Benchmark Papers in Acoustics Series

Editor: R. Bruce Lindsay — Brown University

NOISE CONTROL

Edited by

MALCOLM J. CROCKER
Auburn University

A Hutchinson Ross Benchmark® Book

 VAN NOSTRAND REINHOLD COMPANY

To my mother, Alice D. Crocker

Copyright © 1984 by **Van Nostrand Reinhold Company Inc.**
Benchmark Papers in Acoustics, Volume 17
Library of Congress Catalog Card Number: 83-18453
ISBN: 0-442-21749-8

Manufactured in the United States of America.

Published by Van Nostrand Reinhold Company Inc.
135 West 50th Street
New York, New York 10020

Van Nostrand Reinhold Company Limited
Molly Millars Lane
Wokingham, Berkshire RG11 2PY, England

Van Nostrand Reinhold
480 Latrobe Street
Melbourne, Victoria 3000, Australia

Macmillan of Canada
Division of Gage Publishing Limited
164 Commander Boulevard
Agincourt, Ontario MIS 3C7, Canada

15 14 13 12 11 10 9 8 7 6 5 4 3 2 1

Library of Congress Cataloging in Publication Data
Main entry under title:
Noise control.
    (Benchmark papers in acoustics ; 17)
    "A Hutchinson Ross Benchmark book."
    Includes indexes.
    1. Noise control. I. Crocker, Malcolm J. II. Series.
TD892.N653   1984   620.2'3   83-18453
ISBN 0-442-21749-8

CONTENTS

Contents

SERIES EDITOR'S FOREWORD

The "Benchmark Papers in Acoustics" constitute a series of volumes that make available to the reader in carefully organized form important papers in all branches of acoustics. The literature of acoustics is vast in extent and much of it, particularly the earlier part, is inaccessible to the average acoustical scientist and engineer. These volumes aim to provide a practical introduction to this literature, since each volume offers an expert's selection of the seminal papers in a given branch of the subject, that is, those papers that have significantly influenced the development of that branch in a certain direction and introduced concepts and methods that possess basic utility in modern acoustics as a whole. Each volume provides a convenient and economical summary of results as well as a foundation for further study for both the person familiar with the field and the person who wishes to become acquainted with it.

Each volume has been organized and edited by an authority in the area to which it pertains. In each volume there is provided an editorial introduction summarizing the technical significance of the field being covered. Each article is accompanied by editorial commentary, with necessary explanatory notes, and an adequate index is provided for ready reference. Articles in languages other than English are either translated or abstracted in English. It is the hope of the publisher and editor that these volumes will constitute a working library of the most important technical literature in acoustics of value to students and research workers.

This volume, *Noise Control*, has been edited by Malcolm J. Crocker, who since 1983 has been professor and head of the department of mechanical engineering at Auburn University. Professor Crocker was formerly assistant director for acoustics and noise control of the Ray W. Herrick Laboratories at Purdue University. He is also editor-in-chief of the *Noise Control Engineering Journal,* the refereed journal of the Institute of Noise Control Engineering/USA, and was the 1981 president of that institute. For many years he has been intimately concerned with noise control problems.

In its 27 carefully selected seminal articles the present volume covers thoroughly the development of noise control methods from 1886 to the present. Each of the seven groups of articles into which the volume is divided is prefaced by a commentary explaining the significance of the papers in the group and their relation to other articles in the book. Each is also accompanied

Series Editor's Foreword

by a detailed bibliography of other important writings. No important aspect of noise and its control is omitted. Both in its coverage and its method of presentation this volume should command the enthusiastic attention of all workers in applied acoustics.

R. BRUCE LINDSAY

PREFACE

This book is an attempt to gather together a group of important papers in the field of noise control engineering. Noise is experienced in many aspects of human life. I have attempted to make the book as comprehensive as possible by including papers on individual noise problems in the following broad areas: surface transportation, aircraft, buildings, human response, industrial machinery, measurements, and communities.

Many readers will recognize the papers reprinted in this book as being either the first paper or else a very important paper in the development of literature on different noise topics. These papers obviously do not represent the current state-of-the-art in each topic, but reading these early important papers usually will help the reader understand better the most recently available literature. The literature is extensive on some topics and sparse on others. In some topic areas the choice of papers was obvious; in others it was more difficult. For these reasons, the papers are of differing levels of sophistication.

A deliberate attempt has been made to select the earliest important paper within each area. Important papers published in the last fifteen years have been excluded for three main reasons. First, papers published very recently are normally obtained fairly easily. Second, it is more difficult to judge the importance of very recent publications. Older important papers are still well referenced, while unimportant papers are not. Third, space limitations do not allow inclusion of some excellent recent papers.

Commentaries precede each group of papers to provide the reader with a discussion of recent developments and lists of references that, in the writer's opinion, represent some of the more important recent publications on different topics. Another goal of the commentaries is to discuss the rapid changes that have occurred in surface transportation, aviation, industrialization, and communities in industrialized societies since the 1800s. These social and other changes will continue to influence our present noise problems in many different areas. The noise problems will continue to change and evolve; it is hoped that the commentaries will put both the papers reprinted in this book and the present noise situation in some historical perspective.

Most of the papers selected for this volume were first printed in English. There are a few exceptions, including the paper by Gutin translated from the Russian and the two papers by Meyer et al. translated from the German especially for this volume. Obviously some developments occurred simultaneously in different locations, and some papers have been written on similar topics more or less concurrently in different languages. In such cases, papers in English have been chosen.

Readers may wonder why the two very important 1952 and 1953 papers by Lighthill on jet noise (Part II, Refs. 40, 41) and the 1917 paper by Wente (*Phys. Rev.* **10**:39-63) on the condenser microphone are not included in this book. The reason is that Lighthill's two papers have already been included in the earlier Benchmark volume on Underwater Acoustics and Wente's important paper is included in the recently published Benchmark volume, *Acoustical Measurements: Methods and Instrumentation.*

ACKNOWLEDGMENTS

During the several years it has taken me to research and prepare this volume, I have received invaluable assistance from various colleagues. This assistance has been in two forms. First, I am grateful for the many suggestions of important papers that should be included in this volume. Second, I am very grateful for the detailed comments of many colleagues on my initial selection of papers for this volume. This assistance and these comments have been invaluable, and the book would have been poorer without them. However, the decision on the final selection of papers and the thoughts expressed in the editor's commentaries are entirely my own.

Colleagues who have given assistance and/or comments include: W. I. Acton, R. L. Bannister, L. L. Beranek, W. J. Cavanaugh, Raymond Cohen, Lothar Cremer, P. E. Doak, R. M. Ellis, T. F. W. Embleton, D. R. Flynn, W. Gruner, M. Grutzmacher, J. M. Guinter, R. Hickling, H. H. Hubbard, J. J. Johnson, G. W. Kamperman, R. J. Koenig, Heinrich Kuttruff, W. W. Lang,, A. H. Marsh, W. R. Morgan, P. M. Morse, K. S. Nordby, T. D. Northwood, A. H. Odell, H. F. Olson, A. G. Piersol, Alan Powell, Theodore Priede, R. W. Procunier, M. Rettinger, E. J. Richards, A. Rimsky Korsakov, H. C. Roberts, G. A. Russell, T. J. Schultz, L. C. Sutherland, R. B. Tatge, G. J. Thiessen, E. E. Ungar, W. A. Utley, I. L. Ver, H. E. Von Gierke, and R. W. Young.

My thanks are also expressed to: R. Bruce Lindsay, Norbert Kaemmer, Thomas Reinhart, Ruth C. Crocker, and Peter K. Baade for assistance in translating the two German articles by Meyer et al., and to Joy Meade for contacting each author and publisher to obtain permissions.

My wife, Ruth C. Crocker, kindly made valuable suggestions on obtaining sources of historical information on the development of transportation, industry, and urbanization, which helped me put such developments into better context in the commentaries.

MALCOLM J. CROCKER

CONTENTS BY AUTHOR

NOISE CONTROL

INTRODUCTION

If we use the broad definition of *noise* as *unwanted sound,* then it seems that people have suffered from noise for thousands of years. We use our ears mainly for listening to conversation and for sounds of danger. As people began to live in larger communities, they gradually became troubled by the sounds made by their neighbors' voices, animals, construction activities, and finally, means of transportation. City authorities began to try to control noise and to punish and fine those responsible [1].

About 2000 B.C. the "Epic of Gilgamesh," which describes the life of a Sumerian prince, was written on clay tablets. This story includes a description of the great flood and considers that mankind brought this recrimination from God on itself by making too much noise. In 720 B.C. the Greek city of Sybaris was founded in southern Italy. Apparently it became very prosperous. The authorities required that potters, tinsmiths, and other tradesmen carry on their businesses outside the city walls because of the noise they made [1, Paper 24.] This early form of "industrial" noise ordinance and the Sybarites' sensitivity to discomfort was scorned by the Romans, who told the tale of a Sybarite who could not sleep because of a petal crumpled in his bed of roses.

However, attitudes changed. We find several references in Roman literature to noise. Horace gives us a graphic account of the noise in ancient Rome:

> A contractor rushes along in hot haste with his mules and bearers. A huge machine is moving here a stone and there a beam. Sad funeral processions jostle with large wagons. Here a mad dog runs. There a mud bedaubed sow falls down. Go now and try to meditate on your melodious verses! [2]

During the night the noise was enough to disturb sleep, as Juvenal describes:

> For what sleep is possible in a lodging in Rome? Only those with great wealth can sleep in the city. Here is the reason for the trouble. The movement of four wheeled wagons through the narrow, winding streets, the clamorous outcries of the cattle drovers when brought to a standstill (traffic jam!) would be enough to deprive even General Drusus of his sleep, not to mention seals! [3]

Martial [4] complained that the noise on the streets at night sometimes sounded as though the whole of Rome were traveling through his bedroom. Apparently the noise made by chariot wheels on the cobbled streets was mainly to blame, and the Roman authorities responded with an ordinance banning delivery vehicles from the Forum because of noise and congestion. We know that there were also complaints about noise in Pompeii. On the wall of one house is written the graffiti: "Macerior begs the magistrate to prevent the people from making a noise disturbing the good folks who are asleep."

It is said that in sixteenth century England, Henry VIII forbade wife-beating *at night,* not on humanitarian grounds but because the cries might disturb the neighbors. The fine for offenders was seven shillings, a considerable sum of money in those days [1].

Complaints about noise became more and more frequent in the last two centuries. Most prominent learned people complained that noise disturbed their concentration and other activities. In the 1820s Schopenhauer, the German philosopher, wrote an essay on noise [5]. He complained in particular of the noise of cracking whips. Schopenhauer made the interesting claim that people "of great intellect," as he termed them, were particularly disturbed by noise. He cited the great writers Kant, Goethe, Lichtenberg, and Johann Paul Friedrich Richter who also disliked noise. Schopenhauer's claim that people of great intelligence are more severely affected by noise seems to have found some recent experimental support. Bryan and Colyer (*Acustica* **29** [1973]:228–233) present results suggesting that the performance of very intelligent people can suffer in noisy conditions, while that of less intelligent people can improve. The *arousal* theory may explain this result. Intelligent people may be working at a high degree of arousal, and noise increases this arousal and overloads their capacity for intellectual work. Less intelligent people may be working at a lower level of arousal, and noise can increase their arousal, thus improving their efficiency and performance.

Literature from later in the century contains even more complaints from great writers. Thomas Carlyle apparently dreaded noise, and in particular the sound of a cockcrow, so much that he had a "sound-proof" room built [Paper 24]. Herbert Spencer used to plug his ears with wool to keep out noise. Mark Twain was very much disturbed by city noise. In 1871 and 1889 the weekly journal edited by Charles Dickens featured two articles on noise [6, 7], and he organized a petition to Parliament in London complaining of the noise of street musicians [8]. Among those who signed the petition were Tennyson, Millais, Holman Hunt, John Leech, Wilkie Collins, and Thomas Carlyle.

As U.S. and European society became more urbanized and mechanized at the end of the nineteenth and beginning of the twentieth centuries, noise problems and complaints increased dramatically [8–22]. We no longer read just of the noise caused by people's voices, street musicians, animals (barking dogs), and carriage wheels on cobbled streets, but also of the noise of construction (circular saws) [6], railways, both surface and elevated in London and New York [9, 11], steam whistles, and finally, by 1901, automobiles [11]. Several people in the United States in the 1890s began campaigning for the creation of a Society for the Suppression of Noise [11, 13]. The subject received wide attention and was discussed in hundreds of newspapers and in the lay and medical press in the United States and Europe [11]. The Board of Aldermen of New York City passed an ordinance, first enforced in 1897, making it unlawful to haul iron or steel beams through the streets unless they were prevented from banging against each other; the fine was twenty-five dollars [11]. In 1904–1905 a Mrs. Rice campaigned to reduce the noise produced by the unnecessary use of steam whistles in tugboats in New York harbor. After much effort and in spite of the resistance of the tugboat captains, she managed to mobilize public opinion and, after gathering thousands of signatures, succeeded in getting the noise abated [14].

In the late nineteenth century the cobblestones in the streets of London were replaced with creosoted wood blocks or asphalt. This development occurred later in the big cities of North America so that cab and wagon noise was reduced. The change was probably instituted for reasons other than noise, such as reduced wear on vehicles, ease of maintenance, and street cleanliness. However, by 1911 we find that there were many new complaints in London about the hooters and engines of motor cars, and a Motor Traffic (Street Noises) Bill was introduced into the House of Commons [15].

As already mentioned, city ordinances in the United States in the late nineteenth century began to include regulations against unnecessary noises. By 1910–1911 most large cities, such as Chicago, New York, Cleveland, Louisville, and Milwaukee, had established quiet zones near hospitals and prohibited "unnecessary noise" [16]. However, such regulations were normally not effectively enforced. One exception was in Baltimore, where in 1912 an Anti-Noise Committee of the City Medical Society was formed [17]. It helped produce an ordinance to ban factory steam whistles, and hospital quiet zones were also established. A special anti-noise police officer was appointed to enforce the noise regulations, probably the first such example in history.

Complaints and concern about noise increased between 1910 and 1930 as urbanization continued and the numbers of mechanized vehicles in use increased [8–22]. Imogen Oakley, who in 1915 was chairman of the committee on noise of the American Civic Association, recounts asking an audience of poor women what they considered the greatest evil in their crowded tenement life in Philadelphia. One woman rose and said, "I speak for every woman here. What we cannot stand is noise. It never stops. It is killing us. We work all day and need sleep and rest at night. No one can sleep till midnight and all the noise begins again at five. Many of us have husbands who work at night and must get sleep during the day, but they get no sleep with all the noise that goes on about us" [8].

An increasing number of people in the period 1900–1930, in both the lay and the medical press, began to claim that noise was having a severe and adverse effect on city dwellers [20, 22]. Additional sources of noise mentioned include the use of cut-out mufflers on automobiles, the increasing use of motorcycles (often with no mufflers at all), and streetcar noise caused by bad junction-points and flat wheels [22].

Most people who had visited both European and U.S. cities concluded that European cities were quieter than their U.S. counterparts and that New York was

3

probably at that time the noisiest city in the world [22]. This situation was probably caused by the elevated railways, the continued use of cobblestones, and blaring radio advertisements. In contrast, most city railways in London had been built underground in "tubes" since the middle of the nineteenth century. To make matters worse, the high buildings confined the noise in the streets and caused it to reverberate in the city "canyons" and between the high buildings. In the 1920s some effort was expended to reduce railroad and streetcar noise [23–28, and Paper 1]. However, much of this work was hampered by the lack of good measuring equipment and lack of an agreed-upon way to present the results. Much time that could have been devoted to finding noise reduction solutions was spent on developing noise measurement instrumentation [24, 25]. Nevertheless, some progress was made and noise reduction was achieved in some cases when enlightened management allowed [23, 24, 27, and Paper 1].

In the late 1920s several noise surveys were made in U.S. cities. In 1926 Free, then the science editor of *The Forum,* conducted a noise survey of New York City. Soon after, similar limited surveys were conducted in Chicago (by Norris) and in Boston, St. Louis, London, Paris, and Berlin. In 1930 a much more extensive noise survey was again carried out in New York, this time by a special noise abatement commission appointed by the New York Commissioner of Health [Paper 24]. The surveys made by Free and by the Noise Abatement Commission both concluded that the main source of noise in New York was trucks, and the next most important sources were elevated railways and surface street cars. It is interesting to note that fifty years later the predominant source of community noise in the United States is still truck noise, although it is now more intense and extensive because of a sixfold increase in the numbers of cars and trucks in use and in the number of miles of paved roads.

In 1930 commercial aviation was, of course, in its infancy, and it was hardly a major noise source. Since then, commercial aircraft have largely replaced railroads in carrying passengers over long distances and now transport the vast majority of intercity passengers in the United States. The introduction of the passenger jet aircraft in the early 1960s brought with it much more intense noise. Scheduled passenger aircraft are probably now the second most important source of noise annoyance in U.S. communities.

So far we have been describing sources of noise annoyance. During the last hundred years or more, writers have blamed noise for many other ills: increased mental illness, stress, heart attacks, and absenteeism, as well as reduced productivity. However, attempts to prove conclusively these additional effects (except at very high noise levels) have usually produced controversial results. The human ear and nervous system are very complex and still poorly understood in some respects. However, one effect has been recognized for at least two centuries—that is, the power of either intense, impulsive noise or less intense but chronic loud noise to produce permanent hearing damage. Noise sources in the community are not usually intense enough to cause damage. However, noise and explosions in warfare were known to have produced deafness in the eighteenth century. Parry recounts several cases of temporary and permanent hearing loss [33]. Admiral Lord Rodney is stated to have been made almost completely deaf in 1782 for fourteen days after the firing of eighty broadsides from his ship, H.M.S. *Formidable,* and guncrews and an officer are said to have been made permanently deaf after the battles of Trafalgar and Copenhagen.

Some occupations involving metalwork, such as blacksmithing, had been known for centuries to cause permanent hearing damage. The advent of the Industrial Revolution in Britian about two hundred years ago made intense noise even more common in working places such as the weaving industry [32]. Other occupations such as riveting of boilers for steam engines were even more noisy, and occupational deafness was sometimes known as boilermakers deafness. One of the first quantitative studies of industrial deafness was made by Barr in 1886 [Paper 18]. He compared the hearing of boilermakers, iron-moulders, postmen, and people not exposed to intense noise and showed that the boilermakers' hearing was much worse than that of the other groups. Extensive industrialization in Western countries has exposed many workers to intense continuous noise, and over a period of years many millions have become permanently deaf.

It is interesting to note in the early literature on noise the mixture of correct and fallacious statements. Although the annoyance and damage caused by noise has been recognized for centuries, and the structure of the ear began to be fairly well understood in the middle of the nineteenth century, we find serious scientific debate one hundred years ago in the literature [34] as to whether noise was sensed with the same nerves as sound. Several investigators, including Exner, Brucke, and Helmholtz, conducted experiments and finally concluded that noise and sound were sensed with the *same* nerves in the ear. Even more recently, physicians made the incorrect claim that over a long time intense noise thickened the eardrum and made it insensitive. Now we know, of course, that it destroys the hair cells in the inner ear.

Although the problems of noise have been recognized for many centuries, it seems that scientific and engineering papers on noise control have appeared only in the twentieth century. Of course, increasing industrialization, mechanization, and urbanization have all served to make the noise problem much more severe in many countries in recent years. Since noise is such a severe problem now in some industrialized countries, it is somewhat surprising that there has not been more scientific interest in noise control. Some scientists seem to feel that it is "beneath their dignity" to write about such practical problems. Many methods of controlling noise—use of thicker walls and acoustic absorbing materials—were known in the last century; they have become even better understood in the twentieth century and have been applied to a variety of problems. The papers in this book reflect the variety of noise sources and noise-control techniques that have been applied to them, the theories that have been developed for new noise problems, and the development of instrumentation and techniques for measurement and rating of noise.

Since noise disrupts and interferes with many different human activities, evolving a classification scheme for papers in this book has been difficult. Any

5

classification scheme is bound to be somewhat arbitrary. Also, it had to be decided whether or not to include sections on human response to noise, noise measurements and instrumentation, and propagation of noise in such a book on noise control. In the end, these sections were included, although Part III was kept short; it concentrates on mechanical noise sources in buildings and their control because the important topic of sound transmission is already covered in the companion Benchmark volume, *Architectural Acoustics*, edited by Thomas D. Northwood. One might expect to find the papers by Lighthill on aerodynamic noise in Part II. However, because these papers have already been reprinted in the companion Benchmark volume, *Underwater Sound*, edited by Vernon M. Albers, they were not included in this volume.

It might be argued that some of the papers in Part I might more properly appear in Part IV; others from Part VII in Part I or V, and so on. However, it is hoped that the arbitrariness of the classification scheme is reduced by the editorial comments at the beginning of each section and the references to papers in other sections.

REFERENCES

1. Embleton, T. W. B., 1977, Noise Control from the Ancient Past and the Near Future *Noise/News* **6**(2):26-31.
2. Horace, Epistle II, 2, lines 72-76.
3. Juvenal, Satire III, lines 234-238.
4. Quoted in Embleton, T. W. B., 1977, Noise Control from the Ancient Past and the Near Future, *Noise/News* **6**(2):26.
5. Durant, W., 1962, *The Works of Schopenhauer,* Frederick Unger, New York.
6. Anonymous, 1871, Noises, in *All the Year Round,* C. Dickens, ed., vol. VII, no. 159 (n.s.), Dec. 16, pp. 55-59.
7. Anonymous, 1889, Noise, in *All the Year Round,* C. Dickens, ed., 3rd series, no. 46, Nov. 16, pp. 474-476.
8. Oakley, I. B., 1915, Public Health versus the Noise Nuisance, *Natl. Munic. Rev.* **4**:231-237.
9. Sully, J., 1878, Civilization and Noise, *Fortnightly Rev.* **25**(o.s.):704-720.
10. Ross Browne, W., 1880, Social Plagues, II. Noise, sec. I and II, *Good Words Sunday Mag.* **21**:525-527, 606-608.
11. Girdner, J. H., 1897, To Abate the Plague of City Noises, *North Am. Rev.* **165**:460-468.
12. Girdner, J. H., 1901, Noise and Health, *Munsey's Mag.* **25**:323-326.
13. Hubert, P. G., 1894, For the Suppression of City Noise, *North Am. Rev.* **159**:633-635.
14. Rice, J. L., 1905/1906, An Effort to Suppress Noise, *Forum* **37**:554-570.
15. Anonymous, 1911, Noise, *Spectator* **107**(22):133-134.
16. Bostwick, A. L., 1913, Quiet Zones Near Hospitals, *Natl. Munic. Rev.* **2**:519-520.
17. Watson, W. T., 1914, Baltimore's Anti-Noise Crusade, *Natl. Munic. Rev.* **3**:585-589. See also: Anonymous, 1914, Anti-Noise Policeman, *Outlook* **107**:438-439.
18. Batterson, E. S., 1917, Progress of the Anti-Noise Movement, *Natl. Munic. Rev.* **6**:372-378.
19. Anonymous, 1921, Noise, *Spectator* **127**(12):624-625.
20. Spooner, H. J., 1922, Health Problems Involved in Noise and Fatigue, *Nation's Health,* February, 1922, pp. 91-95.

21. Noble, D. H., 1923, Silencing "Tube" Trains, *Electr. Traction,* March, 1923, pp. 112–113.
22. Nance, W. O., 1922, Gains Against the Nuisances: II. Noise and Public Health, *Natl. Munic. Rev.* **11:**326–332.
23. Anonymous, 1924, Reducing Noise in Train Operation, *Electr. Railw. J.* **63**(4):135–138.
24. Anonymous, 1924, Noise in London Cars Photographed, *Electr. Railw. J.* **63**(13):506.
25. Tobey, S. H., 1926, Reducing Noises in Street Cars, *Electr. Railw. J.* **79**(9):338.
26. Anonymous, 1926, Track Crossings in Cleveland Reduce Noise, *Electr. Railw. J.* **67**(12):511–512.
27. Williams, H. S., 1926, Car Noises from Inadequate Maintenance Demand Attention, *Electr. Railw. J.* **67**(12):476–478.
28. McKelway, G. H., 1926, The Electrical Department Can Help Reduce Noise, *Electr. Railw. J.* **67**(12):481–482.
29. Anonymous, 1928, Reducing Street Car Noises, *Electr. Railw. J.* **72**(6):225–226.
30. Egger, R. A., 1931, Quieting the City's Bedlam, *Natl. Munic. Rev.,* May, 1931, pp. 304–305.
31. Egger, R. A., 1931, The Paris Anti-Noise Ordinance, *Natl. Munic. Rev.,* October, 1931, pp. 616–617.
32. Thompson, E. P., 1966, *The Making of the English Working Class,* Vintage Books, New York, p. 307.
33. Parry, C. H., 1825, *Collections from the Unpublished Medical Writings of the late C. H. Parry,* 1, Underwood, London, p. 554.
34. Anonymous, 1885, On the Recognition of Noises, *J. Sci. Ann. Astron. Biol. Geol. Ind. Arts Manuf. Tech.* **22:**336–341.

Part I

SURFACE TRANSPORTATION NOISE

Editor's Comments on
Papers 1 Through 5

As already discussed in the Introduction, surface transportation noise is not a new problem. The Romans banned wagons and chariots from the Forum during the day because of the congestion and noise [1]. From being probably only a minor irritant in such ancient societies, surface transportation noise has gradually grown to become one of the major social problems in industrialized societies. In order to put the articles reprinted in this section into some perspective, we briefly review the growth of transportation in Europe and North America and the noise made by different forms of transportation.

The development of transportation in the last two or three centuries is intimately related to two other developments, industrialization and urbanization, which have also led to increasing industrial and community noise problems. These problems will be dealt with later in this book. The development first of canals and later of railroads for carrying heavy freight allowed industrialization to occur in Britain around 1780 and later in other European countries and the United States [2, 3, 4]. The development of public transportation in cities allowed the rapid growth of European cities such as London, Berlin, and Paris and U.S. cities such as New York, Detroit, and Chicago in the late nineteenth century. Cheap public transportation transformed the often compact "walking" city of one or two miles diameter into a much larger one, since people could now ride to work from the rapidly developing suburbs [3, 4]. However, higher-speed and then mechanized transportation in dense cities inevitably brought the related problem of noise. In a few cases the resulting noise was serious enough to prevent or curtail the development of some forms of transportation in the last century, and these cases will be discussed. But there were many other, more serious problems facing municipal authorities then, and reduction of noise usually received a low priority. In the twentieth century, however, surface transportation noise has become so intense and pervasive that many countries have introduced regulations to curtail the noise of new vehicles and of those in use. Environmental impact statements must now be written for the planning of major new roads in the United States. In Britain there are strict planning controls for residential developments in noisy areas, and insulation of houses is mandatory if they are subjected to traffic noise from new or improved roads exceeding an A-weighted sound level of 68 dB on the L_{10} (18 hour) index. L_{10} [18 hour] is the level exceeded for 10 percent of the time; an average is taken over an eighteen-hour period.) By 1978, 6,000 homes in Britain had been insulated, and many new roads now incorporate earth mounds and noise barriers [5].

During the last two centuries in Europe and the United States, one form of transportation has dominated in one country or another and then has been replaced by another more economical or faster form. Sometimes several different forms of transportation have coexisted, and some forms developed in the last century have persisted to the present day. Only in the last fifty years have reliable quantitative measurements been made of transportation noise. Accounts of noise in previous centuries are generally only qualitative. However we can be sure that people were disturbed. For example, in Boston iron-wheeled carts, wagoners' shouts, cracking whips, and clanking chains in King Street so annoyed the Great and General Court of Massachusetts that in 1747 it prohibited all vehicles from traveling on either side of the State House during its sessions.

In Europe in the mid- to late eighteenth century, intercity passenger and freight traffic was mainly by stagecoach and horsedrawn wagons, and the system of turnpike roads was gradually expanded [2]. At this time road systems in the United States were generally bad, although some roadways such as the Philadelphia-to-Lancaster turnpike (opened in 1794) were in use. Of much more importance in this period was transportation by water on the Atlantic seaboard and river system, initially by sailing boats and in the first half of the nineteenth century by steamboat [6]. Here, of course, the noise problem was minimal.

In Europe many canals were cut to carry heavy freight in the mid- to late eighteenth century to supplement the river system [2]. An extensive system was also built in the United States in the early 1800s throughout Massachusetts, Maryland, Pennsylvania, Virginia, Ohio, Indiana, and Michigan. However, this system was soon superseded by the railroad [6, 7].

The first really successful railroad was the Stockton to Darlington railway in Britain, which opened in 1825 using George Stephenson's *Locomotion*. The state of Maryland chartered the Baltimore and Ohio Railroad in 1827; it started making regular runs in 1830. Railroads expanded rapidly throughout the United States and Europe during the late nineteenth century. By this time the maps of railroads in many industrialized countries looked very similar to those of today [2, 6]. However, in most countries railroad transportation remained primarily intercity, and with only a few exceptions did it provide transportation for city commuters. Although railroads were undoubtedly noisy, much of this noise was made away from urban areas, and complaints in the literature of railroad noise in the last century are scarce, although they do exist. For example, Carlyle wrote that: "That which the world torments me in most is the awful confusion of noise. It is the devil's own infernal din all the blessed day long confounding God's works and His creatures. A truly awful hell-like combination, and the worst of it is a railway whistle like the screech of ten thousand cats and every cat of them as big as a cathedral." [8] Nathanial Hawthorne sensed the impending encroachment of technology on the quiet Concord woods when he wrote in 1844; "But Hark! There is the whistle of the locomotive—the long shriek, harsh above all harshness . . . it brings the noisy world into the midst of our slumbrous peace." [9] The South London Line opened in 1867, it was the first underground railway in England [2], and it was soon followed by others. These first underground railways were powered by steam locomotives and were dirty and noisy (reasons why they had been built underground in the first place). The weekly journal edited by Charles Dickens described the situation well: "The railway whistle fiend is perhaps responsible for the most aggravating form of noise that goes to swell the general uproar of the metropolis (London). Those who are fated to live where the whizz and rattle of the underground trains are within audible distance, find them quite sufficient to try the strongest nerves" [8].

An attempt to build a similar underground system in New York at this time was prevented at first by the state legislature through the opposition of the horse streetcar interests [3]. Despite this opposition, an elevated railway was started in 1867 and completed in 1870. In 1878 a second elevated railway was added. In the 1880s similar "L's" were built in Brooklyn and Kansas City and in Chicago in the 1890s. These systems were very noisy and dirty. They often dropped oil and hot ashes on pedestrians below, and vibration and noise from them carried a considerable distance. People were always afraid that accidents would occur and that the trains would crash off their tracks and fall to the street below [3]. Girdner described the situation as follows: "The railroads of New York City are the greatest single source of noise in the streets in which they are operated, and for half a block on either side. The elevated road is the worst offender of all. Conversation in a street through which it runs is practically impossible owing to the roar and rattle of passing trains, and it seems impossible that anyone should be able to sleep in a room anywhere near it. . . . Rubber gaskets properly placed under the rails of the elevated roads, and the substitution of electricity or compressed air for steam as a motive power would greatly lessen the noise which these roads now make" [9]. By the 1890s the elevated railways in existence had reached their capacity since trains were running as close together as safety permitted. Municipal authorities refused to allow new lines or extensions to old ones mainly because of public opposition to the noise and dirt created [3]. About this time New York was reputed to be the noisiest city in the world. It is interesting to note that for noise, pollution, and safety reasons elevated railways were not built in large numbers in Europe.

As already mentioned, much more urban traffic was carried by city streets than by railroads in the nineteenth century. In the eighteenth century most cities were small and most points could be reached by walking [4, 12]. Urban public transportation with fare-paying passengers did not develop until the nineteenth century. The important development of the horse omnibus began in Nantes, France, in 1826 and rapidly spread throughout the rest of Europe and the United States. In the 1850s rails were flush-laid in the streets of New York and the horse-drawn street car (horse-car) came into existence. By the 1860s horse-cars generally began to replace horse omnibuses in most U.S. cities. Their advantage was that a smoother, quieter ride was obtained for the passengers and many more passengers could be transported by a train of horses. Horse-cars were much more slowly adopted in Europe, where they often coexisted with omnibuses until the beginning of the twentieth century [12]. By the 1860s and 1870s the present-day phenomenon of the rush hour had developed in large European and U.S. cities with the consequent increase in noise levels. Sir Norman Moore, a British physician, gave a graphic description of the noise in a London street in 1869: "Most of the streets were paved with granite sets and on them the wagons with iron-tired wheels made a din that prevented conversation while they passed by. The roar of London by day was almost terrible—a never varying deep rumble that made a background to all other sounds" [13].

Engineers experimented with various mechanical devices to replace the horse to increase speed and reduce cost. Cable cars were introduced in San Francisco early in the 1870s and in other U.S. cities. These systems mainly used centrally located steam engines and continuous cables [4, 12].

In 1872 a steam locomotive was built in London to pull streetcars (trams). A considerable number were supplied for use in Paris and London. For the next twenty years, steam-powered streetcars (trams) were used in several European and U.S. cities, but they were never very successful [12]. In Europe, public authorities imposed severe antipollution requirements to restrict the smoke, sparks, cinders, and noise produced by the steam engines. For instance, the *Use of Mechanical Power on Tramways Act* of 11 August, 1879, set very stringent operating requirements to be enforced by the Board of Trade in Britain [12]. Steam-powered tramways (streetcars) were not permitted to emit visible smoke or steam; the engine had to operate without noise produced by blast or from the clatter of machinery; all working parts had to be concealed; and the tram had to be governed automatically to a top speed of ten miles per hour [14]. Probably such regulations reflected the desires of society. Similar strict regulations were enforced elsewhere in Europe, such as in Hamburg, Leipzig, and Paris, and partly explain the failure of the steam tram. However, they also seemed to be unable to show a clear economic advantage over the horse-drawn tram. For example, an 1876 shareholder's report of the General French Tramways Company stated: "The problem of mechanical traction does not consist solely in eliminating the serious drawbacks of locomotives—smoke, noise, etc.—but above all in obtaining an economical engine" [12, 15].

In the 1880s and 1890s a revolution in urban transportation occurred: electrification of street railways [3, 4, 12]. Mechanization of urban transportation was already considerably advanced in the United States when electrification arrived. In 1890 steam street railways were carrying 287 million passengers annually over 527 miles of line, while cable cars were carrying 373 million passengers over 283 miles of line, compared with the 1,230 million passengers carried by the horse streetcars over about 5,000 miles of track. In the two short years from 1888 to 1890, 914 miles of track were electrified [12, 16]. By 1893 total street railway track in the United States increased to 12,200 miles (of which 60 percent was electrified), and by 1903 it had reached 30,000 miles (fully 98 percent of which was electrified) [12]. Horse cars became a relic of the past.

Electrification followed quickly in some other countries, such as Germany, but more slowly in others, such as Britain. The much greater economy of electric streetcars (compared with horse-cars and omnibuses) and the consequent reduction in fares brought about a vast increase in passenger use in the period from 1890 to 1920 in the United States and Europe. The increased speed and use of electric streetcars, however, did bring increased community noise problems. There is some evidence that the noise from the metal wheels on the rails and from the crossing points and the gears and trollies and so on was very troublesome. The noise was sufficient to make arterial building lots in Boston less attractive in the 1890s [4]. Serious attempts were made to reduce electric street car and subway noise in the early 1900s [17–23 and Paper 1]. It was felt by some that noise was a serious hindrance to continued expansion of the street-car service because of objections by both passengers and nearby communities. Paper 1 by Williams is typical of several others published in the 1920s. It is interesting to note the rather primitive measurement techniques used, but that despite this, several of the noise reduction methods in use today in vehicles were also known in the 1920s. Electric streetcars are still in considerable use today in some countries (Germany, the Netherlands, and the USSR) although in other countries, (e. g., the United States) they have largely disappeared [12].

In the early 1900s increasing use of internal combustion engines in motor vehicles again revolutionized transportation systems in the United States and Europe. By the late 1880s automobiles were being commercially produced in France and Germany [12]. However, commercial production of cars in large numbers was first undertaken in the United States, since industry there relied heavily on interchangeable and standardized parts and was thus highly suited for mass production [12]. By 1912 yearly production of automobiles in the United States reached almost 500,000. Production of automobiles was initially much slower in Europe. Production of cars, trucks, and buses expanded rapidly after World War I, and by 1930 transportation powered by internal combustion engines had become the predominant form of transportation and producer of urban noise, as is shown by the 1930 New York Noise Survey [Paper 24]. In this survey it was found that road traffic (trucks, horns, brakes, buses, and motorcycles) was the source that received the most complaints (36 percent), with motor trucks alone accounting for 10 percent. The next major source was public transportation (elevated railways, street cars, and subways), totaling 16 percent. Since that time road traffic has expanded vastly. In 1978 there were about 120 million cars, trucks, and buses in use in the United States, and annual production was about 10 million. In Britain the total number of motorized road vehicles in use rose from about 150,000 in 1910 to 800,000 in 1920, to 2,200,000 in 1930 and to 3,100,000 in 1938 [12]. The well-known Wilson Committee report on noise issued in 1963 summed up the situation in Britain: "We conclude that . . . road traffic is, at the present time, the predominant source of annoyance, and no other single noise is of comparable importance" [19].

Road traffic is probably also the predominant source of urban noise annoyance in most other industrialized countries today. In the United States the Environmental Protection Agency (EPA) has stated that "motor vehicles are the principal source of annoyance in urban areas" [25]. The EPA has identified trucks as a major noise source, and in 1975, issued in-use regulations for interstate motor carriers [26]. More recently, it issued limits for new medium-duty and heavy-duty trucks to become effective after January 1978. The reason why medium and heavy trucks were singled out for regulation by EPA is not hard to see. The medium and heavy trucks make about 10 or 15 dB more noise than an automobile and generate about three or four times as much acoustic energy per day as all the cars in use [27]. Railroad noise continues to be a problem, although of less magnitude than traffic noise, and in 1975 the EPA issued noise emission standards for locomotives and railroad cars [28].

In such other countries as Japan, West Germany, France, and Britain, where very high speed railways are in use or being introduced, railroad noise is becoming an increasingly important problem. The energy shortfall that developed in the mid-1970s may produce less reliance on the internal combustion engine in the future, and there may be increased use of electric cars, Stirling-engine vehicles, and public mass transportation systems. This could change the traffic noise situation and even by chance reduce it.

Since traffic noise from motorized road vehicles has been and probably will remain the main noise problem for many years, the remaining papers reprinted in Part I reflect this situation. Apps made a comprehensive survey of car, bus, truck, and motorcycle noise sources in 1957 [29]. Several other authors have reviewed noise sources in vehicles since [30]. The main noise sources include exhaust noise, engine radiated noise, tire noise (at high speeds), and cooling fan noise. Papers on the first three of these four topics are included in Part I.

Exhaust noise is the predominant noise source on an unmuffled internal combustion engine and is usually a major source even on a muffled vehicle. Since the introduction of the internal combustion engine there has been a search for a satisfactory method of muffler (silencer) design. Stewart in the early 1920s was the first to develop theory for acoustic filter design [31] using the lumped impedance approach; Mason in 1927 and Lindsay in 1938 extended this theory to distributed systems [32, 33]. Papers by Stewart, Mason, Lindsay, and White are reprinted in the companion Benchmark volume on Physical Acoustics.

The next landmark development in acoustic filter (muffler) design was the work by Davis and his coworkers at NACA in the 1950s. Davis et al. studied theoretically and experimentallly the transmission loss of many different designs of expansion chamber and side-branch resonator mufflers and obtained good agreement between theory and laboratory experiments. His classic work is reprinted here as Paper 2. Davis obtained disappointing results when he designed a helicopter muffler, which he blamed on finite amplitude effects. However, it seems more likely that the fault lies with his neglect of mean flow and the fact that he compared the measured *insertion loss* of the muffler with the predicted *transmission loss* and thus neglected engine source impedance effects. Some other developments in muffler design that should be noted since then are the four-pole transmission matrix approach developed in the late 1950s by Igarashi [34], the papers by Alfredson and Davies [35, 36], which showed that acoustic filter theory *was* adequate to give good predictions of muffler attenuation on engines provided that mean flow effects were included, and by Young and Crocker [37, 38, 39] and later by Craggs [40], Ross [41], and others on the use of finite elements in predicting the attenuation of mufflers. Several writers, such as Munjal [42], Sullivan [43], and Crocker [44], have also reviewed the state of muffler design for internal combustion engines. The Proceedings of the EPA, Exhaust-System Noise Symposium are a useful source of information on the knowledge of muffler design in 1977 [45]. Methods to predict the acoustic performance of complete engine exhaust systems, including the effect of engine source and tailpipe radiation, have also been described recently [46, 47, 48]. Surprisingly, although theories are now highly developed, many manufacturers still use a very empirical approach to muffler design.

If a vehicle's exhaust muffler and cooling system are well designed, the dominant source of noise is often engine-radiated noise. Diesel engines have been widely used in heavy trucks because of their efficiency, which is higher than that of gasoline engines. Unfortunately, their noise is also higher because of the detonation (or diesel knock) that occurs just before top dead center. Priede and his coworkers have worked on diesel engine noise for twenty-five years, first at C. A. V. and then at the Institute of Sound and Vibration Research at Southampton University in England. They were probably the first to study diesel engine noise systematically. The first of a series of papers on diesel engine noise authored by Priede and his colleagues is reprinted here as Paper 3. Since then several manufacturers have studied methods of reducing diesel engine noise. The results of this work are reported in several books of conference proceedings of the Society of Automotive Engineers (SAE) and of the Institution of Mechanical Engineers (IME) [49–52]. Thien in Austria has reduced diesel engine noise by building engines with integral enclosures [53]. Chung, Seybert, Crocker, and Hamilton have shown that a coherence approach may be used to model diesel engine noise [54, 55, 56]. Intensity techniques that have recently been introduced can greatly speed the localization of sound sources on the surfaces of engines [57–60].

At high speed (above 80 or 90 kilometers/hour) the tires become the dominant noise source on vehicles. The 1960 paper by Wiener reprinted here as Paper 4 was one of the first systematic experimental studies published on tire noise. Since then other developments have included Hayden's [61] theoretical model of tire noise generation, and much more detailed experimental investigations by Tetlow [62] and Leasure [63]. Recent experimental and theoretical tire noise studies are discussed in papers contained in two recent books of proceedings [64, 65].

The last paper in Part I, by Mills and Robinson (Paper 5), addresses the problem of the subjective rating of vehicle noise. It is interesting to note that in 1954 the Automobile Manufacturers Association in the United States rejected the use of A- and B-weighted sound levels and adopted the use of sones for rating loudness of trucks. They specified a maximum loudness of 125 sones at fifty feet for a truck [66]. The use of sones, however, proved to be too complicated in practice for most manufacturers. Young commented on this situation in 1958 (see Paper 17) and suggested that A- and B-weighted sound levels should be brought back into favor. In the 1961 paper by Mills and Robinson (Paper 5) it is seen that A-weighted sound levels are back in use. This paper is interesting because the authors are here already using the drive-by and peak A-weighted sound level method of evaluating vehicle noise, which was later adopted by ISO (and in similar form by SAE). Also, the authors compare these measurements of peak vehicle drive-by noise with the subjective reactions of panels of people and from these reactions attempt to set acceptable limits for vehicle noise. For some alternative methods of measuring vehicle noise indoors in a semi-anechoic environment, see the recent paper by Crocker and Sullivan [67]. Two recent SAE books contain a number of interesting papers on vehicle noise [68] and its measurement [69]. Crocker has also reviewed sources of surface transportation noise, its effects on people, and the technical basis for regulations in a recent book chapter [70].

REFERENCES

1. Grimal, P., 1968, The World of Caesar, in *Greece and Rome*, National Geographical Society, Washington, D.C., See also *Natl. Geogr.*, October, 1968, p. 563.
2. Dyos, H. J., and D. H. Aldcroft, 1969, *British Transport*, Leicester University Press, Leicester, England.
3. Glaab, C. N., and A. T. Brown, 1967, *A History of Urban America*, Macmillan, New York.
4. Warner, S. B., 1962, *Streetcar Suburbs, The Process of Growth in Boston, 1870–1900*, Harvard University Press, Cambridge, Mass. (2nd ed., Atheneum, New York, 1968.)
5. Taylor, R., 1978, *Progress in the Reduction of Noise from Road Transport in the United Kingdom, Soc. Automot. Eng. Paper 780386.*
6. Taylor, G. R., 1951, *The Transportation Revolution 1815–1860*, Harper & Row, New York.
7. Tanner, H. S., 1840, *A Description of the Canals and Railroads of the United States*, T. R. Tanner and J. Disturnell, New York.
8. Quoted in Rice, Mrs. Isaac L., 1906, An Effort to Suppress Noise, *Forum* **37**(April):552–570.
9. Quoted in Marx, L. 1964, *The Machine in the Garden: Technology and the Pastoral Idea in America*, Oxford University Press, New York, pp. 13, 16–32.
10. Anonymous, 1889, Noise, in *All the Year Round*, C. Dickens, ed., 3rd series, no. 46, Nov. 16, pp. 474–476.

11. Girdner, J. H., 1897, To Abate the Plague of City Noises, *North Am. Rev.* **165**:460–468.
12. McKay, J. P., 1976, *Tramways and Trolleys: The Rise of Urban Mass Transportation in Europe,* Princeton University Press, Princeton, N. J.
13. Winter, B., 1971, *Past Positive,* Chatto and Windus, London.
14. Whitcombe, H. A., 1954, *History of the Steam Tram,* South Godstone, Surrey, p. 18.
15. AN, 65 AQ, Q138[1], 1876, Cie. Générale Francaise de Tramways, Annual Report.
16. U.S. Census, *Transportation, 1890,* pp. 681–682.
17. Anonymous, 1924, Reducing Noise in Train Operation, *Electr. Railw. J.* **63**(4):135–138.
18. Anonymous, 1924, Noise in London Cars Ph'tographed, *Electr. Railw. J.* **63**(13):506.
19. Tobey, S. H., 1926, Reducing Noises in Street Cars, *Electr. Railw. J.,* **79**(9):338.
20. Anonymous, 1926, Track Crossings in Cleveland Reduce Noise, *Electr. Railw. J.* **67**(12):511–512.
21. Williams, H. S., 1926, Car Noises from Inadequate Maintenance Demand Attention, *Electr. Railw. J.* **67**(12):476–478.
22. McElway, G. H., 1926, The Electrical Department Can Help Reduce Noise, *Electr. Railw. J.* **67**(12):481–482.
23. Anonymous, 1928, Reducing Street Car Noises, *Electr. Railw. J.* **72**(6):225–226.
24. Wilson Committee, 1963, *Noise—Final Report,* Her Majesty's Stationery Office, London.
25. Environmental Protection Agency, 1974, Motor Carriers Engaged in Interstate Commerce, *Fed. Regist.* **39**(209):38208–38216.
26. Environmental Protection Agency, 1974, Transportation Equipment Noise Emission Controls—Proposed Standards for Medium and Heavy Duty Trucks, *Fed. Regist.* **39**(210):38338–38362.
27. Wyle Laboratories, 1971, *Transportation Noise and Noise from Equipment Powered by Internal Combustion Engines,* Environmental Protection Agency Rep. NTID 300.13.
28. Environmental Protection Agency, 1976, Railroad Noise Emission Standards, *Fed. Regist.* **41**(9):2184–2195.
29. Apps, D. C., Automobile Noise, In *Handbook of Noise Control,* C. M. Harris, ed., McGraw-Hill, New York, pp. 31-1-31-50.
30. Staadt, R. L., 1974, *Truck Noise Control, Soc. Autmot. Eng. Paper SP-386.* Reprinted in *Reduction of Machinery Noise, Revised Edition,* 1975, M. J. Crocker, ed., Purdue University, W. Lafayette, Ind., pp. 158–190.
31. Stewart, G. W., 1922, Acoustic Wave Filters, *Phys. Rev.* **20**:525; Stewart, G. W., 1924, *Phys. Rev.* **23**:520; Stewart, G. W., 1922, *Phys. Rev.* **25**:258; Stewart, G. W., and R. B. Lindsay, 1930, *Acoustics,* Van Nostrand, New York.
32. Mason, W. P., 1927, A Study of the Rectangular Combination of Acoustic Elements, Tapered Acoustic Filters and Horns, *Bell Systems Tec. J.* **6**:258–275.
33. Lindsay, R. B., 1938, Filtration of Sound, *J. Appl. Phys.* **9**:612–622; Lindsay, R. B., 1939, *J. Appl. Phys.* **10**:680–687.
34. Igarashi, J., et al., *Fundamentals of Acoustical Silencers,* I. Rep. No. 339, 1958, pp. 223-241; II. Rep. No. 344, 1959, pp. 67-85; III. Rep. No. 351, 1960, pp. 17-31, Aeronautical Research Institute of Tokyo.
35. Alfredson, R. J., and P. O. A. L. Davies, 1970, The Radiation of Sound from an Engine Exhaust, *J. Sound Vib.* **13**(4):389–408.
36. Alfredson, R. J., and P. O. A. L. Davies, 1971, Performance of Exhaust System Components, *J. Sound Vib.* **15**(2):175–196.
37. Young, C. I. -J., and M. J. Crocker, 1975, Prediction of Transmission Loss in Mufflers by the Finite Element Method, *Acoust. Soc. Am. J.* **57**(1):144–148.

38. Young, C. I. -J., and M. J. Crocker, 1976, Acoustical Analysis, Testing and Design of Flow-Reversing Muffler Chambers, *Acoust. Soc. Am. J.* **60**(5):1111–1118.

39. Young, C. I. -J., and M. J. Crocker, 1977, Finite Analysis Acoustical Analysis of Complex Muffler Systems With and Without Wall Vibrations, *Noise Control Eng.* **9**(2):86–93.

40. Craggs, A., 1976, A Finite Element Method for Damped Acoustic Systems: An Application to Evaluate the Performance of Reactive Mufflers, *J. Sound Vib.* **38**(3):377–392.

41. Ross, D. F., 1981, A Finite Element Analysis of Perforated Component Acoustic Systems, *J. Sound Vib.* **79**(1):133–143.

42. Munjal, M. L., and A. V. Sreenath, 1973, Analysis and Design of Exhaust Mufflers—Recent Developments, *Shock Vib. Dig.* **5**(11):2-14; Munjal, M. L., 1977, Exhaust Noise and Its Control—A Review, *Shock Vib. Dig.* **9**(8a):21-32.

43. Sullivan, J. W., Modelling of Engine Exhaust System Noise, in *Noise and Fluids Engineering,* Proceedings of 1977 Winter Annual Meeting of Am. Soc. Mech. Eng., pp. 161–169.

44. Crocker, M. J., 1977, Internal Combustion Engine Exhaust Muffling, *Noise Control Proc.,* pp. 331–358; Crocker, M. J., 1978, Review of Internal Combustion Engine Exhaust Muffling, in *Proceedings of a Symposium on Surface Transportation Exhaust System Noise,* October 11-13, Chicago, Ill., 1977, Environmental Protection Agency, pp. 295–357.

45. Anonymous, 1978, *Proceedings of a Symposium on Surface Transportation Exhaust System Noise,* October 11-13, 1977, Environmental Protection Agency.

46. Prasad, M. G., and M. J. Crocker, 1981, Evaluation of Four-Pole Parameters for a Straight Pipe with a Mean Flow and a Linear Temperature Gradient, *Acoust. Soc. Am. J.* **69**(4):916–921.

47. Prasad, M. G., and M. J. Crocker, 1981, Insertion Loss Studies on Models of Automotive Exhaust Systems, *Acoust. Soc. Am. J.* **70**(5):1339-1344; Prasad, M. G., and M. J. Crocker, 1981, A Scheme to Predict the Sound Pressure Radiated from an Automotive Exhaust System, *Acoust. Soc. Am. J.* **70**(5):1345-1352.

48. Prasad, M. G., and M. J. Crocker, 1983, Studies of Acoustical Performance of a Multi-Cylinder Engine Exhaust Muffler System, *J. Sound Vib.* **90**(4):491-508. See also: Prasad, M. G., and M. J. Crocker, 1983, Acoustical Source Characterization Studies on a Multi-Cylinder Engine Exhaust System, *J. Sound Vib.* **90**(4):479-490.

49. Anonymous, 1975, Diesel Engine Noise Conference, *Soc. Automot. Eng. Publ.* SP-397.

50. Anonymous, 1979, Proceedings of Diesel Engine Noise Conference, *Soc. Automot. Eng. Publ. P-80.*

51. Anonymous, 1982, Proceedings of a Symposium on Engine Noise, *Soc. Automot. Eng. Publ. P-106.*

52. Anonymous, 1979, Noise and Vibration of Engines and Transmissions, *Inst. Mech. Eng. (London) Proc.,* 176p.

53. Thien, G. E., 1973, The Use of Specially Designed Covers and Shields to Reduce Engine Noise, *Soc Automot. Eng. Paper 730244.*

54. Chung, J. Y., M. J. Crocker, and J. F. Hamilton, 1975, Measurement of Frequency Responses and the Multiple Coherence Function of the Noise-Generation System of a Diesel Engine, *Acoust. Soc. Am. J.* **58**(3):635-642.

55. Seybert, A. F., and M. J. Crocker, 1976, The Use of Coherence Techniques to Predict the Effect of Engine Noise, *J. Eng. Ind.* **97**(13):1227-1233.

56. Crocker, M. J., and J. F. Hamilton, 1980, Modelling of Diesel Engine Noise Using Coherence, *Soc. Automot. Eng. Trans.* **88**:1263-1273.

57. Chung, J. Y., J. Pope, and D. A. Feldmaier, 1979, Application of Acoustic Intensity Measurement to Engine Noise Evaluation, *Soc. Automot. Eng. Paper 790502.*

58. Reinhart, T. E., and M. J. Crocker, 1980, A Comparison of Source Identification Techniques on a Diesel Engine, *Inter-Noise 80 Proceedings,* Noise Control Foundation, New York, pp. 1129–1132.
59. Reinhart, T. E., and M. J. Crocker, 1982, Source Identification on a Diesel Engine Using Acoustic Intensity Measurements, *Noise Control Eng.* **18**(3):84–92.
60. McGary, M. C., and M. J. Crocker, 1981, Surface Intensity Measurements on a Diesel Engine, *Noise Control Eng.* **16**(1):26–36.
61. Hayden, R. E., 1972, Roadside Noise from the Interaction of a Rolling Tire with the Road Surface, in *Noise and Vibration Control Engineering,* M. J. Crocker, ed., Purdue University, West Lafayette, Ind., pp. 59–64.
62. Tetlow, D., Truck Tire Noise, *Sound Vib.,* August 1971, pp.17–23.
63. Leasure, W. A., Truck Tire Noise—Results of a Field Measurement Program, in *Noise and Vibration Control Engineering,* M. J. Crocker, ed., Purdue University, West Lafayette, Ind., pp. 38–45.
64. Anonymous, 1977, SAE Highway Tire Noise Symposium Proceedings, *Soc. Automot. Eng. Publ. P-70,* 324p.
65. *Proceedings from the 1979 International Tire Noise Conference,* National Swedish Board of Technical Development, Stockholm, 1980.
66. Apps, D. C., 1956, AMA-125-Sone-New Vehicle Noise Specifications, *Noise Control* **2**(3):13–17.
67. Crocker, M. J., and J. W. Sullivan, 1978, Measurement of Truck and Vehicle Noise, *Soc. Automot. Eng. Paper 780387.* Also published in *Soc. Automot. Eng. Trans.* **87:**1829–1845.
68. May, D., M. Osman, and E. Rose, eds., 1980, Vehicle Noise Regulation and Reduction, *Soc. Automot. Eng. Publ. SP-456,* 140p.
69. Anonymous, Surface Vehicle Sound Measurement Procedures, *Soc. Automot. Eng. Publ. HS-184.*
70. Crocker, M. J., 1978, Noise of Surface Transportation to Nontravellers, in *Handbook of Noise Assessment,* D. May, ed., Van Nostrand Reinhold, New York, pp. 39–81.

Reprinted from *Electr. Railw. J.* **73**(13):513–516 (1929)

REDUCING NOISE
of Car Operation

By

H. S. WILLIAMS
Assistant Superintendent of Equipment
Department of Street Railways
Detroit, Mich.

A noisy car has actually been made quiet. Tests show efficiency of various schemes for wheel silencing and results from consistent use of high standards of maintenance. Wheel noise can be reduced considerably by use of grooves filled with lead

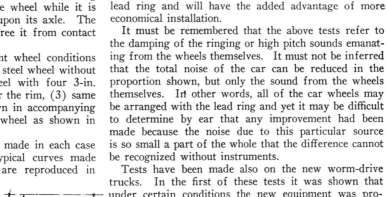

Results in eliminating noise plotted to logarithmic scale

TO DETERMINE the efficiency of various methods of wheel silencing numerous tests have been made with the noise-measuring apparatus of the committee on noise reduction of the American Electric Railway Engineering Association. The set-up for the wheel test is shown in an accompanying sketch (Fig. 1). The essential feature of this test consists of striking a uniform blow on the tread of the wheel while it is mounted in the customary manner upon its axle. The wheel under test is blocked up to free it from contact with the rail.

Tests were made of four different wheel conditions as follows: (1) Conventional forged steel wheel without any quieting device, (2) same wheel with four 3-in. wooden blocks bolted to the web near the rim, (3) same wheel with lead ring insert as shown in accompanying sketch (Fig. 2), and (4) resilient wheel as shown in another sketch (Fig. 3).

A sufficient number of tests were made in each case to obtain reliable results. Four typical curves made by the noise evaluating apparatus are reproduced in Fig. 4. From this the silencing value of the various schemes is readily apparent, their ratio being: (1) Conventional forged steel wheel, sound value, 100 per cent; (2) conventional wheel with wood blocks added, 85.9 per cent; (3) conventional wheel with lead insert, 49.7 per cent, and (4) resilient wheel, 16.4 per cent.

Since these tests were made another scheme has been proposed for accomplishing the same purpose. It consists of the use of a ring of iron welded in four spots to the under side of the wheel tread. This plan has been used successfully on gears. Tests are under way to fix a definite value to this scheme. It is believed that its effectiveness will equal that of the lead ring and will have the added advantage of more economical installation.

It must be remembered that the above tests refer to the damping of the ringing or high pitch sounds emanating from the wheels themselves. It must not be inferred that the total noise of the car can be reduced in the proportion shown, but only the sound from the wheels themselves. In other words, all of the car wheels may be arranged with the lead ring and yet it may be difficult to determine by ear that any improvement had been made because the noise due to this particular source is so small a part of the whole that the difference cannot be recognized without instruments.

Tests have been made also on the new worm-drive trucks. In the first of these tests it was shown that under certain conditions the new equipment was productive of greater noise than the standard apparatus, but the defect causing this has now been corrected and tests made on the new equipment. The results of these tests are summarized in Fig. 5, which shows a very definite improvement of sound with the use of the worm-drive truck and light-weight, high-speed motors. With this truck, gear noise is nearly zero and the balance of the improvement is largely due to decreased weight and closer fitting parts.

The question has been raised as to the effect of composition flooring as compared with ordinary wood flooring on noise

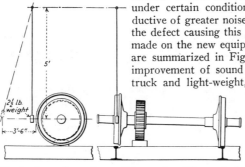

Fig. 1—Set-up for testing various sound muffling methods for car wheels

within the car. To secure information on this point tests were made on a car with wood floor as shown in Fig. 6. After this test was made the car floor was changed to composition floor as shown in Fig. 7, and this

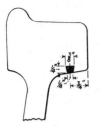

Fig. 2 — Wheel silencing by use of lead ring method

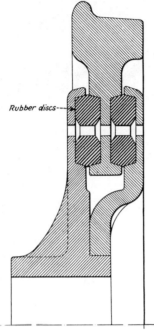

Rubber discs

Fig. 3—Type of resilient car wheel on which tests were made

test was repeated under the same conditions. The result of the several tests is shown in the graph which is reproduced as Fig. 8, which shows that the composition flooring has a slightly beneficial effect upon noise within the car.

There is need for greater accuracy in the manufacture of gears and pinions in the interest of noise reduction. Changes in construction of gearing, such as muffling, to produce quiet operation are entirely nullified if accuracy is lacking. Gear defects consist chiefly of warpage and eccentricity due to heat-treatment.

In checking over gear discard gages approved by the association in 1926 it appears that too great wear is allowed by the use of these gages. A layout to enlarged scale made of a section of a four-pitch gear and pinion

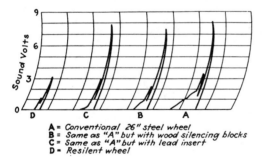

A = *Conventional 26" steel wheel*
B = *Same as "A" but with wood silencing blocks*
C = *Same as "A" but with lead insert*
D = *Resilent wheel*

Fig. 4—Graphical results of wheel silencing tests

is presented herewith as Fig. 9. This shows new teeth and the same teeth when worn to the limit allowable by the discard gage. The amount of back lash under this condition is $\frac{1}{4}$ in., which is too great. Moreover, by the use of the present recommended design for gear discard gage, gearing which just fails of condemnation by the gage would be kept in service for a considerable period, with the result that by the time the motor is again taken down the wear of the teeth will be much

greater than allowable. Noise values mount rapidly when the gear centers are spread. With a spread of gear centers amounting to $\frac{5}{32}$ in., the noise value more than doubles. A spread of this amount means a back lash of approximately $\frac{5}{64}$ in., or about one-fourth of the amount allowable by the discard gage. In view of this, it appears that the gear defect gage should be altered and limits of warpage and eccentricity established to cover the inspection of new gears.

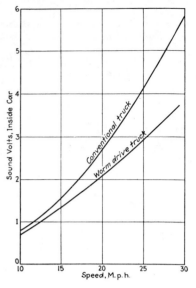

Fig. 5—Comparison of noise of standard and worm-drive trucks

Gear noise is caused by vibration in the material of the gear. The reasons for this may be summarized as follows:

1. Unequal loading of individual teeth, i.e., the load is borne by more teeth at one period than at another, and the increased compressive strains produce vibration.
2. Actual concussion caused by one tooth disengaging before another takes up the load. This allows the following tooth to strike a hammer-like blow with resultant noise and is apt to

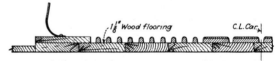

Fig. 6—Wooden floor construction used with tests

occur with gearing of low pitch or with spread gear centers. It may also occur during idling.
3. Inaccurate cutting or distortion due to heat treatment may cause the load to be borne by one tooth at a time whereas two or more should always be in engagement
4. Faulty alignment, causing the load to be carried on part of the tooth instead of being distributed over the whole width of the tooth.
5. Bottoming or actual interference of gear teeth.
6. Damaged gearing.

As a further step toward the manufacture of quieter spur gearing, it is suggested that attention be given to the use of gears of greater diametral pitch and consequently smaller teeth.

Tests were made on two stretches of track of differing construction but with identical car operation. One type of track was built with steel ties and compressed concrete, as shown in Fig. 10, while the other had granite nose block and compressed concrete pavement with rails

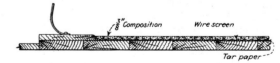

Fig. 7—Type of composition car floor used in tests

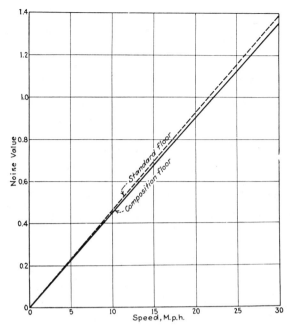

Fig. 8—Comparison of noise with wood and composition floor cars

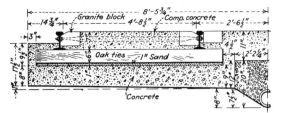

Fig. 11—Granite nose block and compressed concrete pavement with rails on wooden ties

without special selection. This change showed much improvement.

Following this test the gear cases were covered all over with ¾-in. felt and a third set of tests made. This was found to have resulted in still further reducing the noise. Next, the underside of the car floor was covered with ¾-in. soft felt and tests under this condition again showed material betterment.

The next step was to equip the trolley base with a resilient mounting. Rubber was used for this purpose,

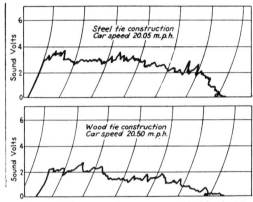

Fig. 12—Comparison of noise measured inside car while operating over track with wood and steel ties.

the type of support being identical with that used by the noise reduction committee in its air compressor tests of 1927. The entire weight of the base was carried on four rubber supports, one of which is shown in detail in Fig. 14. This mounting effectively absorbs the trolley vibration.

The final experiment included changing the car wheels

on wooden ties, as shown in Fig. 11. The sound was measured inside and outside the car, and typical curves showing the results are given in Figs. 12 and 13. These tests indicate that the wooden tie construction is materially quieter than the monolithic concrete type. It is hoped that additional tests on track noise can be made during the coming year, as it is felt that this is important and has been inadequately studied.

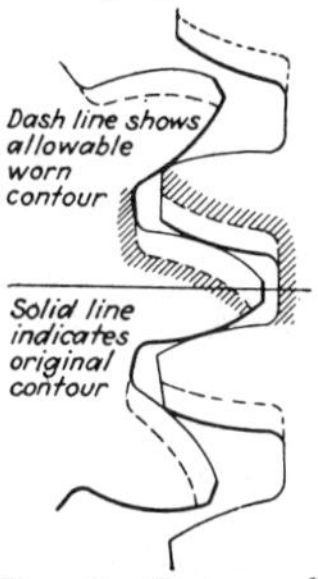

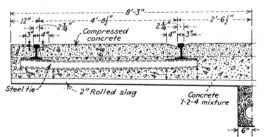

Fig. 9—Gear and pinion, showing original outline and worn condition allowable with present standard gear discard gage

Much has been done through tests to show what can be accomplished in the way of making present car equipment quiet. A modern type car seating 52 passengers and equipped with four 35-hp. motors was selected, it being at the time a noisy car. Progressive changes were made in the car to make it quiet, and by measurements taken the noise value of each step was determined.

Tests were first made with the car as found, to measure the amount of noise which it was creating. It was apparent to the ear that the gearing was very noisy so the car had its gears and pinions changed and the test repeated. In changing the gearing no special gears were used, merely new stock gearing

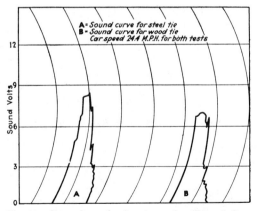

Fig. 13—Comparison of noise measured at the curb line while car was operating over track with steel and wood ties

Fig. 10—Track construction with steel ties and concrete

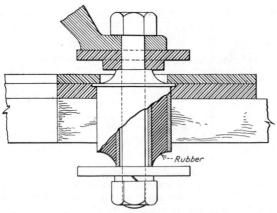

Fig. 14—Method of mounting trolley base to eliminate noise

and equipping with resilient wheels as shown in Fig. 3. One pair of these wheels uses heat-treated aluminum for the central section. Moreover, the construction of this wheel permits not only vertical motion but radial motion as well. Consequently, in starting the sudden jar on the gear is cushioned. In addition, this type of wheel improves the riding qualities of the car very materially.

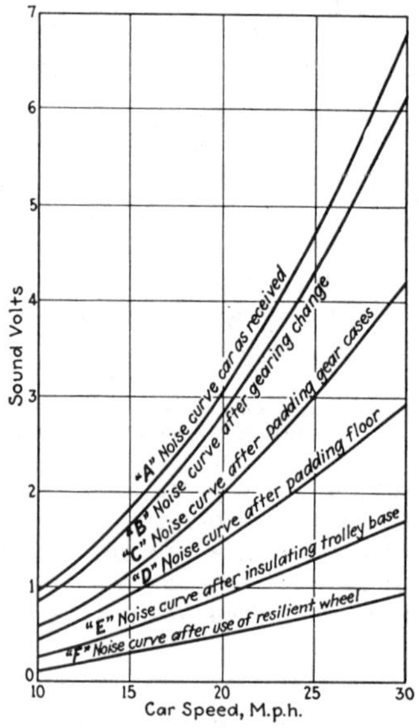

Results of these progressive tests are shown in Fig. 15, from which it will be seen at a glance that each step has been productive of definite results. It must be remembered that these curves represent definite scientific measurements made with all possible care and the personal element is eliminated. Sound vibration amplitude measurements are made with a recording meter while speed measurements are taken from an indicating tachometer. The same car was used throughout the tests and they were made over the same track with as nearly identical weather conditions as possible.

In Fig. 15 the results are plotted exactly as taken from the instruments. However, we are chiefly concerned with the effect of this sound upon the ear. The ear does not interpret sound in a direct ratio to the amplitude of the wave which produces it. The ear is affected in a nearly logarithmic ratio. So, in order to get a correct picture of the situation it is necessary to change Fig. 15

Fig. 15—Progressive results obtained in eliminating noise

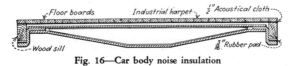

Fig. 16—Car body noise insulation

to a logarithmic scale, which has been done in the chart appearing at the beginning of this article, which shows the true view of the situation.

As a result of this investigation, it has been demonstrated that with moderate expense the noise inside the car—that is, the noise which is heard by the passenger—can be reduced by 31 per cent. This refers to the progressive changes mentioned up to the last step, which is the use of the resilient wheel. The last step is a much more expensive item, but with it included the total noise within the car can be reduced 47 per cent. The noise as measured at the curb is reduced 18 per cent by the whole procedure mentioned in this article.

As so much effort is being expended upon noise reduction, special attention is called to one of the basic principles involved in making sound insulation of maximum effect. This principle is to treat sound insulation exactly as if it were electrical insulation. Resilient padding is effective to damp the noise, but so long as there is metal contact between the parts a large amount of vibration will be transmitted. For example, with the resilient wheel the rim is entirely carried on rubber and no rigid metallic contact of any kind is permitted between the rim and the hub. If the bolts which assemble the parts of the wheel permitted metallic contact between the parts the effectiveness of the wheel would be reduced. The same is true of the trolley base mounting.

REPORT 1192

THEORETICAL AND EXPERIMENTAL INVESTIGATION OF MUFFLERS WITH COMMENTS ON ENGINE-EXHAUST MUFFLER DESIGN [1]

By Don D. Davis, Jr., George M. Stokes, Dewey Moore, and George L. Stevens, Jr.

SUMMARY

Equations are presented for the attenuation characteristics of single-chamber and multiple-chamber mufflers of both the expansion-chamber and resonator types, for tuned side-branch tubes, and for the combination of an expansion chamber with a resonator. Experimental curves of attenuation plotted against frequency are presented for 77 different mufflers with a reflection-free tailpipe termination, and the results are compared with the theory. The experiments were made at room temperature without flow; the sound source was a loud-speaker.

A method is given for including the tailpipe reflections in the calculations. Experimental attenuation curves are presented for four different muffler-tailpipe combinations, and the results are compared with the theory.

The application of the theory to the design of engine-exhaust mufflers is discussed, and charts are included for the assistance of the designer.

Noise spectrums are presented for a helicopter with each of the four muffler-tailpipe combinations installed. These spectrums are compared with the noise spectrum of the unmuffled helicopter. The results show that the overall noise level of the helicopter was reduced significantly by even the smallest of the four mufflers tested.

INTRODUCTION

A theoretical and experimental investigation of the methods of muffler design has been conducted at the Langley full-scale tunnel of the National Advisory Committee for Aeronautics as part of a general research program directed toward the reduction of airplane noise. The acoustic theory and muffler literature were studied with the aim of obtaining a method of predicting muffler characteristics. The theory of acoustic filters is discussed in reference 1. Sections of particular interest in connection with muffler design are the chapters on change in area of wave front, transmission through a conduit with an attached branch, and the filtration of sound, as well as the appendix which gives the branch-transmission theory of acoustic filtration. Experimental checks have been found in the literature which demonstrate that the theory of reference 1 is reasonably accurate for small

filters with stationary air at room temperature as the sound-conducting medium. When the derivation of the equations of the acoustic-filter theory is studied, however, certain assumptions are found which limit the maximum filter dimensions and also the maximum sound pressures for which these equations are applicable. Only limited data are available regarding the accuracy of the theory when applied to filters as large as engine-exhaust mufflers.

The British have studied the problem of aircraft mufflers with limited model experiments and with engine tests (refs. 2, 3, and 4). The model experiments show fair agreement with theory as to attenuation for a particular multiple resonator low-pass filter of the type described in reference 1 and for a multiple-expansion-chamber silencer. The experiments also showed a definite tendency for increasing flow velocity to increase the attenuation at low frequencies of expansion-chamber silencers. Air flow had little effect on the attenuation of the multiple resonator. In both cases, however, the flow velocities investigated were much lower than those which are found in engine-exhaust pipes. Muffler design has also been studied by the Germans with particular emphasis on mufflers for single-cylinder engines (refs. 5, 6, and 7). Ground tests of a large number of different mufflers on an actual engine are reported in references 8 and 9. The experimental results of reference 8 showed that, for the particular muffler discussed, both the low-frequency cutoff and the first high-frequency cutoff were near the calculated frequencies, which was encouraging. Unfortunately, however, the data of references 8 and 9 were not suitable for detailed verification of the theory because of interfering engine noise from sources other than the exhaust.

Although the literature indicated that certain acoustic theories could be useful in the design of engine-exhaust mufflers, neither the range of validity of the various theories with respect to muffler size nor the accuracy of the theories in predicting the attenuation of mufflers installed on actual engines could be deduced from the available data. It became apparent that, before more detailed information regarding the validity of the equations could be obtained, a test method was needed which would allow conditions to be

---

[1] Supersedes NACA TN 2893, "Theoretical and Measured Attenuation of Mufflers at Room Temperature Without Flow, With Comments on Engine-Exhaust Muffler Design" by Don D. Davis, Jr., George L. Stevens, Jr., Dewey Moore, and George M. Stokes, 1953 and NACA TN 2943, "The Attenuation Characteristics of Four Specially Designed Mufflers Tested on a Practical Engine Setup" by George M. Stokes and Don D. Davis, Jr., 1953.

closely controlled and which would reduce the number of variables involved. A relatively simple and fundamental approach seemed to be to develop a suitable apparatus and then to measure the attenuation characteristics of various types of mufflers in still air at room temperature. In order to eliminate the effects of tailpipe resonance, a termination with the characteristics of an infinite pipe was indicated. Such an attenuation-measuring apparatus was developed for the first part of this investigation.

The objective of this part of the investigation was to obtain from theoretical considerations equations for the attenuation of various types of mufflers and then to investigate the validity of these equations experimentally throughout a rather large range of muffler size in order to determine the limitations of the various equations with respect to muffler types, muffler dimensions, and sound frequencies. Because it is important in airplane-engine muffling to avoid excessive back pressures, only those types of mufflers which permit the exhaust gas to flow through the muffler without turning have been considered in this investigation.

Of course engine mufflers must be terminated with a tailpipe of finite length in actual practice. The influence of the finite tailpipe was studied in the second part of this investigation. A method for including the effect of the tailpipe in the muffler calculations was proposed, and an experiment was then conducted to investigate the validity of this method.

The problem of practical muffler design is discussed in Part III, and families of calculated attenuation curves for three types of mufflers are presented therein for the assistance of the designer.

The final part of this report describes an application of the theory to the design of four mufflers for a particular aircraft engine and the tests of these mufflers installed on the engine. The purpose of this part of the investigation was to study the practicality of the design methods and equations which had been developed and, also, to obtain some idea of the size of muffler which is required in practice to provide a significant noise reduction. Of particular interest was the question whether certain factors which had not been studied in the previous parts of the investigation would affect seriously the performance of the mufflers. Factors of primary concern were the very large sound pressures in the engine exhaust pipe and the flow velocity of the exhaust gas. In order to make possible a comparison of experimental data, the same mufflers were used for the finite tailpipe study and for the engine tests.

For an investigation of this nature, it is desirable to have an engine dynamometer stand; however, in this case, a helicopter was used for the engine tests because it was readily available. This was believed permissible because the helicopter rotor noise was expected to be lower than the engine noise, at least for the unmuffled engine.

SYMBOLS

a radius of connector between exhaust pipe and branch chamber

A displacement amplitude of an incident wave

B displacement amplitude of a reflected wave

c velocity of sound

c_0 conductivity of connector between exhaust pipe and branch chamber, $\dfrac{\pi a^2}{l_c + \beta a}$

d diameter of expansion chamber

f frequency

f_c cutoff frequency

I sound current

k wave-length constant, $2\pi f/c$

l' length of conical connector, measured along surface

l_1 length of pipe between connectors of two successive branches in a multiple resonator or length of pipe between two chambers of a combination muffler

l_2 length of resonant chamber

l_c one-half of effective length of connector between two expansion chambers or length of connector between exhaust pipe and branch chamber

l_e length of expansion chamber

l_t effective length of tailpipe

m expansion ratio; ratio of chamber cross-sectional area to exhaust-pipe cross-sectional area

M number of chambers in multiple-resonator muffler

n number of orifices or tubes which form connector between exhaust pipe and branch chamber

p sound pressure

R resistive component of impedance

S cross-sectional area

t time

V volume of resonant chamber

x distance coordinate measured along pipe

X reactive component of impedance

Z impedance

Z_0 characteristic impedance, acoustic resistance to transmission of a plane wave in a pipe, $\rho c/S$

β constant in conductivity equation

λ wave length, c/f

μ coefficient of viscosity of sound-conducting medium

ρ average density of sound-conducting medium

$\sigma = 4\pi \dfrac{l'}{\lambda}$

ξ instantaneous displacement of a particle of the medium in which a plane acoustic wave is transmitted

$\dot{\xi}$ instantaneous velocity of a particle of the medium in which a plane acoustic wave is transmitted

ω circular frequency, $2\pi f$

Subscripts:

b branch

c connector

i incident wave

r resonant

re reflected wave

t tailpipe

tr transmitted wave

Note: Bars ‖ are used to denote the absolute value (modulus) of a complex number.

I. INFINITE TAILPIPE

THEORY

The equations that have been used in the calculation of attenuation for the mufflers discussed in this report are derived and presented in the appendixes. Mufflers of the expansion-chamber type are treated in appendix A. The method used throughout the derivation of attenuation equations for single expansion chambers, double expansion chambers with external connecting tubes, and double expansion chambers with internal connecting tubes is that of plane-wave theory. In this theory the sound is assumed to be transmitted in a tube in the form of one-dimensional or plane waves. At any juncture where the tube area changes, part of the sound incident on the juncture is transmitted down the tube and part of it is reflected back toward the source. An expansion-chamber muffler consists of one or more chambers of larger cross-sectional area than the exhaust pipe, which are in series with the exhaust pipe. This type of muffler provides attenuation by taking advantage of the reflections from the junctures at which the cross-sectional area changes. A three-dimensional sketch of a typical double expansion chamber with an internal connecting tube is shown in figure 1 (a). The theory shows that below a certain frequency, which is called the cutoff frequency, the muffler is relatively ineffective. An approximate equation for determining this cutoff frequency has been derived and is presented in appendix A.

Mufflers of the resonator type are treated in appendix B. A typical single-chamber resonator is shown in figure 1 (b). This type of muffler consists of a resonant chamber which is connected in parallel with the exhaust pipe by one or more tubes or orifices. In certain frequency ranges the impedance at the connector is much lower than the tailpipe impedance. The resonant chamber then acts as an effective short circuit which reflects most of the incident sound back toward the source; thus, the amount of sound energy that is permitted to go beyond the muffler into the tailpipe is reduced. The attenuation equation for the single-chamber resonator is first derived by the method of lumped impedances; that is, phase differences between the two ends of the connector and between different points in the chamber are considered negligible. For this case, attenuation equations are developed first by considering the resistance in the connector and then by omitting this resistance; then, two additional equations, both of which omit the resistance, are developed. The first equation considers the effect of phase differences in the connector, whereas the second equation considers the effect of phase differences inside the chamber.

A typical multiple-chamber resonator is shown in figure 1 (c). For mufflers of this type, the equation given in reference 8 is used. In the derivation of this equation resistance is neglected, the connector and chamber are considered as lumped impedances, and the central tube between the resonators is treated as a distributed impedance. The sound in this central tube is considered to be transmitted in the form of plane waves. The multiple resonators, like the multiple expansion chambers, have a cutoff frequency. An

approximate equation for this cutoff frequency is also given in appendix B.

The conductivity c_0 is a very important physical quantity which enters into the determination of both the resonant frequency and the amount of attenuation for resonator-type mufflers. The quantity ρ/c_0 is, as is explained in reference 1, the acoustic inertance that is associated with a physical restriction in an acoustic conduit. Because this quantity is determined by the acoustic kinetic energy that is associated with the presence of the restriction and because this energy is a function of the conduit configuration on either side of the restriction as well as of the physical dimensions of the restriction itself, the conductivity is physically a rather elusive quantity which is predictable in only certain special cases, such as that of a circular orifice in an infinite plane. In most practical cases, it is therefore necessary to base an estimate of c_0 on past experimental evidence.

The prediction of c_0 is discussed in reference 1. In the case of a single connector, with diameter not too large in comparison with the exhaust-pipe diameter, the equation given is

$$c_0 = \frac{\pi a^2}{l_c + \beta a}$$

where β is an empirical constant, which has been found to be usually between $\pi/2$ and $\pi/4$. If the connector is composed

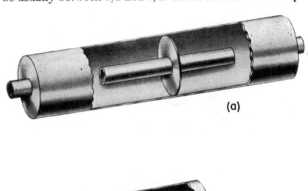

(a)

(b)

(c)   L-77028

(a) Double expansion chamber with internal connecting tube (muffler 19).
(b) A typical single-chamber resonator.
(c) Double-chamber resonator (muffler 54).

FIGURE 1.—Sketches showing internal details of several mufflers.

of several orifices, a further uncertainty is introduced since the interference effects among the orifices are not known. In this report, the calculated curves will be based on the experimentally measured conductivity in those cases where the calculated and experimental conductivities show significant differences. In a section immediately following the presentation of the single-resonator and multiple-resonator results, the problem of conductivity prediction is discussed with the assistance of the experimental results.

Equations are derived in appendix C for two types of combination mufflers. The first is a combination of two resonators tuned at different frequencies and the second is a combination of an expansion chamber and a resonator. Combinations of these types are shown in figures 1(d) and 1 (e).

L-77029

(d) Combination of a resonator and an expansion chamber (muffler 71).
(e) Two resonators tuned to different frequencies (muffler 73).
(f) Combination of several quarter-wave resonators (muffler 74).

FIGURE 1.—Concluded.

MUFFLERS

The mufflers used in the "infinite" tailpipe part of the experimental investigation were constructed of 18-gage sheet steel (0.049-in. thickness) and unless otherwise specified were of circular cross section. Seam welds were used throughout to prevent leakage between the adjacent chambers of the mufflers. In all cases, the exhaust-gas flow is from left to right. Three-dimensional sketches showing internal details of several of the mufflers are given as figure 1. Photographs of some of the mufflers are shown as figure 2. Results are presented for 74 mufflers that were built to fit a 3-inch-diameter exhaust pipe. These mufflers varied in diameter from 4 inches to 24 inches and in length from 1 inch to 96 inches. In addition, results are presented for three mufflers that were built to fit a 12-inch-diameter exhaust pipe.

The types of mufflers on which the most extensive tests were made are the single expansion chamber, the multiple expansion chamber, the single resonator, and the multiple resonator. The single-expansion-chamber mufflers were empty cylindrical tanks with inlet and outlet tubes centrally located at the two ends. Multiple expansion chambers were constructed by placing two or more expansion chambers in series and connecting them with either internal or external tubes. These connecting tubes varied in length from 0.05 inch (the thickness of the central baffle in the muffler) to 42 inches and had a diameter of 3 inches. Each of the single-resonator mufflers consisted of an enclosed volume connected to the exhaust pipe by either tubes or circular orifices. The resonant chamber was located either as a branch projecting from the side of the exhaust pipe or as an annular chamber concentric with the exhaust pipe. In this type of muffler and in others in which the muffling element is located in "parallel" with the exhaust pipe, the exhaust gas, as a whole, is not required to flow through the volume chamber as it is in the expansion-chamber type of muffler. The multiple-resonator mufflers consist of two or more identical resonators spaced at equal intervals along the exhaust pipe. A few mufflers were constructed of combinations of the above types. In addition, side-branch tubes with one end closed were investigated.

APPARATUS

The test apparatus used in this investigation is shown schematically in figure 3 and a photograph of the equipment used for testing the mufflers with 3-inch inlet diameter is shown as figure 4. The sound was generated by the 15-inch coaxial loud-speaker shown at the left and was conducted through a 3-inch tube to the muffler, which was attached to the tube by rubber couplings. The sound which passed through the muffler continued down a 3-inch tube to the termination, which consisted of several feet of loosely packed cotton. The section of the tube between the loud-speaker and the muffler is called the exhaust pipe in this report, and the section of the tube beyond the muffler is called the tailpipe.

(a) Mufflers for 3-inch-diameter exhaust pipe.

FIGURE 2.—A group of mufflers investigated.

(b) Central-tube diameter, 12 inches; muffler 74.

FIGURE 2.—Concluded.

Measuring stations at which microphones could be inserted were installed in the exhaust pipe and the tailpipe. These measuring stations had the same cross-sectional area as the tube and were so designed that the microphone, when inserted, produced only a slight restriction in the acoustic tube. Because of the interaction between the incident sound wave traveling toward the muffler and the wave reflected by the muffler traveling back to the loud-speaker, the sound pressure varied with distance along the exhaust pipe. A sliding measuring station was, therefore, installed in the exhaust pipe. Three stationary measuring stations, unevenly spaced, were inserted in the tailpipe between the muffler and the cotton termination. With a 3-inch pipe in the muffler position instead of a muffler, the cotton was adjusted until the reflections from the termination were minimized. Reflections were detected by differences in the sound pressures at the various tailpipe measuring stations. With the termination used in this investigation, the pressures at these three stations varied by a maximum of about $\pm \frac{3}{4}$ decibel for frequencies between 120 and 700 cycles per second and about $\pm 1\frac{1}{2}$ decibels for frequencies between 40 and 120 cycles per second.

A General Radio Company type 759–B sound-level meter was used to determine the sound-pressure levels at the measuring stations. The crystal microphone of this meter produced an electrical signal proportional to the sound pressure when it was inserted at the measuring stations. The meter indicated the sound-pressure level in decibels, defined as $20 \log_{10} \frac{p}{p_0}$ where p_0 is the standard base-pressure level of 0.0002 dyne per square centimeter. An oscilloscope and a sound analyzer were used as auxiliary equipment to make periodic checks of the wave form (freedom from harmonics) of the sound at the measuring stations.

The power supply for the loud-speaker consisted of the output of an audio oscillator feeding into a 50-watt amplifier. No harmonics were detectable within 40 decibels of the fundamental level in the input to the loud-speaker at the operating conditions used in this investigation. An electronic voltmeter was used to determine the input voltage supplied to the loud-speaker.

Part of the investigation involved the testing of three large mufflers in a 12-inch-diameter tube. A photograph of the apparatus used is shown in figure 5. In general, the

apparatus was similar in principle to the 3-inch apparatus. Traversing microphones operated by a pulley and cable arrangement were used in both the exhaust pipe and the tailpipe. In order to simplify the apparatus, the microphones were placed inside the 12-inch pipes, as shown in figure 6, where they imposed less than a 4-percent maximum area restriction. A cotton termination was again used, although it was not quite as effective as was the termination of the 3-inch apparatus.

FIGURE 5.—Apparatus used for testing mufflers designed for a 12-inch exhaust pipe.

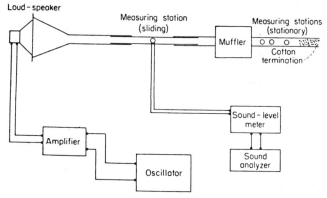

FIGURE 3.—Schematic diagram of experimental apparatus for infinite tailpipe investigation.

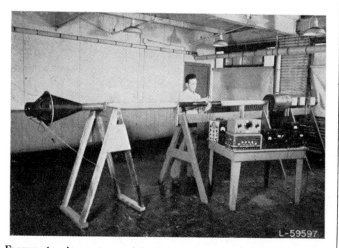

FIGURE 4.—Apparatus used for testing mufflers designed for a 3-inch exhaust pipe.

FIGURE 6.—Movable-microphone arrangement in the 12-inch tailpipe.

METHODS AND TESTS

In the tests of each of the muffler configurations, the maximum sound-pressure level obtainable at the sliding measuring station in the exhaust pipe and the sound-pressure levels at the three stationary measuring stations in the tailpipe were recorded. The data from the three tailpipe stations provided a running check on the absence of reflections in the tailpipe. The attenuation is defined as $20 \log_{10} \frac{p_i}{p_{tr}}$, where p_i is the incident-wave pressure in the exhaust pipe and p_{tr} is the transmitted-wave pressure in the tailpipe. The tailpipe data obtained in these tests give the true transmitted sound-pressure levels in the tailpipe, but the exhaust-pipe readings do not give the incident-wave sound-pressure levels in the exhaust pipe; instead, they give the maximum sound-pressure levels in the exhaust pipe. This maximum pressure is due to the superposition of the incident wave and the wave which is reflected from the muffler. In some cases, it is possible to calculate precisely the difference between the true attenuation and the quantity measured in these tests. This measured quantity is the maximum drop in sound-pressure level between the exhaust pipe and the tailpipe. The calculated difference can be applied as a correction to the experimental data. The corrected experimental data can then be compared with the calculated attenuation curves. Although this method provides an exact correction for the experimental data, it has certain disadvantages. It becomes quite tedious because separate correction calculations must be made for each separate muffler. Also, each time the muffler type is altered slightly, new equations must be derived. This process would become quite difficult and time-consuming for some of the more complicated muffler types. For these reasons a much simpler method of correction was devised, although at some sacrifice in terms of accuracy. This approximate correction was obtained as follows:

Assume that all sound reflection takes place from a single point and that the incident sound pressure is unity. If five percent of the incident wave is reflected, the maximum pressure in the exhaust pipe, which occurs at that point where the incident and reflected waves are exactly in phase, is 1.05. Then the sound-pressure level in the exhaust pipe will be $20 \log_{10} \frac{1.05}{1.00}$ or 0.42 decibel higher than the incident-wave sound-pressure level. Ninety-five percent of the incident pressure will be transmitted, so that the true attenuation will be $20 \log_{10} \frac{1.00}{0.95}$ or 0.45 decibel. The maximum drop which would be measured experimentally would be 0.42+0.45 or 0.87 decibel. By this procedure, table I was compiled, from which the approximate correction curve shown in figure 7 was plotted. This correction has been applied to all experimental data presented in Part I of this report. Some idea of the magnitude of the error introduced by using this approximate correction instead of the exact correction

may be obtained from figure 8, which was calculated for an expansion-chamber muffler. The top curve is the calculated difference between the maximum sound-pressure level in the exhaust pipe (at the point where the incident and reflected waves are in phase) and the sound-pressure level in the tailpipe (see eq. (A13)). The top curve is labeled "measured" because this is the quantity which, in the tests, was determined directly from experimental measurements. The lower curve shows the true attenuation of the muffler, based on the difference between the incident-wave pressures in the exhaust and tailpipes (eq. (A10)). The middle curve was obtained by applying the approximate corrections (fig. 7) to the measured attenuation curve. Note that the difference between the exact and approximately corrected attenuation curves is quite small at the higher values of attenuation.

Insofar as was practicable, the attenuation was calculated for each muffler tested by the theory of the appendixes, and the calculated attenuation curves and corrected experimental attenuation data were plotted. A maximum frequency of 700 cycles per second was chosen for the experiments because most of the exhaust noise energy is contained in the range below this frequency (ref. 9).

TABLE I.—CALCULATED CORRECTIONS TO MEASURED ATTENUATION VALUES (PART I)

Percent reflection	Rise in exhaust pipe=Correction, db	True attenuation, db	True attenuation +Rise=Measured attenuation, db
5	0.42	0.45	0.87
10	.83	.92	1.74
20	1.58	1.94	3.52
30	2.28	3.10	5.38
40	2.92	4.24	7.16
50	3.52	6.02	9.54
60	4.08	7.96	12.04
70	4.61	10.46	15.07
80	5.11	13.98	19.09
85	5.34	16.48	21.82
90	5.58	20.00	25.58
95	5.80	26.03	31.83
97	5.90	30.46	36.35
99	5.98	40.00	45.98
99.5	6.00	46.02	52.02
99.9	6.02	60.00	66.02
100	6.02	∞	∞

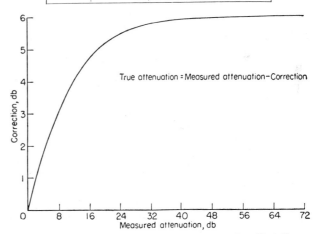

True attenuation = Measured attenuation-Correction

FIGURE 7.—Corrections to measured attenuation (Part I).

The leakage of room noise into the microphone at the tail-pipe measuring stations limited the minimum measurable noise level. Consequently, the maximum measured attenuation for any muffler tested was limited to about 50 decibels and at the higher frequencies was somewhat less. If the tailpipe measuring stations and the microphone had been better isolated from external noise and if the muffler walls had been rigid and nonconducting to sound, higher values of attenuation could have been measured. No attempt was made to obtain such measurements because values of attenuation higher than 50 decibels did not seem important to this investigation. In practice, noise transmission through the muffler walls prevents the attainment of even a 40-decibel attenuation with the usual thin-wall sheet-metal construction. Furthermore, other noise sources on an airplane are normally loud enough so that an exhaust noise reduction of the order of 50 decibels is not warranted.

RESULTS AND DISCUSSION

The results of this part of the investigation are presented in the form of curves of attenuation in decibels plotted against frequency in cycles per second. The curves have been calculated by the theory of the appendixes and they are accompanied by experimental points. The validity of the theory is examined by comparing the theoretical and experimental results. A sketch of each muffler is shown beside the corresponding attenuation curve. The unit of length for the dimensions or constants given below the individual sketches is 1 foot. The results for the various types of mufflers are presented in the following order:

    Expansion chamber (figs. 9 to 11)
    Resonator (figs. 12 to 14)
    Side-branch tube (fig. 15)
    Combinations (fig. 16)
    Large-diameter mufflers (fig. 17)

The speed of sound was about 1,140 feet per second and this number has been used to determine the wave lengths corresponding to the frequencies presented $\left(\lambda = \dfrac{c}{f}\right)$.

SINGLE EXPANSION CHAMBER

The attenuation in decibels of a muffler which consists of a single expansion chamber is given by the following formula (appendix A, eq. (A10)):

$$\text{Attenuation} = 10 \log_{10}\left[1 + \frac{1}{4}\left(m - \frac{1}{m}\right)^2 \sin^2 kl_e\right]$$

This equation indicates that the attenuation increases as the ratio m of the chamber area to the exhaust-pipe area increases and that the attenuation curve is cyclic, repeating itself at frequency intervals determined by the length of the muffler l_e and the velocity of sound inside the muffler c $\left(k = \dfrac{2\pi f}{c}\right)$.

Effect of expansion ratio.—The effect of varying the expansion ratio is shown in figure 9 (a) where m is varied from

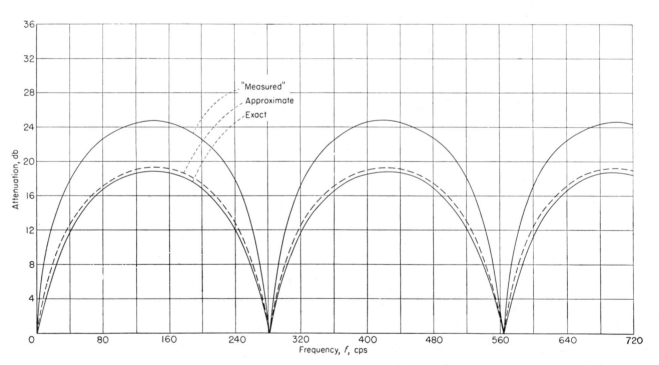

FIGURE 8.—Computed comparison of exact and approximately corrected attenuation curves for a single expansion chamber.

4 to 64. This figure shows clearly that the requirement for high attenuation is that the muffler have a large expansion ratio. Although the experimental points show some scatter, it appears that the theory is valid for muffler diameters as large as the wave length of the sound. This region of validity includes the region of practical interest in airplane-muffler design. The failure of the theory to predict the large loss of attenuation for muffler 4 at 700 cycles per second is believed to be due to the fact that the theoretical assumption of plane sound waves is no longer valid.

The complete solution for the velocity potential inside a circular tube shows that there are an infinite number of possible vibrational modes for the transfer of sound energy. Equation (A10) is based on the plane-wave mode, which may exist at any frequency. Other modes, which contain angular and radial nodes, are also possible at sufficiently high frequencies. Because the tubes and chambers which make up these mufflers are concentric, no vibrational modes which involve angular nodes would be expected. If these modes are eliminated, the lowest frequency at which any mode other than the plane wave can be transmitted without atten-

uation is given by $f = 1.22 \dfrac{c}{d}$. (The basic limiting condition is that $J_1\left(k\dfrac{d}{2}\right) = 0$, where J_1 is the Bessel function of the first kind of order 1, which has $\dfrac{2\pi f}{c}\dfrac{d}{2} = 3.83$ for its lowest root.) In terms of the wave length, this expression can be rewritten as $\lambda = 0.82d$. Thus, the assumption of plane waves is valid for wave lengths down to somewhat less than the chamber diameter. For muffler 4 the critical frequency given by this formula is 694 cycles per second. The experimental results show a sudden loss of attenuation between 650 and 700 cycles per second, which indicates that the appearance of this undamped higher vibrational mode has reduced seriously the muffler effectiveness.

Effect of length.—The effect of varying the length of the muffler is shown in figure 9 (b). The peak attenuation, about 20 decibels, is essentially unaffected by the length change and is a function only of the expansion ratio. The frequency at which this peak occurs is reduced, however, as the length of the muffler is increased. The frequency at which the peak attenuation occurs is inversely proportional to the muffler length. The cyclic nature of the attenuation

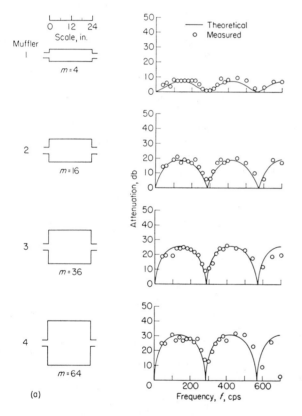

(a) Effect of expansion ratio m.

FIGURE 9.—Comparison of theoretical and experimental attenuation characteristics for single-expansion-chamber mufflers. Equation (A10).

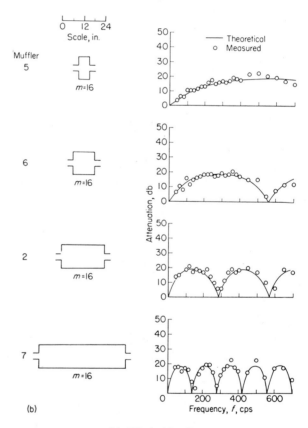

(b) Effect of length.

FIGURE 9.—Continued.

curve is evident with the attenuation dropping to zero for frequencies at which the muffler length equals an integral multiple of one-half the wave length $\lambda/2$ $\left(f=\dfrac{nc}{2l_e}\right.$, where n is any integer$\Big)$. The experiment and theory agree throughout the range tested which includes wave lengths as short as 0.4 of the muffler length in the case of the longest muffler. The theory contains no assumptions which directly limit this length. From the scale sketch of muffler 5, however, which has a diameter twice its length, it might appear that the sound waves inside the chamber would hardly be plane waves. Nevertheless, the experimental points are in good agreement with the plane-wave theory. Inasmuch as agreement is shown throughout the frequency range investigated ($\lambda=0.4l_e$ to $57l_e$), there appears to be no practical length limitation on the plane-wave theory for expansion chambers.

Effect of shape.—The effect of shape variations is shown in figure 9 (c). Tapering either or both ends of the chamber

has little effect on the muffler performance except for some loss of attenuation near 700 cycles per second. The acoustical length of these mufflers was measured from the longitudinal center of the tapered sections. Although the mufflers are relatively insensitive to the steep tapers tested, it is probable that long slender tapers would act as horns and would tend to reduce the muffler effectiveness severely at the high frequencies. This effect is demonstrated in figure 10, which shows the attenuation for conical connectors as a function of the wave length, taper length, and expansion ratio. The curves of figure 10 were calculated from the equation

$$\text{Attenuation}=10\log_{10}\left\{\left[1+\frac{(\sqrt{m}-1)^2}{\sqrt{m}}\left(\frac{1-\cos\sigma}{\sigma^2}\right)\right]^2+\left[\frac{(\sqrt{m}-1)^2}{\sqrt{m}}\left(\frac{\sigma-\sin\sigma}{\sigma^2}\right)\right]^2\right\}$$

where $m=\dfrac{S_2}{S_1}$ and $\sigma=4\pi\dfrac{l'}{\lambda}$. This equation was derived from equation (3.97) on page 86 of reference 1.

Changing from a circular to approximately elliptical cross section with the cross-sectional area held constant resulted in a loss of attenuation above 600 cycles per second (muffler 11, fig. 9 (c)). At this frequency the wave length is slightly less than the length of the major axis of the ellipse. The loss of attenuation is probably due to the appearance of a higher-order vibrational mode as was found in the case of muffler 4. The solution of the wave equation in elliptic coordinates (ref. 10) shows that the critical frequency for the mode which was found to limit the circular muffler 4 (the H_o mode in electrical terminology) is actually increased as the chamber becomes elliptic, whereas the measured critical frequency for muffler 11 is much lower than for a circular muffler of the same perimeter. Thus, some other vibrational mode, with a lower critical frequency, must be responsible for the loss of attenuation of muffler 11 above 600 cycles per second. The lack of circular symmetry in

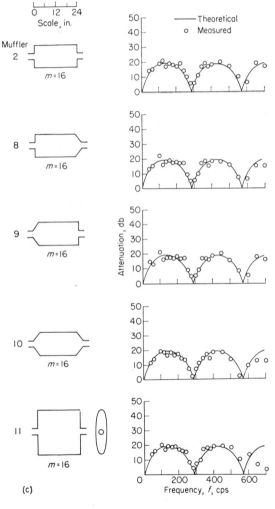

(c) Effect of shape.

FIGURE 9.—Concluded.

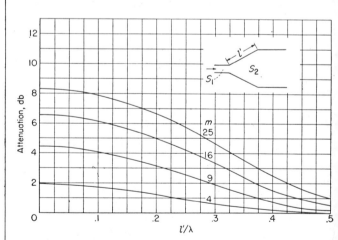

FIGURE 10.—Acoustical characteristics of truncated cone. (See ref. 1, p. 86.)

the elliptic case suggests consideration of the elliptical modes comparable to the unsymmetrical circular modes. Reference 10 describes two such modes oriented at right angles to each other. The mode which most closely matches the measured critical frequency is the odd H_1 mode.

In connection with the effect of changes of shape, reference 9 shows that large flat walls should be avoided wherever possible because of their tendency to vibrate and thus transmit exhaust noise energy to the atmosphere.

MULTIPLE EXPANSION CHAMBER

Equations are developed in appendix A for the attenuation of double expansion chambers with external connecting tubes and with internal connecting tubes. The method used in appendix A may also be used to develop equations for three or more expansion chambers connected in series. The data to be presented include calculated attenuation curves for the double expansion chambers.

Effect of number of chambers.—The effect of increasing the number of expansion chambers in a muffler is shown in figure 11 (a) where data are presented for mufflers of one, two, and three chambers. The maximum attenuation is shown to increase as the number of chambers is increased, although the addition of the third expansion chamber results in only a small increase in the measured attenuation. Because the attenuation of the three-chamber muffler was found to be quite similar to that of the two-chamber muffler, it appears that the addition of a third chamber will result in little increased attenuation for mufflers of practical construction. For this reason and because of the increased complexity of the calculations, the theoretical attenuation of muffler 13 was not calculated. A region of low attenuation is encountered at the lower frequencies with the multiple

expansion chambers. This region is predicted theoretically and will be discussed further. In the case of muffler 12, the calculated values agree fairly well with experiment down to a wave length about equal to the length of one of the chambers.

Effect of connecting-tube length with an external connecting tube.—Figure 11 (b) shows the effect of changing the length of the tube connecting the expansion chambers when this connecting tube is external to the chambers The frequency at which the low-frequency pass region (region of relatively low attenuation) occurs is shown to decrease as the length of the connecting tube is increased. An approximate formula for the upper-frequency limit, which is called herein the cutoff frequency, of this low-frequency pass region has been developed and is included as equation (A18) in appendix A. Cutoff frequencies determined from this equation are compared with those determined from the more exact equation (A17) in table II. The maximum attenuation in the first attenuating band above the low-frequency pass region is shown to increase as the connecting-tube length is increased. With the longer connecting tubes, regions of low attenuation, with a width of 50 cycles per second or more, occur between the large loops of the attenuation curves. These pass bands would be objectionable in a muffler if a significant amount of exhaust

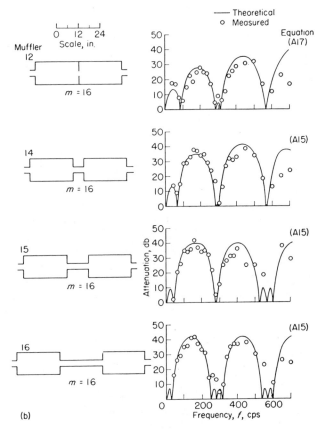

(b) Effect of connecting-tube length with an external connecting tube.

FIGURE 11.—Continued.

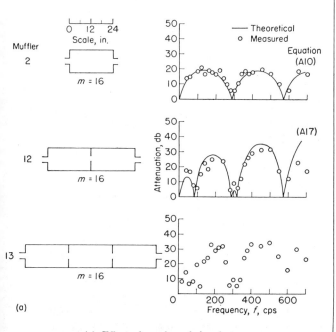

(a) Effect of number of chambers.

FIGURE 11.—Multiple-expansion-chamber mufflers.

TABLE II.—CUTOFF FREQUENCY FOR DOUBLE EXPANSION CHAMBERS

[$c = 1140$ fps]

Muffler	m	l_e, ft	l_c, ft	f_c, cps	
				Approximate (eq. (A18))	Exact (eq. (A17))
12	16	2	0.10	85.8	86.1
14	16	2	.25	59.9	59.1
15	16	2	.50	44.0	43.3
16	16	2	1.00	31.7	30.6
17	16	2	.25	59.9	60.6
18	16	2	.50	44.0	43.8
19	16	2	1.00	31.7	31.7
20	16	2	1.50	26.1	25.9
21	4	1	.50	123.2	122.4
22	9	1	.50	84.0	83.9
23	16	3	1.50	21.2	21.2

noise was present within these bands. The calculations and experiment are in agreement down to a wave length about equal to the length of one of the chambers.

Effect of connecting-tube length with an internal connecting tube.—Figure 11 (c) shows the effect of connecting-tube length when the connecting tube is symmetrically located inside the expansion chambers. The low-frequency pass region is again present and the frequency at which it occurs is again lowered as the connecting-tube length is increased. The cutoff frequency may be found approximately by using the same formula as in the case of the external connecting tubes (appendix A). The maximum attenuation in the first attenuating band above the low-frequency pass region is again increased as the connecting-tube length is increased. Also, pass regions are again encountered at the higher frequencies. The calculations again seem valid throughout most of the range investigated. When extremely high values of attenuation are calculated, the measurements are not accurate because of limitations of the apparatus. (See section entitled "Methods and Tests.") Very interesting results were obtained with muffler 19, for which the connecting-tube length was the same as the chamber length. The pass frequency at about 280 cycles per second, which is due to half-wave resonance of the expansion chambers, was eliminated. Although the attenuation did decrease in this region, the minimum attenuation measured was 27 decibels. The elimination of this pass region could prove quite useful in the design of a muffler which is required to attenuate over a wide frequency band. Further calculations and experiments have been made to investigate this phenomenon.

Effect of having the internal-connecting-tube length equal to the chamber length.—Results are shown in figure 11 (d) for four mufflers of different expansion ratios and lengths which had the common feature of an internal connecting tube of the same length as one of the expansion chambers. The results show, in all cases, that the pass region which normally occurs when the length of the expansion chamber is one-half the wave length is eliminated. This region is replaced by a region of reduced attenuation. The calculations for muffler 23 show that this phenomenon again occurs when the muffler length is 3/2 times the wave length. The pass region which occurs when the muffler length is equal to the wave length is not affected. The calculations show

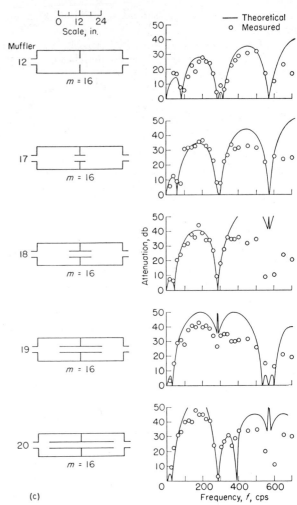

(c) Effect of connecting-tube length with an internal connecting tube. Equation (A17).

FIGURE 11.—Continued.

regions where the attenuation increases rapidly to infinity. Except for some discrepancy shown by the lower attenuation points in these regions, the calculations agree moderately well with the measurements down to a wave length about equal to the length of one of the chambers. The experiments, which were performed in advance of the detailed attenuation calculations, do not show points of extraordinarily high attenuation in these regions. A careful experimental survey which has since been made on another muffler of this general type, however, revealed in each such region a point of very high attenuation which was so sharply tuned that it appears to have no practical value.

Figure 9 (b) shows that, if a broad attenuation band is desired with a single expansion chamber, the chamber length should be reduced, but this reduction lowers the attenuation at low frequencies. If a longer double expansion chamber of the type shown in figure 11 (d) can be used, a broad attenuation band may be obtained without the loss of low-frequency attenuation, if the cutoff frequency is not too high. (Compare mufflers 6 and 19.)

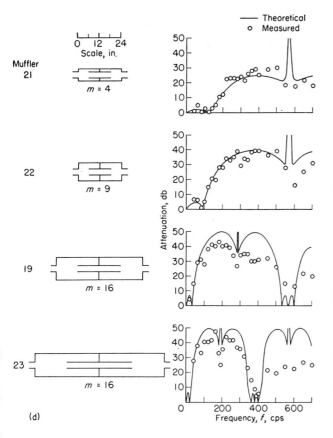

(d) Mufflers with internal connecting tubes equal in length to the individual chamber lengths. Equation (A17).

FIGURE 11.—Concluded.

PRINCIPLES OF SINGLE-CHAMBER RESONATORS

Figure 1 (b) is a sketch of a typical resonator-type muffler which consists of an enclosed volume surrounding the exhaust pipe, the volume being connected to the exhaust pipe through two short tubes. The pressure fluctuations in the exhaust pipe are transmitted to the volume chamber through the two small connecting tubes. Since these tubes are short compared to the wave length of the sound, the phase differences between the two ends of the tubes can be neglected. Thus, the gas in the tubes can be considered to move as a solid piston of a certain mass upon which the tube walls exert a certain viscous or friction force. As this effective piston of gas moves in and out, the gas inside the volume chamber undergoes alternate compression and expansion. The attenuation of such a resonator can be computed by substituting equations (B7) and (B8) into equation (B4) of appendix B. In a large number of practical cases, the friction force between the air and the walls of the connecting tube is sufficiently small that it can be neglected in comparison with the mass forces acting on the air in the connecting tube and the compression forces within the volume chamber. Because of this fact, the equation for the attenuation of a frictionless resonator is also presented in appendix B (eq. (B10)).

Single resonators of two very different physical configura-tions were investigated. The first configuration consisted of a resonant chamber located as a branch from the exhaust pipe. These resonators were, in general, relatively small and the calculations included viscous forces in the connecting tubes. In the second configuration, the resonator was an annular chamber surrounding the exhaust pipe (fig. 1 (b)). The resonators of this configuration were generally somewhat larger than those of the first configuration and viscous forces were omitted from the calculations.

BRANCH RESONATORS

Effect of varying resonator volume.—The effect of varying the chamber volume of a resonator is shown in figure 12(a). The calculated and experimental curves are in general agree-ment although there is here, as in the succeeding data of figure 12, a general tendency for the muffler to give a higher than calculated attenuation at frequencies above resonance and a lower than calculated attenuation at the resonant fre-quency. As the calculations indicate, decreasing the volume V raises the resonant frequency. These resonators are quite effective at the resonant frequency but the attenuation falls off rapidly at lower or higher frequencies.

Effect of varying c_0 and V with the ratio $\sqrt{c_0/V}$ constant.—Figure 12(b) shows the effect of varying c_0 and V together while keeping their ratio constant. The resonator equation states that the resonant frequency of a group of mufflers should be constant if the ratio $\sqrt{c_0/V}$ is constant. This ratio will be called the resonance parameter. Mufflers 27 and 25 are found to have the same resonant frequency, but muffler 27 has a broader region of attenuation. This broader attenu-ation region is predicted by the theory and is due to the

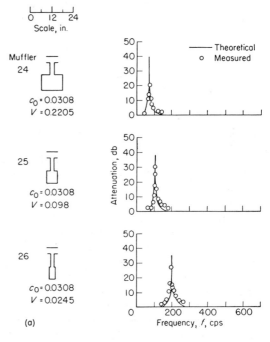

(a) Effect of volume V. Equation (B4).

FIGURE 12.—Single-chamber resonators with resonator chambers separate from tailpipe.

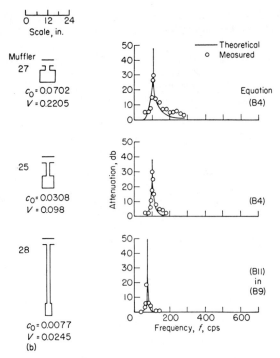

(b) Effect of varying volume V and conductivity c_0 together, with $\sqrt{c_0/V}$ constant.

FIGURE 12.—Continued.

larger values of both c_0 and V for muffler 27. The value of the parameter $\sqrt{c_0\overline{V}}/2S$, which will be called the attenuation parameter, is increased to more than twice that for muffler 25. The data for muffler 28 show a decrease in the resonant frequency. This apparent contradiction of the theory is due to the fact that the connecting tube in muffler 28 is not negligibly short compared to the wave length. The calculated attenuation curve for muffler 28 was obtained by taking into account the wave nature of the sound flow in the connecting tube. (See appendix B, eq. (B11).) At the resonant frequency of this muffler, the length of the connecting tube is of the order of one-fifth of the wave length.

Effect of varying cross-sectional area of the connecting tube.—Increasing the connecting-tube area increases c_0; thus, the values of both the resonance and attenuation parameters $\sqrt{c_0/V}$ and $\sqrt{c_0\overline{V}}/2S$ are increased. Consequently, the resonant frequency is increased and the attenuation region becomes broader (fig. 12(c)). A comparison of mufflers 29 and 30 shows that, if an attempt is made to obtain low-frequency attenuation simply by decreasing c_0, the result may be very disappointing. Both the magnitude of the attenuation and the width of the attenuation region decrease as c_0 decreases.

Effect of varying length of connecting tube.—Increasing the connecting-tube length decreases c_0 and, therefore, has the opposite effect from an increase of the connecting-tube area. This is shown in figure 12(d). Note again that, when the resonant frequency is decreased without changing the volume, the attenuation region becomes narrower.

Effect of changing connecting-tube configuration with c_0 held constant.—Although the conductivity c_0, is an important quantity in the attenuation equation, the physical configuration of this conductivity enters into only the viscous resistance term which is very small for most of the

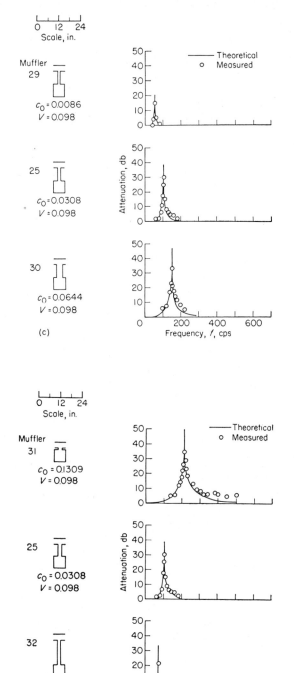

(c) Effect of area of connecting tube S_2. Equation (B4).
(d) Effect of length of connecting tube. Equation (B4).

FIGURE 12.—Continued.

resonators tested. Thus, the characteristics of a resonator are theoretically nearly independent of the manner in which the conductivity is obtained. The actual effect of changes in the physical configuration of the conductivity was investigated by testing three mufflers which had different connecting tubes but the same c_0 and V (fig. 12 (e)). Although mufflers 26 and 33 give about the same results, muffler 34, which has the smallest connecting tube, gives less attenuation than either of the other mufflers. In this connection, a definite, though often unrecognized, limitation of the linearized acoustic theory is of interest. If the three resonators in figure 12(e) are to have the same attenuation, it is necessary that the mass flow in the connecting tubes be the same. But this condition requires a higher velocity as the tube diameter is reduced. Inasmuch as the linearized theory requires that the changes in velocity, pressure, and density be small, it follows that for a given pressure in the exhaust pipe a limiting tube diameter exists below which the velocity is so high that the theory is not valid. This phenomenon has an important bearing on the design of engine-exhaust mufflers. The velocity in a connecting tube of fixed diameter will increase as the sound-pressure level in the exhaust pipe increases. Inasmuch as the sound pressures inside an engine exhaust pipe are extremely high, care must be exercised to avoid a connecting tube which is too small to permit the required flow into and out of the chamber. Apparently this muffler limitation has never been investigated on an actual engine. Muffler configuration 30 of reference 9, however, is interesting in this connection. The performance of this muffler was initially disappointing, but when additional conductivity holes were added (con-

figuration 31, ref. 9) the attenuation was markedly improved, even though the c_0 was much larger than was desired. Perhaps this muffler would have been even better if the ¼-inch orifices had been replaced by a few tubes of ¾-inch to 1-inch diameter which had the same c_0 as the ¼-inch orifices.

CONCENTRIC RESONATORS

In general, the resonators so far discussed have had relatively narrow attenuation bands. They would be useful in quieting a fixed-frequency noise source but are inadequate for use on a variable-speed engine or even on a fixed-speed engine with objectionable noise spread over a wide frequency band. For engines of these types a much broader attenuation band is desired. Basically, a broader band requires increased chamber volume and conductivity. Results are presented in figure 13 for single-chamber resonators of larger volume than those presented in figure 12. The mufflers shown in figure 13 are of conventional arrangement with the chamber located concentric with the exhaust pipe.

Effect of varying $\sqrt{c_0 V}/2S$ with the resonance parameter constant.—The data of figure 13(a) show the expected

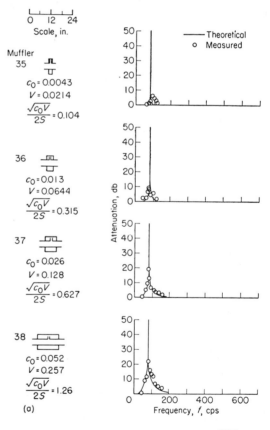

(a) Effect of varying the attenuation parameter $\dfrac{\sqrt{c_0 V}}{2S}$ with the resonance parameter $\sqrt{\dfrac{c_0}{V}}$ constant. Equation (B10).

FIGURE 13.—Single-chamber resonators with resonator chambers concentric with tailpipe.

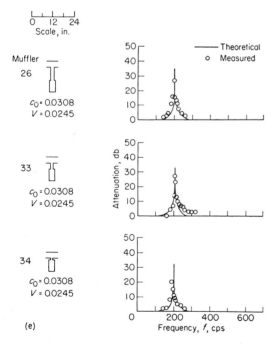

(e) Effect of varying connecting-tube area and length together, with c_0 constant. Equation (B4).

FIGURE 12.—Concluded.

broadening of the attenuation region as the value of the attenuation parameter is increased while the resonance parameter $\sqrt{c_0/V}$ is kept constant. The resonant frequency was constant as predicted by the theory. Viscous forces were omitted from the calculations for these and all other mufflers shown in figure 13.

A similar investigation was made with the resonators tuned for a higher frequency and with orifices used for the connector instead of tubes (fig. 13 (b)). All four mufflers were designed for a resonant frequency of 280 cycles per second, but mufflers 40 and 41 resonate at higher frequencies. In each of these two mufflers the conductivity was much higher than was expected. This result illustrates a serious problem in muffler design—that of predicting the conductivity of a group of orifices. This problem is considered further after the multiple-resonator data have been discussed. The calculated curves for mufflers 40 and 41 were obtained by using the c_0 as determined from the measured resonant frequency and the chamber volume. No definite resonant frequency was observed for muffler 42.

The measured attenuation of mufflers 41 and 42 falls below the calculated curves in the region near 600 cycles per second. The chamber is about one-half wave length long at this frequency and thus violates the theoretical assumption that the dimensions of the chamber are small compared to the wave length of the sound. Muffler 40, however, does not show this loss of attenuation at 600 cycles per second.

Effect of varying the chamber length and connector location with the chamber volume constant.—A group of mufflers was investigated in which the length and diameter of the resonator chamber and the location of the connector were varied while holding the chamber volume and the connector configuration constant (fig. 13 (c)). The measured attenuation of muffler 41 agrees with the calculated values except for the previously mentioned dip at 600 cycles per second. The resonator theory gives the same calculated attenuation for all of the mufflers shown in figure 13 (c). Actually no two of the five mufflers have the same measured attenuation. The explanation is found in the length of these mufflers. At the frequency at which muffler 41 resonates, the length of muffler 43 is about two-thirds of the sound wave length; therefore, it seems necessary to consider the wave nature of the sound. With this consideration, it is found that, when the distance from the connector to the end of the chamber is approximately one-fourth wave length, the reflection from the closed end of the chamber is 180° out of phase with the incoming pressure wave at the connector location. This results in high attenuation. For the configuration of muffler 43

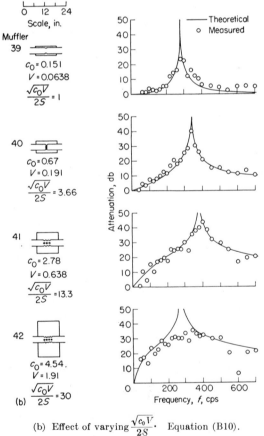

(b) Effect of varying $\dfrac{\sqrt{c_0 V}}{2S}$. Equation (B10).

FIGURE 13.—Continued.

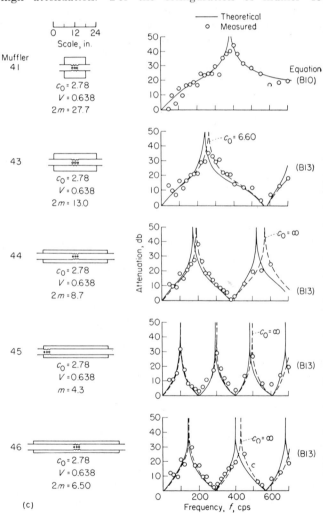

(c) Effect of chamber length and connector location with constant chamber volume.

FIGURE 13.—Continued.

(centrally located connector), this condition occurs when the muffler length is one-half the sound wave length. Inasmuch as this condition is satisfied at a frequency lower than the resonant frequency predicted from the values of c_0 and V, the fact that the resonator calculations fail to predict the characteristics of this muffler is not surprising.

Because the resonator theory was inadequate for mufflers 43 to 46, it was necessary to develop a different theory, based on the distributed impedance of assumed plane waves in the chambers. An equation derived for this case (appendix B, eq. (B13)) was used to calculate the attenuation of mufflers 43 to 46. In applying equation (B13) to the mufflers with the c_0 in the center (mufflers 43, 44, and 46), the chambers were considered to be the equivalent of chambers of twice the cross-sectional area and half the length of the actual chambers. Thus m was replaced by $2m$, S_2 by $2S_2$, and l_2 by $\frac{1}{2}l_2$ in making the calculations. The value of c_0 for these mufflers was first assumed equal to the measured c_0 value for muffler 41, because the hole configurations were identical. The resulting attenuation curves are shown by the solid lines in figure 13 (c). The calculated curves (solid lines) did not give the correct resonant frequencies. Consideration of the sketches of these mufflers indicated that it was probably incorrect to assume a constant c_0 for this group of mufflers.

A simple consideration can be used to show that the c_0 is a function not only of the connector but also of the objects which it connects. Consider a thin baffle, containing a small orifice, placed in a tube of very large diameter. The c_0 of the orifice then equals the orifice diameter. If, now, the diameter of the large tube be continuously decreased until it reaches the orifice diameter, the same orifice will simply form part of the tube and the c_0 will be infinite. In figure 13 (c), the effective area ratio between the exhaust pipe and the outer chamber varies from 27.7 to 4.3, and it seems reasonable to expect that as this ratio decreases and the pipe and chamber areas become better matched the c_0, for the same orifices, will increase. As a test of this reasoning, the attenuation was calculated for muffler 43 by using $c_0 = 6.60$ and for mufflers 44, 45, and 46 by using the limiting value $c_0 = \infty$. Comparison of the dashed and solid curves with the experimental data shows that the c_0 must be much higher for these mufflers than for muffler 41.

A comparison of the simple resonator theory with the more exact plane-wave theory will help to define the limitations of the simple theory, which is a "lumped impedance" theory. The impedance of the volume chamber is given as

$$-i \frac{\rho c}{S_2} \cot k l_2$$

by the plane wave or "distributed impedance" theory (note second term of eq. (B12)). If the assumption is made that $\tan k l_2 = k l_2 = \frac{\omega}{c} l_2$, the chamber impedance becomes

$$-i \frac{\rho c^2}{\omega S_2 l_2} = -i \frac{\rho c^2}{\omega V}$$

This is the value used in the lumped-impedance theory, and the difference in chamber impedance is the only difference between the two theories. When l_2 is one-eighth of the sound wave length, this difference is about 10 percent of the chamber impedance, and the error increases as the ratio l_2/λ increases. Because $\cot k l_2$ is a cyclic function, the distributed-impedance theory predicts a series of resonant frequencies, whereas only a single resonant frequency is predicted by the lumped-impedance theory. The experimental results show that with the appropriate value for c_0 the distributed-impedance theory is valid throughout the frequency range for mufflers 43 to 46, inclusive.

Comparison of the two theories indicates that the lumped-impedance theory is valid in the region near and below the first resonant frequency if l_2 is less than one-eighth of the wave length at the resonant frequency. In order to compare further the two theories, the attenuation of muffler 41 has been calculated by both methods. The value of c_0, computed by the distributed-impedance theory from the measured resonant frequency, was found to be almost double the value that was used in the lumped-impedance calculation (fig. 13 (c)). The attenuation calculated with this higher value of c_0 in the distributed-impedance equation (B13), however, differed from that calculated with the lumped-impedance equation by a maximum of only 1.4 decibels at a frequency of 700 cycles per second. Thus, in the case of muffler 41, the lumped-impedance theory has been extended to a case where l_2 is 0.175 times the resonant wave length by the expedient of using a fictitious value of c_0 that is much lower than the actual c_0 as given by the distributed-impedance theory. This fictitious value of c_0 was determined by using the equation

$$k_r = \sqrt{\frac{c_0}{V}}$$

from the lumped-impedance theory and by using the measured resonant frequency to determine k_r.

A comparison of the results for mufflers 44 and 45 shows that the attenuation region between two consecutive pass regions is wider when the conductivity is in the center of the muffler than when it is at one end, because of a decrease in the effective chamber length and an increase in the effective area ratio. The effect of the difference in chamber length, which changes the resonant frequency, can be eliminated by dividing the width of the attenuation region for a particular muffler by the resonant frequency of that muffler. A comparison on this basis shows that in the first attenuation band muffler 44 provides 10 decibels or more of attenuation over a frequency range of about 1.2 times the resonant frequency, while muffler 45 provides this attenuation over a range of only 0.8 times the resonant frequency. This difference in relative width of the attenuation bands is due to the difference in the effective area ratios. Mufflers based on this phenomenon of plane-wave resonance of the chambers are discussed further in a subsequent section of this report.

Venturi-shaped central tube.—The data that have been presented show that the width of the attenuation band can be increased by increasing the value of the attenuation parameter $\sqrt{c_0 V}/2S$. It is obviously possible to increase the value of this parameter without increasing the external size of a muffler if the area S is reduced. A significant reduction of the exhaust-pipe and tailpipe area is, however, impractical for most aircraft engines because of the resultant increase in engine back pressure. An idea for avoiding this difficulty has nevertheless been devised. It was believed that a significant decrease in the central-tube area at the connector location might be obtained without excessive back pressure if the central tube of the muffler were built in the shape of a venturi with the connector located at the throat. The acoustics of such a muffler were investigated by designing and testing a muffler with the same external dimensions as muffler 40 but with a venturi-shaped central tube which reduced the area at the connector by a factor of four. The data of figure 13 (d) show that the modified muffler 47 provides much more attenuation than muffler 40. This increase is particularly striking in the region above the resonant frequency. For comparative purposes, a theoretical curve is shown which gives the attenuation of a muffler having the same values of c_0 and V as muffler 47 but which has an exhaust pipe of constant diameter equal to the minimum diameter (1.5 in.) of the pipe of muffler 47. For frequencies above about 70 percent of the resonant frequency, muffler 47 provides approximately the attenuation of such a muffler. Thus, in cases where some additional back pressure is permissible, the venturi-shaped central tube is a powerful means for increasing the attenuation of a muffler of fixed external dimensions. Design curves based on equation (B10) show that a significant attenuation increase is obtained if the area is reduced by a factor of two.

MULTIPLE RESONATORS

If it is desired to increase the amount of attenuation from a resonator-type muffler, one obvious possibility is to combine two or more resonators in a single muffler. A muffler of this type with two consecutive identical resonators is discussed in reference 8. An equation for the attenuation is included along with other approximate equations useful in the preliminary design of such mufflers. The attenuation equation of reference 8 has been modified in appendix B (eq. (B15)) to emphasize the important parameters. In addition to the attenuation parameter $\sqrt{c_0 V}/2S$ and the resonance parameter $\sqrt{c_0/V}$, the distance between connectors l_1 is found to be a third important parameter. The attenuation is directly proportional to the number of resonant chambers in the muffler. The validity and range of application of this attenuation equation have been investigated by testing a group of mufflers of the multiple-resonator type (fig. 14).

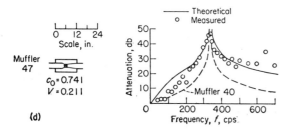

(d)

(d) Effect of venturi-shape contraction in central tube. Equation (B10).

FIGURE 13.—Concluded.

Effect of number of chambers.—The calculated and measured attenuation characteristics of mufflers composed of one, two, and three consecutive resonators are shown in figure 14 (a). For the single-chamber resonator, muffler 48, the attenuation has been calculated by both the multiple-resonator equation and the equation used in the preceding section for single resonators. The single-resonator equation is fairly accurate for wave lengths longer than $4l_2$ but is considerably in error for shorter wave lengths (higher frequencies). As would be expected, however, it does predict the resonant frequency. The multiple-chamber equation is inaccurate through most of the range but predicts the resonant frequency and the pass frequencies accurately. Inasmuch as the multiple-resonator formula is derived for an infinite filter of identical chambers, the experimental results show that a single resonator produces less attenuation than is predicted for one resonator of an infinite filter.

The data for mufflers 49 and 50 show that the attenuation increases with the number of chambers. Limitations of the apparatus prevented the measurement of the extremely high peak attenuation of these mufflers. General agreement with the theory is found except at the higher frequencies. There is some question as to the cause of the loss of attenuation at high frequencies. Since the attenuation, even though less than predicted, is still quite high, it is not certain that failure of the attenuation equation is responsible. Vibration of the muffler walls may be transmitting high-frequency sound into the tailpipe. Also, the leakage of external noise into the microphone at the measuring stations, which limited the maximum measurable attenuation at lower frequencies to about 50 decibels, may have increased at the higher frequencies, so that the measurable attenuation is limited to somewhat less than 50 decibels.

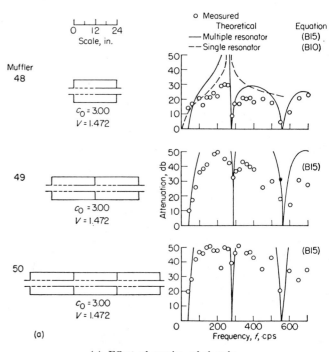

(a)

(a) Effect of number of chambers.

FIGURE 14.—Multiple-chamber resonators.

Effect of diameter with resonance parameter constant.— If the diameter of the muffler is increased while the resonance parameter remains constant, the value of the attenuation parameter will increase. The experimental data of figure 14 (b) confirm the theoretical prediction that this increase in the value of the attenuation parameter will result in an increase in both the magnitude of the attenuation and the width of the first attenuation band. The low frequency cutoff occurs at lower frequencies as the diameter is increased. The cutoff frequency for these three mufflers has been computed in three different ways. The results are shown in the following table:

| Muffler | Values of f_c | | |
	Exact (eq. (B15))	Approximate (eq. (B16))	Equation (B3) of reference 8
51	87.0	87.3	146.5
52	63.2	63.5	93.8
53	40.1	40.4	45.7

This table shows that for these particular mufflers equation (B16) is sufficiently accurate for preliminary design calculations. The assumption made in obtaining equation (B3) of reference 8, however, is not permissible for these mufflers.

Although both mufflers 51 and 52 show a sharp drop in measured attenuation at the predicted cutoff frequency, the attenuation does not drop to zero until well below the predicted cutoff frequency. This lack of agreement may be due to the fact that the mufflers had only two chambers, whereas the theoretical cutoff frequency was based on an infinite number of chambers. It is known that for a single chamber the cutoff frequency is zero, and it seems plausible that f_c may approach the predicted value only as the number of chambers becomes large.

Effect of length.— Mufflers 53 and 54 differ in both length and volume but the resonance parameter has been kept constant (fig. 14(c)). Comparison of these two mufflers shows that increasing the length decreases the frequency at which the first upper pass band occurs. The attenuation characteristics of muffler 55 are of an altogether different type. It has been pointed out in connection with single resonators that an attenuation curve of this type is characteristic of mufflers in which the plane-wave nature of the sound in the chamber is predominant. Muffler 55 is so long that the plane-wave resonance occurs in the chambers at a lower frequency than the volume resonance. Consequently, it has been necessary to consider the wave nature of the sound field in the chambers in making the calculations. This was accomplished by making use of equation (B12) for the branch impedance. The dashed curve shows the attenuation of two chambers of an infinite filter and was obtained by substituting equation (B12) for Z_b in equation (B14). The solid curve shows the attenuation of a two-chamber muffler terminated by an infinite tailpipe and was obtained by using equation (C8) in equation (C7). The branch impedance was again obtained from equation (B12). The attenuation of muffler 54 has also been computed by using equation (B12) for the branch impedance. The results, shown by the dashed curve, indicate that the sudden increase in attenuation at frequencies of 320 and 600 cycles per second is due to length resonance in the chambers.

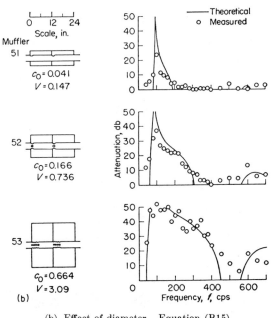

(b) Effect of diameter. Equation (B15).

FIGURE 14.—Continued.

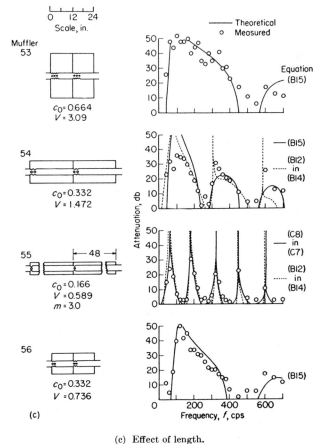

(c) Effect of length.

FIGURE 14.—Continued.

Muffler 56 differs physically from muffler 54 in length alone. This decrease in length, however, affects all three muffler parameters. The result is an increase in the cutoff frequency, an increase in the resonant frequency, an increase in the width of the first attenuation band, and an increase in the width of the first upper pass band.

Effect of conductivity.—Figure 14(d) presents results for a group of mufflers identical except for values of c_0. In all cases, tubes were used to obtain the conductivity. In general, the effects of increasing the conductivity are correctly predicted by the theory. For instance, both the experiment and the theory show that the cutoff and the resonant frequencies are raised, the first attenuation band is widened, and the first upper pass band is narrowed. The attenuation, however, did not drop completely to zero at the calculated cutoff frequency. Muffler 57, which had a very low conductivity, failed to produce the high attenuation predicted near the resonant frequency. This is believed to be due to viscous effects. Another indication of the effect of viscosity is obtained by comparing mufflers 59 and 60. Although both mufflers had the same values of c_0, muffler 60, which had

larger diameter tubes, gave more attenuation at frequencies near resonance. The first attenuation band extended to higher frequencies than were predicted for both of these mufflers, although the attenuation was less than 10 decibels at these higher frequencies.

Figure 14(e) shows results from a group of mufflers similar to those shown in figure 14(d). In this case, however, orifices were used to obtain the conductivity. The trends are quite similar to those shown in figure 14(d). Note from the experimental data that, if the value of c_0 is sufficiently high, the first upper pass band is narrowed until it is almost eliminated. At the same time, however, the cutoff frequency is continually increased.

Elimination of the first upper pass band.—Consideration of equation (B15) indicates that it might be possible to eliminate the first upper pass band ($\sin kl_1 = 0$) by choosing the resonant frequency such that $\frac{f}{f_r} = 1$ when $\sin kl_1 = 0$. A case of this type is shown in the design curve for $k_r l_1 = \pi$. In

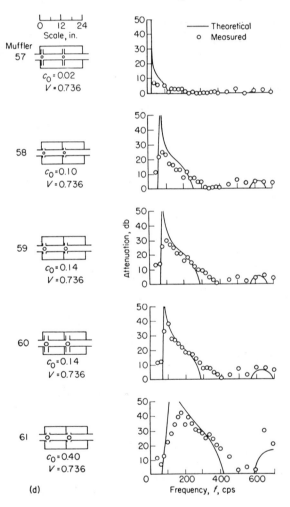

(d) Effect of conductivity c_0 using tubes. Equation (B15).

FIGURE 14.—Continued.

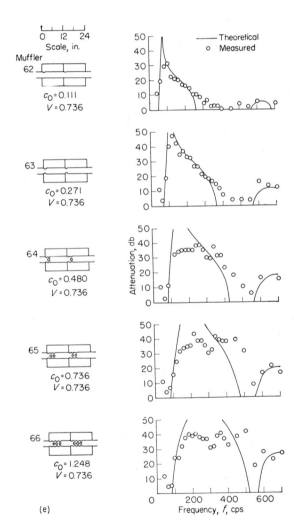

(e) Effect of conductivity c_0 using orifices. Equation (B15).

FIGURE 14.—Continued.

the usual construction of mufflers of this type, however, the chamber length is equal to l_1. Then when $k_r l_1 = \pi$, $k_r l_2 = \frac{\pi}{2}$ and the wave length is one-half the chamber length. Therefore, the chamber cannot properly be considered as a lumped impedance at the resonant frequency. If plane-wave motion is assumed in the chamber, k_r will approach $\frac{\pi}{l_1}$ $\left(= \frac{\pi}{2l_2} \right)$ only as the value of c_0 approaches infinity (see eq. (B12)). In order to determine whether it is possible in practice to eliminate the first upper pass band, muffler 67 was built. This muffler was tested after most of the data presented herein had been analyzed. In order to allow the measurement of higher values of attenuation than those in the previous tests, the experimental apparatus was reassembled in another location with the loud-speaker outside the room in which the measurements were made. The exhaust pipe entered the room through a hole in the wall which was sealed with sponge rubber. The tailpipe extended out the other end of the room through a similar hole. With this arrangement, it was possible to measure an attenuation of 65 decibels.

Two theoretical curves are presented in figure 14 (f). The solid curve, which shows the complete elimination of the pass band, was calculated for $c_0 = \infty$. The dashed curve, which shows a very narrow pass region, was calculated for $c_0 = 9.95$. The experimental points follow the solid curve up to about 340 cycles per second. In the critical first upper pass region, however, the measured attenuation drops from 65 decibels at 340 cycles per second to 29 decibels at 360 cycles per second, then rises sharply to 51 decibels at 380 cycles per second, drops again to 24 decibels at 400 cycles per second, and then begins to rise again. Both the initial drop and the final rise parallel the dashed curve ($c_0 = 9.95$), but the theory gives no explanation for the intermediate peak attenuation of 51 decibels which occurs at the point where the dashed curve goes to zero. Of course the actual behavior of a muffler in this very critical region cannot be accurately predicted without including viscous terms in the impedances. When $\frac{f}{f_r} = 1$ the branch reactance is zero, and when $\sin k l_1 = 0$ the pipe reactance is zero. Since these events both occur at nearly the same

frequency for muffler 67, only the resistances are left to control the sound flow. Therefore, it is inaccurate to neglect them in this region.

The points at about 360 and 400 cycles per second were determined by careful survey to be points of minimum attenuation. Thus, the experimental results prove that it is possible to obtain significant attenuation in a frequency region which is normally a pass band. The second upper pass band, however, was not eliminated.

CONDUCTIVITY PREDICTION

The results that have been presented show that the conductivity is a very important physical quantity which enters into the determination of both the resonant frequency and the amount of attenuation for resonator-type mufflers. It is unfortunate, therefore, that the conductivity should be, as has been mentioned in the section entitled "Theory," a somewhat elusive quantity to predict. In an attempt to eliminate some of the uncertainty regarding the prediction of c_0, it was computed by the following equation for those volume-controlled resonators which showed a well-defined resonant frequency:

$$c_0 = \frac{\pi a^2}{l_c + \beta a}$$

Two values for β, $\pi/2$ and $\pi/4$, were used. Where more than one connecting element was used, the calculated conductivity of a single element was multiplied by n, the number of elements. The results of this calculation are tabulated in table III along with the values of c_0, that are listed beside the corresponding attenuation curves. In each case, the listed c_0 was used in calculating the theoretical curve.

The data of table III indicate that, within the range of this investigation, when tube connectors are used, β may be

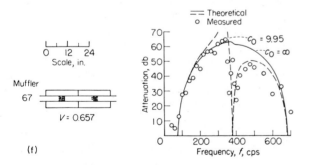

(f) Effect of setting resonant frequency equal to first pass frequency ($k_r l_1 = \pi$). Equation (B12) in (B14).

FIGURE 14.—Concluded.

TABLE III.—COMPARISON OF CALCULATED c_0 VALUES WITH c_0 VALUES LISTED IN FIGURES 12 TO 14

Muffler	Number of connectors per chamber	Number of chambers	l_c, in.	$2a$, in.	Calculated c_0, ft		Listed c_0, ft
					$\beta = \frac{\pi}{4}$	$\beta = \frac{\pi}{2}$	
Tube connector to chamber							
24	1	1	6.8	2.0	0.034	0.031	0.0308
25	1	1	6.8	2.0	.034	.031	.0308
26	1	1	6.8	2.0	.034	.031	.0308
27	1	1	2.16	2.0	.089	.070	.0702
29	1	1	6.8	1.0	.009	.009	.0086
30	1	1	6.8	3.0	.074	.064	.0644
31	1	1	.43	2.0	.215	.131	.1309
32	1	1	13.6	2.0	.018	.017	.0172
33	1	1	3.0	1.4	.036	.031	.0308
34	1	1	1.28	1.0	.039	.032	.0308
37	1	1	1.75	1.0	.031	.026	.026
38	1	1	1.66	1.5	.066	.052	.052
58	4	2	.25	.5	.147	.102	.100
59	4	2	1.00	1.0	.188	.147	.140
60	4	2	3.00	1.5	.164	.141	.140
Orifice connector to chamber							
39	2	1	0.05	1.0	0.296	0.157	0.151
40	6	1	.05	1.0	.887	.470	.670
47	10	1	.05	.50	.664	.369	.741
51	1	2	.05	.50	.066	.037	.041
52	4	2	.05	.50	.265	.148	.166
56	8	2	.05	.50	.529	.296	.332
62	3	2	.05	.50	.198	.111	.111
63	7	2	.05	.50	.463	.259	.271

taken as $\pi/2$ with sufficient accuracy for design purposes. In the case of orifice connectors, the results are not so conclusive. In general, however, it appears that β can be taken as $\pi/2$ if only a few orifices are used. The experiments indicate that, as the number of orifices is increased, the conductivity per orifice tends to increase (compare mufflers 39 and 40, or mufflers 51 and 47). The determination of an accurate method of predicting the value of c_0 for a group of orifices would require a study of such parameters as the number, diameter, and spacing of the orifices as well as the diameter of the central tube. Until the results of such research become available, however, the designer should, wherever possible, use only a few tubes or orifices, unless he has available the relatively simple equipment required to determine the resonant frequency experimentally after construction of a sample muffler.

TUNED TUBES

Two acoustical circuit configurations have been considered which make use of the velocity at which plane sound waves travel to obtain interference and resulting attenuation.

Side-branch tubes.—The first of these configurations consists of a side branch of constant area with the end closed. At a frequency for which such a tube is, for instance, one-quarter wave length long, a wave traveling from the exhaust pipe to the closed end and back to the exhaust pipe will arrive in phase opposition to the incoming wave in the exhaust pipe. The interference between the two waves results in attenuation. Appendix B gives the equation for the attenuation of mufflers of this type (eq. (B13)). The attenuation characteristics of three of these mufflers are presented in figure 15. For each of these mufflers the tube

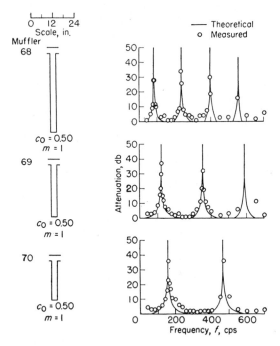

FIGURE 15.—Side-branch tubes with same diameter as exhaust pipe. Equation (B13).

diameter is equal to the exhaust-pipe diameter so that the area ratio m is one. Although attenuation above 20 decibels can be obtained, this high attenuation is limited to very narrow frequency bands. Consequently, the mufflers shown in figure 15 would not be suitable for variable-speed engines.

The analysis of the results obtained with resonator-type mufflers has shown that several of these mufflers with high ratios of length to diameter exhibit the characteristic behavior of tuned-tube mufflers (mufflers 43, 44, 45, and 46 of fig. 13 (c) and muffler 55 of fig. 14 (c)). These mufflers had much wider attenuation bands than the tuned tubes of figure 15. The calculations show that this increase in the width of the attenuation band is a direct result of the increased area ratio m.

Quincke tubes.—The second type of tuned-tube muffler is commonly known as the Quincke tube. It consists of two tubes of different lengths connected in parallel, with the combination inserted in series with the exhaust pipe. This arrangement is discussed in reference 1. Because of the characteristics of sharp tuning and narrow attenuation bands, an arrangement of this type seems unsuitable for an engine-exhaust muffler. Consequently, no mufflers of this type were included in this investigation.

COMBINATIONS

After investigating several types of mufflers, a few mufflers were tested which either combined two of the types or combined two or more sections of different size but of the same type. Mufflers 71 and 72 combined a resonator with an expansion chamber (fig. 16 and fig. 1 (d)). The results show the importance of the location of the conductivity for, although the mufflers are identical in all other respects, the attenuation of muffler 71 is much higher than that of muffler 72. Apparently the entrances to the two chambers are too close together in the case of muffler 72. The theory (appendix C, eq. (C18)) correctly predicts the better effectiveness of muffler 71.

It appeared probable that the requirement of a very broad attenuation region could best be satisfied by combinations of resonators which were tuned to different frequencies. Consequently an attenuation equation was developed for a combination of two resonators (appendix C, eqs. (C3) to (C7)), and one such combination was investigated experimentally (muffler 73). This muffler is shown in figure 16 and in figure 1 (e). The results show an attenuation of more than 10 decibels over an uninterrupted frequency band of width equal to about six times the lowest frequency of the band, in spite of the fact that this muffler is relatively small (12-inch diameter and 12-inch length). Muffler 73 was also tested in the reverse position (muffler 73R), with the high-frequency chamber to the front. The results show no appreciable difference except in the region below the first resonant frequency.

Muffler 74 is effectively a combination of four tuned tubes. The internal details of this muffler are shown in figure 1 (f). Although some attenuation is obtained over a wide frequency band, the attenuation spectrum consists of a series of very sharp peaks and hollows.

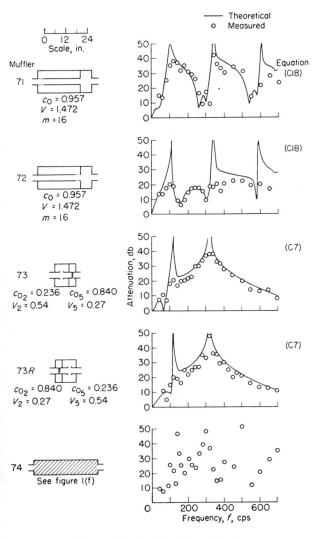

FIGURE 16.—Combination mufflers.

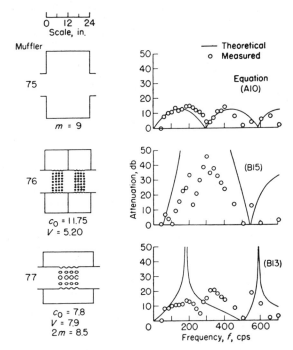

FIGURE 17.—Mufflers for large-diameter (12 in.) exhaust pipes.

MUFFLERS FOR A 12-INCH EXHAUST PIPE

All equations which have been presented include, in one manner or another, the assumption that the dimensions of certain elements are small compared to the sound wave length. In order to determine the effect of violating this assumption, three mufflers were designed for installation in a 12-inch-diameter exhaust pipe (fig. 17). Muffler 75 is a large expansion-chamber-type muffler. Inasmuch as the wave motion is accounted for in the expansion-chamber equation, it might seem, at first, that no size assumption has been made. The discussion of expansion chambers, however, showed that the plane-wave assumption carried an implicit assumption regarding the diameter. For muffler 75 the critical frequency for the first radial mode of vibration is 463 cycles per second. The experimental results show a loss of attenuation between 400 and 500 cycles per second. Below 400 cycles the calculations and experiment are in fair agreement, except that the effective length of the chamber seems to be somewhat shorter than the actual length.

Muffler 76 is a double resonator and muffler 77 is a single resonator. For both of these mufflers, the lack of agreement between calculations and experiment is quite pronounced. The results show that it is possible to obtain attenuation in pipes of this size but they also show that, because of the assumptions made, the equations used in this report are not adequate to predict this attenuation. Calculations for such mufflers must include consideration of other vibrational modes in addition to the plane-wave mode.

II. FINITE TAILPIPE

For the first part of this investigation, a reflection-free muffler termination (an effectively infinite tailpipe) was used in order to reduce the number of variables involved. In some cases a muffler or filter in a long duct or pipe line may have an essentially reflection-free termination. Engine mufflers, however, must be terminated in a tailpipe of finite length in actual practice. In Part II of this report, therefore, a method is presented which permits the tailpipe to be included in the muffler calculations. The validity of this method has been investigated experimentally by testing four muffler-tailpipe combinations. The dimensions of these mufflers are within the limits for which the basic muffler theory has been shown to be valid in Part I. The apparatus used in the tailpipe investigation is described briefly and the results of the tests are discussed.

THEORY

The problem of radiation from an unflanged circular pipe has been solved for the case where the incident sound is of the plane-wave mode (ref. 11). It is possible by use of this information to compute the tailpipe impedance and, thus, to introduce the tailpipe into the muffler calculations. A less accurate, but somewhat simpler, method is to add an end correction of 0.61 times the pipe radius to the length of the pipe and to assume that the pipe is terminated in a zero impedance (total reflection) with a phase shift of 180° between the incident and reflected waves. This method is justified at sufficiently low frequencies, because the reflection coefficient approaches unity as the frequency approaches zero. In order to determine the frequency range within which this approximation is applicable, the attenuation of a single-chamber resonator with a tailpipe has been calculated by both methods. The results (table IV) show that the approximation gives results within less than 0.1 decibel for frequencies up to 520 cycles per second. The attenuation curve is plotted in figure 18 (a). Note that the calculations have been made for $c=2000$ fps. This value is typical of the speed of sound in the hot exhaust gas from an aircraft engine. The attenuation has been based on the ratio of the absolute values of the incident-wave pressure just ahead of the conductivity openings and the incident-wave pressure in the tailpipe. The equation used for the approximate calculation is developed in appendix D (eq. (D10)).

Before proceeding further with the consideration of tailpipe effects, some discussion is necessary concerning this basis for calculating the attenuation. The user of a muffler ordinarily thinks of the attenuation due to a muffler as being the difference, at some point in the open air, between the sound level from an open exhaust pipe and the sound level after a muffler has been installed. The sound pressure in the open air due to an open exhaust pipe or a tailpipe is, at a given frequency, directly proportional to the pressure of the incident wave traveling in the pipe. Therefore, the attenuation can also be defined as the difference between the sound-pressure levels of the incident waves inside the open exhaust pipe and the tailpipe. It has been shown that the reflection coefficient from the end of an open exhaust pipe is nearly unity for the frequency range of this investigation. Also, for frequencies at which the attenuation of a muffler is high, there is a very strong reflection from the conductivity location back into the exhaust pipe. (See table I.) Now consider an engine to which are attached alternately an open exhaust pipe and another exhaust pipe of the same length as the open pipe but one that is terminated in a muffler and tailpipe. The reflected waves in the exhaust pipes are very strong in both cases; furthermore, the same sound source is feeding the two exhaust pipes and the pipes have the same length; therefore, it follows that the incident waves will have about the same strength. Thus, it is possible, in approximation, to calculate the attenuation as the difference between the sound-pressure levels of the incident wave entering the muffler in the exhaust pipe and the incident wave leaving the muffler in the tailpipe. This approximation should be valid in the frequency range for which the open-pipe reflection coefficient is near unity and for which the muffler also provides attenuation of the order of 15 decibels or more. Although the exhaust-pipe length has a very definite effect on the sound characteristics of a complete engine-exhaust system, it is possible by this method to separate the effect of the exhaust pipe length from the rest of the system. Since the open exhaust pipe itself reflects a large part of the sound, it is entirely possible that under certain conditions a muffler could permit more sound to escape than does the open exhaust pipe, with a resultant negative attenuation. A negative attenuation value, under the present definition of attenuation, does not imply that sound energy has been created inside the muffler; it means simply that the percentage of the sound energy which reaches the atmosphere is greater with the muffler installed than it is without the muffler.

Consideration of equation (D10) (appendix D) has led to an idea which may permit the elimination of the first upper tailpipe pass band of a single-chamber-resonator muffler. If the resonator is tuned to the usual pass frequency, then, when $kl_t=\pi$, both the tailpipe impedance and the resonator impedance will equal zero. In this event the pass frequency may be eliminated. A calculation has been made for a muffler identical with the muffler of figure 18 (a), except for the large change in conductivity required to tune the resonator to the frequency at which $kl_t=\pi$. The results shown in figure 18 (b) indicate that the width of the attenuation band is nearly doubled. At the same time, however, the cutoff frequency is increased slightly and the magnitude of the attenuation is lowered in the low-frequency region. Although no experimental data are available for this muffler, it seems possible, in view of the experimental results for muffler 67 (fig. 14 (f)), that some attenuation may be obtained near the resonant frequency, with the resultant elimination of the first upper tailpipe pass band.

The case of a single expansion chamber with a finite tail

TABLE IV.—COMPARISON OF TWO METHODS FOR CALCULATING THE ATTENUATION OF A SINGLE-RESONATOR MUFFLER WITH TAILPIPE

[Muffler constants: $c_0=0.261$ ft; $V=0.338$ ft^3; $c=2000$ fps; Tailpipe length $=20$ in.; $S=0.0247$ sq ft]

Frequency, f, cps	Attenuation, db	
	Calculations using exact tailpipe impedance	Calculations using approximate tailpipe impedance
20	−0.81	−0.81
40	−4.10	−4.10
60	−9.63	−9.65
80	−2.24	−2.24
100	5.09	5.10
120	9.86	9.87
140	13.67	13.67
160	17.09	17.10
180	20.28	20.29
200	23.37	20.37
220	27.08	27.09
240	31.38	31.39
260	37.35	37.36
280	∞	∞
300	38.48	38.50
320	33.14	33.16
340	29.35	29.37
360	26.63	26.67
380	24.45	24.48
400	22.39	22.41
420	20.30	20.33
440	18.27	18.30
460	16.00	16.03
480	13.37	13.39
500	10.32	10.34
520	6.34	6.33
540	−.42	−.56
560	−14.29	−17.06
580	−.45	−.40
600	4.55	4.65

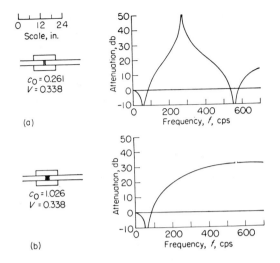

(a) Single-resonator muffler with tailpipe. $k_r l_t = 0.48\pi$. (See also table IV.)

(b) Single-resonator muffler with tailpipe. $k_r l_t = \pi$.

FIGURE 18.—Theoretical attenuation characteristics of single-chamber resonators with tailpipes. $c = 2000$ fps. Equation (D10).

pipe has also been considered, and an equation is presented in appendix D for the attenuation of such a muffler.

MUFFLERS

Sketches of the muffler-tailpipe combinations that were used in the experimental investigation of the effect of tailpipe length are shown in figure 19. These mufflers were designed for use on a particular aircraft engine. The design of these mufflers will be discussed later in the report in connection with a test of these mufflers on the engine for which they were designed. The mufflers were made from $\frac{1}{16}$-inch mild steel.

APPARATUS AND TESTS

The experimental investigation of the effect of the tailpipe was conducted outdoors in an open area and in calm air. In these tests, as in the previous experiments, the air inside the mufflers was at the ambient temperature and there was no steady air flow. Hence, these tests will be referred to as "cold tests." The apparatus that was used is shown schematically in figure 20. The electronic equipment included an audio oscillator, a power amplifier, a speaker, an oscilloscope for monitoring the wave form, and a sound-level meter.

The cold-test data were obtained by sending sound waves at a single frequency alternately into a muffler and into an open exhaust pipe and by taking the difference between the sound-pressure level observed in the open air at a distance of 20 inches from the outlet of the muffler tailpipe and that observed at a distance of 20 inches from the outlet of the open exhaust pipe. In order to insure that the mufflers were tested for the same wave lengths in the cold test as in the subsequent engine test, the cold-test frequencies were adjusted to produce the wave lengths for which the mufflers were designed. In the presentation of the cold-test results, the experimental frequencies are multiplied by the ratio of

the sonic velocity in the actual exhaust gas to the sonic velocity in the cold test in order to correct for the temperature difference between the two conditions. For the cold test, the frequency range was from 30 cycles per second to 400 cycles per second; for the engine test, the frequency range having equal wave lengths is 52 to 700 cycles per second. The ambient noise level for the cold tests was about 60 decibels.

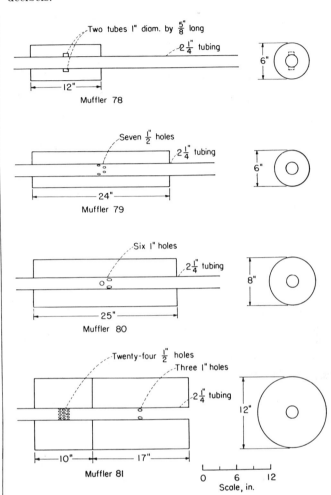

FIGURE 19.—Sketches of muffler-tailpipe combinations tested.

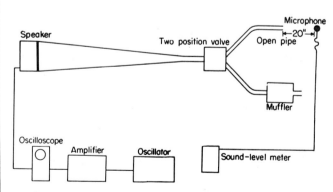

FIGURE 20.—Schematic diagram of the arrangement of apparatus for cold test of mufflers.

49

RESULTS

The experimental results obtained from the muffler cold tests and the theoretical attenuation predicted for each muffler are shown in figure 21. The theoretical curves for mufflers 78, 79, and 80, which were computed from equation (D10), show that these mufflers were designed to have chamber resonances (points of maximum attenuation) at approximately 280 cycles per second and tailpipe resonances at about 400, 580, and 580 cycles per second, respectively.

A comparison of the experimental and theoretical data shows good agreement for these three mufflers. For example, the higher frequency cutoff points, which are a function of tailpipe length, are seen to fall very close to the predicted frequencies; furthermore, the measured attenuation

falls near that computed theoretically at all frequencies except those near the computed chamber resonance. The mufflers were not expected to provide the infinite attenuation calculated at the chamber resonant frequency; the calculated infinite values occurred only because the viscous forces were neglected in order to simplify the calculations. With this limitation, it may be concluded that equation (D10) is valid for predicting the attenuation characteristics for muffler-tailpipe combinations under the cold-test conditions.

The double-chamber resonator curve computed for muffler 81 shows two chamber resonant frequencies and no high-frequency tailpipe pass bands. The difference between the curve shapes for the single-chamber and double-chamber mufflers is, of course, due to the changes in the acoustical circuit. The attenuation for muffler 81 was computed by substituting the tailpipe impedance iX_t from appendix D for the impedance Z_e in the equations given for a combination of two resonators in appendix C and working out the expression for the attenuation.

For the cold tests, the two largest mufflers (mufflers 80 and 81) were wrapped with several layers of felt. In the absence of the felt wrappings, the maximum attenuation was limited to about 25 to 30 decibels by the radiation from the 1/16-inch-thick outer walls. Reduction of this radiation would be an important factor in the design of a muffler from which a higher attenuation is desired.

III. APPLICATION TO MUFFLER DESIGN

VARIABLES DEPENDENT ON OPERATING CONDITIONS

Under the conditions of the investigations discussed in Parts I and II of this report, acoustic theory has been shown to predict the performance of several types of mufflers within a frequency range which is governed by the dimensions of the muffler elements. These investigations were designed to allow the study of several of the dimensional variables involved in exhaust muffling.

In order to isolate the effects of these variables, it was necessary to eliminate certain other variables dependent on operating conditions which could be separately investigated at some future time. The three major variables which have not been discussed are exhaust-gas temperature, exhaust-gas velocity, and exhaust-pipe sound pressure. A discussion of these variables follows.

TEMPERATURE

The preceding investigations were made at room temperature or at atmospheric temperature and the velocity of sound was about 1,140 feet per second. The higher temperature in the engine exhaust gas will result in a higher sonic velocity. From the data of figure 8 of reference 9 and from temperature measurements made during the engine tests described in Part IV of the present report, the sonic velocity inside the tailpipe is estimated to be about 2,000 feet per second. It is believed that the primary effect of a change in the exhaust-gas temperature is the corresponding change in the velocity of sound. It is necessary in the de-

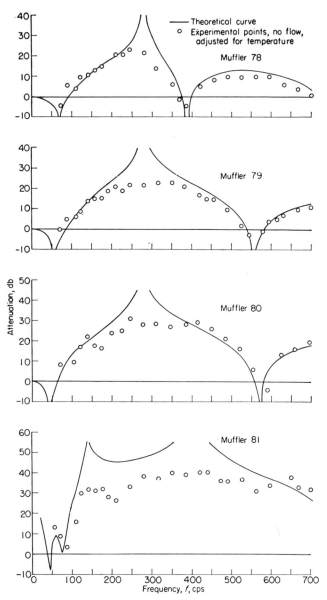

FIGURE 21.—Cold-test data and theoretical curves for mufflers tested.

sign of mufflers to use the actual sonic velocity of the exhaust gas. If the exhaust-gas temperature is known, the approximate velocity of sound may be determined by using the relation which has been found for air $c = 49\sqrt{T}$ feet per second, where T is the absolute temperature on the Fahrenheit scale.

The calculations that have been presented have included the tacit assumption that the temperature and average density in the muffler chambers are the same as those in the exhaust pipe. If significant differences are found in practice, they can be accounted for by using the most accurate available values for ρ and c at each element in calculating the impedance of that element. In this connection, it is interesting to note that the impedance of a resonant chamber is proportional to ρc^2 (eq. (B5)). But since c^2 is proportional to T and ρ is proportional to T^{-1}, the chamber impedance is independent of T. The connector impedance is a function of T, but, unless it is a long tube, the connector will be at the exhaust-gas temperature. Thus, for resonator-type mufflers, a temperature difference between the exhaust pipe and the chamber would be expected to have little effect on the performance of the muffler.

EXHAUST-GAS VELOCITY

In an actual engine-exhaust-muffler installation the exhaust gas which transmits the sound is in motion, whereas in the preceding investigations there was no net flow of air. The actual case may be considered to consist of an alternating, or sound, flow superimposed on a steady exhaust-gas flow. A theoretical approach to the problem of determining the effect of the steady flow on the acoustic characteristics of an exhaust system has been made in reference 12. No experimental data, however, are included. The conclusion of the theory is that the velocity effect is a function of $\sqrt{1 - M^2}$, where M is the Mach number of the exhaust flow. If the theory is assumed to be essentially correct, the following results are obtained.

Consider first the characteristics of the muffler itself. In the useful range of expansion ratios, the exhaust-gas velocity inside an expansion chamber is much lower than that in the exhaust pipe. Because the permissible engine back pressure limits the Mach number in the exhaust pipe to a value considerably less than 1, the Mach number inside the expansion chamber will be so low that M^2 is negligible when compared with 1. Thus the exhaust velocity will have no appreciable effect on the attenuation of a single expansion chamber. In the case of multiple expansion chambers, however, the exhaust-gas velocity in the connecting tubes may be high enough to alter the muffler characteristics significantly. (See ref. 2 for experimental data.) In the resonant chamber of a resonator-type muffler there is no steady exhaust-gas flow; therefore, the single resonator will not be affected by exhaust-gas flow. In the case of multiple resonators, as in multiple expansion chambers, the impedance of the connecting tubes will be affected by the exhaust-gas velocity.

Consider next the tailpipe characteristics. The tailpipe impedance will vary with the flow velocity. This will, of course, affect the attenuation of any practical muffler installation. According to the theory, the main effect of increased exhaust velocity is to lower the resonant frequencies of the tailpipe and to reduce the attenuation due to the tailpipe at those frequencies for which the tailpipe impedance reaches a maximum. On the whole, these effects are probably relatively small, inasmuch as the tailpipe resonant frequency is reduced by only 9 percent at a Mach number of 0.3, which corresponds to an exhaust velocity of 600 feet per second when c is 2,000 feet per second.

Note that most of the preceding conclusions regarding the effect of exhaust-gas velocity must be regarded as tentative, because they have been based on an unproved theory. Furthermore, the experimental data of reference 2 tend to cast some doubt on the validity of the theory. This uncertainty shows the need for additional research on the effects of exhaust-gas velocity.

INCREASED SOUND PRESSURE

In the derivation of the classical acoustic theory it is assumed that the sound pressures are very small in comparison with the static pressure of the medium (ref. 1). This assumption is made in order to permit the linearization of the differential equation of motion. However, in connection with engine tests previously made at this laboratory (ref. 9) certain nonlinear effects were observed, particularly the buildup of sharp wave fronts in long exhaust pipes as evidenced by the explosive character of the sound from such pipes. The detection of such nonlinear effects indicates that the exhaust sound pressure inside the pipes is high enough so that the classical linearized theory may give results which are somewhat in error. Further study of the behavior of acoustic elements—resonators, orifices, and tubes—in the presence of nonlinear sound fields is required before the effects of very high sound pressures on the performance of an acoustic system will be quantitatively known.

RELATIVE MERITS OF MUFFLER TYPES INVESTIGATED

None of the muffler types discussed should have excessive back pressures if the exhaust pipe is the proper size because the exhaust gas is not forced around sharp 180° turns. The expansion chambers will probably have the highest back pressures of the types tested because of the energy losses in the expansion and contraction processes but, at least for the single expansion chambers, this back pressure should be within allowable limits.

In general, single-chamber mufflers are useful where the required frequency range is small; whereas, for high attenuation over a very wide frequency range, two or more different chambers will be required in order to obtain attenuation at the pass frequencies of the individual chambers and the tailpipe.

Reference 7 indicates that, in the case of engine exhausts having large sound pressures, mufflers of the expansion-chamber type must be used, because the attenuation of a resonator is dependent on the existence of small sound pressures. The experiments of reference 9, however, have shown

51

that resonator mufflers can be quite effective in an engine-exhaust system, even though the theoretical assumption of small sound pressures is violated. (This assumption is actually made also in deriving the equations for the attenuation of expansion-chamber mufflers.) The muffler designer is, therefore, not necessarily restricted to expansion chambers. The answer to the question as to which type, for a given muffler size and a given back pressure, is the more effective depends in part upon the relative magnitudes of the effects of high sound pressure and of exhaust-gas velocity on the two types.

In case the adverse effects of high sound pressures are found to be excessive for resonators, it is suggested that a combination muffler, with the expansion chamber first in order to reduce the sound pressures entering the resonator, may be most effective. (See muffler 67 of ref. 9.)

MUFFLER-DESIGN PROCEDURE

On the basis of the theory which has been presented, a muffler-design procedure was developed. Because some of the important variables have not been investigated as yet, the procedure must be judged by the results obtained in practical applications. Modifications of the procedure are to be expected as a result of experience gained in the applications. This procedure begins with the determination of a required attenuation spectrum, which defines the noise reduction that the muffler is expected to produce.

REQUIRED ATTENUATION SPECTRUM

The first step in muffler design is to determine, at a known distance from the exhaust pipe, the sound-level spectrum of the engine which is to be quieted. This should be done at several speeds and loads within the operating range or, at the very least, at the maximum and minimum speeds of the normal operating range. In estimating the critical operating conditions likely to be encountered from the standpoint of noise, it is useful to recall that for a particular engine the magnitude of the noise is controlled largely by the engine torque, whereas the frequencies are controlled by the engine speed (refs. 8 and 9).

After the engine-noise spectrum has been determined, an allowable spectrum should be established, consisting of the maximum allowable sound-pressure level as a function of frequency. The fact that other noise sources (such as engine air intake, engine clatter, and the propeller) place a practical limit on the attainable reduction in overall airplane noise will influence the choice of the allowable spectrum. As the desired noise reduction increases, it becomes necessary to treat more of these other noise sources. In particular, it was necessary to treat both the engine exhaust and the propeller to obtain significant noise reduction for the liaison airplane of reference 8.

The difference between the measured and allowable spectrums will establish the minimum attenuation which is required at each frequency; this difference will be called the required attenuation spectrum.

MUFFLER SELECTION

Compare the required spectrum with the design curves (figs. 22 to 24) and select from these curves a muffler design which will provide somewhat more than the required attenuation throughout the frequency range. (The use of these design curves will be discussed.) In the case of a single expansion chamber or resonator, the tailpipe must be carefully selected. From the required cutoff frequency compute the necessary tailpipe length by using the approximate equations which have been presented (eq. (D6) or (D12)). Next, by use of this tailpipe length, determine the location of the high-frequency pass bands. If the first pass frequency is too low, it will be necessary to choose a larger muffler in order that the tailpipe may be shortened or else to add another chamber which will provide attenuation at the tailpipe pass frequency. If a double expansion chamber or multiple resonator has been selected, the approximate equations or the design curves may be used to determine the cutoff and pass frequencies. Several of the muffler types may be considered in this manner in order to determine which will result in the smallest muffler that will provide the required attenuation in a particular case. It is usually not necessary to carry out detailed attenuation calculations until the final configuration has been closely approached. The detailed calculations will then provide a final check on the theoretical suitability of the selected muffler.

A test of the chosen muffler installation on the engine may show that modifications are required, owing to the influence of factors which have not been investigated as yet (in particular, the high exhaust-pipe sound pressures). Even with the assistance of the information presented in this report, it is likely that a certain amount of trial and error will be necessary in muffler design when the goal is a very highly efficient muffler in terms of attenuation per unit of weight or volume.

DESIGN CURVES

Three sets of design curves, showing the attenuation of mufflers terminated with the characteristic pipe impedance Z_0, are presented in figures 22, 23, and 24; these curves have been calculated from equations (A10), (B10), and (B15), respectively, of the appendixes. Simple examples will be given to indicate how these charts can be used to eliminate the need for detailed attenuation calculations in the preliminary stages of muffler design.

SINGLE EXPANSION CHAMBER

Figure 22 shows the attenuation of single expansion chambers in terms of nondimensional parameters. The parameter kl_e is a combination length and frequency parameter. The other parameter is the expansion ratio m.

Suppose that a muffler is desired to provide a minimum attenuation of 10 decibels between frequencies of 100 and 300 cycles per second. An expansion ratio of 9 will provide 10 decibels at $kl_e=0.8$. At three times this value of kl_e (i. e., $kl_e=2.4$), it will also provide about 10 decibels. Thus, $m=9$

is satisfactory. The length of the muffler is determined by the fact that 100 cycles per second corresponds to $kl_e = 0.8$, so that $0.8 = \frac{2\pi l_e}{c} \times 100$ $\left(\text{let } c = 2000 \text{ fps, then } l_e = \frac{2000 \times 0.8}{2\pi \times 100} = 2.54 \text{ ft}\right)$. If the exhaust-pipe diameter is 2 inches, the expansion-chamber diameter will be $2\sqrt{9}$ or 6 inches.

If this muffler is too long, another procedure is possible. Let $m = 25$; thus, the diameter is increased to 10 inches. The design curve shows 10-decibel attenuation at $kl_e = 0.25$. At a kl_e of $\frac{300}{100} \times 0.25 = 0.75$, the attenuation is more than adequate. The length of this muffler will be $l_e = \frac{2000 \times 0.25}{2\pi \times 100}$ or 0.795 foot.

SINGLE-CHAMBER RESONATOR

Figure 23 shows the attenuation of single-chamber resonators in terms of nondimensional parameters. The attenuation is plotted against f/f_r which is the ratio between the sound frequency and the resonant frequency of the resonator. Curves are plotted for several values of the attenuation parameter $\sqrt{c_0 V}/2S$.

Suppose again that the muffler is desired to provide a minimum attenuation of 10 decibels between $f = 100$ and 300 cycles per second. In terms of the chart this means that the frequency at which the right leg of the attenuation curve crosses the 10-decibel line must be three times the frequency at which the left leg crosses the 10-decibel line. The chart shows that this requires a value somewhat higher than 3.16, say approximately 4, for the attenuation parameter. The value of f/f_r corresponding to 100 cycles per second will be about 0.55

$$f_r = \frac{100}{0.55} = 182 \text{ cps}$$

Therefore,

$$\sqrt{\frac{c_0}{V}} = \frac{2\pi f_r}{c} = \frac{2\pi \times 182}{2000} = 0.57 \text{ ft}^{-1}$$

The exhaust pipe is 2 inches in diameter so that

$$\sqrt{c_0 V} = 2 \times S \times \frac{\sqrt{c_0 V}}{2S}$$

$$= 2 \times \left(\frac{\pi}{4} \frac{(2)^2}{(12)^2}\right) \times 4 = 0.174 \text{ ft}^2$$

$$c_0 = 0.57 \times 0.174 = 0.099 \text{ ft}$$

$$V = \frac{0.174}{0.57} = 0.305 \text{ ft}^3$$

Any combination of length and diameter which will give this volume is permissible, as long as the dimensions are not too large in comparison with the 300-cycle wave length at the exhaust-gas temperature (see experimental results). If a length of 1 foot is selected, the diameter becomes 0.645 foot or 7¾ inches.

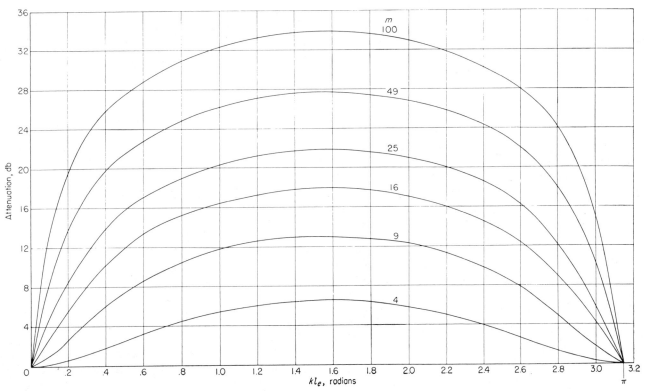

FIGURE 22.—Expansion-chamber design curves.

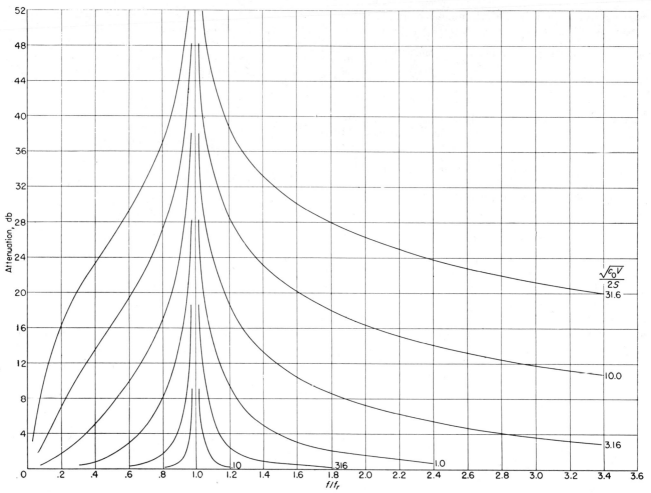

FIGURE 23.—Single-chamber-resonator design curves.

MULTIPLE-CHAMBER RESONATOR

Figure 24 shows the attenuation per chamber of multiple-chamber resonators in terms of nondimensional parameters. Because three parameters are involved (appendix B), several charts are required to describe fully the possible configurations. Three such charts are presented.

As an example of the use of these charts, assume that for a particular engine spectrum the sound level at the fundamental frequency (100 cycles per second) is to be reduced 13 decibels. The levels at the other frequencies are to be reduced to the point where the speech interference is nowhere greater than at the fundamental frequency. This criterion results in a required attenuation of 13 decibels at 100 cycles per second, 4 decibels at 200 cycles per second, and zero at higher frequencies. The top chart of figure 24 $\left(k_r l_1 = \frac{1}{2}\right)$ shows that this objective could be met with $\frac{\sqrt{c_0 V}}{2S} = 1$ for a two-chamber muffler with $f_r = 100$ cycles per second. By using these values the muffler dimensions are found as follows:

$$k_r = \frac{2\pi \times 100}{2000} = 0.314 \text{ ft} = \sqrt{\frac{c_0}{V}}$$

$$l_1 = \frac{0.5}{0.314} = 1.59 \text{ ft} = 19 \text{ in.}$$

$$\sqrt{c_0 V} = 2 \times \frac{\pi}{4} \frac{(2)^2}{(12)^2} \times 1 = 0.0436 \text{ ft}^2$$
$$c_0 = 0.314 \times 0.0436 = 0.0137 \text{ ft} = 0.164 \text{ in.}$$

$$V = \frac{0.0436}{0.314} = 0.139 \text{ ft}^3$$

In order to obtain this volume with a concentric resonant chamber 19 inches long, a chamber diameter of 4.5 inches is required. The overall length of the two-chamber muffler is 38 inches. The use of a tube connector seems advisable in order to obtain the low c_0 required without creating excessive sound velocities in the connector.

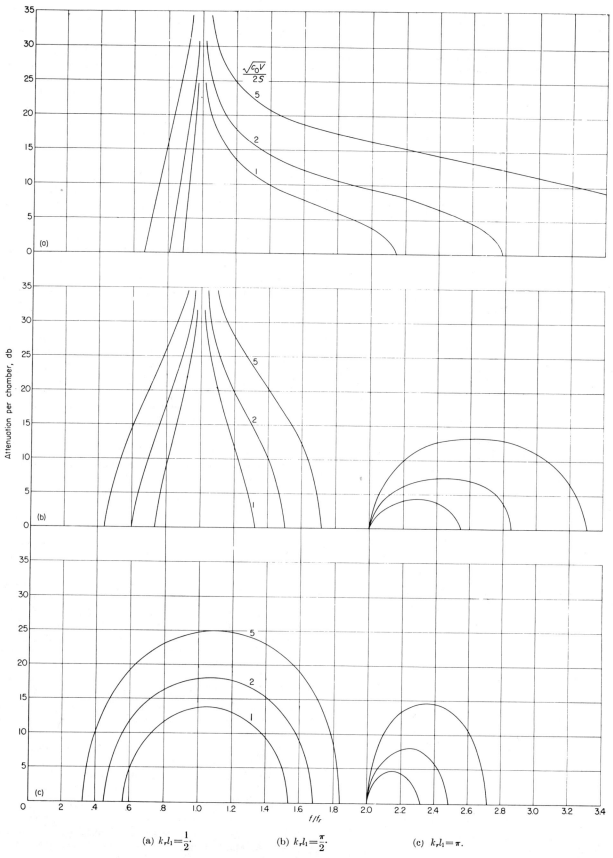

(a) $k_r l_1 = \frac{1}{2}$.   (b) $k_r l_1 = \frac{\pi}{2}$.   (c) $k_r l_1 = \pi$.

FIGURE 24.—Multiple-chamber-resonator design curves.

IV. ENGINE TESTS

In order to study the practicality of the design methods which have been described and, also, to obtain some idea of the size of muffler which is required in practice to provide a significant reduction of engine noise, four mufflers were designed for and tested on an actual aircraft engine. As was stated in the introduction, the engine of a helicopter was used for these tests. The design of the mufflers followed, in general, the design procedure which has been presented.

BASIC DATA

The first step in the design procedure is the determination of the engine-exhaust noise spectrum of the unmuffled engine. In the case of the helicopter the spectrum of the exhaust noise alone could not be determined. Instead the overall noise spectrum of the helicopter was measured. This spectrum, which includes an unknown amount of extraneous noise from such sources as the engine air intake, rotor blades, and engine clatter, is presented in figure 25.

Temperature measurements showed the speed of sound in the exhaust pipe to be approximately 2,000 fps.

MUFFLERS AND DESIGN

In order to insure that an adequate test range would be covered in the investigation, four resonator-type mufflers were designed and constructed. Three of the mufflers had single resonant chambers, whereas the fourth had two resonant chambers. The double-chamber muffler was designed with the intent to provide enough exhaust-noise attenuation so that the extraneous noise level could be measured. Figure 19 shows schematic drawings of these mufflers.

The mufflers were designed to give successive increases in attenuation and to have the acoustical properties shown in the following table:

Muffler	Chamber resonant frequency, cps	Tailpipe resonant frequency, cps	Attenuation parameter, $\frac{\sqrt{c_0 V}}{2S}$
78	280	400	4.33
79	280	580	6.03
80	280	580	12.00
81	{140, large chamber {400, small chamber	Undetermined	{9.5, large chamber {16.15, small chamber

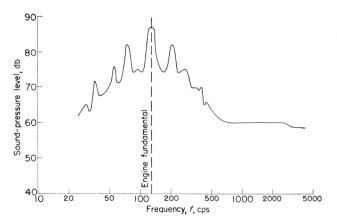

FIGURE 25.—Unmuffled-helicopter-noise frequency analysis.

Mufflers 78, 79, 80, and 81 were made from 1/16-inch mild steel and weighed 12, 17, 21, and 32 pounds, respectively. Figure 26 shows the mufflers installed on the test helicopter.

It may be of interest at this point to indicate the method used in the design of these mufflers with a specific example included for muffler 79. The fact that the test helicopter had two exhaust systems, one exhausting three cylinders and the other exhausting four cylinders, did not require the design of different mufflers for the two exhaust pipes. Although the exhaust-pressure pulse from each cylinder contains components at the individual cylinder firing frequency and at harmonics of this frequency, the phase relationships are such that, when the pressure pulses of all seven cylinders are combined in the atmosphere, the components at the cylinder firing frequency and at many of the harmonics are partially canceled. The mufflers must attenuate those frequency components in both exhaust pipes which combine to cause undesirably high noise levels in the atmosphere. Consequently, the mufflers are designed on the basis of the noise in the atmosphere, rather than that inside the individual exhaust pipes, and, as a result, the two mufflers are identical. The seventh harmonic of the cylinder firing frequency is referred to as the engine fundamental frequency. The prominence of this harmonic in the unmuffled engine noise (see fig. 25) is due to the fact that this frequency is the lowest at which the components of all seven cylinders are nearly together in phase.

(1) The noise spectrum of the unmuffled helicopter (fig. 25) showed that most of the disturbing noise fell in the frequency range from 70 to 350 cycles per second and that 10 decibels of overall attenuation would reduce the noise to a desired level. The muffler must be made to resonate within this frequency band in order to obtain maximum quieting; thus, 280 cycles per second was chosen for the muffler resonant frequency. In order to provide a 10-decibel reduction from 70 to 350 cycles per second, a muffler having an attenuation parameter $\frac{\sqrt{c_0 V}}{2S}$ value of approximately 6.0 was selected from the design curves.

(2) A tube for conducting the exhaust gases through the muffler for filtering must be chosen. The engine-exhaust back pressure should be kept small; consequently, a tube used for this purpose must be large enough to keep the back pressure within acceptable limits. The tubing selected for muffler 79 was 2¼ inches, the same size as the existing exhaust ducting on the test helicopter. It should be noted that the attenuation parameter $\frac{\sqrt{c_0 V}}{2S}$ shows that the internal-tube area governs the muffler size for a given attenuation; for this reason, the tube should be selected as small as practicable.

(3) In order to obtain the length for this central tube, a desired tailpipe length is computed and added to the length necessary to conduct the exhaust gases to the conductivity holes. The conductivity holes mark the origin of the tailpipe for single-chamber mufflers. Before the tailpipe length can be computed, however, some specific frequency for tailpipe resonance must be selected. This frequency must fall within a range in which little or no attenuation is needed because, as the tailpipe resonant frequency is neared, the

Muffler 78

Muffler 79

Muffler 80

L-77948

FIGURE 26.—Muffler installation.

muffler attenuation drops to a negative value over a narrow band. The tailpipe resonant frequency selected for muffler 79 was 580 cycles per second. The effective tailpipe length is computed as follows:

$$l_t = \frac{\lambda}{2} = \frac{c}{2f} = \frac{2000 \times 12}{2 \times 580} = 20.68 \text{ inches}$$

By applying the end correction $\Delta l_t = 0.61r$ where r is the tailpipe inside radius, the resulting true tailpipe length is

$$20.68 - 0.61(1.125 - 0.063) = 20.03 \text{ inches}$$

Inasmuch as the tailpipe length also affects the low-frequency cutoff of the muffler, a check is required to see whether this cutoff falls within the desired attenuation band. The cutoff frequency is determined from equation (D12):

$$f_c = \frac{f_r}{\sqrt{1 + \frac{\sqrt{c_0 V}}{2S} k_r l_t}} = \frac{280}{\sqrt{1 + 6 \frac{2\pi 280}{2000} \frac{20.68}{12}}} = 88 \text{ cps}$$

Since the cutoff frequency is within the frequency band in which muffling was desired, a decision must be made as to whether it is beneficial to increase the tailpipe length and thereby lower the high frequency cutoff or to increase the chamber size in order to obtain a small attenuation gain in the low-frequency range. The tailpipe length was not changed, and the resulting loss of low-frequency attenuation was accepted because all available criteria for judging the effects of noise agree that somewhat higher noise levels are tolerable at low frequencies than at higher frequencies.

(4) The conductivity factor c_0 determines the muffler resonant frequency for a given volume. The equation

$$f_r = \frac{c}{2\pi}\sqrt{\frac{c_0}{V}}$$

shows the relationship that exists among the conductivity, volume, and resonant frequency. With the use of this expression and for the values of the parameters chosen, the volume and conductivity for muffler 79 can be determined as follows:

$$f_r = \frac{c}{2\pi}\sqrt{\frac{c_0}{V}} = 280 \text{ cps}$$

Solving for $\sqrt{\frac{c_0}{V}}$ yields

$$\sqrt{\frac{c_0}{V}} = \frac{280 \times 2\pi}{2000} = 0.880$$

$$\sqrt{\frac{c_0 V}{2S}} = 6$$

and solving for $\sqrt{c_0 V}$ gives

$$\sqrt{c_0 V} = 6 \times 2 \times \frac{\pi}{4} \frac{(2.25 - 0.125)^2}{144} = 0.295$$

$$c_0 = \sqrt{\frac{c_0}{V}} \sqrt{c_0 V} = 0.880 \times 0.295 = 0.260 \text{ ft}$$

By substitution

$$V=\frac{0.260}{0.880^2}=0.336 \text{ cu ft}$$

This volume and a chosen muffler length of 2 feet were used to calculate the muffler diameter, 5.9 inches. For the sake of construction simplicity, the diameter was chosen to be 6.0 inches. This diameter change required small adjustments to be made in the values of volume and conductivity; the new values calculated were 0.338 cubic foot and 0.261 foot for volume and conductivity, respectively.

(5) The required muffler conductivity was obtained by drilling several ½-inch holes in the central tube of the muffler. In determining the number of holes required, a value of $\frac{\pi}{2}$ was used for the constant β in the conductivity equation. The calculation follows:

$$n=\frac{c_0 \text{ chosen for muffler 79}}{c_0 \text{ per ½-inch hole}}$$

$$=\frac{0.261}{\frac{\pi\left(\frac{1}{4}\frac{1}{12}\right)^2}{\frac{1}{16}\frac{1}{12}+\frac{\pi}{2}\frac{1}{4}\frac{1}{12}}}=7.27 \text{ or 7 holes}$$

Experience has shown that there are some effects on the conductivity caused by the close spacing of holes which often require the number of holes to be changed in order to obtain the desired conductivity c_0, or resonant frequency. The actual conductivity c_0 can be determined by experimental tests.

(6) After all dimensions for the muffler have been deter-mined, the theoretical attenuation characteristics of the resonator should be computed and analyzed with the use of equation (D10). This equation may also be written in the following form:

$$\text{Attenuation}=10 \log_{10}\left(\frac{p_1}{p_t}\right)^2$$

$$=10 \log_{10}\left[1+\frac{c}{S}\frac{1}{2\pi f\,\frac{1}{c_0}-\frac{c^2}{2\pi fV}}\sin\left(\frac{4\pi f}{c}\right)l_t+\left(\frac{c}{S}\right)^2\left(\frac{1}{2\pi f\,\frac{1}{c_0}-\frac{c^2}{2\pi fV}}\right)^2\sin^2\left(\frac{2\pi f}{c}\right)l_t\right]$$

If the predicted attenuation does not conform to the desired conditions, small changes in the originally selected design values may be made to achieve the desired results.

APPARATUS

The test helicopter (fig. 27) was used as the muffler test bed in this investigation. The tail rotor was removed for the tests to prevent its noise from interfering with the sound measurements. The noise emanating at the main rotor fundamental frequency (13 cps) was known to be of little significance in these tests. However, a possibility that the higher harmonics of the rotor might interfere with the exhaust noise measurements was recognized.

The helicopter was powered by a R–550–1, 180-horsepower, 7-cylinder engine having twin exhaust stacks. One stack exhausted three cylinders and the other, four. Figure 28 shows a diagrammatic sketch of the field-test setup and surrounding terrain.

FIGURE 27.—Muffler 79 installed on helicopter with tail rotor removed.

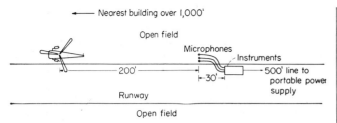

FIGURE 28.—Engine-test arrangement.

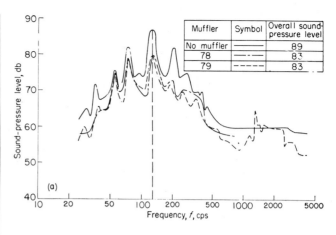

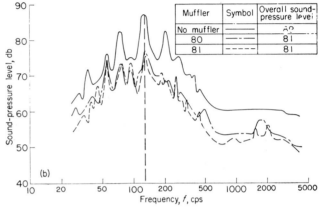

(a) Mufflers 78 and 79.
(b) Mufflers 80 and 81.

FIGURE 29.—Comparison of recorded frequency analyses of helicopter noise with and without mufflers.

The sound measuring equipment used in the field tests consisted of a General Radio Company type 759–B sound-level meter, a General Radio Company type 760–A sound analyzer, and a Western Electric type 700–A sound-level meter and filter set. Both the frequency analysis and the overall sound pressure level were recorded on a twin recorder. This equipment gave an overall measuring accuracy of about 2 decibels when operating under field conditions. The response of the equipment was found to drop rapidly for frequencies below 40 cycles per second. A water-cooled crystal pressure pickup was utilized to obtain a time history of the pressure variation inside the exhaust pipe ahead of the muffler. Indications of the exhaust-gas temperatures were obtained through use of chromel-alumel thermocouples and a Lewis potentiometer.

TESTS

The field tests were conducted before sunrise on the Langley landing field. The ambient field noise level was approximately 62 decibels at the start of the field tests. Changes that may have occurred in the ambient field noise after the helicopter engine was started could not be determined. The muffler field test included the investigation of the four mufflers of different size on the modified helicopter to determine the attenuation characteristics of the mufflers at an engine speed of approximately 2,200 revolutions per minute. In order to determine more fully the conditions under which the mufflers were operating, internal exhaust-gas sound pressures and temperatures were measured during one of the test runs.

As a further check on the practicality of the muffler design, the helicopter was flown with the first three mufflers attached. The pilot, who had considerable flying experience with the test helicopter, reported no noticeable change in performance.

RESULTS AND DISCUSSION

The results of these muffler tests, which are discussed in the following sections, show the effectiveness of the muffler in reducing the exhaust noises along with the merits and shortcomings of the theoretical equation under investigation (eq. (D10)). The muffler experimental results are presented in the form of tables and curves.

MUFFLER-ENGINE TESTS

The muffler-engine-test results are shown in figure 29 and table V. These data describe the manner in which the amplitude of the exhaust noise varies with frequency.

Figure 29 (a) shows the unmuffled-exhaust noise spectrum in addition to the noise spectrums for both mufflers 78 and 79. Similarly, the spectrums for mufflers 80 and 81 are shown in figure 29 (b).

Frequency analysis.—The curve describing the envelope for the unmuffled-exhaust noise frequencies shows that the fundamental firing frequency (noted by the dashed line) is by far the largest noise-producing harmonic and, thus, is the frequency which should be given the greatest attenuation. The peaks occurring at 75 and 205 cycles per second are the next largest sound-producing frequencies of the engine noise. These two peaks, along with the fundamental peak mentioned previously, define the frequency band where most of the annoying noise is found to exist and, consequently, the range which should be given the greatest attention. When the noise spectrum from each of the four mufflers is compared with that of the unmuffled engine, it becomes obvious that considerable muffling was obtained in the 75 to 205 cycles per second frequency band. In general, the curves are seen to have the same characteristic shape.

Suppose now that a comparison is made between the cold tests and the engine tests. (See figs. 21 and 29.) Figure 21 shows that mufflers 78, 79, and 80 should have yielded their

TABLE V.—BAND-PASS ANALYSIS OF HELICOPTER NOISE AT 200 FEET

Muffler	Overall sound level, db	Sound pressure level, db, from—								
		0 to 50 cps	50 to 100 cps	75 to 150 cps	100 to 200 cps	150 to 300 cps	200 to 400 cps	300 to 600 cps	400 to 800 cps	600 to 1200 cps
78	85	72	80	81	81	77	75	70	60	54
79	85	70	79	82	81	76	73	69	60	58
80	83	70	78	80	79	74	72	67	60	57
81	83	70	78	80	78	73	72	67	59	58
No muffler	91	74	84	88	87	83	81	71	61	58

greatest attenuation at 280 cycles per second and no attenuation in one lower and one higher frequency cutoff band. A point-by-point comparison between the data of these two figures showed that the helicopter noise spectrum was not reduced by the amount predicted for the muffler in the cold test. For instance, the cold-test data for muffler 79 showed about 11 decibels of attenuation was obtained at 128 cycles per second; the engine test, however, showed that 7 decibels of attenuation was realized when the muffler was tested on the helicopter. Similarly, at 200 cycles per second, approximately 20 decibels of attenuation may have been expected but only 11 decibels were measured during the engine test. After inspecting the data for all mufflers tested it was concluded that, although effective muffling was received, no muffler reduced the helicopter noise by the amounts predicted from the muffler cold tests.

Band-pass analysis.—In order to provide a rough check on the frequency-analyses data, certain band-pass analyses were made. These band-pass data (table V) give sound pressure levels with overlapping octaves for frequency bands, ranging from 0 to 1,200 cycles per second. Before further discussion of these data, it should be pointed out that the meter used in taking these measurements was of a different type from that used for the frequency analysis. A constant 2-decibel calibration difference was found to exist between the two meters used. For identical sound signals, the meter used to record the band-pass analysis always read 2 decibels more than the meter used to record the frequency spectrum.

Good agreement between these data was found in the frequency range of 75 to 400 cycles per second. This range is most important in the present study because most of the annoying noise falls within these limits. The band-pass analysis generally is not as useful for analyzing the data as the frequency spectrums; nevertheless, it can be used profitably to check other data and to find regions of large sound energies.

Tailpipe characteristics.—The theoretical data previously discussed (fig. 21) showed that certain pass bands occurred at frequencies both above and below the muffler resonant frequency. For muffler 78, these bands are from 0 to 93 cycles per second and from 375 to 400 cycles per second. Although the theoretical data showed no attenuation should have been obtained in the frequency range from 0 to 93 cycles per second, the frequency analysis of figure 29 (a) indicates that some effective quieting was received. Some muffling also was obtained in the predicted high-frequency pass band. In this band, however, the attenuation is very small, ranging from ½ to 2 decibels. The marked decrease in attenuation in the frequency range from 375 to 400 cycles

per second is sufficient to indicate that the tailpipe resonance must have occurred in this frequency band; this result agrees with the theory. The cold tests also showed this attenuation decrease. It may therefore be concluded that the theoretical expression is valid for predicting the tailpipe resonance of the muffler under engine test conditions and that some slight attenuation may be realized during such resonances. Further evidence of these tailpipe resonances may be found by checking the data for mufflers 79 and 80.

Internal sound pressures of the exhaust system.—As stated previously, the test engine had two separate exhaust manifolds, one exhausting three cylinders and the other, four. A schematic drawing showing this arrangement appears in figure 30. Sound-pressure data, as signaled by a crystal pickup gage placed in the left exhaust manifold, are presented in figure 31. The curve of figure 31 (a) describes one cycle of this sound variation. The curve of figure 31 (b), having 4 humps, shows the exhaust-pressure variation for the 4-cylinder exhaust. This curve was not obtained directly from recorded data but was synthesized with the aid of the measured 3-cylinder exhaust curve.

Close examination of the plot showing the 3-cylinder exhaust pressure reveals that the sound pressure in the system did not go as high when the second consecutive exhaust valve opened as when the first valve opened. An examination of the exhaust system reveals that the first cylinder exhaust valve remains open for a considerable time after the second cylinder valve opens; thus, the volume of the system is increased. This increased volume allows, in effect, an additional expansion of the exhausting gases and provides a damping of the peak sound pressures.

The maximum peak exhaust pressure measured is shown to be approximately 7 pounds per square inch. This value corresponds to a sound-pressure level of 189 decibels. This pressure is far greater than both the pressure assumed in theory and the sound pressure used for the cold tests. The peak pressures measured entering the mufflers attached to the cold-test setup were of the order of 141 decibels or 0.028 pound per square inch. In order to reduce large peak sound pulses, collector rings may be employed. The pressure records of figure 31, for example, indicate that, if a complete circular collector ring had been installed on the engine, the magnitude of the pressure peaks might have been reduced by over 50 percent. In addition, only one muffler would have been required.

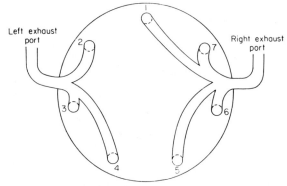

FIGURE 30.—Schematic drawing of helicopter-engine-exhaust system. Firing order: 1, 3, 5, 7, 2, 4, 6.

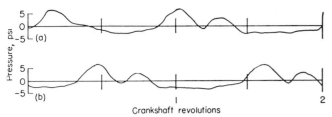

(a) Variation of sound pressure measured in the 3-cylinder exhaust of the test helicopter.

(b) Variation of sound pressure in the 4-cylinder exhaust of the test helicopter as estimated from 3-cylinder data.

FIGURE 31.—Exhaust-pipe sound pressure.

POSSIBLE REASONS FOR DISCREPANCIES BETWEEN COLD TESTS AND ENGINE TESTS

Some reasons may be given to account for the discrepancies that exist between the attenuations obtained from the cold tests and those obtained from the engine tests. These reasons include (1) the large differences in operating conditions, and (2) the prevailing extraneous noises of the engine tests.

Differences in operating conditions.—In the section entitled "Variables Dependent on Operating Conditions" in Part III, the possible effects of three variables were discussed. The effect of temperature was taken into account in the design of the engine mufflers by using the speed of sound in the hot exhaust gas. The exhaust-gas velocity was not taken into account, but it is doubtful whether this factor could have caused a loss of attenuation, inasmuch as the available evidence indicates that the exhaust-gas velocity either causes an increase in attenuation or has little influence on attenuation. The exhaust-gas velocity for this engine is estimated to be about 500 feet per second. The cold-test experiments were conducted with peak sound pressures of the order of 141 decibels (0.028 psi); whereas, the peak sound pressures from the engine entering the mufflers were about 189 decibels (7.0 psi). This sound-pressure increase of 250 times in the muffler system raises the sound pressure to a point where it is no longer small with respect to the static (atmospheric) pressure. An original assumption made in the development of the theoretical equation was that the sound pressure would be small in comparison with the static pressure. It is obvious that this assumption was not satisfied during the engine tests, and this fact may be responsible for some loss of attenuation.

Extraneous noise.—Another factor which may account for some of the discrepancies between data is extraneous noise. The influence of this factor on the exhaust noise spectrum presented is difficult to determine. No pure extraneous noise spectrum could be obtained whereby a quantitative point-by-point comparison could be made. The extraneous noise, as discussed herein, is made up of all noises which originate from sources other than the exhaust gas. These noises include engine air intake, engine blower, engine clatter, vibrating fuselage, main rotor, and distant aircraft. The combination of these noises, when integrated with those from the exhaust gases, yields all the curves described in figure 29. If the exhaust-gas noise is the most pronounced noise in a system and if it is reduced continuously, some

point will be passed where the exhaust and extraneous noises will be equal. At this point the extraneous noise will be equally as important as the exhaust in determining the noise spectrum. Thus, the spectrum will stop defining the shape of exhaust noise in detail and begin to show some characteristics of the extraneous noises. A reduction of the exhaust noise well below that of the extraneous noise will leave a spectrum containing principally extraneous noise. With this fact in mind, the large, two-chamber muffler (81) was designed to attenuate the exhaust noise so much that the extraneous noise spectrum could be approximately determined. The spectrum for muffler 81 (fig. 29 (b)) has practically the same shape as that for muffler 80. This observation indicates that muffler 80 must have reduced the exhaust noise to a point where the extraneous noise became prevalent and that muffler 81 could have only further reduced the exhaust noise; consequently, only slightly more overall noise reduction was provided. Overall sound-pressure measurements showed the same sound energy (81 decibels) was present at the microphones when both mufflers 80 and 81 were installed. Thus, the exact attenuation provided by the mufflers could not be determined because of the extraneous noise level. It is of interest to note here that, as the extraneous noise level is approached, the mufflers must reduce the exhaust noise in greater increments to reduce the overall noise level by equal amounts. For instance, if the extraneous noise is 85 decibels and the exhaust noise is 100 decibels, the overall noise will be 100.1 decibels. If a muffler reduces the exhaust noise by 12 decibels, the overall noise will be reduced by 10.4 decibels to 89.7 decibels. If the exhaust noise is reduced another 12 decibels (to 76 decibels), the overall noise level is reduced by only 4.2 decibels to 85.5 decibels. This explanation shows very clearly that the amount of overall noise reduction which can be gained by the use of a given muffler is dependent upon the relative intensities of the extraneous and exhaust noises. It may be concluded, therefore, that a muffler used to attenuate a noise level which considerably exceeds that of the extraneous noise can provide much more overall noise reduction than if it were working in a noise range close to the extraneous noise.

SIGNIFICANCE OF MEASURED NOISE REDUCTION

In order that the significance of the noise reductions obtained may be interpreted, some comparisons and comments are made on the basis of the information contained in reference 13 regarding the sound levels of aircraft traffic. For those familiar with the noise of various types of airplanes on takeoff, figure 27 of this reference provides a meaningful comparison. The noise of the unmuffled 180-horsepower helicopter has about the same intensity level as that of the 150-horsepower Stinson Voyager or the 165-horsepower Beech Bonanza. The smallest muffler tested on the helicopter reduced the intensity to about that of the quietest airplane of figure 27 of reference 13, a 65-horsepower Piper Cub. These comparisons are made at takeoff power at a distance of 200 feet. The three airplanes mentioned were all equipped with standard production mufflers.

As a further indication of the significance of the sound levels measured in this investigation, a comparison in terms of relative loudness is made. Relative loudness is defined herein, as in reference 13, as the perceived loudness of sound heard by the average ear relative to the loudness of the normal conversational voice at a 3-foot distance. The variation in perceived loudness with the loudness level (in phons or decibels) is taken from the American Standards Association Standard Z24.2–1942. Relative loudnesses of the five configurations of this investigation, based on the overall sound levels given in table V, are approximately 5.3 for the unmuffled helicopter, 2.9 with mufflers 1 and 2, and 2.5 with mufflers 3 and 4, all at a distance of 200 feet at takeoff power. Thus, muffler 78, for example, reduces the loudness of the noise as perceived by the average ear by about 45 percent. This example gives an indication of the magnitude of the noise reduction obtained although, of course, the human mind takes into account other factors besides loudness in judging the annoyance due to a particular noise. On the basis of the data in reference 13, the distances at which the helicopter noise would have the same loudness as the reference conversational voice are estimated at about 1,800 feet for the unmuffled helicopter, 800 feet with mufflers 78 and 79, and 630 feet with mufflers 80 and 81. It is evident from this discussion that the mufflers produced a very significant reduction in the noise of the helicopter.

CONCLUDING REMARKS

Attenuation curves have been calculated for a large number of mufflers, all of which are designed to permit the exhaust gas to flow through the mufflers without turning. Comparison of the calculated curves with experimental data has shown that it is possible, by means of the acoustic theory, to predict the attenuation in still air at room temperature of mufflers of the size required for aircraft engines. There are, however, certain limits to the muffler size and the frequency range within which these equations are applicable. These limits include:

(a) For expansion chambers, the acoustic wave length must be greater than about 0.82 times the chamber diameter.

(b) For resonators, if the connector is longer than about one-fifth of the wave length at the desired resonant frequency, the wave nature of the sound flow in the connector must be taken into account.

(c) For resonators, if the acoustic path length from the connector to the closed end of the chamber is of the order

of one-eighth wave length or more, the wave nature of the flow in the chamber must be accounted for.

The conductivity was predicted with reasonable accuracy for connectors composed of a small number of holes or tubes. Where large numbers of holes in close proximity were used, the conductivity was not accurately predictable. In such cases, the designer must rely on an experimental determination of the conductivity through measurement of the resonant frequency.

Methods have been found which, in theory, will eliminate pass bands in three specific cases. The pass bands that can be eliminated are:

(a) The odd-numbered upper pass bands of a double-expansion-chamber muffler.

(b) The first upper pass band of a multiple-resonator muffler.

(c) The first upper tailpipe pass band of a single-resonator muffler.

A method has been presented which permits the effect of the tailpipe to be included in the muffler calculations. Specific equations have been developed for the attenuation with tailpipes of single expansion chamber and single-chamber-resonator mufflers. Experimental verification of the equation for the single-chamber resonator was obtained under cold-test conditions.

Four resonator-type mufflers have been tested on a helicopter engine. Even the smallest of these mufflers reduced the overall noise by a significant amount. Because this overall noise included a considerable amount of extraneous noise, an accurate determination of the exhaust-noise reduction was not possible. The experimental results seem to indicate, however, that the exhaust-noise reduction may have been considerably less than that which was obtained in the cold tests of these same mufflers. The theory is handicapped severely by the fact that the sound pressures inside the exhaust pipe were found to be much larger than those assumed in the basic theory. In order to isolate the effects of large sound pressures and exhaust-gas flow velocities on the attenuating properties of mufflers, further tests are necessary in which extraneous noises are held to a low level.

Langley Aeronautical Laboratory,
National Advisory Committee for Aeronautics,
Langley Field, Va., *October 6, 1952.*

APPENDIX A

ATTENUATION OF EXPANSION-CHAMBER MUFFLERS

ASSUMPTIONS AND GENERAL METHOD

In the derivation of the equations for the attenuation of expansion-chamber mufflers, the following conditions are assumed:

(1) The sound pressures are small compared with the absolute value of the average pressure in the system.

(2) The tailpipe is terminated in its characteristic impedance (no reflected waves in the tailpipe).

(3) The muffler walls neither conduct nor transmit sound energy.

(4) Only plane pressure waves need be considered.

(5) Viscosity effects may be neglected.

By definition, the attenuation in decibels due to a combination of acoustic elements placed in a tube is

$$10 \log_{10} \left(\frac{\text{Average incident sound power}}{\text{Average transmitted sound power}} \right) \quad \text{(A1)}$$

In the manner of reference 1 (p. 72), let the displacements and particle velocities of the incident and reflected waves at an arbitrary point x be written as

$$\left.\begin{array}{ll} \xi_i = A e^{i(\omega t - kx)} & \dot{\xi}_i = i\omega A e^{i(\omega t - kx)} \\ \xi_{re} = B e^{i(\omega t + kx)} & \dot{\xi}_{re} = i\omega B e^{i(\omega t + kx)} \end{array}\right\} \quad \text{(A2)}$$

where the positive x-direction is taken as the direction of propagation of the incident wave and the constants A and B are, in general, complex numbers. For plane waves the acoustic pressure p is equal to $\mp \rho c^2 \frac{\partial \xi}{\partial x}$, where ρ is the average density of the gas. The incident and reflected pressures can therefore be written as

$$\left.\begin{array}{l} p_i = i\omega\rho c A e^{i(\omega t - kx)} \\ p_{re} = i\omega\rho c B e^{i(\omega t + kx)} \end{array}\right\} \quad \text{(A3)}$$

The average sound power in the incident wave is

$$\frac{\omega}{2\pi} \int_0^{2\pi/\omega} p_i \dot{\xi}_i S \, dt$$

where, since this is a calculation of actual power, only the real parts of p_i and $\dot{\xi}_i$ can be considered. After the integration is performed, the average sound power is obtained as

$$\frac{1}{2} \rho c \omega^2 S |A|^2$$

If the attenuation between two points located at cross sections of equal area is desired, the formula is

$$\text{Attenuation} = 10 \log_{10} \left| \frac{A_1}{A_2} \right|^2 \quad \text{(A4)}$$

provided there are no reflected waves at the point 2.

SINGLE EXPANSION CHAMBER

A schematic diagram of a single expansion chamber is shown below:

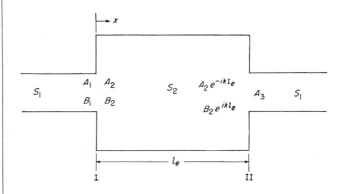

The origin of x is taken at junction I. If constant factors are omitted, the equations for continuity of pressure and flow volume at junction I can be written, with the assistance of equations (A2) and (A3), as

$$A_1 + B_1 = A_2 + B_2 \quad \text{(A5)}$$

$$S_1(A_1 - B_1) = S_2(A_2 - B_2)$$

or

$$A_1 - B_1 = m(A_2 - B_2) \quad \text{(A6)}$$

Similarly, at junction II, the expressions are

$$A_2 e^{-ikl_e} + B_2 e^{ikl_e} = A_3 \quad \text{(A7)}$$

$$m(A_2 e^{-ikl_e} - B_2 e^{ikl_e}) = A_3 \quad \text{(A8)}$$

If, now, equations (A5), (A6), (A7), and (A8) are solved simultaneously for the ratio A_1/A_3, the result is

$$\left.\begin{aligned} \frac{A_1}{A_3}&=\cos kl_e+i\,\frac{1}{2}\left(m+\frac{1}{m}\right)\sin kl_e\\ \left|\frac{A_1}{A_3}\right|&=\sqrt{1+\frac{1}{4}\left(m-\frac{1}{m}\right)^2\sin^2 kl_e} \end{aligned}\right\} \qquad (A9)$$

and the attenuation (see eq. (A4)) is

$$\text{Attenuation}=10\log_{10}\left[1+\frac{1}{4}\left(m-\frac{1}{m}\right)^2\sin^2 kl_e\right] \quad (A10)$$

The design curves of figure 22 were obtained by plotting this equation against kl_e.

If the equations are solved for B_1/A_3, the result is

$$\frac{B_1}{A_3}=-i\,\frac{1}{2}\left(m-\frac{1}{m}\right)\sin kl_e \qquad (A11)$$

When measurements are taken in the manner described in the section entitled "Methods and Tests" in Part I, the maximum pressure measurable in the exhaust pipe to the left of junction I will be proportional to

$$\left|\frac{A_1}{A_3}\right|+\left|\frac{B_1}{A_3}\right|$$

and will be found at the station x at which the incident and reflected waves are in phase with each other. The maximum measured attenuation will thus be given by

$$10\log_{10}\left(\left|\frac{A_1}{A_3}\right|+\left|\frac{B_1}{A_3}\right|\right)^2 \qquad (A12)$$

Substitution of equations (A9) and (A11) into equation (A12) results in

$$\text{Maximum measured attenuation}=10\log_{10}\left[1+\frac{1}{2}\left(m-\right.\right.$$
$$\left.\frac{1}{m}\right)^2\sin^2 kl_e+\left(m-\frac{1}{m}\right)\sin kl_e\sqrt{1+\frac{1}{4}\left(m-\frac{1}{m}\right)^2\sin^2 kl_e}\right]$$
$$(A13)$$

The upper curve of figure 8 was computed from this equation.

DOUBLE EXPANSION CHAMBER WITH EXTERNAL CONNECTING TUBE

A schematic diagram of a double-expansion-chamber muffler with the connecting tube external to the chambers is shown below, with the symbols to be used also included:

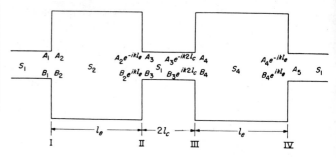

The effective length of the connecting tube $2l_c$ is equal to the physical length plus an end correction. If the same basic method is used as for the single expansion chamber, the equations of continuity of pressure and flow at the four indicated junctions are:

At junction I

$$A_1+B_1=A_2+B_2$$
$$A_1-B_1=m(A_2-B_2)$$

At junction II

$$A_2e^{-ikl_e}+B_2e^{ikl_e}=A_3+B_3$$
$$m(A_2e^{-ikl_e}-B_2e^{ikl_e})=A_3-B_3$$

At junction III

$$A_3e^{-ik2l_c}+B_3e^{ik2l_c}=A_4+B_4$$
$$A_3e^{-ik2l_c}-B_3e^{ik2l_c}=m(A_4-B_4)$$

At junction IV

$$A_4e^{-ikl_e}+B_4e^{ikl_e}=A_5$$
$$m(A_4e^{-ikl_e}-B_4e^{ikl_e})=A_5$$

The simultaneous solution of these equations results in

$$\frac{A_1}{A_5}=\frac{1}{16m^2}\left[(m+1)^4e^{2ik(l_e+l_c)}-(m^2-1)^2e^{-2ik(l_e+l_c)}-\right.$$
$$2(m^2-1)^2e^{2ikl_c}+2(m^2-1)^2e^{-2ikl_c}-$$
$$\left.(m^2-1)^2e^{2ik(l_e-l_c)}+(m-1)^4e^{-2ik(l_e-l_c)}\right]$$

This equation can be written, in terms of trigonometric functions, as

$$\frac{A_1}{A_5}=\frac{1}{16m^2}\{[4m(m+1)^2\cos 2k(l_e+l_c)-4m(m-1)^2\cos 2k(l_e-l_c)]$$

$$+i[2(m^2+1)(m+1)^2\sin 2k(l_e+l_c)-$$

$$2(m^2+1)(m-1)^2\sin 2k(l_e-l_c)-4(m^2-1)^2\sin 2kl_c]\} \quad (A14)$$

The attenuation is

$$\text{Attenuation}=10\log_{10}\left\{\left[R\left(\frac{A_1}{A_5}\right)\right]^2+\left[I\left(\frac{A_1}{A_5}\right)\right]^2\right\} \quad (A15)$$

when R and I are used to denote the real and imaginary parts, respectively.

DOUBLE EXPANSION CHAMBER WITH INTERNAL CONNECTING TUBE

A schematic diagram of a double-expansion-chamber muffler with the connecting tube internal to the expansion chambers is shown below, with the symbols to be used also indicated:

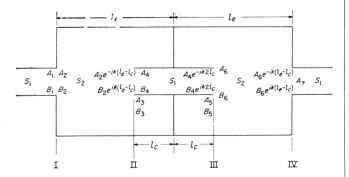

The basic equations of continuity of pressure and flow at the four indicated junctions are:

At junction I

$$A_1+B_1=A_2+B_2$$

$$A_1-B_1=m(A_2-B_2)$$

At junction II

$$A_2e^{-ik(l_e-l_c)}+B_2e^{ik(l_e-l_c)}=A_3+B_3=A_4+B_4$$

$$m(A_2e^{-ik(l_e-l_c)}-B_2e^{ik(l_e-l_c)})=A_4-B_4+(m-1)(A_3-B_3)$$

At junction III

$$A_4e^{-2ikl_c}+B_4e^{2ikl_c}=A_5+B_5=A_6+B_6$$

$$A_4e^{-2ikl_c}-B_4e^{2ikl_c}+(m-1)(A_5-B_5)=m(A_6-B_6)$$

At junction IV

$$A_6e^{-ik(l_e-l_c)}+B_6e^{ik(l_e-l_c)}=A_7$$

$$m(A_6e^{-ik(l_e-l_c)}-B_6e^{ik(l_e-l_c)})=A_7$$

In addition, because of the total reflection from the bulkhead separating the two chambers,

$$B_3=A_3e^{-2ikl_c}$$

$$B_5=A_5e^{2ikl_c}$$

The simultaneous solution of these equations results in

$$\frac{A_1}{A_7}=\cos 2kl_e-(m-1)\sin 2kl_e\tan kl_c+$$

$$\frac{i}{2}\left\{\left(m+\frac{1}{m}\right)\sin 2kl_e+(m-1)\tan kl_c\right.$$

$$\left.\left[\left(m+\frac{1}{m}\right)\cos 2kl_e-\left(m-\frac{1}{m}\right)\right]\right\} \quad (A16)$$

The attenuation is

$$\text{Attenuation}=10\log_{10}\left\{\left[R\left(\frac{A_1}{A_7}\right)\right]^2+\left[I\left(\frac{A_1}{A_7}\right)\right]^2\right\} \quad (A17)$$

CUTOFF FREQUENCY

In the design of double-expansion-chamber mufflers, it is important to be able to predict the low-frequency limit of the first effective attenuation region. This frequency is called the cutoff frequency f_c. It may, of course, be found from a plot of equation (A17) but a more rapid method of estimating f_c is desirable for use in the preliminary design of a muffler. The semiempirical equation

$$f_c\approx\frac{c}{2\pi}\frac{1}{\sqrt{ml_cl_c+\frac{l_e}{3}(l_e-l_c)}} \quad (A18)$$

has been found quite satisfactory for this purpose within the range of variables covered in this investigation (see table II).

APPENDIX B

ATTENUATION OF RESONATOR MUFFLERS

SINGLE RESONATORS

In the derivation of the equation for the attenuation due to a single resonator in a side branch, assumptions (1), (2), and (3) of appendix A are required. Assumptions (4) and (5) are modified as follows:

(4) Only plane pressure waves are propagated in the exhaust pipe and the tailpipe.

(5) The influence of the viscosity of the fluid may be neglected everywhere except in the tubes or orifices which form the connector between the exhaust pipe and the volume chamber of the resonator.

The following two additional assumptions are necessary:

(6) The boundary-layer thickness is small compared to the diameter of the tube or orifice in which viscosity effects are considered.

(7) The dimensions of the resonator are small relative to the wave length of the sound considered.

Consider the effect of a side branch of impedance $Z_b = R_b + iX_b$ opening into a tube in which plane sound waves are propagated. At the point where the branch joins the tube, the conditions of continuity of pressure and sound current give

$$p_i + p_{re} = p_b = p_{tr} \qquad (B1)$$

$$I_i - I_{re} = I_b + I_{tr} \qquad (B2)$$

where subscripts i and re refer to the incident and reflected waves ahead of the branch, b refers to the branch, and tr refers to the transmitted wave behind the branch. For a plane wave $p = Z_0 I$, where Z_0 is the characteristic impedance of the tube. If the currents are written in terms of pressure and impedance, equation (B2) becomes

$$\frac{1}{Z_0}(p_i - p_{re}) = p_{tr}\left(\frac{1}{Z_b} + \frac{1}{Z_0}\right) \qquad (B3)$$

If, now, equations (B1) and (B3) are solved simultaneously for the ratio p_i/p_{tr}, the result is

$$\frac{p_i}{p_{tr}} = 1 + \frac{Z_0}{2Z_b} = 1 + \frac{Z_0}{2(R_b + iX_b)}$$

Hence the attenuation is

$$\text{Attenuation} = 10 \log_{10}\left|\frac{p_i}{p_{tr}}\right|^2 = 10 \log_{10}\frac{\left(R_b + \frac{Z_0}{2}\right)^2 + X_b^2}{R_b^2 + X_b^2} \qquad (B4)$$

A schematic diagram of a single-resonator muffler is shown below with the symbols that are used indicated:

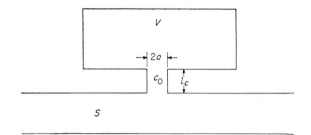

On the basis of the listed assumptions, the impedances of the various components are (ref. 1, p. 118)

$$\text{Volume-chamber impedance} = -i\,\frac{\rho c^2}{\omega V} \qquad (B5)$$

$$\text{Connector impedance} = \frac{l_c}{\pi a^3}\sqrt{2\mu\rho\omega} + i\left(\frac{\omega\rho}{c_0} + \frac{l_c}{\pi a^3}\sqrt{2\mu\rho\omega}\right) \qquad (B6)$$

where c_0 is the conductivity and l_c is the effective length of the connector. Since, in the resonator side branch, the volume chamber and the connector are in series

$$R_b = \frac{l_c}{\pi a^3}\sqrt{2\mu\rho\omega} \qquad (B7)$$

$$X_b = \frac{\omega\rho}{c_0} - \frac{\rho c^2}{\omega V} + \frac{l_c}{\pi a^3}\sqrt{2\mu\rho\omega} \qquad (B8)$$

These values, when substituted into equation (B4), give the attenuation of a single-resonator muffler.

In many cases it is possible to neglect the effect of viscosity without introducing excessive error, except at the resonant frequency. If $\mu = 0$, equation (B4) simplifies to

$$\text{Attenuation} = 10 \log_{10}\left(1 + \frac{Z_0^2}{4X_b^2}\right) \qquad (B9)$$

By inserting the value of X_b and making use of the fact that $f_r = \frac{c}{2\pi}\sqrt{\frac{c_0}{V}}$ it is possible to bring equation (B9) into the form

$$\text{Attenuation} = 10 \log_{10}\left[1 + \left(\frac{\frac{\sqrt{c_0 V}}{2S}}{\frac{f}{f_r} - \frac{f_r}{f}}\right)^2\right] \qquad (B10)$$

The design curves of figure 23 have been obtained from this equation. Since viscosity has been neglected, the predicted attenuation rises to infinity at the resonant frequency $\frac{f}{f_r}=1$.

If the effective length of the connector l_c is not sufficiently short compared to the sound-wave length, assumption (7) of appendix B is violated and the wave nature of the flow in the connector must be considered. Muffler 28 is an example of this case. For a connector of length l_c and area S_c terminated by a volume V, the branch reactance (with viscosity omitted) is

$$X_b = \frac{\rho c}{S_c}\left(\frac{\tan kl_c - \frac{S_c}{kV}}{\frac{S_c}{kV}\tan kl_c + 1}\right) \qquad (B11)$$

This expression can be obtained from equation 5.30, page 125, reference 1 by substituting the volume-chamber impedance $-i\frac{\rho c^2}{\omega V}$ for the impedance which is symbolized by Z_l in the reference. Having obtained the branch reactance, the attenuation, with viscosity neglected, is calculated from equation (B9). The attenuation of muffler 28 was calculated in this manner. Strictly speaking, an end correction is required at both ends of the connector in determining the effective connector length l_c when equation (B11) is used. This correction will reach a maximum of about 0.8 times the connector radius, at each end of the connector, if the connector radius is much smaller than that of the exhaust pipe and the volume chamber.

If the resonant chamber is itself long, the resonance becomes a length-controlled phenomenon instead of a volume-controlled one and the attenuation can be determined by assuming plane-wave motion in both the connector and the chamber.

In case the connector is short and the chamber is long, as in the following sketch, another approach may be used:

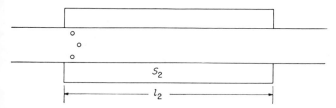

Again, the problem is to determine the branch impedance. For a closed chamber the branch impedance is (again with viscosity omitted)

$$Z_b = i\left(\frac{\rho\omega}{c_0} - \frac{\rho c}{S_2}\cot kl_2\right) \qquad (B12)$$

The attenuation is therefore

$$10\log_{10}\left[1 + \frac{1}{4}\left(\frac{m}{\frac{kS_2}{c_0} - \cot kl_2}\right)^2\right] \qquad (B13)$$

For a muffler in which the connector is located at the center of the resonant chamber, rather than at the end, the effective chamber length is one-half the actual chamber length l_2

and the effective expansion ratio is twice the physical expansion ratio m. These effective values should be used in equation (B12) or (B13). Because of the typical attenuation characteristics of resonators of this type (eq. (B13)), they are called "quarter-wave" resonators.

MULTIPLE RESONATORS

The attenuation of M identical chambers of an infinite filter composed of branch resonators is given by (see ref. 8)

$$\text{Attenuation} = -8.69M\cosh^{-1}\left|\cos kl_1 + i\frac{Z_0}{2Z_b}\sin kl_1\right| \qquad (B14)$$

where

$$\frac{Z_0}{2Z_b} = \frac{\rho c}{2S}\cdot\frac{1}{i\left(\frac{2\pi f\rho}{c_0} - \frac{\rho c^2}{2\pi fV}\right)}.$$

By use of the substitution $\sqrt{\frac{c_0}{V}}=\frac{2\pi f_r}{c}$, this equation may also be written as

$$\frac{Z_0}{2Z_b} = -i\frac{\frac{\sqrt{c_0 V}}{2S}}{\frac{f}{f_r} - \frac{f_r}{f}}$$

Substituting this expression in equation (B14) and making use of the fact that $k_r = \frac{2\pi f_r}{c}$ gives

$$\text{Attenuation} = -8.69M$$

$$\cosh^{-1}\left|\cos\left(k_r l_1\frac{f}{f_r}\right) + \frac{\frac{\sqrt{c_0 V}}{2S}}{\frac{f}{f_r} - \frac{f_r}{f}}\sin\left(k_r l_1\frac{f}{f_r}\right)\right| \qquad (B15)$$

where the inverse hyperbolic cosine is taken with a negative sign. Thus at a given frequency the attenuation, per chamber, of a multiple-resonator muffler is a function of three basic parameters: $\sqrt{c_0 V}/2S$, $k_r l_1$, and $\sqrt{c_0/V}$ (since f_r is controlled by $\sqrt{c_0/V}$). The design curves of figure 24 were calculated from equation (B15).

In reference 1 the cutoff frequency is given as

$$f_c = \frac{c}{\pi}\sqrt{\frac{S}{l_1 V}\left(\frac{1}{1 + \frac{4S}{l_1 c_0}}\right)} \qquad (B16)$$

In terms of the resonant frequency equation (B16) can be written in the form

$$f_c = \frac{f_r}{\sqrt{1 + \frac{c_0 l_1}{4S}}} \qquad (B17)$$

These equations for f_c are, in reality, approximations since lumped impedances were assumed in the derivation (see ref. 1). The approximation should be valid within the range of variables where $\tan k_c l_1$ can be taken as $k_c l_1$ within the permissible limits of accuracy.

In the case of mufflers with long chambers the expression for Z_b given by equation (B12) can be used in equation (B14). Instances where this substitution has been made are pointed out in the text and in the figures.

APPENDIX C

COMBINATIONS

TWO RESONATORS TUNED AT DIFFERENT FREQUENCIES

A schematic diagram of a muffler composed of two resonators tuned at different frequencies is shown below with the subscripts that will be used to indicate various locations also shown:

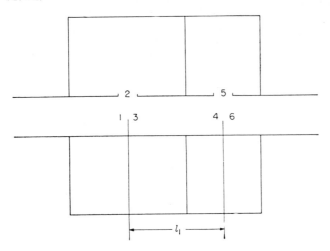

The assumptions made are the same as for the single resonators. At station 1 let

$$Z_1 = \frac{p_1}{I_1} = \frac{i\omega\rho c(A_1+B_1)e^{i\omega t}}{i\omega S(A_1-B_1)e^{i\omega t}} = Z_0\frac{A_1+B_1}{A_1-B_1}$$

From this relationship

$$\frac{B_1}{A_1} = \frac{Z_1-Z_0}{Z_1+Z_0} \qquad\qquad (C1)$$

where Z_1 is the impedance of the first branch and the circuit to the right of this branch in parallel. Similarly

$$\frac{B_3}{A_3} = \frac{Z_3-Z_0}{Z_3+Z_0} \qquad\qquad (C2)$$

The equation for continuity of pressure at the junction is

$$A_1+B_1 = A_3+B_3$$

$$\frac{A_1}{A_3} = \frac{1+\dfrac{B_3}{A_3}}{1+\dfrac{B_1}{A_1}}$$

Substituting from equations (C1) and (C2) gives

$$\frac{A_1}{A_3} = \frac{Z_3}{Z_1}\frac{Z_1+Z_0}{Z_3+Z_0}$$

Similarly

$$\frac{A_4}{A_6} = \frac{Z_6}{Z_4}\frac{Z_4+Z_0}{Z_6+Z_0} = \frac{Z_4+Z_0}{2Z_4}$$

since $Z_6=Z_0$. Now

$$A_4 = A_3e^{-ikl_1}$$

so that

$$\frac{A_1}{A_6} = \frac{A_1}{A_3e^{-ikl_1}}\frac{A_4}{A_6} = \frac{e^{ikl_1}}{2}\left(\frac{Z_1+Z_0}{Z_1+\dfrac{Z_0Z_1}{Z_3}}\right)\left(\frac{Z_4+Z_0}{Z_4}\right) \qquad (C3)$$

The values of the impedances in this equation are

$$Z_4 = \frac{Z_0Z_5}{Z_0+Z_5} \qquad\qquad (C4)$$

$$\frac{Z_3}{Z_0} = \frac{Z_4\cos kl_1 + iZ_0\sin kl_1}{Z_0\cos kl_1 + iZ_4\sin kl_1} \qquad (C5)$$

$$Z_1 = \frac{Z_2Z_3}{Z_2+Z_3} \qquad\qquad (C6)$$

The attenuation is determined by inserting the values given in equations (C4), (C5), and (C6) into equation (C3) and working out the expression for

$$\text{Attenuation} = 10\log_{10}\left|\frac{A_1}{A_6}\right|^2 \qquad (C7)$$

If the branch impedances have no resistive components, the result obtained is

$$\frac{A_1}{A_6} = \frac{1}{2}\frac{[R_3X_2{}^2+Z_0R_3{}^2+Z_0(X_2+X_3)^2]+i[R_3{}^2X_2+X_2X_3(X_2+X_3)]}{[R_3X_2{}^2\cos kl_1 + Z_0X_2R_3\sin kl_1]+i[R_3{}^2X_2+X_2X_3(X_2+X_3)\cos kl_1 - Z_0X_2(X_2+X_3)\sin kl_1]} \qquad (C8)$$

where

$$R_3 = \frac{Z_0 X_5^2 (Z_0^2 + X_5^2)}{[(Z_0^2 + X_5^2) \cos kl_1 - Z_0 X_5 \sin kl_1]^2 + X_5^4 \sin^2 kl_1}$$

$$X_3 = \frac{Z_0^2 X_5 (Z_0^2 + X_5^2) \cos 2kl_1 + \frac{1}{2}(Z_0^5 + Z_0^3 X_5^2) \sin 2kl_1}{[(Z_0^2 + X_5^2) \cos kl_1 - Z_0 X_5 \sin kl_1]^2 + X_5^4 \sin^2 kl_1}$$

(C9)

These equations were used to calculate the attenuation of muffler 73 (see fig. 16). It has been found necessary to include the length l_1, even though it may be much less than the sound wave length under consideration.

A RESONATOR AND AN EXPANSION CHAMBER

A schematic diagram of a muffler composed of a resonator in combination with an expansion chamber is shown below:

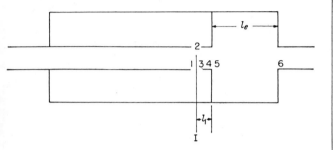

The boundary conditions to be satisfied at station I are

$$A_1 + B_1 = A_2 + B_2 = A_3 + B_3 \qquad (C10)$$

$$S_1(A_1 - B_1) = S_2(A_2 - B_2) + S_1(A_3 - B_3) \qquad (C11)$$

From equations (C10) and (C11)

$$A_1 = \frac{1}{2}\frac{S_2}{S_1}(A_2 - B_2) + A_3 \qquad (C12)$$

For the side branch

$$Z_2 = \frac{p_2}{I_2} = \frac{i\omega \rho c (A_2 + B_2) e^{i\omega t}}{i\omega S_2 (A_2 - B_2) e^{i\omega t}} = \frac{\rho c}{S_2} \frac{A_2 + B_2}{A_2 - B_2}$$

from which

$$A_2 - B_2 = \frac{\rho c}{S_2 Z_2}(A_2 + B_2) = \frac{\rho c}{S_2 Z_2}(A_3 + B_3) \qquad (C13)$$

If equation (C13) is substituted into equation (C12), the result is

$$A_1 = \frac{S_2}{2S_1}\frac{\rho c}{S_2 Z_2}(A_3 + B_3) + A_3$$

Since $A_3 = A_4 e^{ikl_1}$, $B_3 = B_4 e^{-ikl_1}$, and $\frac{\rho c}{S_1} = Z_0$, the preceding equation can also be written as

$$A_1 = \left(1 + \frac{Z_0}{2Z_2}\right) A_4 e^{ikl_1} + \frac{Z_0}{2Z_2} B_4 e^{-ikl_1} \qquad (C14)$$

Let the subscripts 1 and 3 of equations (A9) and (A11) be replaced by 4 and 6, respectively. Then the ratios A_4/A_6 and B_4/A_6 can be written as

$$\frac{A_4}{A_6} = \cos kl_e + i\frac{1}{2}\left(m + \frac{1}{m}\right)\sin kl_e \qquad (C15)$$

$$\frac{B_4}{A_6} = -i\frac{1}{2}\left(m - \frac{1}{m}\right)\sin kl_e \qquad (C16)$$

By using equations (C14), (C15), and (C16), the ratio A_1/A_6 can be written as

$$\frac{A_1}{A_6} = \left(1 + \frac{Z_0}{2iX_b}\right)\left[\cos kl_e + i\frac{1}{2}\left(m + \frac{1}{m}\right)\sin kl_e\right]e^{ikl_1} +$$
$$\frac{Z_0}{2iX_b}\left[-i\frac{1}{2}\left(m - \frac{1}{m}\right)\sin kl_e\right]e^{-ikl_1}$$
$$= \left\{\left(1 - i\frac{Z_0}{2X_b}\right)\left[\cos kl_e + i\frac{1}{2}\left(m + \frac{1}{m}\right)\sin kl_e\right] -\right.$$
$$\left.\frac{Z_0}{4X_b}\left(m - \frac{1}{m}\right)\sin kl_e \, e^{-2ikl_1}\right\}e^{ikl_1}$$
$$= \left\{\cos kl_e + \frac{Z_0}{4X_b}\left(m + \frac{1}{m}\right)\sin kl_e -\right.$$
$$\frac{Z_0}{4X_b}\left(m - \frac{1}{m}\right)\cos 2kl_1 \sin kl_e + i\left[\frac{1}{2}\left(m + \frac{1}{m}\right)\sin kl_e -\right.$$
$$\left.\left.\frac{Z_0}{2X_b}\cos kl_e + \frac{Z_0}{4X_b}\left(m - \frac{1}{m}\right)\sin 2kl_1 \sin kl_e\right]\right\}e^{ikl_1}$$

(C17)

The attenuation is given by

$$\text{Attenuation} = 10\log_{10}\left|\frac{A_1}{A_6}\right|^2$$
$$= 10\log_{10}\left\{\left[\cos kl_e + \frac{Z_0}{4X_b}\left(m + \frac{1}{m}\right)\sin kl_e -\right.\right.$$
$$\left.\frac{Z_0}{4X_b}\left(m - \frac{1}{m}\right)\cos 2kl_1 \sin kl_e\right]^2 +$$
$$\left[\frac{1}{2}\left(m + \frac{1}{m}\right)\sin kl_e - \frac{Z_0}{2X_b}\cos kl_e +\right.$$
$$\left.\left.\frac{Z_0}{4X_b}\left(m - \frac{1}{m}\right)\sin 2kl_1 \sin kl_e\right]^2\right\}$$

(C18)

APPENDIX D

ATTENUATION OF MUFFLERS WITH FINITE TAILPIPES

SINGLE EXPANSION CHAMBER

Consider a muffler composed of an expansion chamber with expansion ratio m terminated with a tailpipe of effective length l_t. At the upstream end of the muffler,

$$A_1 + B_1 = A_2 + B_2 \tag{D1}$$

$$A_1 - B_1 = m(A_2 - B_2) \tag{D2}$$

At the downstream end, assuming total reflection from the end of the tailpipe

$$A_2 e^{-ikl_e} + B_2 e^{ikl_e} = A_3 + B_3 = A_3\left(1 - e^{-i2kl_t}\right) \tag{D3}$$

$$m\left(A_2 e^{-ikl_e} - B_2 e^{ikl_e}\right) = A_3\left(1 + e^{-i2kl_t}\right) \tag{D4}$$

These four equations, when solved simultaneously for A_1/A_3, give

$$\frac{A_1}{A_3} = \frac{1}{4m}\{[4m \cos kl_e - 2(m^2-1)\sin 2kl_t \sin kl_e] +$$
$$i\,[2(m^2+1)\sin kl_e - 2(m^2-1)\cos 2kl_t \sin kl_e]\}$$

The attenuation is $10 \log_{10} \left|\dfrac{A_1}{A_3}\right|^2$ where

$$\left|\frac{A_1}{A_3}\right|^2 = 1 + \frac{(m^2-1)^2}{2m^2}\sin^2 kl_e - \frac{m^2-1}{2m}\sin 2kl_t \sin 2kl_e -$$
$$\frac{m^4-1}{2m^2}\cos 2kl_t \sin^2 kl_e \tag{D5}$$

The approximate cutoff frequency is found by setting the preceding expression equal to zero and solving for k, with the approximations that

$$\sin kl_e = kl_e$$
$$\sin 2kl_e = 2kl_e$$
$$\sin 2kl_t = 2kl_t$$
$$\cos kl_t = 1$$

The result is

$$f_c \approx \frac{c}{2\pi}\sqrt{\frac{4 + \dfrac{2l_e}{ml_t}}{\left(m + \dfrac{1}{m}\right)l_e l_t}} \tag{D6}$$

SINGLE RESONATOR

A schematic diagram of a single-resonator muffler with a finite tailpipe is shown below:

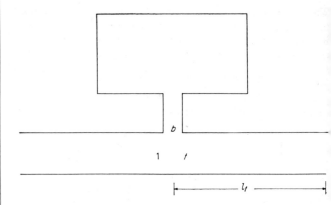

The method of appendix C gives

$$\frac{A_1}{A_t} = \frac{Z_t}{Z_1}\frac{Z_1 + Z_0}{Z_t + Z_0}$$

If now the substitution

$$Z_1 = \frac{Z_b Z_t}{Z_b + Z_t}$$

is made, the result is

$$\frac{A_1}{A_t} = 1 + \frac{Z_0/Z_b}{\dfrac{Z_0}{Z_t} + 1} \tag{D7}$$

If the correct values for Z_0/Z_b and Z_0/Z_t are inserted in this equation, the attenuation may be calculated from equation (A4). As an example, the attenuation equation will be developed for the case where Z_b is a pure reactance and total reflection is assumed at the open end of the tailpipe. In this case

$$\frac{A_1}{A_t} = 1 + \frac{Z_0/iX_b}{\dfrac{Z_0}{iX_t} + 1} \tag{D8}$$

Upon reduction this gives

$$\left|\frac{A_1}{A_t}\right|^2 = 1 + \frac{2(Z_0/X_b)(Z_0/X_t)}{(Z_0/X_t)^2 + 1} + \frac{(Z_0/X_b)^2}{(Z_0/X_t)^2 + 1} \tag{D9}$$

inally, for the single-branch resonator with a tailpipe substitute

$$\frac{Z_0}{X_t}=\cot kl_t$$

where l_t includes the end correction mentioned in Part II of this report under the heading "Theory." Substitute also

$$kl_t=k_r l_t \frac{f}{f_r}$$

and

$$\frac{Z_0}{X_b}=\frac{\sqrt{c_0 V}}{S\left(\frac{f}{f_r}-\frac{f_r}{f}\right)}$$

with the result

$$\text{attenuation}=10\log_{10}\left[1+\frac{\sqrt{c_0 V}}{S}\frac{\sin 2k_r l_t \frac{f}{f_r}}{\left(\frac{f}{f_r}-\frac{f_r}{f}\right)}+\frac{c_0 V}{S^2}\frac{\sin^2 k_r l_t \frac{f}{f_r}}{\left(\frac{f}{f_r}-\frac{f_r}{f}\right)^2}\right] \tag{D10}$$

Note that in equation (D10) the parameters which determine the attenuation characteristics are $\sqrt{c_0 V}/S$, $k_r l_t$, and f_r (or $\sqrt{c_0/V}$).

The pass frequencies can be found by setting the sum of the second and third terms of equation (D10) equal to zero, with the result

$$\tan kl_t=-2S\left(\frac{k}{c_0}-\frac{1}{kV}\right) \tag{D11}$$

The attenuation will be zero for any value of k which satisfies this equation. For the cutoff frequency this equation can be simplified by the use of the approximation

$$\tan kl_t=kl_t$$

with the result

$$f_c=\frac{f_r}{\sqrt{1+\frac{\sqrt{c_0 V}}{2S}k_r l_t}}=\frac{f_r}{\sqrt{1+\frac{c_0 l_t}{2S}}} \tag{D12}$$

Use of this equation gives a value of approximately 88 cycles per second for the cutoff frequency of the muffler of figure 18 (a). The more exact calculation gives $f_c=85$ cycles per second. Note the similarity in form between equation (D12) and equation (B17).

REFERENCES

1. Stewart, George Walter, and Lindsay, Robert Bruce: Acoustics. D. Van Nostrand Co., Inc., 1930.
2. Davis, A. H. (With Appendix by N. Fleming): Further Model Experiments Concerning the Acoustical Features of Exhaust Silencers. Rep. No. N.108, British N.P.L. (Rep. No. 1421, A.R.C.), Feb. 1935.
3. Davis, A. H., and Fleming, N.: The Attenuation Characteristics of Some Aero-Engine Exhaust Silencers. Rep. No. N.125, British N.P.L. (Rep. No. 2249, A.R.C.), Feb. 1936.
4. Morley, A. W.: Progress in Experiments in Aero-Engine Exhaust Silencing. R. & M. No. 1760, British A.R.C., 1937.
5. Martin, Herbert: Muffling Without Power Loss in the Four-Stroke-Cycle Engine. Translation No. 328, Materiel Div., Army Air Corps, Aug. 3, 1938.
6. Buschmann, H.: Noise in Motor Vehicles. R.T.P. Translation No. 2584, British Ministry of Aircraft Production.
7. Martin, H., Schmidt, U., and Willms, W.: The Present Stage of Development of Exhaust Silencers. R.T.P. T.I.B. Translation No. 2596, British Ministry of Aircraft Production. (From MTZ, No. 12, 1940).
8. Czarnecki, K. R., and Davis, Don D., Jr.: Dynamometer-Stand Investigation of the Muffler Used in the Demonstration of Light-Airplane Noise Reduction. NACA TN 1688, 1948.
9. Davis, Don D., Jr., and Czarnecki, K. R.: Dynamometer-Stand Investigation of a Group of Mufflers. NACA TN 1838, 1949.
10. Chu, Lan Jen: Electromagnetic Waves in Elliptic Hollow Pipes of Metal. Jour. Appl. Phys., vol. 9, no. 9, Sept. 1938, pp. 583–591.
11. Levine, Harold, and Schwinger, Julian: On the Radiation of Sound From an Unflanged Circular Pipe. Phys. Rev., vol. 73, no. 4, Second ser., Feb. 15, 1948, pp. 383–406.
12. Trimmer, John D.: Sound Waves in a Moving Medium. Jour. Acous. Soc. Am., vol. 9, no. 2, Oct. 1937, pp. 162–164.
13. Field, R. L., Edwards, T. M., Kangas, Pell, and Pigman, G. L.: Measurement of Sound Levels Associated With Aircraft, Highway and Railroad Traffic. Tech. Dev. Rep. 68, CAA, U. S. Dept. Commerce, July 1947.

3

Reprinted by permission of the Council of the Institution of Mechanical Engineers from pages
19-32 of *Proceedings of a Symposium on Engine Noise and Noise Suppression,* Institution
of Mechanical Engineers, London, 1958, 46p.

ORIGINS OF DIESEL ENGINE NOISE

By A. E. W. Austen, B.Sc., Ph.D.*, and T. Priede, Ph.D., A.M.I.Mech.E.*

INTRODUCTION

THIS PAPER DESCRIBES the first part of a study of the noise emitted by reciprocating internal-combustion engines in the automotive size range, though some measurements on a petrol engine are included for comparison.

The grounds on which it was decided to put the work in hand were: because of its high efficiency, particularly in road transport applications, in which the engine runs most of its time at part load, it is clearly in the national interest that the Diesel engine should be more widely adopted. Certain of the shortcomings tending to prevent wider use—lower power than the petrol engine for a given swept volume and engine weight, and greater first cost—were within sight of being to a large extent overcome. It was thought that eventually the greater noise emission of the Diesel engine might prove decisive, that adequate reduction of the noise might prove difficult, and that this was a problem to which a fuel-injection equipment manufacturer should make a contribution.

The work described has been carried out over several years; the objects were to establish fully the mechanism of the generation and radiation of the various components of the noise, to assess their relative importance, and to establish, in principle, means of reducing them.

The approach has been to identify the source of the predominant noise, reduce it, identify the source of the then predominant noise, and so on. At the same time the general aim was kept in mind and each noise source was explored in detail as opportunity offered.

In the earlier stages of the work measurements were made as engines became available and it was only in the later stages that engines and test facilities specially for noise work were obtained. Techniques and apparatus have been under continuous development during the work. For these reasons the experimental data are, in places, less complete and consistent than could be wished.

The paper is of the nature of an interim report. It is shown that the noise from certain particular sources must first be reduced, and means for doing so are outlined. Much progress has been made in establishing the mechanism of generation of the remaining, basic, noise but little work has

The MS. of this paper was received at the Institution on 28th November 1957.
* *C.A.V. Ltd, Acton, W.3.*

been done to reduce it. There are, however, clear pointers to the possible lines of attack.

TEST CONDITIONS

The noise discussed here is the airborne noise emitted from the engine installed on a test bed but supported on rubber mountings as in a vehicle installation. Except where otherwise stated the microphone was placed 3 ft from the engine on the injection-pump side of the engine.

Details of the measuring equipment and the way it was used are given in Appendix III.

In some cases, measurements have been used to estimate the subjective loudness on the sones scale. This is a scale on which the numerical value is intended to be proportional to the magnitude of the auditory sensation experienced by the observer. The method of loudness estimation used is given in Appendix IV. The procedure has been consistent and is satisfactory for comparisons throughout this paper. For reasons given in the Appendix, however, comparisons should not be made with other loudness estimations.

Loudness is a physical quantity which, although it involves subjective responses, can nevertheless be measured given a large enough group of observers. The method of estimating loudness used here is known to give results accurate enough for most purposes. Loudness, however whether measured directly, or estimated from octave-band measurement, does not evalute the annoyance caused by a noise. Experience with loudness estimates gained during the course of this work tends to show that subjective impressions of 'noisiness', which clearly include some evaluation of offensiveness as well as loudness, may sometimes conflict with loudness estimations when the subjective impression is based on a short time exposure to the noise. When the observer is exposed to the noise for extended periods, however, the agreement with loudness estimations generally appears to be quite good.

EXHAUST AND INLET NOISE

Exhaust noise is produced by the sudden release of gas into the exhaust system by the opening of the exhaust valve; the closing of the valve produces only a very minor effect Means of silencing the exhaust are well known and in practical installations silencing is generally adequate so far as persons in the vehicle are concerned.

72

Inlet noise is produced both by the opening and closing of the inlet valve, as shown in the noise oscillogram of Fig. 8. At opening, the pressure in the cylinder is generally above atmospheric pressure and a sharp positive pulse sets the air in the inlet passage into oscillation at the natural frequency of the column. This oscillation is rapidly damped by the change of volume of the system caused by the downward movement of the piston. During this period, high-frequency noise is generated by high-velocity air-flow across

frequency region (at about 120 and 250 c/s respectively) which are due to air inlet noise. The dotted curves show the effect of adequate silencing.

Inlet noise commonly decreases with decrease of load in Diesel engines as shown by the chain-dotted curve of Fig. 9a. The reduction is not so marked as in petrol engines because the reduction of overall pressures due to intake throttling in the petrol engine does not occur in Diesel engines.

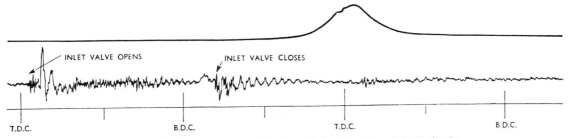

Fig. 8. *Oscillogram of Noise from Single-cylinder Engine and Cylinder Pressure*

Engine running at 1100 rev/min.

the valve seat. The closing of the inlet valve produces a similar oscillation, and it continues for some time. As the oscillogram shows, the sound pressure due to combustion is in this case small compared with that due to the inlet noise.

In practical installations, inlet noise is generally not fully silenced, and in some vehicles it is the predominant source of noise. The octave noise spectra of two engines given in Fig. 9 (full curves) show pronounced peaks in the low-

Inlet noise is markedly affected by the timing of both valves and the flow properties of the exhaust valve and exhaust system; late inlet valve timing, late exhaust valve closing, and large exhaust valve flow-area tend to reduce inlet noise.

The authors used and found fully effective, Stewart low-pass silencers (1)[*] of volume about 5 times the cylinder volume, of five sections, and lined with sound-absorbing material. Such silencers were used to obtain the lower curves in Fig. 9.

NOISE EMITTED

Characteristics of Noise

Noise measurements on a large number of automotive Diesel engines (with inlet and exhaust silenced) have shown a striking similarity in shape of noise spectrum. All the spectra so far obtained show a broad peak in the frequency range 800–2000 c/s similar to that of the octave band spectrum of Fig. 9. Narrow band spectra for 2-, 4-, and 10-l. engines are reproduced in Fig. 10, which shows that the peak contains a number of discrete components. Subjectively, this noise predominates both because of its high intensity and because the frequency is in the range to which the ear is most sensitive. It is emitted in impulses coinciding with the rapid increase of cylinder pressure, and is the objectionable hard knock characteristic of Diesel engines.

The spectrum of the petrol engine, Fig. 10d, is different. The components in the frequency range 800–2000 c/s are of lower intensity and the peaks of greatest intensity are in the frequency range 400–600 c/s. It is shown below that there is reason to ascribe this, not to any differences of structure, but to differences in the excitation due to cylinder pressure in petrol and Diesel engines.

[*] *A numerical list of references is given in Appendix V.*

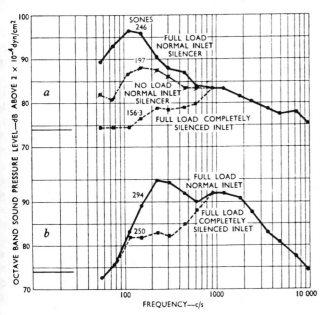

Fig. 9. *Octave Band Spectra of Noise of Two Diesel Engines with Normal and with Fully Silenced Air Inlets*

a 2-litre engine, 2 000 rev/min.
b 4·2-litre engine, 2 000 rev/min.

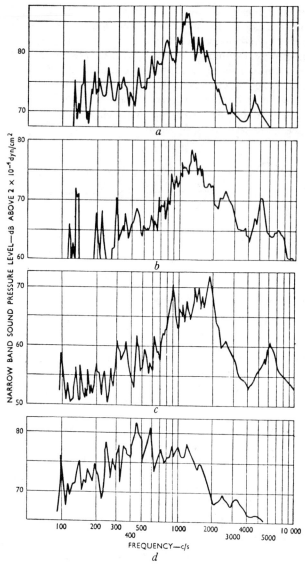

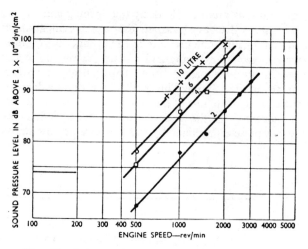

Fig. 11. *Calculated Overall Sound Pressure Level Against Speed for Four Diesel Engines*

Calculated overall sound pressure level is plotted against speed on a logarithmic scale for four Diesel engines of different sizes in Fig. 11. Straight lines of nearly the same slope give a good fit in all cases. The slope corresponds to a thousandfold increase in sound intensity for a tenfold increase in speed (30 dB/decade); that is, sound-intensity $\propto$ (speed)^3, or sound-pressure $\propto$ (speed)$^{\frac{3}{2}}$.

Effect of Engine Load

Overall sound pressure level is plotted against load for a Diesel engine and for a petrol engine, at 2000 rev/min, in Fig. 12. In the Diesel engine the sound pressure level at no load differs only very slightly from that at full load, whereas in the petrol engine the sound pressure level at no load is less than at full load by some 10 dB.

Table 5 shows that whereas maximum pressure is not greatly reduced in the Diesel engine at no load, there is a

Fig. 10. *Narrow Band Spectra of Three Diesel Engines and a Petrol Engine*

a 10-l., 6-cylinder Diesel engine, 1000 rev/min, full load.
b 4-l., 6-cylinder, direct-injection Diesel engine, 1000 rev/min, full load.
c 2-l., 4-cylinder, indirect-injection Diesel engine, 1000 rev/min, full load.
d 3-l., 6-cylinder petrol engine, 2000 rev/min, full load.

Effect of Engine Speed

In this and the following two sections the results are presented in terms of total sound pressure level. The quantity used (referred to as calculated overall sound pressure level) has been calculated by summing the intensities in the six octave bands (measured at a distance of 3 ft from the engine) covering the range from 150 to 9600 c/s. In certain cases, the noise in the frequency range below that covered was of considerable intensity but was due to a poorly silenced air intake, or background noise; it was not desired to take this noise into account.

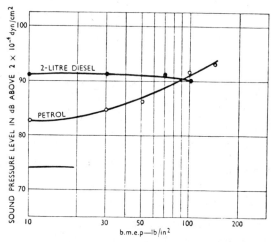

Fig. 12. *Effect of Load on Overall Calculated Sound Pressure Level for a Diesel Engine and a Petrol Engine*

great reduction in the petrol engine owing to the reduced air charge.

Table 5

			Load	Maximum pressure, lb/in²
Diesel	.	.	0 Full	900 1080
Petrol	.	.	0 Full	130 900

Effect of Engine Size

Overall sound pressure level is plotted against engine swept-volume on a logarithmic scale in Fig. 13 for the engines of Fig. 11, together with three other small engines, including engine D in the quietened condition, to indicate the magnitude of the variation between engines of about the same size. The speed in all cases was 2000 rev/min.

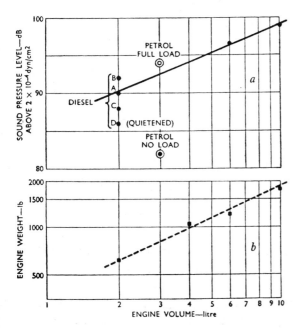

Fig. 13. Comparison of Diesel and Petrol Engines

a Calculated sound pressure against size at 2000 rev/min, full load.
   o Petrol, full load.   ● Petrol, no load.

b Engine weight (logarithmic scale) against engine size.

The causes of the variation among the small engines are considered below.

The line drawn is of slope 13·3 dB/decade. Anticipating data given later in the paper, this is what would be expected if vibration velocities of engine surfaces do not vary with engine size, the increase in intensity of sound radiated then being due to the increase in size of radiating surface.

The weight of the engines, on a similar logarithmic scale, is plotted against swept volume in Fig. 13*b*. The slope of the line shows that the weight varies more slowly than the

volume, the slope corresponding to a fivefold increase of weight for a tenfold increase of volume; that is, weight $\propto$ (volume)$^{0.7}$.

The observed variation of sound pressure level with speed and size makes it possible to express quantitatively the familiar result that, power for power, a large engine running slowly is quieter than a smaller one running faster; e.g. from the above results, if the comparison is between an engine A and another, B, of twice the swept volume, running at half the speed, B may be expected to be quieter to the extent of 5 dB; i.e. the overall noise would be of $\frac{1}{3}$ the intensity, and the loudness might be about 70 per cent that of A.

COMPARISON OF DIESEL AND PETROL ENGINES

The points for the petrol engine inserted in Fig. 13 show that at full load the overall sound pressure level of the petrol engine is comparable with that of Diesel engines of similar size. At no load, however, the sound pressure level of the petrol engine is considerably smaller.

As indicated earlier, however, the spectrum of the petrol engine is of different character from that of the Diesel engine and it is therefore of interest to compare loudnesses. Octave band spectra are plotted in Fig. 14*a* for the petrol engines and for four Diesel engines of comparable size, at full load 2000 rev/min, and in Fig. 14*b* for idling. The corresponding total loudnesses, estimated from the octave band spectra, are given in the block diagrams, Fig. 14*c* and *d* respectively, the shading indicating that part of the loudness due to noise in the octave bands up to 600 c/s. At full load the loudness of the petrol engine is comparable with that of either of the normal Diesel engines A and B. At idling, however, it is notably quieter, the loudness being about $\frac{1}{3}$ that of Diesel engines A and B. The specially treated Diesel engine D is rather less than half as loud as the petrol engine at full load, but it is still louder—about $1\frac{1}{2}$ times—at idling.

SOME PARTICULAR NOISE SOURCES

In this section the more readily identifiable sources of noise in an engine are evaluated, and means are outlined of reducing the noise they produce to a level below that of the remaining noise, which is then referred to as basic noise.

Fuel-injection Equipment

The noise from the fuel-injection equipment will not be observable against engine noise if the intensity of the fuel-injection equipment noise is substantially lower than that of the whole engine in all frequency bands. It has been shown that engine noise increases in intensity with increase in size of engine, and markedly with increase in speed. The noise of the conventional type of fuel-injection equipment increases only slightly with increase in speed and with increase in pump and injector size in the automotive range. Injection equipment noise is therefore most likely to be observable compared with engine noise at low engine speed and on small engines.

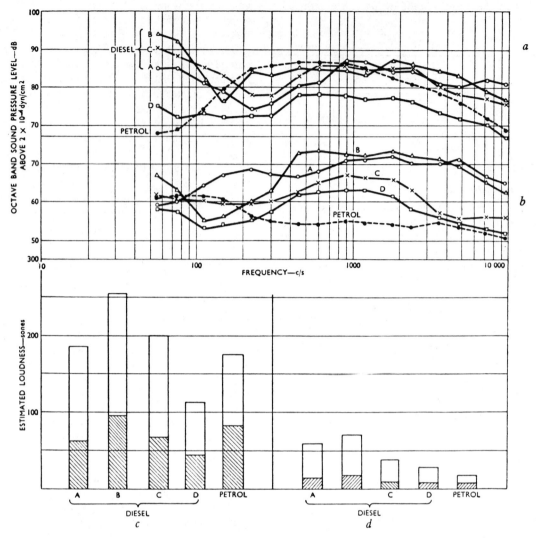

Fig. 14. Octave Band Spectra and Loudness of Petrol Engine and Diesel Engines at Idling and Full Load

a 2000 rev/min, full load. *b 500 rev/min, idling.* *c 2000 rev/min, full load.* *d 500 rev/min, idling.*

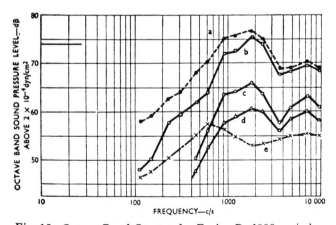

Fig. 15. Octave Band Spectra for Engine D, 1000 rev/min, and for Injection Equipment Only with Different Mountings

a Engine. b Injection equipment. c Injectors in head.
d Injectors in neoprene. e Pump only.

Fig. 15 shows octave band spectra for a 2-l. Diesel engine at 1000 rev/min, for the injection equipment mounted on the engine but driven separately, for the same injection pump mounted on a dead structure, and for the injectors alone mounted in a separate cylinder head and also mounted in neoprene holders. The noise emitted by the injection pump (spectrum e) and injectors themselves (spectrum d) is very small. The noise is considerably enhanced when the injection equipment is mounted on the engine (spectrum b) owing to the transmission of vibrations from the injection equipment to the engine structure and radiation from the considerably larger engine surfaces. The noise, however, compared with that of the engine itself, is still of no more than comparable intensity in some frequency bands, and it is found (see below and Fig. 17) that silencing the injection equipment produces little overall quieting. Of the engines tested this is the one in which engine noise is of smallest intensity and in which, therefore, injector noise is most likely to be detectable; thus, the conclusion also applies to all the other engines.

It has, however, been found that relatively minor engine modifications, which are dealt with below, can produce enough reduction of engine noise to leave the noise produced by the present fuel-injection equipment of higher intensity in some frequency bands. A more detailed consideration of fuel-injection equipment noise and means of reducing it is therefore given.

Pump Noise

The contribution of the pump to total injection equipment noise is indicated in the spectrum of Fig. 15e as a peak around 650–800 c/s. This arises from bending vibrations of the pump camshaft excited by changes of fluid pressure in the pump elements. The corresponding frequency in 6-cylinder pumps is in the range 400–600 c/s. In both cases the addition of a flywheel to the pump may produce a second component of about 200 c/s. The torque reaction of the pump produces a bodily movement of base-mounted pumps at a frequency in the range 100–150 c/s.

Any of these components may become of significant magnitude if the pump is mounted too flexibly, or mounted on a large unstiffened engine surface. There is no difficulty, however, in keeping the noise intensity negligibly small by making the mounting rigid, flange mounting is very effective.

Injector Noise

A narrow band spectrum of noise from a single injector (as in engine D) is shown in Fig. 16.

By means of oscillograms of needle lift, noise, and nozzle spring vibration, it has been shown that the peak labelled S in Fig. 16 is due to spring surge, that labelled N, in the region from 5000 c/s upward, is noise of distributed frequencies resulting from the impact of the nozzle needle with the seat, on closing. The intensity of both components varies (generally together) in an apparently capricious manner with pump speed and fuelling, according to whether the needle seats on a void (that is, fuel is absent), when the noise intensity is high, or on to fuel, when the impact is less violent and the noise intensity may be lower by up to 10 dB.

The sudden closing of the needle is essential to the proper functioning of the injection equipment, and it was considered to be impracticable to ensure that the needle always closed on to fuel. The noise components due to spring surge can be eliminated by replacing the steel nozzle spring with a spring consisting of a cylinder of synthetic rubber in compression; but as shown by the dotted curve in Fig. 16 this leaves the component due to needle impact substantially unchanged.

Both components may be reduced by more than enough by a design of injector with reduced needle mass and spring force. This is shown by the octave band spectra in Fig. 17 in which (a) shows the noise, at idling, with the original covers and injectors; (b) is for a 'dead' valve (and timing) cover and normal injectors—the noise from 1000 c/s upward is now due mainly to the injectors; (c) shows the effect of the original covers with special injectors which gives a reduction only in the 5000–10 000 c/s region; (d) is for 'dead' covers and special injectors—the *combination* gives a

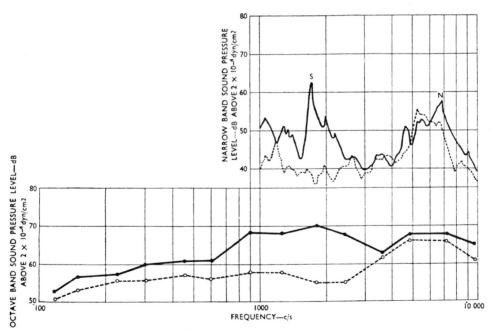

Fig. 16. Narrow and Octave Band Spectra of Noise from Set of Injectors with Steel Springs and with Synthetic Rubber Springs

———— Steel springs.   — — — — Synthetic rubber springs.

Pump speed 500 rev/min. Microphone 2 ft. from injectors.

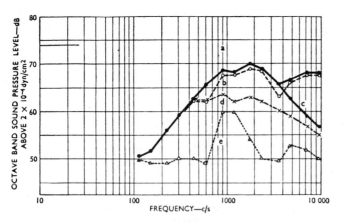

Fig. 17. Octave Band Spectra for Engine D Showing Effect of Injectors and Mountings

Idling 500 rev/min.

(a) Original covers and injectors.
(b) 'Dead' valve and timing cover.
(c) Original covers and special injectors.
(d) 'Dead' covers and special injectors.
(e) Pump driven separately, special injectors.

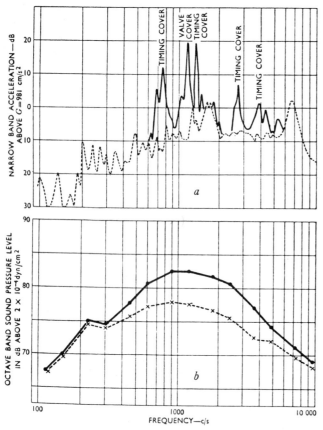

Fig. 18. Octave and Narrow Band Spectra of Engine D.
Special

a Narrow band spectra of vibration of covers and other engine surfaces.
   Octave band spectra of engine noise with normal and 'dead' valve and timing covers. Special injectors, 2000 rev/min.

   ● Normal valve and timing covers.
   × 'Dead' valve and timing covers.

substantial quietening; (e) for the same combination but with pump driven separately, shows that injector noise is now well below even the reduced engine noise.

Engine Covers

On most Diesel engines so far examined either the valve cover, or the timing cover, or both, have been a major source of noise. Fig. 18b (full line) shows an octave band spectrum for engine D. Exploration with a vibration pick-up attached to the surfaces of valve and timing cover showed the prominent peaks labelled in Fig. 18a. The dotted curve is drawn through points representing the peak values of vibration levels found in other parts of the structure.

The covers were light, highly resonant structures of cast aluminium. They were replaced by 'dead' structures: rubber reinforced with steel strip for the valve cover, and a double steel sheet construction with the interspace packed with asbestos wool for the timing cover. The resulting noise spectrum is shown dotted in Fig. 18b. The overall sound pressure level is reduced by 5 dB and the loudness from 133 to 103 sones. The high noise level produced by the original covers is due to their high Q-factor*; the values, determined by observing oscillographically the decay of the vibrations, are 23 and 54 for valve and timing covers, respectively.

Noise radiated from the engine sump did not exceed the noise from other engine surfaces in engine D. Noise from the sump has not yet been investigated in detail in other engines.

Flywheel and Front Pulley

In a case where it was necessary to test an engine with clutch housing removed, it was found that the flywheel was a major source of noise. Fig. 19 shows narrow band spectra for the engine with clutch housing removed and with clutch housing in position. The components labelled were identified, by observing the endwise displacement of the centre and top edge of the flywheel, as due to tilting and endwise displacement of the flywheel respectively. Since the noise radiated extends up to quite high frequencies it is important that no apertures should be left in the clutch housing.

There are also endwise vibrations of the front pulley which is large enough to radiate effectively at the high frequencies (around 1000 c/s). The sound intensity due to this source has not been explored.

BASIC ENGINE NOISE

Some of the noise sources considered above are due to forces on the engine structure resulting from cylinder pressure. The remaining noise, radiated from the main surfaces of the engine structure, is due mainly, often wholly, to cylinder pressure.

Narrow band analysis of cylinder pressure (the electrical output of a 'catenary' strain-gauge cylinder pressure gauge

* *The Q-factor is a number which is a measure of the frequency selectivity of an oscillatory system. For a simple system it is equal to the ratio of the amplitude produced by a sinusoidal force of frequency equal to the natural frequency to that produced by an equal force of very low frequency.*

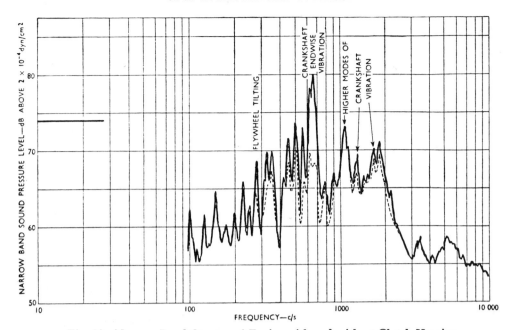

Fig. 19. Narrow Band Spectra of Engine with and without Clutch Housing
——————— Without clutch housing.   — — — — With clutch housing.

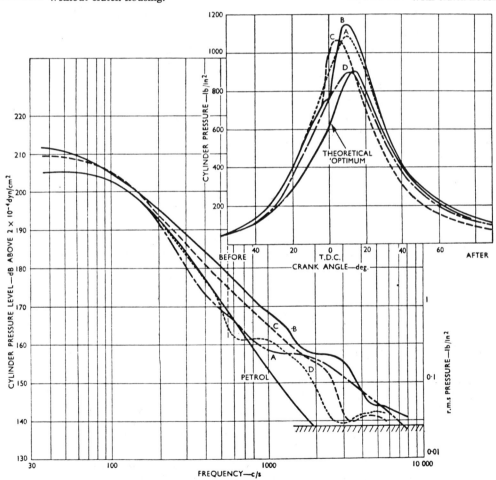

Fig. 20. Cylinder Pressure Spectra, and Diagrams, for Diesel Engines and a Petrol Engine
2000 rev/min, full load.   Cylinder pressure level curves are drawn through peaks of individual harmonics (Figs. 22 and 27).

79

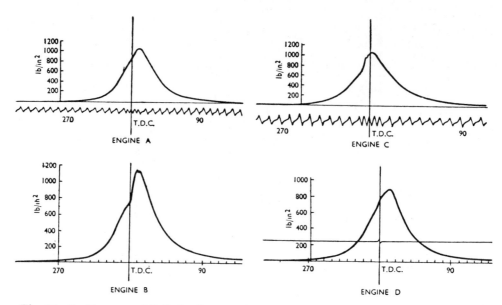

Fig. 21. Oscillograms of Cylinder Pressure for Diesel Engines at 2000 rev/min, Full Load

(2)) has been used as a measure of the exciting propensities of cylinder pressure. For convenience, the cylinder pressure analysis is expressed in the usual noise units, that is, dB relative to 2×10^{-4} dyn/cm². Then the quantity

attenuation (dB)

= cylinder pressure level (dB) − sound pressure level (dB)

may be used as a measure of the 'noisiness' of the engine structure. By this means, in making comparisons between engines, the amount of the difference in noise ascribable to different forms of cylinder pressure, and the amount due to differences in structure, can be clearly distinguished. Fig. 20 shows the cylinder pressure analyses (or spectra) for Diesel engines A, B, C, and D and for a petrol engine,

at full load. In Fig. 21 oscillograms of the corresponding cylinder pressure diagrams are reproduced. Up to about 500 c/s the curves in Fig. 20 are drawn through peaks (drawn in full in Figs. 22 and 27) corresponding to harmonics of the cylinder firing frequency ($\frac{1}{2}$ engine speed); at higher frequencies, individual harmonics are no longer resolved.

Both Diesel and petrol engines show a steady decrease of cylinder pressure level, after the first few harmonics, at a rate of about 40 dB/decade. In the petrol engine spectrum this decrease persists. With the Diesel engines, however,

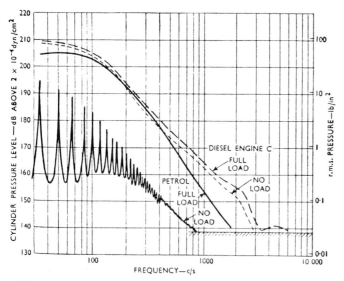

Fig. 22. Cylinder Pressure Spectra for Diesel Engine C and a Petrol Engine at 2000 rev/min

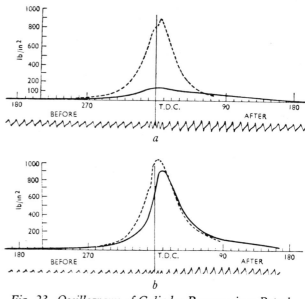

Fig. 23. Oscillograms of Cylinder Pressure in a Petrol Engine and a Diesel Engine

——————— Petrol engine.　　— — — — Diesel engine.

a 2000 rev/min, no load.　　b 2000 rev/min, full load.

the rate of decrease becomes less steep from about 800 c/s; this is ascribed to the rapid pressure rise in the Diesel engine following ignition, which causes harmonics of considerable magnitude of period about equal to the time of pressure rise and less.

Figs. 22 and 23 show cylinder pressure spectra and oscillograms respectively for Diesel engine C and the petrol engine at full load and no load. As with engine noise, there is a large reduction of cylinder pressure level with reduction of load in the petrol engine (some 20 dB), but a small decrease in the Diesel engine.

The effect of speed has not been fully investigated but, to a first approximation, the spectra for an engine at different speeds are geometrically similar, with a shift parallel to the frequency axis corresponding to the change of speed. This is consistent with the tendency of the cylinder pressure diagram of a Diesel engine to remain of similar form (on a degree base) with change of speed.

'Attenuation' values (that is, cylinder pressure level—octave-band sound pressure level) are plotted in Fig. 24 for

speeds, etc., is found with the Diesel engines and noises due to different mechanisms of generation sometimes intervene when the cylinder pressure is particularly smooth. Values of the attenuation are, however, generally repeatable within a total scatter of 2 dB though occasionally the scatter is up to 5 dB. The attenuation curve for the petrol engines lies mainly within the group of curves for the Diesel engine, which indicates that its structure is not dissimilar, as regards noise, from that of the Diesel engines.

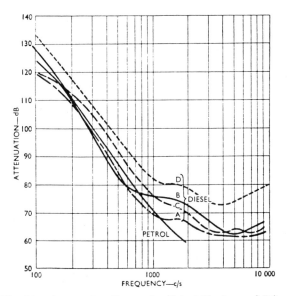

Fig. 25. Attenuation Curves for Petrol Engine and Diesel Engines A, B, C, and D

Fig. 25 shows that the variation of attenuation among the Diesel engines is not very large. Comparison with Fig. 20 shows that there is rather more variation of cylinder pressure level in the critical region from 800 to 2000 c/s than of attenuation, 15 dB as compared with 10 dB. The combination of low cylinder pressure level with high attenuation in engine D is particularly favourable. The high attenuation in this case is largely due to the provision (experimentally) of dead valve and timing covers.

It is of some importance to establish how far it is practicable to go in smoothing the cylinder pressure in a Diesel engine without appreciable sacrifice of efficiency, and to what extent excitation due to cylinder pressure is then still greater than if smoothing were taken to the limit. It is not, as yet, possible to answer these questions with certainty.

There is some indication from a theoretical consideration (private communication by W. T. Lyn) that with a 15/1 compression ratio engine limited to a maximum cylinder pressure of 1000 lb/in² a rate of pressure rise of 30 lb/in² per degree crankangle can be achieved with less than 1·6 per cent (of the efficiency) sacrifice of efficiency. This is in comparison with the practicable maximum for this peak pressure if there is no restriction on rate of pressure rise. The 2-l. engines considered here are of high compression

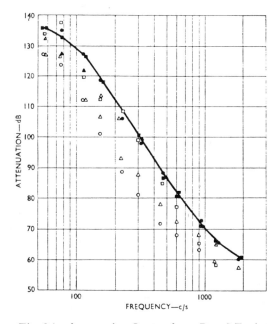

Fig. 24. Attenuation Curves for a Petrol Engine

○ 2000 rev/min.⎫
△ 3000 rev/min. ⎬ No load.
□ 4000 rev/min.⎭

● 2000 rev/min.⎫
▲ 3000 rev/min. ⎬ Full load.
■ 4000 rev/min.⎭

the petrol engine. The full-load points lie fairly closely on a line, but the no-load values are scattered in the direction of lower attenuation. This is taken to mean that at full load the predominant noise is due to excitation of the engine surfaces by forces due to cylinder pressure. At no load, however, some other source of noise intervenes so that the noise level measured is spuriously high and the attenuation spuriously low.

Attenuation curves for Diesel engines A, B, C, and D (D with dead covers) are given in Fig. 25 together with that of the petrol engine. Some scatter of points for different

ratio (in the range 17–21) and this finding may not apply. The calculated cylinder pressure which results is, however, reproduced in Fig. 20. Visual inspection suggests that the cylinder pressure of engine D is comparable as regards 'noisiness' with the theoretical curve. This conforms with the finding that the fuel consumption of engine D is not appreciably inferior to that of the other engines despite its low noisiness.

As a working hypothesis it is suggested that the theoretical curve, and a rate of pressure rise of about 30 lb/in²/deg should be regarded as the target. It would appear that the cylinder pressure level, and hence sound pressure level, in the critical frequency range 800–2000 c/s, will not be more than about 5 dB greater than the lowest level achievable without relation to efficiency.

Current experience suggest that this target can probably be achieved under most engine conditions, but that it may be very difficult to do so at idling with complete absence of smoke and smell.

MECHANISM OF GENERATION OF BASIC ENGINE NOISE

It is instructive to attempt to correlate the frequency analyses of cylinder pressure, the possible orders of magnitude and observed amplitudes of vibration of external engine surfaces, and the intensity of the noise radiated.

To deal first with the relation between the vibration of a radiating surface and the sound intensity produced, rough estimates may be made from standard solutions quoted in convenient form by Beranek (3). Sound pressure levels at a point on the axis of a vibrator 1 m from the surface have been calculated for a disc in free space, and for a piston in the end of a tube and for a piston in an infinite baffle, for three sizes of radiator. It is convenient for calculation to use, instead of r.m.s. vibration velocity of the surface, the quantity $\rho_0 cu$, where ρ_0 is the density of air and c the velocity of sound in air. This quantity $\rho_0 cu$, has the dimensions of pressure and may be plotted as sound pressure level relative to 0·0002 dyn/cm². (For a plane surface of dimensions much greater than the wavelength $\rho_0 cu$ is the sound pressure close to the surface). The results are given in Fig. 26 plotted as

$$\text{radiation attenuation} = 20 \log_{10} \frac{\rho_0 cu}{2 \times 10^{-4}} - \text{S.P.L. at 1 m}$$

The very large effect of size of radiator in relation to the wavelength should be noted.

In Fig. 27, curve (a) is the cylinder pressure level spectrum for engine D; curve (c) is the corresponding narrow band noise spectrum with the engine in the 'quietened' condition. The difference between these two curves is the 'attenuation' as defined earlier (except that the narrow band spectrum has been used here whereas octave band levels are normally used).

The curves (b₁), (b₂), and (b₃) represent expected values of surface vibration velocity (plotted as $\rho_0 cu$) (pressure level) calculated from the measured sound pressure level and the radiation attenuation as defined above. Curve (b₁)

was calculated using the data of Fig. 26 assuming the whole engine to be vibrating, with a bodily displacement, and to be comparable, as a radiator, to a disc of area equal to the projected area of the engine. It is estimated that if the engine vibrates in a rotary mode about a horizontal or vertical axis it will behave as if it were about ⅓ the effective area; this represents about 10 dB additional attenuation (curve (b₂)).

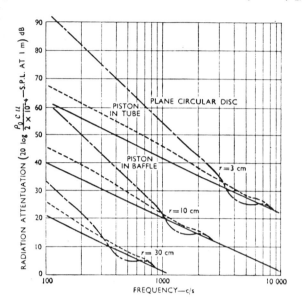

Fig. 26. Calculated 'Radiation Attenuation' for Plane Circular Disc, Piston in Tube, and Piston in Baffle

At frequencies of about 1000 c/s and above, radiation appears to occur mainly from the crankcase. It is supposed that bending vibrations of the crankcase wall are responsible, and that at any given frequency the radiators are of such size as would resonate (as a circular plate clamped at the edges) at that frequency. On these assumptions the lowest frequency at which resonance is possible is 700 c/s for a plate of diameter 10 in.; the diameter decreases with increase of frequency (as $1/\sqrt{n}$), giving a radiation attenuation, independent of frequency, of 26 dB. It is now supposed that there are two such resonant areas at 700 c/s, and that the number varies inversely as the area of the individual resonator, that is, as frequency. This gives an attenuation of 23 dB at 700 c/s decreasing to 13 dB at 7000 c/s. This result has been used to plot curve (b₃).

An extensive, but not exhaustive, exploration of vibration levels on engine surfaces gave the results indicated by the points in Fig. 27. There is rough agreement between the experimental and calculated values; and the agreement might be improved if vibration measurements were made at many more points closely spaced over the engine surfaces.

The frequency range 800–2000 c/s in which the characteristic Diesel engine noise is radiated is of particular importance and it is of interest to try to account for the surface vibrations in this region. The 'structure attenuation', that is, cylinder pressure level – vibration pressure level (Fig. 27) is 60–70 dB. A rough calculation on an idealized

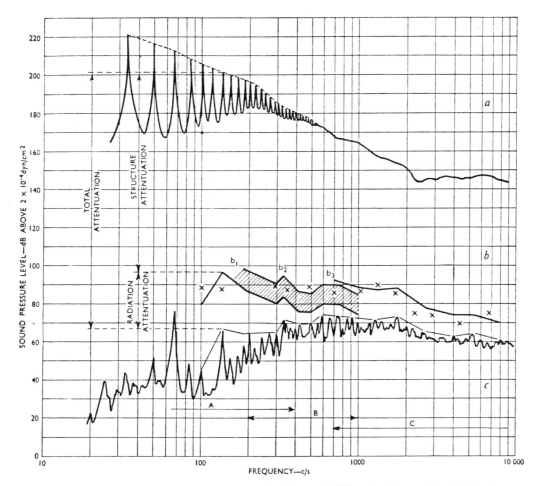

Fig. 27. Cylinder Pressure, Narrow Band Noise, and Vibration Spectra for Engine D

a Cylinder pressure.   *b* Vibration velocity × $\rho_0 c$.   *c* Sound pressure.

model of the cylinder head showed that the bending deflection of the double-walled parts of the structure under the force due to cylinder pressure was too small to account for the observed vibrations; the value of structure attenuation obtained was 105 dB. Displacements due to strain in the cylinder-head bolts and adjacent metal, and also due to bodily stretch of the whole crankcase, appeared to be even smaller. The cylinder pressure force applied in a direction normal to the wall of the crankcase, however, would produce bending vibrations of considerably more than the observed magnitude (structure attenuation 56 dB, independent of frequency as is observed), but it is not easy to see how the force can be so applied. The cylinder pressure force applied in such a manner to the edge of the crankcase wall as to cause a bending moment, even though with only a very small moment distance, could, with a dynamic magnification such as it observed, still produce the observed vibration. It would appear from the typical construction indicated in the sketch of Fig. 28 that the occurrence of such a bending moment is probable, and this hypothesis has been adopted provisionally.

Noise in the frequency ranges marked in Fig. 27 is ascribed tentatively to different mechanisms as follows:

(A) Bodily vibration of whole engine, as displacement, caused by reaction of forces due to vibration of engine flywheel–crankshaft system.

(B) Rotational vibrations about both horizontal and vertical axes of whole engine caused by reactions of forces due to resonant crankshaft bending vibrations and torsional oscillations.

(C) Vibration of resonating sections of engine surfaces, generally crankcase, resulting from transmission of force due to cylinder pressure, both directly, and *via* crankshaft vibrations.

CONCLUSIONS

For the sake of continuity, discussion of the results has been included in the various sections above. Some observations of a more general nature are made here.

The mechanism of the generation and radiation of engine noise is complex, and overwhelming quietening cannot be

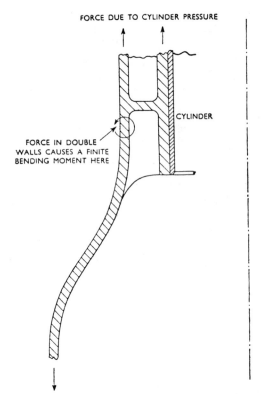

FORCE DUE TO CYLINDER PRESSURE

CYLINDER

FORCE IN DOUBLE
WALLS CAUSES A FINITE
BENDING MOMENT HERE

Fig. 28. Typical Construction of Cylinder–Crankcase Junction

produced by giving attention to any single factor. Air intake noise, and valve and timing covers are liable to be strong sources of noise. Reduction in the noise from them may produce a major quietening, and mechanical noise, or in extreme conditions injection-equipment noise, may then be prominent. In some engines a contribution should be obtainable by smoothing the cylinder pressure.

All of the remaining sources (which together give rise to what has been termed basic noise) are believed to involve resonant vibrations of the crankshaft–flywheel system, the crankshaft or the crankcase. Stiffening, to the extent usually contemplated for other engineering reasons, is not likely to have much effect on the magnitude of the noise, though its 'quality' may be altered by displacement towards higher frequencies. Stiffening to the extent of carrying the double-walled structure of the cylinder region right round the crankcase, with closely spaced cross-connections, might give a major reduction of noise. Since the mechanical amplifications involved are high, at least 15, a major reduction of noise should be obtainable by introducing damping great enough to approach critical damping. This could probably be done with the crankcase but it is not easy to see how a 'dead' crankshaft could be made.

This paper is a study of *airborne* engine noise and the importance of the high-frequency part of the noise has been stressed. Current experience is tending to show that in quietening the driving compartment in Diesel versions of light commercial-vehicles, airborne engine noise does not

present the major problem; this applies still more to passenger cars, where sound absorbing material tends to be used freely. The high-frequency components, however, remain important as regards the noise heard by bystanders, and may be objectionable inside the vehicle when driving near walls with windows down.

The work is being continued by completing the exploration of the possibility of producing further structural quietening, in particular by reducing the vibration of crankcase surfaces.

At the same time a study is being made of the relation between the rate of injection of fuel and the rate of heat release and hence, the form of the cylinder pressure. The aim is to determine the injection-rate–time relation which will give the optimum form of cylinder-pressure diagram. There is some indication that achievement of the optimum cylinder pressure diagram may not be possible in all modern high-speed engines over the whole speed and load range by adjustment of the injection characteristic alone. If this should prove to be so, it will be necessary to modify also the air flow in the combustion space. Since there are earlier combustion systems which can be made quiet without sacrifice of efficiency up to a limited maximum speed, there is good hope that quietening may become possible up to the highest useful speeds.

Thus, with sufficient effort applied to the development of the combustion system and the fuel-injection system to go with it, there is reason to believe that high-speed Diesel engines will be achieved which will be as quiet as equivalent petrol engines are now.

ACKNOWLEDGEMENTS

The authors wish to thank the many firms in the motor industry who assisted by the loan of engines and provision of test facilities, and the many colleagues who contributed to the work, particularly Mr. Grover and Mr. Gardiner who carried out much of the testing and analysis, and Mr. Stockwell and Mr. Coleman for the development of special measuring apparatus.

Thanks are also due to C.A.V. Ltd, for permission to publish the paper.

APPENDIX III

MEASURING APPARATUS

All measurements were made with a calibrated moving coil or condenser microphone, filter and either Solartron measuring amplifier or Bruel and Kjaer level recorder.

A set of ST and C octave filters, or Bruel and Kjaer third-octave filters, were used for analysis for noise assessment, and Muirhead wave analyser set to 'in tune, high' for all narrow band analyses for diagnosis of noise sources, etc. The wave analyser was also used for cylinder pressure analysis, and a special amplifier (d.c. so as to permit static calibration), with low noise level and low impedance output, was developed for this purpose.

In all cases r.m.s., as distinct from peak voltage (representing sound pressure), was measured despite the impulsive nature of the noise, since it is generally the case that the r.m.s. value represents more closely the subjective effect of the noise.

Vibration measurements were made with a Bruel and Kjaer barium titanate accelerometer, type 4307, which has only recently become available and permitted vibration measurements over the whole audio-frequency range.

Noise measurements were generally made on engines in cubicles with walls covered with sound-insulating material, and of size such that the noise at the chosen microphone position, 3 ft from the engine on the near side, was not appreciably affected by room reflections.

APPENDIX IV

The method used for estimating loudness from octave-band spectra is that proposed for pure tones by Fletcher and Munson (4), and for octave-band measurements by Gates (5), but using other data.

The loudness level (phons) corresponding to the sound pressure level in each octave band is obtained using the data of Churcher and King (6). Each loudness level is then converted into loudness (sones) using the formerly internationally agreed loudness-level–loudness relation (7). The loudnesses of all the octave bands are then summed to give total loudness.

Recently, the agreed loudness-level–loudness relation has been replaced by a new one (8) so that the loudness estimates made in this paper should not, strictly, be labelled 'sones'. The accuracy of the procedure, however, is not great enough to justify recalcula-tion of all earlier loudness estimations and it was thought that too much confusion would be caused by substituting a different term for 'sones'.

APPENDIX V

REFERENCES

(1) DAVIS, A. H. and FLEMING, N.   1934 A.R.C. Report Nos. 105, 108.
(2) DRAPER, C. S. and LI, Y. T.   1949 *J. Aero. Sci.*, vol. 16, p. 593, 'A New High Performance Engine Indicator of the Strain Gauge Type'.
(3) BERANEK, L. L.   1954 'Acoustics' (McGraw-Hill Book Co., New York and London).
(4) FLETCHER, H. and MUNSON, W. A.   1933 *J. Acoust. Soc. Amer.*, vol. 5, p. 82, 'Loudness, its Definition, Measurement and Calculation'.
(5) GATES, B. G.   1937 *J. Instn elec. Engrs*, vol. 81, p. 57, Contribution to discussion.
(6) CHURCHER, B. G. and KING, A. J.   1937 *J. Instn elec. Engrs*, vol. 81, p. 57, 'The Performance of Noise Meters in Terms of the Primary Standard'.
(7) FLETCHER, H.   1935 *J. Frank. Inst.*, vol. 220, p. 405, 'New Concepts of the Pitch, the Loudness and the Timbre of Musical Tones'.
(8) STEVENS, S. S.   1955 *J. Acoust. Soc. Amer.*, vol. 27, p. 815, 'The Measurement of Loudness'.

Reprinted from *Noise Control* **6**(4):13–16 (1960)

Experimental Study of the Airborne Noise Generated by Passenger Automobile Tires[*]

FRANCIS M. WIENER

BOLT BERANEK AND NEWMAN INC., CAMBRIDGE, MASSACHUSETTS

SUSTAINED driving at high speeds and the increasing loads under which modern automobiles are operated have resulted in a search for improved tire materials and constructions. At the same time, the noise generated by tires has come in for closer scrutiny. Airborne tire noise is only one of the many sources of noise in an automobile. Further, one must remember that the tire is not only a radiator of airborne sound, but also an important link in the suspension and steering systems of the automobile. Thus, there exists a multiplicity of transmission paths carrying stimuli from the tire-road interface to the car occupant. In the aggregate, the occupant's subjective response to the performance of a given tire may be determined not only by the level and character of the airborne tire noise, but also, and perhaps much more so, by the seat vibrations, the steering performance, and other related fac-

tors. This paper is not concerned with the many factors that influence subjective appraisals of tire performance. Rather, it attempts to shed light on how automobile tires perform as radiators of airborne sound under typical operating conditions.

Noise Measurement Procedure

To measure the airborne noise generated at the tire-road interface, a condenser microphone was placed on the outside of the rear fender of a fully loaded, medium-heavy passenger car of recent manufacture. The microphone diaphragm was located in a vertical plane through the rear axle of the vehicle, about one foot from the road. To reduce the wind noise below the levels to be measured, the microphone was equipped with a cloth-covered cylindrical windscreen of efficient design. The output signal was recorded on magnetic tape while the car coasted from a speed of about 60 mph down to about 15 mph, with the engine off and the transmission in neutral. Suitable markers were recorded on the tape at several points during a run to indicate the actual car speed at that time as indicated by a calibrated speedometer.

Selected samples from the recordings, of about 10-sec duration, were rerecorded on a loop of magnetic tape and analyzed in one-third-octave bands using a square-law detector and integrating circuit,

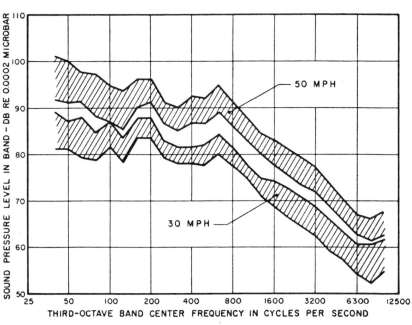

Fig. 1. Typical tire noise spectra on rough road.

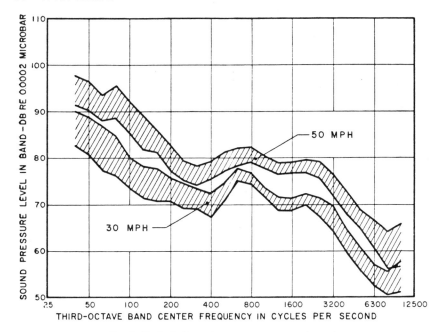

Fig. 2. Typical tire noise spectra on smooth road.

four-ply construction with a standard all-purpose tread. The tire mold was such that the spacing between the cleats of the tread was not constant but varied along the circumference.

Figure 1 shows the noise spectra measured for a standard tire on the rough road for two road speeds. On the abscissa are plotted the center frequencies of the one-third-octave bands used in the analysis of the data. On the ordinate are plotted the integrated sound-pressure levels in each band in db *re* 0.0002 microbar. It is seen that the measured tire noise encompasses a wide frequency spectrum, with approximately constant band levels up to about 800 cps and decreasing levels toward the higher audio frequencies. The analysis was generally restricted to frequencies above 40 cps. In the few cases where the analysis was carried down to the 20-cps filter band, the noise levels at 20 cps were found to be about the same as those at 40 cps. The shaded areas on the graph give an indication of the spread of the data when the measurements were repeated (four to six runs) under nominally identical conditions. The results show that as the road speed is decreased, the levels are shifted downward by a nearly constant amount, that is

with due regard to the frequency response of the microphone and the recording system. Absolute sound-pressure levels were obtained by recording a 400-cycle calibration tone at frequent intervals in the field, using an acoustic calibrator which produced a known sound pressure level at the microphone. The self-noise of the system, as determined by the aerodynamic noise about the windscreen, was carefully checked in the laboratory by rotating a microphone and windscreen on a centrifuge [1] and also by immersing the microphone and the windscreen in an air stream of low turbulence. Both methods yielded similar results for equal air speeds.

The test tires were mounted on standard wheels which were dynamically balanced and then mounted on the car for tests. Prior to the tests the tires had been stored at constant room temperature and inflated to standard pressure. No attempt was made to change the air pressure during the experiments.

Two roads were selected for the tests. One was an interstate highway with a relatively smooth macadam surface ("smooth road"). The other was a paved highway surfaced with solidly imbedded gravel and crushed rock ("rough road"), the peak-to-peak roughness of

which was estimated at about ¾ in. Tests were conducted only on straight and level stretches and only when the road was dry and traffic was light.

Results

A considerable number of tires of various constructions were tested. The results reported here pertain to tires of size 8.00-15, of standard

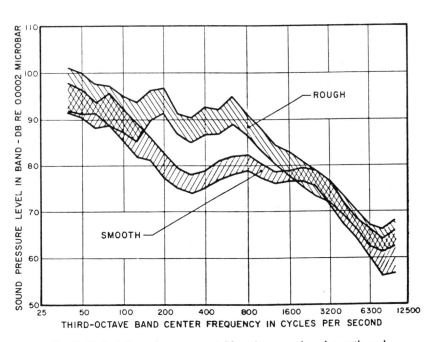

Fig. 3. Typical tire noise spectra at 50 mph on rough and smooth road.

without appreciable change in spectrum shape.

Reducing the level of tire excitation by reducing the road roughness results in lower noise levels radiated by the tire in the 100- to 1500-cps range, as shown in Fig. 2. Again, reduction of the speed from 50 mph to 30 mph results in approximately uniformly decreased noise levels over the entire frequency range investigated. There is no evidence of pronounced peaks in the spectrum due to the tire tread, a finding which is in agreement with the data obtained by Luetgebrune [2] inside an automobile coasting on a cobblestone pavement with the windows open.

It is interesting to examine Fig. 3 which shows the tire noise spectra for a constant road speed of 50 mph as a function of the level of excitation provided by the different road surfaces. Only in the frequency range between about 100 cps and 1500 cps do the sound-pressure levels on the rough road appreciably exceed those for the smooth road. This condition was found to hold also for lower road speeds, as shown in Fig. 4. One possible explanation for this finding is that at the very low frequencies noise from other sources, e.g., the rear wheel drive, might exceed the noise generated by the tire-road interface. (However, the possibility

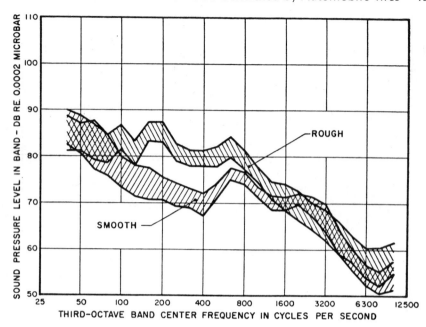

Fig. 4. Typical tire noise spectra at 30 mph on rough and smooth road.

cannot be entirely excluded that locally generated low-frequency turbulence, not present in the smooth flow in which the windscreens were tested in the laboratory, masks the tire noise at the low frequencies.) At the high frequencies, the excitation of the tire by the two different road surfaces may possibly not be greatly different.

The testing of tires of various construction revealed, within the limits of this test series, a remarkable constancy of the noise spectrum with respect to changes in the fiber material of the tire reinforcing fabric and the tread rubber composition for a given set of operating conditions. For example, Fig. 5 shows two spectra obtained on the rough road at 50 mph for tires containing two different synthetic fibers in their reinforcing fabric. The differences in the noise levels shown are significant, but rather small. Tests with an experimental tire with a smooth tread also showed only relatively small changes in the noise spectrum, as compared with that of a similar tire with normal tread. Figure 6 shows similar results obtained on the smooth road.

To see how much the tire noise contributed to the total noise inside the automobile, noise measurements were made inside the test car using a microphone placed in the ear canal of a dummy head. The dummy was placed in the seat next to the driver and the car was operated at constant speed (60 mph) with the windows closed and the engine operating normally and the transmission in gear. In Fig. 7 two spectra are shown for the rough and smooth roads, respectively. It can be seen that, under the conditions tested, the road noise, consisting of airborne tire noise and noise due to vibrations of the wheels and suspension system, con-

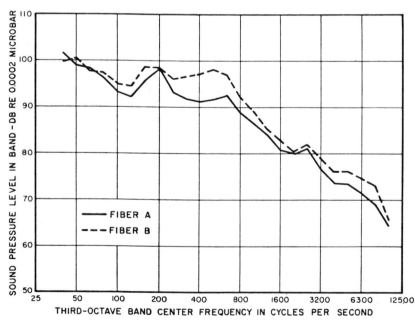

Fig. 5. Tire noise spectra at 50 mph on rough road for different tires.

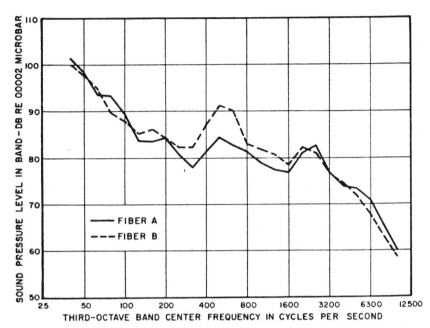

Fig. 6. Tire noise spectra at 50 mph on smooth road for different tires.

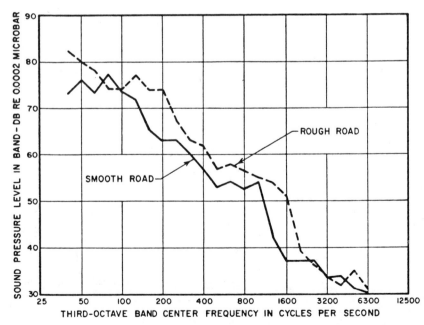

Fig. 7. Noise spectra inside an automobile at 60 mph with windows closed.

tributed significantly to the noise levels inside the automobile when driven over the rough road.

Conclusions

The conclusions drawn from this work can be summarized as follows:

1. The airborne noise generated at the tire-road interface by typical passenger automobile tires under normal operating conditions has a continuous spectrum that extends over the entire audio frequency range.

2. Increased excitation by a rough road surface results in an increase of the noise levels, primarily in the frequency range from about 100 to 1500 cps. With changes in car speed within the limits of this test series, the levels change in such a manner that the shape of the noise spectrum remains essentially constant.

3. For the types of tires tested, the tire noise does not vary greatly with variations in tire construction.

4. Road noise, including airborne tire noise, may contribute significantly to the total noise levels inside the automobile.

5. In assessing the over-all performance of a tire as evaluated from the user's standpoint, airborne tire noise is only one of several factors relating to tire performance which must be taken into account. Any over-all evaluation must necessarily involve psychophysical considerations.

Acknowledgments

The author is indebted to Dr. J. J. Baruch of Bolt Beranek and Newman, Inc. for his interest and helpful suggestions. Mr. C. M. Gogos ably assisted with the road tests. ▲ ▲ ▲

References

* Some of the material contained in this paper was the subject matter of a talk entitled "On the airborne noise generated by passenger automobile tires," presented at the Third International Congress on Acoustics, September 1959, Stuttgart, Germany.
1. L. L. Beranek, J. Acoust. Soc. Am. 25, 313 (1953).
2. H. Luetgebrune, Kautschuk u. Gummi 8, WT-91 (1955).

Reprinted from *Engineer,* June 30, 1961, pp. 1070–1074

The Subjective Rating of Motor Vehicle Noise

By C. H. G. MILLS* and D. W. ROBINSON†

A subjective experiment is described, which was designed to establish a relationship between the subjective rating of noise emitted by motor vehicles, and objective measurements made with a sound-level meter employing " A " weighting. The noise sources employed were nineteen production vehicles driven in a number of different conditions, and the subjective ratings were made by a panel of fifty-seven observers. The results show that in the case of private cars and commercial vehicles satisfactory correlation is obtained between the subjective and objective measurements. The results of motor-cycles as a group show a greater dispersion which is largely caused by shortcomings of the sound-level meter when measuring motor-cycle noise ; the dispersion is not significantly improved by sub-classification into two-stroke and four-stroke or into single-cylinder or twin-cylinder machines. Guidance is given on interpretation of results by means of numerical examples.

DURING the past two years much experimental work has been carried out in order to specify a test procedure, and a method of measurement, that can be applied successfully to the statutory measurement of noise emitted by motor vehicles. The background to this work has been described in an article by D. W. Robinson et al,[1] recently published in this journal, in which it was shown that a sound-level meter employing " A " weighting, gave better correlation between objective measurements and subjective ratings of motor-vehicle noise than did a similar meter which employed " B " weighting. The use of a sound-level meter employing " A " weighting has been proposed subsequently as a suitable instrument for making statutory noise measurements on motor vehicles, and, as an obvious extension of the investigation, it was necessary to determine the relationship between the subjective rating of motor-vehicle noise and the corresponding meter indications in sound level (A).

The subjective experiments carried out by the National Physical Laboratory in 1959, and referred to above, also yielded information on the relationship between sound level (A) indications and subjective ratings, but, because the experiment was carried out using a sample of traffic on a public road, the distribution of types of vehicles was not wholly satisfactory and only a very small sample of motor-cycles was available during these tests. Comparison of the results of the above tests with the results of superficially similar tests carried out in Switzerland[2] and the U.S.A.[3] showed that there was significant disagreement between the results of all three investigations regarding the sound level (A) corresponding to a subjective rating of " acceptable." It was therefore decided that the National Physical Laboratory and the Motor Industry Research Association should jointly carry out another subjective experiment under controlled conditions, and the results of that experiment are presented in this paper.

The broad object of the experiment was to make objective measurements (sound level

(A)) and subjective measurements on a wide range of noises emitted by motor vehicles, and to determine the relationship between the subjective ratings and the objective meter indications. It was also required that the experiment should be so designed that it yielded information on the relative acceptability of the noises emitted by different classes of vehicles.

THE DESIGN OF THE EXPERIMENT

It was necessary to carry out the subjective experiment employing the widest range of noises and the largest number of subjects that could be accommodated, taking into account the practical difficulties of organisation. Unfortunately, the tests had to be

* The Motor Industry Research Association.
† National Physical Laboratory.

made during late autumn in 1960, and the attendant difficulties caused by the weather at that time of year imposed some limitations. The tests were carried out using " live " motor-vehicle noise, in the open air, on one of the test tracks at the Proving Ground of the Motor Industry Research Association. The number of vehicles employed, and the total number of noises which were rated, were limited to some extent by the time for which the subjects could be exposed to the weather conditions which were likely to prevail.

When the tests were actually carried out

fifty-seven subjects were available who rated the noises emitted by nineteen vehicles, each vehicle operating under six different conditions. The number of subjective ratings which each observer made was 150, representing 114 different vehicle conditions and thirty-six repeats. The test was carried out between the hours of 12 noon and 3 p.m. in two parts, separated by a lunch interval. The observers were seated back to back in two lines parallel to the track of the test vehicles, and the vehicles were driven past them in alternately opposite directions. Half the observers made " sighted judgments," facing the test vehicles, and half made " unsighted judgments " sitting with their backs to the test vehicles.

One of the test surfaces at the Proving Ground of the Motor Industry Research Association was employed for the experiment and a plan of the test site is shown in Fig. 1. The track, which was surfaced with a smooth, porous, asphalt carpet, was a little over 1 km in length, with ample space at each end for parking and turning vehicles. One side of the track was bounded by young, widely spaced conifers, and the observers sat in line with the trees in a wide gap approximately half-way along the track. The test site was approximately in the centre of the Proving Ground, with no buildings or other objects capable of causing an acoustic disturbance within a radius of ¾ mile. All other traffic on the Proving Ground was stopped throughout the test.

The vehicles proceeded up and down the marked centreline of the road and the observers sat at a mean distance of 7·5m from the centreline. Two rows of chairs, thirty chairs in each row, were placed alongside the track at the listening position, arranged back to back. The row facing away from the test track, was raised 30cm above the forward facing row, to avoid the forward facing heads casting a sound shadow on the

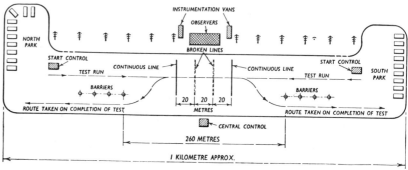

Fig. 1—Test site

other row. Each row of chairs was 12m long with a gap of one chair width in the centre of the row to accommodate the measuring microphones. The mean distance of the observers' heads from the centreline of the track was 7·5m, " sighted " subjects being 15cm less than the mean distance and " unsighted " subjects 15cm more than the mean distance from the centreline of the track.

Two instrumentation vans were parked, one at each end of the rows of observers, about 5m from the nearest observer. The " unsighted " observers had only a very restricted view of the test vehicles even if they turned their heads, the instrumentation vans and the small conifers acting as efficient sight screens. Fig. 2 shows the test in operation and the disposition of observers, instrumentation vans and a test vehicle. It should be noted that the theoretical difference in loudness of a sound caused by rotation of the head is insignificant in the case of sound spectra such as motor vehicles emit. For this reason it was preferable to orientate half the listeners facing away

Fig. 2—The test site showing disposition of the listeners and a test vehicle

rather than to employ blindfolds, which would have been inconvenient in view of the duration of the trials.

The choice of test vehicles was based upon obtaining a representative range of noise emission from each class of vehicle. Vehicles were selected from four classes, private cars, commercial vehicles, and four-stroke and two-stroke motor-cycles. Within each class, vehicles were chosen, on the basis of previous measurements, to represent extremes of noise emission, plus a few vehicles reasonably distributed between the extremes. The nineteen vehicles finally selected included three private cars, three high-performance cars, one moped, one motor-scooter, two two-stroke motor-cycles, four four-stroke motor-cycles and five commercial vehicles. All vehicles employed were new or in virtually new condition. No attempt was made to modify vehicles to produce either higher or lower noise levels than standard production types.

Although the actual vehicles were chosen to give a representative range of sound levels when tested under the proposed I.S.O.‡ test conditions, each vehicle was tested under two other driving conditions, in order to extend the range of sounds to be judged by the subjects. Each vehicle passed the observers in two directions, employing each of three distinct driving procedures, thus providing, in general, six different noises at the listening positions.

The three vehicle operating conditions were as follows:

(1) *Proposed I.S.O. Procedure.*—The vehicle approached the test area at an engine speed of three-quarters of the r.p.m. at which, according to the manufacturer, the engine developed maximum b.h.p. A gear was

‡ International Organisation for Standardisation Technical Committee 43, Draft Recommendation No. 419, October, 1960.

chosen such that the road speed approached 50 km.p.h. as closely as possible, but first gear was excluded with vehicles having more than three forward gears.

As the front of the vehicle crossed a line 10m from the measuring position (the centre of the line of observers in this case) the throttle was fully opened and held open until the rear of the vehicle crossed a line 10m past the measuring position. The throttle was then fully closed and drivers made every attempt to coast until they were at least 100m from the measuring position.

(2) *Braked Full Throttle Tests.*—The engine and road speed of each vehicle was the same as for the approach conditions in test (1), but the vehicle was driven past the observers at full throttle and constant speed, the speed being controlled by steady application of the vehicle brakes. The operating conditions for the test were stabilised from at least 30m before the measuring position to 30m after it.

(3) *Constant Speed Cruising Tests.*—Each vehicle was driven past the observers at a constant indicated speed of 50 km.p.h. in top gear. Test conditions were stabilised over the same distance as in (2) above.

Each vehicle carried out all the above tests at least once in each direction past the observers.

The provision of about sixty subjects presented some difficulties, and only a minimum of selection could be applied. Apart from ensuring that males and females were represented, no further control could be exercised. Fifty-seven subjects actually took part in the tests, twenty-three females and thirty-four males. The male subjects were selected mainly from M.I.R.A. staff, excluding those who worked habitually in a noisy environment, plus seven males from the N.P.L. and two from the Ministry of Transport. The majority of the female subjects were kindly loaned by the National Coal Board West Midlands No. 4 Area Office, and were all office workers. A further three females were selected from the M.I.R.A. staff and one from the N.P.L.

TEST PROCEDURE

Each separate test run, under one of three conditions and in one or other direction past the observers, was treated as a separate "vehicle-condition". The order in which

the "vehicle-conditions" were presented to the observers was randomised, within the limitation that no one vehicle could undertake consecutive runs in opposite directions. Each vehicle operated under each of the six "vehicle-conditions" and many vehicles carried out the same test procedure twice during the experiment for control purposes, resulting in a total of 150 "vehicle-conditions" being presented to the observers. Only one vehicle was permitted to be in the central test area at any one time, but the vehicles followed each other with as little delay as possible. A central controller was in contact with controllers at each vehicle park by means of V.H.F. radio, and by this means it was possible to present a different vehicle to the observers each thirty seconds.

Measuring microphones were set up 7·5m from the centreline of the test surface, in the open space in the centre of the lines of observers. The following objective measurements were made during the test:

(1) Sound level (A) measured on two independent sound-level meters employing I.E.C. weighting.

(2) A continuous record of sound level (A) on a high-speed level recorder, adapted to read r.m.s.

(3) Single track calibrated tape recordings for future play-back on to a sound-level meter and the high-speed level recorder.

The sound level (A) assigned to each "vehicle-condition" was the highest recorded during the passage of the vehicle concerned. The various methods of measurement referred to in (1), (2) and (3) above provided reliable objective results with adequate cross checking.

The form of subjective measurement employed for this experiment was identical to that used in the earlier experiment carried out by the N.P.L. The subjects were asked to rate the noises which were presented to them according to a six-point rating scale, the verbal description of which was printed on the answer sheets as shown in Fig. 3.

NAME: _____ AGE: _____ SEX: _____
SEAT No:

TEST No.	A	B	C	D	E	F
	—	QUIET	ACCEPTABLE	NOISY	EXCESSIVELY NOISY	—
1						
2						
3						
4						
5						
6						
7						
8						
9						
10						
11						
12						
13						

Fig. 3—Form of answer sheet used for the subjective measurements

No descriptions were assigned to the first and last categories, which the subjects were instructed to regard as extremes to provide a reference for the intermediate categories. The subjects were permitted to interpolate between adjacent categories by marking both of them. Instructions were given verbally to the subjects as a group and were kept as brief as possible without reference to any hypothetical environmental conditions.

The subjects were allocated numbered seats, and did not change position during the first seventy-five test runs. For the second group of seventy-five tests the five subjects at each end of the forward facing line changed places with the subjects immediately behind them in the "unsighted" line of observers. The purpose of the interchange was to check the relationship between results obtained with " sighted " and " unsighted " observers in case a marked difference were apparent, but this proved to be unnecessary. All " unsighted " observers were asked to make no attempt to look at the vehicles.

For most of the test the weather was cold, clear, and bright, with a light N.E. wind. Towards the end of the test the wind increased slightly, increasing the discomfort of the subjects but not causing any difficulty with the relevant objective measurements reported herein. All subjects were protected by warm clothing, and blankets and rugs were provided. No clothes were worn which could affect hearing.

RESULTS

For convenience in expressing the results, the verbal categories of the rating scale were first expressed numerically, so that " quiet " became 2, " acceptable " became 4, and so on. Thus, the numerical scale ran from 0 to 10. Each judgment recorded by a listener could in this way be expressed as a number, and the values averaged, either for the whole group, or for various sub-groups of listeners.

The principal results are shown in the form of correlation diagrams (Figs. 4-10), in which

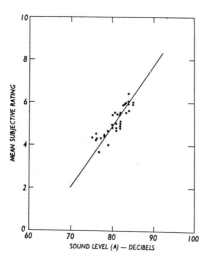

Fig. 4—Diesel vehicles : all observers

the average subjective rating for each vehicle is plotted against the recorded sound level (A) in decibels. Each point in the diagrams represents the passage of one vehicle. Previous experience has shown that the scatter of the points in such diagrams is excessive if all vehicles are included in one group. The vehicles have therefore been sub-classified in various ways in order to determine how many such sub-classes are needed to reduce the scatter to reasonable proportions. Referring to the figures, it is apparent that the scatter of results in the case of petrol-driven and diesel-driven vehicles respectively is quite small (5·5dB and 7dB respectively), and that there seems to be no need of sub-classification within these groups.

Initially, the motor-cycles were divided into two classes, two-stroke and four-stroke respectively, with the results shown in Figs. 6 and 7. The scatter in each of these cases is seen to be larger (two-strokes, 11dB ; four-strokes, 7dB), and, as shown in Fig. 8, the

scatter is scarcely increased if both classes are combined (11·5dB). An alternative classification into single and twin-cylinder machines was therefore made. The results in this case are shown in Fig. 9 and 10, and indicate no marked advantage of this manner of sub-classification (single-cylinder machines, 11dB ; twin-cylinder machines, 10dB). In particular, the most discordant points in Fig. 6 (two-strokes) and Fig. 9 (single-cylinder) respectively, represent the same machines. Clearly, there would be no advantage in further sub-classification.

In view of some conflicting evidence on the relation between sound level and subjective rating, which has been discussed in the previous paper, it was of interest to ascertain, so far as possible within the limitations of the

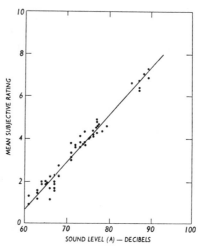

Fig. 5—Petrol vehicles : all observers

present tests, to what extent judgments of noisiness were affected by different listening conditions and different sub-grouping of observers. The average subjective ratings for the vehicles were therefore plotted for various sub-groups of listeners, namely those who faced the vehicle track and who may thus have been influenced by seeing the vehicles, those who were unsighted and faced away from the vehicles, males only, females only, and various age groups.

Since the results for the whole group were substantially unaffected by any of these sub-groupings of the listeners, the detailed results are not reproduced, but the summarised results are shown in Table I.

In order to make detailed comparison possible, a straight line is shown on each of

the correlation diagrams. This is the cal-culated regression line, obtained by regarding the mean subjective rating as an independent variable and the indicated sound level as the dependent variable. There is, of course, no logical reason to assume a linear relation, but it is evident that the scatter of the experimental points does not justify the fitting of higher order curves. By means of these straight lines, it is possible to read off the sound level

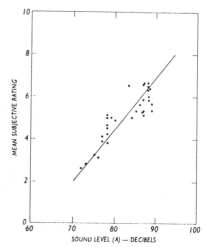

Fig. 6—Two-stroke motor-cycles : all observers

corresponding to steps of the subjective rating scale as judged by the average listener.

Possibly the most significant point on the rating scale is the numerical value 5, which corresponds to the demarcation line between " acceptable " and " noisy." However, some interest also attached to the rate at which the noisiness grows with the objectively-measured sound level, i.e. to the slope of the regression lines. A convenient way of comparing these features for the different vehicle classes and sub-groups of listeners is by a tabulation of the sound levels corresponding to the numerical steps, 2, 5 and 8 of the rating scale. Table 1 shows these levels, and demonstrates clearly that the demarcation line referred to is, for all cases, close to 80 dB (A). From the table small but definite trends can be detected as between sub-groups of listeners. Thus, the age groups thirty plus consistently rate vehicles less noisy than the younger listeners, or otherwise expressed, they are more tolerant of a given objective noise level. For the four vehicle classes the effect amounts to 1·2 dB, 1·6 dB, 0·6 dB and 1·1 dB

TABLE I

The values of sound level (A) read off the regression lines at subjective ratings of Q quiet, EN excessively noisy and D—demarcation between acceptable and noisy

Group	Diesel vehicles			Petrol vehicles			Two-stroke motor-cycles			Four-stroke motor-cycles		
	Q	D	EN	Q	D	EN	Q	D	EN	Q	D	EN
10 to 29 years of age	69·5	80	90·5	65·5	79	92	70	82	94	67·5	82	96·5
30 to 59 years of age	70·5	81	92	66·5	80·5	94	70	82·5	95	69	83	97·5
Males	69	80·5	92	66	79·5	93	69	82	94·5	67·5	82·5	97
Females	71·5	80·5	89·5	66·5	79·5	92·5	71	83	94·5	69	83	97
Facing	69	79·5	90·5	65·5	79	92·5	71·5	82·5	94	68·5	82·5	96·5
Non-facing	70·5	81	91·5	66·5	79·5	92	67·5	81·5	96	67	82	97
All	70	80·5	91	66	79·5	93	70	82·5	94·5	68	82·5	97

(average 1.1 dB), and would perhaps be dismissed as insignificantly small in relation to the dispersion if it were not persistently observed.

Comparing the results for males and females, a different, but equally persistent, effect is observed, namely, that the former compass a wider range in decibels for a given subjective interval than the latter. In terms of the interval " quiet " to " noisy " the respective decibel ranges for the four vehicle classes are : Males 15·1, 18·1, 17·0 and 19·4 (average 17·4) ; Females 12·3, 15·4, 15·4 and 18·5 (average 15·4). Once again, the effect, though probably real, is inconsiderable in absolute magnitude.

No systematic trend is apparent on comparing the results of sighted and un-sighted observers. Moreover, the magnitude of the differences is unimportant.

Comparison with Earlier Results

It is interesting also to compare the present results with those obtained in the previous investigation (1959.) In some important respects, notably the rating scale used and the instructions to the subjects, the two investigations were similar. On the other hand, the physical conditions in which they were carried out were markedly different, the test material for the earlier investigation consisting of normal main road traffic. The nomenclature applied to the classes of vehicles differed slightly between the two experiments, but the effect on the types of vehicle included in each class is insignificant. The earlier results may have been influenced adversely, though only to a small extent, by the fact that the track of the vehicles was not under accurate control. The results of the two investigations are compared in Table II and show a remarkable similarity,

Statistical Discussion

That there are appreciable divergences of opinion by individuals may be seen from the fact that the standard deviations of the judgments (of the whole group) were about 0·97 units of the numerical rating scale (values for individual vehicle tests ranged

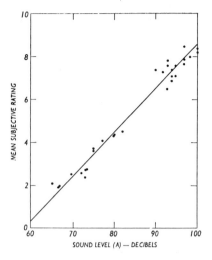

Fig. 7—Four-stroke motor-cycles : all observers

from 0·5 to 1·3). This represents some 4 dB when interpreted on the sound level scale. For most of the tests, judgments were spread over three or four adjacent (numerical) steps and in a few cases, as many as five, meaning that a noise judged " quiet " by some was judged " noisy " by others.

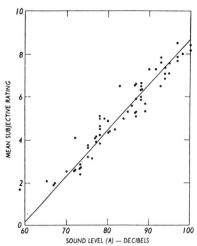

Fig. 8—All motor-cycles : all observers

and (c). To do this, it is necessary to estimate the magnitude of (a) independently by repeat judgments under identical physical conditions. A number of such repeats were included in the tests (thirty-six out of 114), and by analysing the corresponding results the standard deviation associated with the repeatability of listeners' judgments was evaluated to be 0·82 (numerical) units on the rating scale. This component of variance can, in principle, be extracted to estimate the residual scatter due to (c). This net residual scatter corresponds to a standard deviation in decibels of 1·0, 1·3 and 2·4 for diesel-engined vehicles, petrol-engined vehicles and motor-cycles respectively.

The fact that a large part of the scatter is attributable to the shortcomings of the sound-level meter is qualitatively obvious from some of the diagrams. For example, two four-stroke motor-cycles which gave sound level meter readings of 65 and 90 respectively were invariably (i.e. by whatever group of listeners) judged noisier than the meter reading would suggest, and two which gave meter readings of 82 and 100 were invariably judged less noisy. Similar discrepancies are evident with two two-stroke motor-cycles which read 83 and 84 respectively, and a close study of the scatter diagrams reveals a large number of similar examples. Since there can be no correlation whatever between the judgments of individual male and female listeners, it follows that

TABLE II

Comparison of present results with those of an earlier investigation for all listeners. The values of sound level (A) read off the regression lines at subjective ratings of Q = quiet, EN = excessively noisy, and D = demarcation between acceptable and noisy.

	Diesel vehicles			Petrol vehicles			Motor-cycles		
	Q	D	EN	Q	D	EN	Q	D	EN
Present investigation (1960)	70	80·5	91	66	79·5	93	68·5	82·5	96·5
	Commercial vehicles			Private cars			Motor-cycles		
Earlier investigation (1959)	72·5	82	91·5	65·5	80	94	71	83·5	96·5

both as regards the levels in decibels corresponding to the demarcation line between " acceptable " and " noisy " and to the rate of growth of the subjective rating on the noisiness scale with sound level. One feature, for example, observed in 1959 was that the range from " quiet " to " noisy " was compassed in a smaller objective range for commercial vehicles than for other classes. This is clearly exemplified again in the present work.

Bearing in mind that the two investigations have been carried out quite independently, with different people as observers, and noting that no marked differences are apparent with age, sex, aid of visual observation, &c., it seems reasonable to conclude that a level close to 80 dB fairly represents the demarcation line in the opinion of typical British listeners, under the conditions of this type of experiment.

No support can be found for the results of a Swiss investigation[2] that this demarcation line should be set around 73 dB on the sound level (A) scale, but differences of national habit in regard to the attitude to motor-vehicle noise cannot, of course, be discounted. It may be significant, however, that the Swiss observers were instructed to assume a hypothetical listening situation related to their normal daily activities.

One listener recorded an average (numerical) judgment of 5·87 (nearly " noisy ") over the whole 150 tests, whilst at the other extreme a value of 3·76 occurred (quieter than " acceptable "), the grand average being 4·63. Significance should not be attached to these considered as absolute values, of course, since they depend on the particular vehicles used and the manner in which they were driven. The relative attitude of these extreme listeners, however, is significant, corresponding to a difference of the order of 9 dB. Even the existence of these extreme listeners, however, does little to bridge the gap between the Swiss observations and the present results.

The scatter of the points on the diagrams (e.g. Fig. 4) is compounded of a number of factors, namely ;

(a) the uncertainty of individual judgments ;

(b) the fact that such judgments were quantised in units on a scale running from 0 to 10 ;

(c) the inherent lack of correspondence between the action of the meter and that of human listening ; and

(d) errors of objective measurement.

Of the above factors, (b) and (d) are unlikely to be of any consequence, but it is difficult to resolve the importance of (a)

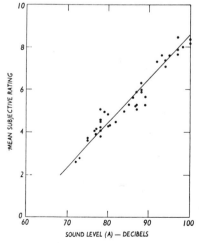

Fig. 9—Single-cylinder motor-cycles : all observers

similarities of the male and female average scatter diagrams must result from factors (*c*) and (*d*) above, and of these (*d*) may be ruled out, in view of the many cross-checks in the sound-level meter readings.

INTERPRETATION OF RESULTS

Subject to the limitations of any experiment of a statistical character, the straight lines in Figs. 4–10 represent the conclusions of this work in so far as the average listener is concerned. To interpret the results fully it must be appreciated that two main variances are associated with the subjective data, one arising from the dispersion of the subjects' individual judgments, and the other from the shortcomings of the meter. A further small uncertainty is associated with the fact that the results refer to a limited sample of listeners (fifty-seven), and there is the possibility of a sampling error, estimated to be not more than 0·8 dB, in the interpretation of absolute values. This, however, may be discounted in view of the magnitude of other uncertainties.

Each point on the graphs represents the average judgment of fifty-seven observers and if, for example, it was required that all

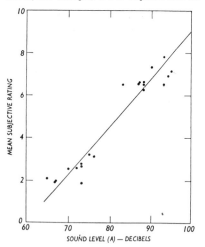

Fig. 10—Twin-cylinder motor-cycles : all observers

observers were "protected" at some predetermined point on the subjective rating scale, rather than only the less susceptible half, then the basic level associated with this point (as read off the diagrams, or obtained from Table III) would have to be adjusted downwards. If three standard deviations are taken as the criterion (the probability of any results lying between ±1 std. deviation (σ) is 68 per cent, within ±2 σ is 95 per cent, and ±3 σ is 99·7 per cent) a downward adjustment of some 12 dB would be necessary to ensure that it would be very improbable that any subject would rate any noise above the predetermined point on the scale. Similarly if 12 dB were added to the basic level, it would be very improbable that any subject would rate any noise below the predetermined point.

On the other hand it has been shown that the sound-level meter has shortcomings in representing the subjective noisiness of motor vehicles, especially motor-cycles, in the sense that it does not place all vehicles in the same rank order for noisiness as does the average listener, i.e., two vehicles judged equally noisy might yield different meter readings. It might be argued that shortcomings of the meter should not be permitted to "penalise" the vehicle giving the higher reading but which is still subjectively acceptable. To offset this, upward adjustments of the basic level would be required of, say, three standard deviations of the residual scatter due to the shortcomings of the meter. The upward adjustments would amount to about 4 dB, 3 dB and 7 dB for petrol-engined vehicles, diesel-engined vehicles and motor-cycles respectively.

It should be appreciated that any upward adjustment that is made in order to avoid the "penalisation" of vehicles by the shortcomings of the meter, must result in a reduction of the proportion of listeners who are "protected." Thus various levels may be set corresponding to "noisy," "acceptable," "quiet," &c., and according to the degree of "protection" required by the listeners, or the vehicles, or both.

By way of illustration the case is considered in which the basic point on the subjective rating scale at which "protection" is required is the demarcation between "acceptable" and "noisy." Referring to Table III, this point is represented by sound-levels of 79·5 dB, 80·5 dB and 82·5 dB (A) for petrol-engined vehicles, diesel-engined vehicles and motor-cycles respectively, and if no vehicle exceeded these sound-levels approximately half the listeners would judge them to be "just acceptable." It is of interest to note that the average listener permits a slightly higher sound-level (A) for motor-cycles than for the other two classes of vehicles.

If it is required to "protect" all listeners then 12 dB (3 σ criterion) should be subtracted from the basic levels quoted above resulting in sound-levels of 67·5 dB, 68·5 dB and 70·5 dB (A) respectively for the three classes of vehicles. On the other hand, it if is required that no vehicle is "penalised" by the shortcomings of the meter, then upward adjustments should be made to the basic levels amounting to 4 dB, 3 dB and 7 dB (A) (3 σ criterion) resulting in sound-levels of 83·5 dB, 83·5 dB and 89·5 dB (A) for petrol-engined vehicles, diesel-engined vehicles and motor-cycles respectively.

ACKNOWLEDGMENTS

The authors wish to acknowledge the assistance of their colleagues, in particular Messrs. W. C. Copeland and J. C. Dixon, who carried out the experimental work, and Mr. L. Verra, Mathematics Division, National Physcial Laboratory, who carried out the numerical analysis. They also acknowledge the assistance of subjects who took part in the listening tests, and of Mr. Rayner of the National Coal Board who made a number of his staff available for this purpose. The work described forms part of the research programme of the Motor Industry Research Association, and of the Applied Physics Division, National Physical Laboratory, and the article is published by permission of the Directors of these bodies.

REFERENCES

[1] Robinson, D. W., Copeland, W. C., and Rennie, A. J. THE ENGINEER, March 31, 1961, Vol. 211, No. 5488, pages 493–97.
[2] Generaldirektion P.T.T. Switzerland, Forschungs und Versuchsanstalt Report No. 22637 and Appendix, October, 1959.
[3] Andrews, B., and Finch, D. M., "Truck Noise Measurement," *Proc. Highways Research Board (U.S.A.)*, December, 1951, pages 456–65.

Part II

AIRCRAFT NOISE AND VIBRATION

Editor's Comments on
Papers 6 Through 13

Aviation is largely a twentieth-century development. As a major source of community noise it has become increasingly significant in the last thirty or forty years.

The Wright brothers are credited with the first successful manned powered flight in 1903. However, most people at first assumed that the heavier-than-air airplane would be used mainly in military and sporting activities, and little thought was given to commercial use of airplanes before World War I.

Development of the airplane was extremely rapid during World War I, and when hostilities suddenly ended in 1918 the air forces of Britain, France, and Germany contained considerable numbers of planes and trained pilots. Attempts were soon made to develop regularly scheduled passenger service. Soon many airlines were in operation in Europe, and in 1923 almost 10,000 passengers were carried on the London-to-Paris route. After a series of mergers, airlines such as

Imperial Airways were formed in Britain in 1924 and Deutsche Luft-Hansa in Germany in 1926, most with some form of government subsidies or support.

Surprisingly, airline service in the United States lagged behind that in Europe by several years. Although an airmail service had been operated by the U.S. Post Office since 1918, it was not until 1926 that scheduled passenger service was begun in the United States by Western Air Express [1]. Many small airlines were formed in the United States in the period 1927 to 1930, and this period also saw (as a result of various mergers) the formation of United, Eastern, TWA, and American Airlines.

Air travel rapidly expanded in the United States in the late 1920s. In 1928, 100,000 passengers were carried by German Airlines (the leader in Europe) and 60,000 by U.S. airlines. One year later the situation was reversed, with 120,000 passengers being carried by the German Airlines and 160,000 by U.S. airlines [1]. Passenger service was stimulated by the 1925 Airmail (Watres) Act and amendments, which gave extra compensation to airmail operators who provided passenger accommodation [1, 2, 3]. During the period 1934 to 1938 there was a steady growth in airline traffic (with an increase of about 300,000 passengers each year). The famed DC-3 made its debut during this period, while European governments became increasingly preoccupied with military aviation. The United States emerged after World War II as operator of 50 percent of the world's airlines and supplier of 90 percent of the world's civilian aircraft [1]. New routes were added and service continued to expand.

Air travel has continued to expand rapidly since World War II. Worldwide passenger miles traveled have doubled about every five or six years since that time, from about 5,000 million in 1945 to about 36,000 million in 1955, to about 120,000 million in 1965, and to almost 500,000 million in 1975. Throughout all this time, domestic U.S. seat-miles have accounted for about 50 percent of this total.

Whether the trend to increased air travel will continue in light of the rapid increase in fuel costs during the 1970s is unclear. However, the aircraft noise problem during the past half-century and the likely problem in the future must be viewed against this story of very rapid growth and changes that have occurred and still are occurring.

Until the late 1950s the dominant aircraft propulsion system was the internal combustion engine and propeller. The main noise sources were the propellers and engine exhaust, with propeller noise probably dominating at high power settings. Work on exhaust muffling (including that of Davies, Paper 2, on helicopter exhaust noise) is discussed in the previous section. Work on propeller noise was conducted in several European countries (for example, England, Germany, and the Soviet Union), beginning soon after World War I. Presumably the main motivation was to reduce the noise of military aircraft both to cut down the risk of detection and to decrease cabin noise. It seems that there was considerable concern during the 1920s, 1930s, and 1940s in several countries, including England, France, Germany, Soviet Union, Sweden, and the United States, that aircraft noise was causing deafness in pilots [4-24]. As the use of civilian aircraft grew, the incentive to reduce noise to the community and passengers of commercial aircraft also grew. Bassett and Zand in 1934 claimed that the growth of air transportation was being limited by passenger objection to excessive cabin noise. They produced impressive noise reductions of 12 dB and more on later models of Condor passenger planes by use of vibration-isolation, sealing, and absorption materials. Their paper is reproduced here as Paper 6.

Lynam and Webb in 1919, in England, were probably the first to investigate propeller noise [26]. However, their assumptions (the second of which was that each element of the propeller acts as a simple omnidirectional source) led to poor experimental agreement with theory. Hart made the same omnidirectional assumptions in his work in 1929 [27]. Paris [28, 29] and Kemp [30] made measurements of propeller noise in the early 1930s showing that the noise has strong directional characteristics at the blade passing frequency and its harmonics. Paris [28], by making many assumptions, did manage to predict the main directional shape of propeller noise and obtain agreement with measurements. However, probably the first paper that satisfactorily predicted propeller noise quantitatively was by Gutin (Paper 7). The paper was first translated from Russian into German in 1936 and then into English as NACA TM 1195 in 1948. Gutin's theory satisfactorily predicts the noise at the blade passing frequency, although higher harmonics are not so well predicted. This fact is discussed, for example, by Applegate and Crocker [31]. Propellor theory was soon improved and refined by later workers such as Ernsthausen and Deming [32, 33, 34]. It is impossible to mention all of the many theoretical and experimental papers published on this topic since the 1940s; however, several recent good reviews of propeller (rotor) theory have been provided by Sharland [35] and Morfey [36].

Large numbers of military and civilian propeller aircraft were built during and after World War II. During this period the first military jet-engine powered aircraft were built in Britain and Germany. Soon after, military aircraft became predominantly jet powered, and in 1958 the first commercial passenger jet aircraft (the Boeing 707 and DH Comet) were introduced into service in the United States and Britain [25]. Most large passenger aircraft are now powered by jet engines, although very large numbers of small passenger propeller-driven engined aircraft and helicopters driven by rotors also exist.

Von Gierke published a good review of aircraft noise sources in 1953, reproduced here as Paper 8. This review is still useful today since all major aircraft noise sources are discussed. However, the aircraft noise situation has changed somewhat since 1953. The early jet-engine passenger aircraft were much noisier than their piston-engine counterparts not only because the sound pressure levels were higher, but also because much higher frequency noise was produced. For more recent reviews of aircraft noise and its assessment, see Refs. 37, 38, and 39.

In the early 1960s, passenger jet aircraft noise soon became the subject of many complaints from communities near airports; much effort has since been expended to reduce it. The first theoretical attempt to explain jet noise was produced by Lighthill in 1952. Lighthill's two famous papers [40, 41] have already been reproduced in the Benchmark book on Underwater Sound [42], and the reader is referred there for these papers. Lighthill modeled jet noise production theoretically and showed that it occurs as a result of the shearing action of the high-speed jet flow mixing with the atmosphere. Many workers, such as Ribner and Ffowces-Williams, have continued to work on jet noise theory, and Powell has given a good recent review of the state of current theories [23].

At the same time that Lighthill's theoretical papers were being written, experimental work on jet noise was being conducted in several institutions. These experiments with both model and full-scale jets produced the basis for both prediction of jet noise and better theoretical understanding. Notable were the contributions of Hubbard and his colleagues at NASA (Langley) and Powell at Southampton University in Britain. Hubbard showed that the peak frequency of jet noise was proportional to the exit velocity and inversely proportional to the diameter, thus exhibiting a Strouhal number relationship. Hubbard demonstrated that there appears to be a relationship between the frequency spectrum of the turbulence in the jet and the noise just outside it. He also showed that the higher-frequency noise appears to be generated in the mixing region near the jet (where the turbulent eddies are small), and that the lower-frequency noise is generated further downstream (where the eddies are larger). These are important results. The paper by Hubbard and Lassiter is reproduced here as Paper 9. This paper is brief and more details are given in a NACA report published about the same time [44]. Powell's research in the same period was also noteworthy. Paper 10 describes his work on the noise of choked jets. When the critical pressure ratio across a jet nozzle is exceeded, sonic conditions exist at the jet exit, and jet screech can be produced. Powell was one of the first to explain this phenomenon. For a paper giving more details of this work (it is too long to be reproduced here), the reader is referred to Ref. 45.

The early theoretical and experimental papers mentioned above were useful in obtaining an understanding of the production of jet noise. However, the first direct attempt to reduce jet noise appears to be by Westley and Lilley, and a condensation of their 1952 report is reproduced here as Paper 11. Westley and Lilley's work is important because it led to the development of jet silencers that were employed on pure jet passenger aircraft during the 1960s. There were many different designs of such silencers; they were usually based on the principle of increasing the jet exhaust nozzle perimeter several times (by using corrugations of some kind), thus ensuring more rapid mixing of the high-speed jet with the atmosphere.

During the 1960s the first fan-jet engines (also known as turbofans or bypass jets) were used on passenger jet aircraft. The original aerodynamic motivation in using fan-jet instead of pure-jet engines was to produce a more efficient engine by reducing the loss of kinetic energy in the high-speed exhaust by producing a larger-diameter, slower-speed jet. However, as is seen by studying Lighthill's results and those of Hubbard and Lassiter, the lower-speed exhaust of the fan-jet was quieter, a happy result from the noise point of view. Although take-off noise is reduced by the use of fan-jet engines, approach or landing noise is unfortunately not reduced because the discrete frequency compressor (and fan) noise of such engines becomes a serious problem. The paper by Tyler and

Sofrin, published in 1961, was perhaps the first to examine compressor noise systematically, and it won the 1961 Manly Memorial Award. Part of the paper has been reproduced here as Paper 12. This paper has formed the basis for much later work on jet-engine compressor noise. Acoustic liners used in the inlets of fan-jet engines are now capable of reducing noise by about 15 PNdB.

Besides being a source of human annoyance, the noise of jet exhausts in aircraft and space vehicles has become a serious problem in the last twenty years. The structure of aircraft and space vehicles can be caused to vibrate so that undesirable noise is radiated into the interior cabins or cavities. In the case of very intense acoustic or turbulent pressures, fatigue or catastrophic structural failure can occur. For these reasons it has become necessary to calculate the response of structures to noise fields. Powell [46] first formulated the response of structures to such fields. Powell's approach requires a calculation of the *joint-acceptance* of the structure and of its response in each of its different modes of vibration. Although suitable for simple structures, this approach becomes unmanageable for complicated structures with ill-defined boundary conditions. In the early 1960s several workers, including Smith, Lyon, and Maidanik, developed a statistical method of estimating structural response that became known as statistical energy analysis (SEA). SEA has the advantage that a manageable analysis results when the simpler question, such as "what is the space-averaged response of a structure in a given frequency band," is asked. SEA has become quite widely used in estimating the response of such structures as aircraft, spacecraft, buildings, ships, trucks and farm tractors to noise. The early papers by Smith, Lyon, and Maidanik [47, Paper 13, and 48] are now the classical ones in the SEA field. Unfortunately there is not room to reproduce them all here in their entirety; an extract of one of them is given here as Paper 13. Lyon and Maidanik's paper (Paper 13) shows that in certain idealized cases the power flow between coupled systems—that is, structures, or structures and air cavities—is proportional to the *modal energy difference*. Several papers have been written since to prove that this result can also be shown for more complicated cases, and this assumption is usually made in practice with SEA. Maidanik's paper includes derivations of the radiation resistance of ideal cases of simply supported beams and plates [48]. The paper also discusses the effect of panel ribs on response and gives comparison between theory and experiment for the radiation resistance, total resistance, and coupling factor of a panel.

Many papers have been written on SEA since its beginnings in the early 1960s. Fahy [49] and Lyon [50] have written recent reviews of SEA. Lyon has written a book [51]. Crocker et al. have shown how SEA can be used to predict sound transmission through single and double panels and double panels connected with ties [52, 53, 54].

REFERENCES

1. Davies, R. E. G., 1964, *A History of the World's Airlines*, Oxford University Press, London.
2. Bigham, T. C., and M. J. Roberts, 1952, *Transportation, Principals and Problems*, McGraw-Hill, New York.

3. Lloyd Wilson, G., 1954, *Transportation and Communications,* Appleton-Century Crofts, Inc., New York.
4. Scott, V. T., 1923, Airplane Deafness and Its Prevention, *Milit. Surg.* **52:**300-301.
5. Troina, F., 1933, Changes in Hearing in Air Pilots, *Valsalva* **9:**337-353.
6. Aubroit, P., 1935, Note on the Pathogenesis of Certain Cases of Occupational Deafness Among Aviators, *Oto-Rhino-Laryng. Int.* **19:**65-68.
7. Foges, G., 1935, Acoustic Apparatus of Flyers and Its Significance for Safety in Aviation, *Arch. Gewerbepathol. Gewerbehyg.* **6:**197.
8. Dickson, E. D. D., A. W. G. Ewing, and T. S. Littler, 1939, The Effects of Aeroplane Noise on Auditory Acuity of Aviators: Preliminary Remarks, *J. Laryng. Otol.* **54:**531-548.
9. Lerouge, J., 1940, Effects of Noise on Aviators, *J. Progres Méd.* (Paris) **68:**132-135.
10. Bunch, C. C., 1941, The Problem of Deafness in Aviators, *War Med.* **1:**873-886.
11. Pastore, P. N., 1941, Hearing Among Experienced Aviators, *Mayo Clinic Proc. Staff Meet.* **16:**214-217.
12. Salem, W., 1942, Does Occupational Deafness Exist in Aviation?, *Rev. Med. Hyg. Milit.* **31**(3):281-286.
13. Box, N. H. E., 1943, Occupational Deafness in Airmen, *Med. J. Australia* **2:**126.
14. Flaiz, J. W., and E. G. Wolf, 1943, The Effect of Noise on Aviation Personnel, *Contact* (Pensacola) **3:**160-167.
15. Meyer zum Gotlesberge, A., 1943, Hearing Injuries in Flying Personnel, *Luft-fahrtmedizin* **8:**256-264.
16. Firestone, C., 1943, Status of Auditory Mechanism in Pilot of Extensive Experience; Evaluation of Factors Contributing to State of Hearing Diminution in Experienced Pilot and Correlation of Them into Otologic Entities, *Laryngoscope* **53:**419.
17. Heyden, R., 1944, Aviator's Deafness, *Acta Oto-Laryng.* Stockh. **32:**164-175.
18. Schmalix, J., 1944, Inner Ear Injury Caused by Airplane Noise, *Luftfahrtmedizin* **8:**368-380.
19. Fowler, E. P., 1945, Causes of Deafness in Flyers, *Arch. Otolaryng.* **42:**21-32.
20. Iwaszkiewicz, 1945, Problem of Hearing in Aviators, *Lek. Wojsk.* **36:**53-56.
21. Graebner, H., 1947, Auditory Deterioration in Airline Pilots, *J. Aviat. Med.* **18:**39-47.
22. Causse, R., and P. Falconnet, 1947, Deafness in Aviators, *Ann. d'Oto-Laryng.* **64:**436.
23. Radziminski, A., 1948, Defects of Hearing in Aviators, *Oto-Laryng. Polska* **2:**25.
24. Senturia, B. H., 1949, Hearing Changes During Pilot Training, *Mil. Surg.* **105:**205.
25. Dyos, H. J., and D. H. Aldcroft, 1969, *British Transport,* Leicester University Press, Leicester, England.
26. Lynam, E. J. H., and H. A. Webb, 1919, *The Emission of Sound by Airscrews,* Advisory Committee for Aeronautics (London), R & M No. 624.
27. Hart, M. D., 1929, *The Aeroplane as a Source of Sound,* Aeronautical Research Communication (London), R & M No. 1310.
28. Paris, E. T., 1932, *Philos. Mag.* **13:**99.
29. Paris, E. T., 1933, A Measurement of the Fundamental Sound Generated by the Airscrew of an Aeroplane in Flight, *Philos. Mag.* **16**(103):50.
30. Kemp, C. F. B., 1932, *Phys. Soc. (London) Proc.* **44**(pt. 2):151.
31. Applegate, S. L., and M. J. Crocker, 1976, Reducing the Noise of a Rotary Lawn Mower Blade, *Noise Control Eng.* **6**(1):30-34.
32. Ernsthausen, W., 1936, On the Origin of Propeller Noise (in German), *Luft-fahrtforschung* **13:**433-440. Translated in *Natl. Advis. Comm. Aeronaut.* TM825, 1937.
33. Ernsthausen, W., 1941, Influence of Aerodynamic Characteristics on the Sound Field and Radiated Power of an Airscrew (in German), *Akust. Z.* **6:**245-261. Also published in *Luftfahrtforschung,* **18:**289-304 (1941).
34. Deming, A. F., 1940, Propellor Rotation Noise due to Torque and Thrust, *Acoust. Soc. Am. J.* **12:**173-182.

35. Sharland, I. J., 1964, Sources of Noise in Axial Flow Fans, *J. Sound Vib.* **1:**302–322.
36. Morfey, C. L., 1973, Rotating Blades and Aerodynamic Sound, *J. Sound Vib.* **28:**587–617.
37. Large, J. B., 1970, Aircraft Noise and Sonic Boom, in *Transportation Noises,* J. D. Chalupnik, ed., University of Washington Press, pp. 3–14.
38. Hubbard, H. H., and D. J. Maglieri, 1974, A Brief Review of Air Transport and Noise, *Noise Control Eng.* **3**(3):16–24.
39. Crocker, M. J., 1978, Noise of Air Transportation to Nontravelers, in *Handbook of Noise Assessment,* D. May, ed., Van Nostrand Reinhold, New York, pp. 82–126.
40. Lighthill, M. J., 1952, On Sound Generated Aerodynamically, I. General Theory, *R. Soc. (London) Proc.* **A221:**564–587.
41. Lighthill, M. J., 1954, On Sound Generated Aerodynamically, II. Turbulence as a Source of Sound, *R. Soc. (London) Proc.* **A221:**1–32.
42. Albers, V. M., ed., 1972, *Underwater Sound,* Benchmark Papers in Acoustics, Dowden, Hutchinson and Ross, Stroudsburg, Pa.
43. Powell, A., 1977, Flow Noise: A Perspective on Some Aspects of Flow Noise and of Jet Noise in Particular, Parts I and II, *Noise Control Eng.* **8**(2):69–80; *Noise Control Eng.* **8**(3):108–119.
44. Lassiter, L. W., and H. H. Hubbard, 1952, Experimental Studies of Noise from Subsonic Jets in Still Air, *Natl. Advis. Comm. Aeronaut. TN 2757.*
45. Powell, A., 1953, On the Noise Emanating from a Two-Dimensional Jet Above the Critical Pressure, *Aeronaut. Q.* **4:**103–122.
46. Powell, A., 1958, On the Response of Structures to Random Pressure and to Jet Noise in Particular, in *Random Vibrations,* S. H. Crandall, ed., M. I. T. Press, Cambridge, Mass., pp. 187–229.
47. Smith, P. W., Jr., 1962, Response and Radiation of Structural Modes Excited by Sound, *Acoust. Soc. Am. J.* **34**(5):640–647.
48. Maidanik, G., 1962, Response of Ribbed Panels to Reverberant Acoustic Fields, *Acoust. Soc. Am. J.* **34**(6):809–826.
49. Fahy, F. J., 1974, Statistical Energy Analysis: A Critical Review, *Shock Vib. Dig.;* L'Analyse Statistique Énergétique—Une Revue Critique, *Rev. d'Acoust.* no. 33, 1975.
50. Lyon, R. H., 1978, Recent Developments in Statistical Energy Analysis, *Shock Vib. Dig.* **10**(2):3–7.
51. Lyon, R. H., 1975, *Statistical Energy Analysis of Dynamical Systems: Theory and Applications,* M. I. T. Press, Cambridge, Mass.
52. Crocker, M. J., and A. J. Price, 1969, Sound Transmission Using Statistical Energy Analysis, *J. Sound Vib.* **9**(3):469–486.
53. Price, A. J., and M. J. Crocker, 1970, Sound Transmission Through Double Walls Using Statistical Energy Analysis, *Acoust. Soc. Am. J.* **47:**683–693.
54. Crocker, M. J., M. C. Battacharya, and A. J. Price, 1971, Sound and Vibration Transmission Through Panels and Tie Beams Using Statistical Energy Analysis, *J. Eng. Ind. Trans. (ASME)* **93**(3):775–782.

6

NOISE REDUCTION IN CABIN AIRPLANES

Preston R. Bassett and Stephen J. Zand
Brooklyn, N. Y.

Any new form of transportation generally brings some new sort of nuisance to the persons using it. Air transportation, the newest and fastest form of travel, is no exception, and the chief deterrent to air travel is the noise to which the passengers are subjected. As it is impossible to discuss any physical phenomena without having measuring units and methods of securing them, a description of the decibel scale as well as a discussion of the commonly used noisemeter is given, including a description of a most inexpensive, simple, and yet very reliable tuning-fork method.

NOISE accompanies all forms of transportation. Over 90 per cent of the noise of a city is caused by transportation—trucks, trolley cars, trains, etc. In general, the amount of noise is proportional to the speed of transportation. The fact that the speed of transportation has been increasing so rapidly in recent years has brought about an increase in noise level in cities and in high-speed transportation to a point where the public has become very conscious of it. The traveling public remained uncomplaining and almost unaware of the noises of

P. R. BASSETT

S. J. ZAND

traveling until, with increased speed, these noises built up to what may be termed the discomfort level. When this discomfort level is reached, the traveling public then becomes acutely aware of it and tends to avoid such forms of transportation.

The airplane has been for some years the most rapid form of transportation, but only recently has it been making a bid for passengers. No attention has been paid to noise, however, until now. Pilots and early passengers took it for granted that the noise was just an unavoidable evil that went with flying and high speed. However, when air lines in this country started to go on schedules and carrying many passengers, who flew as a quick method of transportation rather than merely for a thrill, then the passengers began to complain. They not only complained, but many have decided that, except in emergencies, the discomfort of air transportation is not worth the speed, and therefore have

returned, whenever possible, to the use of more comfortable forms. The fact that air transportation is actually losing passengers because of discomfort is a challenge to the aviation industry. There are now very many people who are air-minded, but the majority cannot be converted to use air transportation until it is brought out of the discomfort level.

It is of course impossible totally to eliminate noise in any form of transportation. It is fortunate that there is no demand for complete elimination of noise. The demand is merely to keep the noise under the discomfort level. There is a very broad region of noise levels which can be termed as comfort level, and all that is necessary is that the general level within the airplane be kept in this region.

Lord Kelvin once said: "If you can measure that of which you speak, you know something of your subject, but if you cannot measure it, your knowledge is unsatisfactory." Only in recent years have means been found to measure noises quantitatively and qualitatively. A new unit of noise, the decibel, has come into common usage. Instruments for the measurement of noise levels in decibels have been developed rapidly to a stage where accurate measurements may be made with portable apparatus.

The measure of the sensation of loudness is not easy to define. The difficulty may be circumvented in a way indicated by Weber's law: "The increase of stimulus to produce the minimum perceptible increase of sensation is proportional to the preexisting stimulus."[3] That is, commencing with a certain sound intensity, the increase of intensity δE which produces a *noticeable* change of sensation δS may be measured. From Weber's law, Fechner derived the relation

$$\delta S = \frac{k\delta E}{E} \quad \text{or} \quad S = k \log E$$

where S is the magnitude of the sensation, E the intensity of the stimulus, and k a constant. While obviously it is not possible to measure S directly, there is no difficulty in determining the ratio $\delta E/E$ as a function of E. The simplest way of measuring this ratio is by means of electrical apparatus where the acoustical energy is transformed into electrical energy. The unit commonly used in this country for such measurement of loudness is the decibel,[4] a unit which may be said to be, approximately, the smallest change in the level of sound which the normal ear can detect. More accurately, as seen from the Weber-Fechner relation, this unit may be defined as a ratio of intensities. Hence, if the intensities of two sounds are in a ratio of 10 to 1, they differ by 10 db; if the intensities are in a ratio of 10^2 to 1, i.e., 100 to 1, the sounds differ by 20 db; and so on. In general, therefore, the number of decibels measuring the difference between two sounds is ten times the Briggs logarithm of the intensity ratio, db = $10 \log (I_1/I_2)$. It is obvious, therefore, that any decibel scale

[1] Vice-President in Charge of Engineering, Sperry Gyroscope Company, Inc. Preston R. Bassett was graduated from Amherst College in 1913, with A.B. and M.A. degrees, and continued graduate work in engineering at Brooklyn Polytechnic Institute. He has been associated with the Sperry Gyroscope Company for 18 years, first as Research Engineer and since 1929 as Chief Engineer.

[2] Aeronautical Research Engineer, Sperry Gyroscope Company, Inc. Stephen J. Zand was graduated from the École Polytéchnique Fédérale, Zurich, Switzerland, in 1920. He did post-graduate work at the École Superieure d'Aéronautique in Paris, France. He came to America in 1925 and worked as airplane designer at the Ford Motor Company, as a research engineer in the Pioneer Instrument Company, and since April, 1932, he has been engaged in research work for the Sperry Gyroscope Company. In 1931, for his paper, "A Study of Instrument Boards and Airplane Structure" (*S.A.E. Journal*, October, 1931), he was awarded the Wright Brothers Gold Medal.

Contributed by the Aeronautic Division and presented at the Semi-Annual Meeting, Chicago, Ill., June 26 to July 1, 1933, of THE AMERICAN SOCIETY OF MECHANICAL ENGINEERS.

NOTE: Statements and opinions advanced in papers are to be understood as individual expressions of their authors, and not those of the Society.

[3] Wood, "A Textbook of Sound," Macmillan, New York, N. Y.
[4] See "The Decibel," by S. J. Zand, *Sperryscope*, vol. 6, no. 11; "The Decibel," by S. J. Zand, *Aeronautical Engineering*, Feb., 1933, vol. 8, no. 2, p. 19.

requires the more or less arbitrary adoption of some definite zero point from which the intensity or power ratio should be measured. Omitting this reference may cause misunderstandings.

The first scale used by the Noise Abatement Committee of New York City fixed the zero at 4.4×10^{-16} watts per sq cm. The scale we use has its zero at 1 millibar sound pressure, which, under normal atmospheric conditions, is approximately 24.4×10^{-16} watts per sq cm.

Recently, the Society of Acoustical Engineers of America has adopted a new zero equal to 1×10^{-16} watts per sq cm, or 0.207 mb of sound pressure.

Thus, 90 db above 1×10^{-16} watts is equivalent to 83 db above 4.4×10^{-16} watts and 76 db above 1 millibar. Throughout this article, the decibel-above-1-millibar scale is used.

While loudness appears to the ear to increase by simple arithmetical progression, the sound energy increases by logarithmic progression, rising from ten to ten billion (10^1 to 10^{10}), while loudness goes from ten to one hundred (10 to 100). This peculiar relation must be clearly understood, and hence we have gone into considerable detail to impress that decibels per se have no significance whatever and are only convenient symbols for expressing a ratio. The difference between 10 and 20 miles is the same as between 90 and 100 miles. But between sounds of 10 and 20 decibels above the threshold of hearing there is an intensity difference of 90, while the difference between sounds of 90 and 100 decibels is 9 billion units of energy. As a consequence of this relation, it is at once seen that doubling the intensity does not double the loudness, but merely increases the loudness by 3 db. Therefore, two sources, each producing 100 db, give a resultant noise of only 103 db instead of 200 db. Conversely, if an airplane cabin has a noise level of 100 db and we take steps to reduce the sound energy to one-half, we will have reduced the noise level to only 97 db, hardly a noticeable achievement. It is this discouraging fact that makes sound reduction so difficult.

While the measurement of acoustical energy by mechanical means is not impossible,[5] it has been found highly impracticable, especially in measuring moving vehicles. This is due to the extremely small amount of energy which produces an auditory sensation. Thus, speech of a level of 60 db would appear to have a lot of energy, but measurements indicate that the average power of a speaker is between 25 and 50 microwatts or thereabouts. Thus, it would require 15 million persons speaking simultaneously to produce the equivalent of a single horsepower of acoustical energy. Obviously, then, the only method of measurement will be the electrical, by which the incoming sound wave produces a corresponding electrical variation. These are amplified in a known and always constant ratio by means of vacuum tubes, and the amplified output is a magnified but faithful reproduction of the acoustical disturbance which is measured.

The action of the noise meter, a simplified diagram of which is shown in Fig. 1, starts with a microphone, generally of the dy-

Fig. 1 Schematic Diagram of a Noise Meter

namic or condenser type, which picks up the sound wave and produces its electrical equivalent. The electrical impulse is then amplified, generally by means of a high-gain and high-grade resistance-coupled amplifier. (Great care must be exercised in order not to introduce distortion.) This magnified output is then rectified, finally actuating a meter causing the pointer or hand of the indicator to move along the meter scale to a point determined

by the intensity of the sound wave. The figure indicates that a noise meter is also supplied with an attenuator and a weighing network. The attenuator enables the observer to control the amplitude of the electrical wave by large steps, so that when a very loud sound is encountered which would drive the needle off the scale of the meter, a known amount of attenuation may be inserted, bringing the needle back into the range of the meter. The actual reading is then the sum of the attenuator setting and the meter reading. The weighing network is designed to render the noise meter more sensitive, and vice versa, so that the pitch of a tone will automatically affect the meter in the same way that it affects the ear. While this is only possible in rough approximation, owing to the complicated characteristics of the ear, it can be done with a reasonable amount of precision for all practical purposes.

In addition, a noisemeter may contain a series of band pass filters which may be thrown in and out of a circuit by means of a convenient selector switch or plug-in arrangement. The action of such filters is to suppress any wave the frequency or pitch of which does not lie in the particular frequency region passed by the filter. One filter passes waves of only very low pitch, having a frequency of, say, 64 cycles or less; a second passes frequencies from 64 to 128, etc. With six or eight such filters an "octave" analysis is possible—very valuable information which will generally facilitate the working out of proper acoustical treatments.

Naturally, noise meters and frequency analyzers are expensive and generally bulky and heavy. In order to use them it is necessary to have the vehicle to be tested at disposal for a certain amount of time. For a rough check, with skilled observers as close as ± 2 db, there is a very simple method which is based on the masking effect of one sound against another.[6] The only tool necessary is a calibrated tuning fork and a watch. The observer strikes a tuning fork (256 or 512 cycles) with a constant force and brings the fork in front of his ear canal, moving the fork slightly to create a warble. At the same time, he observes the second hand of a watch (a stopwatch is preferable). He observes the time until the noise of the fork is masked by the noise to be measured. Suppose the fork held in such a position near the ear canal makes a noise of 90 db, and from the calibration of the fork it is known that it decays at the rate of 1.5 db per sec. Then if the fork remains audible only 20 sec, it will follow that the noise at this frequency is approximately $90 - (1.5 \times 20) = 60$ db. If measurements of this type are made, using a number of forks, a surprisingly good picture of intensity and frequency distribution can be gathered with the simplest instrumentation.

Returning now to the relationship of speed to acoustical comfort, we may now study it on a decibel scale. Fig. 2 has been prepared in a unique way to show both methods of determining noise levels; decibels as measured on the noise meter are plotted against masking time of a tuning fork. All readings on this chart were obtained by the authors on various tests and trips within the last year. The readings include every form of transportation from sailboat to open-cockpit airplane. Included are readings on both American and European subways, trains, and passenger airplanes. The grades of comfort are indicated approximately at the right of the chart.

Although the high readings on the chart are all airplanes, it is to be noted that the newer soundproofed airplanes have moved down into the region of subways and trolley cars as regards noise level.

To show more clearly the relation of sound level to speed, Fig. 3 presents several curves of different transportation units, plot-

[5] Lord Rayleigh, "Theory of Sound," vol. 1, Macmillan, 1926.

[6] V. O. Knudsen, "Architectural Acoustics; Method of Measuring Noise," p. 255.

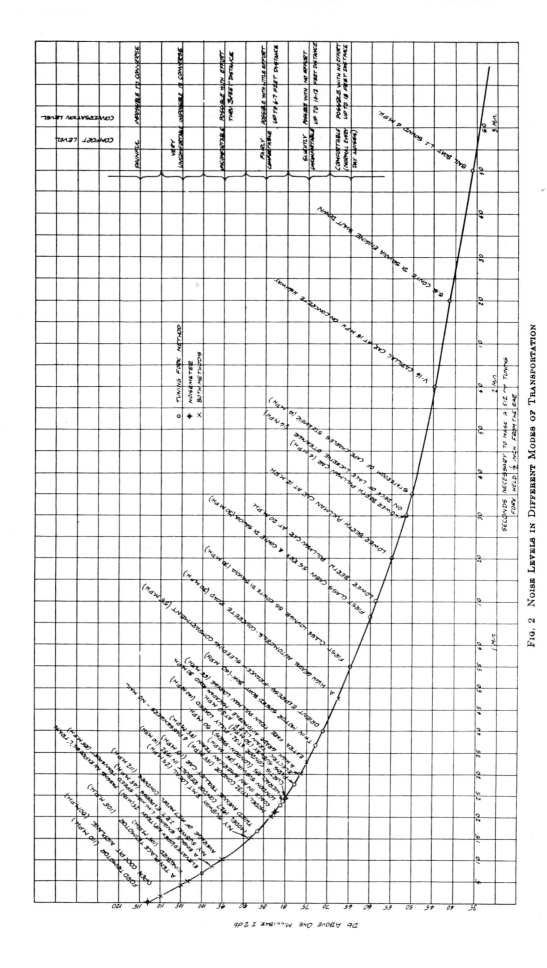

Fig. 2 Noise Levels in Different Modes of Transportation

106

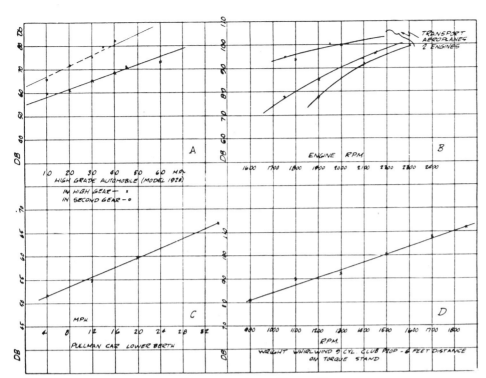

FIG. 3   DIAGRAM OF SPEED VERSUS NOISE

both high and low frequency, with the low frequency predominant

(4) Low-frequency noises are very difficult to deal with because most materials are much better barriers to high-pitch sound than to low-pitch

(5) The power plant is attached directly to the structure which contains the cabin, which allows vibration from the engine and propeller to be transmitted by direct solid conduction into the cabin

(6) On multiengine airplanes, the possibility of beats of low-pitch character and high intensity is very troublesome, with no direct remedy as yet in sight

(7) Weight limitations are very severe, and it is a known fact that sound insulation is a direct function of mass.[7]

ting speed against decibels. *A* shows that the noise of an automobile increases approximately as a straight-line function of its speed. *B* gives the increase of noise level in three different types of transport airplanes with increased engine rpm. It is interesting to note the convergence of the airplane curves toward some point above 110 db. It is suspected that this is due to the propeller-tip speed, which increases in noise output rapidly as it approaches the velocity of sound, and produces so much noise at these high velocities as to completely mask the engine noises. Hence, all airplanes would have about the same maximum noise if they could be pushed to the point of having tip speeds of 1100 fps. Some military and racing ships reach this point, and, as is well known, they make a terrific and penetrating noise, measuring well over 110 db.

Chart *C* gives the increase of noise level in a Pullman car with increased speed. It is to be noticed that in this case the increased noise is not caused by the engine, but by local noises. It is apparent that the train has an advantage over an airplane in that the power plant or engine is so far removed from the passengers that its noise is attenuated to a harmless level in the passenger compartments. Furthermore, if soundproofing is still found desirable, additional weight is easily permissible for the purpose. In the airplane, there are, on the other hand, severe weight limitations.

Chart *D* gives the relationship of noise and engine speed of a standard Wright Whirlwind engine. When it is realized that one, two, or three of these noise-producing units must be placed in the close vicinity of the passengers, it is apparent that noise reduction in airplanes is a most difficult undertaking.

The following list covers the handicaps under which the acoustical engineer must work to reduce the noise in aircraft:

(1) The airplane is the fastest vehicle, and hence inherently the noisiest

(2) The acoustical disturbances are very near the travelers (engine, propeller, exhaust)

(3) Frequency analyses show that the noise of an airplane is of a very complex character, containing sounds of

In April, 1932, the Sperry Gyroscope Company started on the study of noise reduction in airplanes. Before any program of research could be established, it was necessary to determine what had already been done or written in this field, and then to determine for ourselves the sound levels in existing airplanes.

Available data were very few. The work of Dr. W. S. Tucker in England (A.M.I.E.E., January, 1928) and the Bureau of Standards (Research Paper No. 63, 1929) in this country were about the only valuable contributions that we found. The former showed the important part played by the propeller in noise production and the necessity of slow tip speeds. The latter showed the extreme difficulty of obtaining a good sound

[7] Bureau of Standards Research Paper no. 63, 1929.

FIG. 4   NOISE METER REBUILT FOR AIRPLANE USE

107

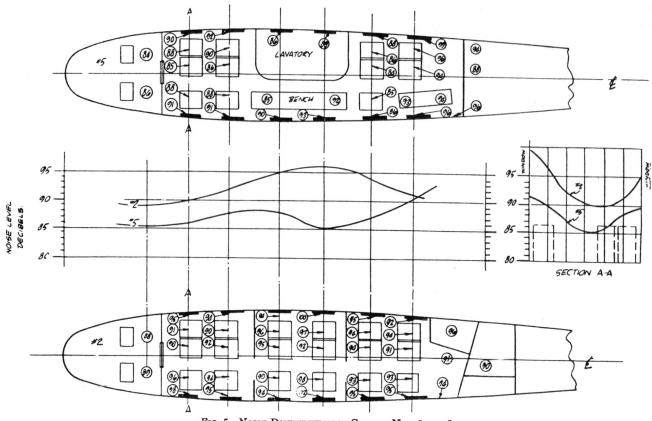

FIG. 5   NOISE DISTRIBUTION IN CONDOR NOS. 2 AND 5

insulator without excessive weight; in fact, it proved in a very disconcerting way that most sound insulators were only as good as they weigh.

The years 1930 and 1931, apparently on account of the depression, produced nothing in this field. So 1932 started from scratch. We first checked up a number of airplanes, paying particular attention to large transport planes. The tests were carried out either by the use of a noise meter (see Fig. 4), system Free, type 123, a commercial apparatus built by Dr. E. E. Free to our specifications for airplane use, or by the tuning-fork method. In many cases both methods were used to form a useful check. Generally, the tests were carried out at cruising speed, but in some cases at varying rpm, the latter check to establish relations of speed versus noise level. We took at least 20 to 40 readings in selected places along the cabin, the arithmetical mean serving as a basis of comparison of the noises of one airplane against the other.

As the research program was started in cooperation with Eastern Air Transport, Inc., we checked their ships first. This system owns a fleet of five Curtiss Condors, built about 1927 and 1928, which were scheduled for a complete overhaul that was to include acoustical treatment. Tests were made before and after this rebuilding. The Condor is an example of a relatively quiet ship. It probably was the quietest transport at the time it was put in service. The following summary of our tests will give a good idea of the noise levels encountered in this type of ship:

The E.A.T. Condor No. 2, an 18-passenger biplane, with Conqueror V-12 engine, geared three-blade metal propeller, speed about 110 mph. The exhaust of the inside bank of cylinders was piped on the near side of the nacelle. The average noise level (arithmetical mean of 31 readings) was 98.5 db; the highest was 102 db, near the fourth window from the front of the ship; the lowest was 88 db, in the pilot's cockpit.

The E.A.T. Condor No. 5, which was the same as No. 2, with the exception that here the exhaust was piped on the far side of the nacelle. The average noise level (arithmetical mean of 36 readings) was 94.5 db; the highest was 99 db, near the last window; the lowest was 86, in the pilot's cockpit.

The pitch in both ships was decidedly in the lower octaves, lower than 150 cycles, but a great variety of high-pitch noise was also present, coming from loose accessories such as ash trays, lamp shades, etc. Hissing sounds, about 2000 to 5000 cycles, came from the individual ventilators.

Before proceeding further with other tests, let us now analyze the two ships. Fig. 5 shows a longitudinal and a transverse curve of noise distribution.

It is at once apparent that one side of either ship is more quiet than the other. The reason is very simple. It is more quiet, as it should be, on the side where there are two seats in a row. The absorptive area is greater; hence this side is less noisy. The fact that No. 5 ship is generally quieter than No. 2 is due mainly to the exhaust manifold arrangement. In No. 5, the exhaust is farther away from the side of the ship and is concealed behind a large and heavy mass, the engine. Therefore the noise cannot reach the walls of the cabin as easily as on No. 2, where it is on the near side. Both ships show a decided drop in noise level in the pilot's cockpit, which is not at all what one would expect. Upon analysis, it is not extraordinary if one considers the fact that the cockpit of the Condor is far forward of the propeller and engine, the chief noise makers.

The Kingbird, with two radial direct-drive engines and a two-blade propeller, a six-place high-wing monoplane, speed about 115 mph. At the time of the tests, this particular ship had double windows. The highest reading was 100 db, in the pilot's cockpit and first right seat; the lowest, 94 db, rear left seat; the average, 99 db.

An interesting observation can be made here regarding the noise distribution. On this particular ship, the propellers and the engines are quite ahead of the cabin proper and rather near the cockpit. Consequently, the center of disturbance is ahead of the passengers and the noise decays about 6 db when measured in the rear of the airplane. Probably the most striking and important observation that was made on this airplane was on a retest of the same airplane after removing the double windows. With single windows, the average noise level was still 99 db. The double windows added weight, but were totally ineffective. At first this seems paradoxical, as it is known from the experience of architectural acoustics that double windows will materially quiet down a room where they are properly installed.

FIG. 6 SHEAR-TYPE SHOCK ABSORBERS AT THE ENGINE MOUNT

Let us now compare the two cases. In a building located on a busy thoroughfare, we have an outside noise level of about 65 db. The transmission loss of the masonry in a well-constructed building may be as high as 40 db, thus bringing the room without windows down to a noise level of 25 db. According to Sabine, a single window will have a transmission loss of about 30 db. So, if a single window is inserted in the wall of the room, more noise would be transmitted through it than through the wall, the room noise level would increase to 35 db, and the window would be properly blamed. If we install a double window of such characteristics that it will have a transmission loss of 40 db, then the wall and window will have a uniform opacity to sound and the room will show a remarkable improvement. Now, with the airplane, the case is reversed; the window is a better sound insulator than the thin cabin walls. The best cabin wall will have a transmission loss of about 25 db, so that no matter how opaque the windows are, the sound will come through the walls. Instead of using double glass for a local sound insulator (weighing as much as 3 lb per sq ft), if the same weight is spent for absorbing materials within the cabin, much better results will be obtained. The result with double windows against single windows has been checked by the authors in the case of another airplane and fully confirmed our idea of the fallacy of this construction.

We tested next the Ford Tri-Motor. This airplane, no longer in production, is an example of what complete neglect of acoustic considerations may produce. The highest reading was 118 db in the pilot's cockpit, the highest reading we have obtained in a passenger ship. The lowest was 104, near the rear. The average was 108.5. Prolonged exposure to this high level of sound can temporarily injure the sense of hearing. Ringing of the ears and physical discomfort persist for several hours after being subjected to this noise level. A long period of subjection to this level causes permanent hardness of hearing.

This airplane had no soundproofing whatsoever. This is especially true in the inside of the cabin, where not a piece of fabric or other fibrous material is introduced, and the noise reverberates constantly. Further, no attempt has been made on this airplane to prevent vibration from the engines reaching the passengers. Thus, noise is transmitted in the cabin by solid conduction. The propellers, being of the two-blade type, create a greater amount of noise than would a three-blade one. The center exhaust manifold runs under the floor of the airplane, where its pulses can be directly transmitted to the passenger compartment.

It has been further observed on another Ford Tri-Motor that the noise level was about the same as on the former. Further, the pitch is between 300 and 700 cycles, thus corresponding to the average pitch of human speech, and consequently no conversation is possible.

A different type of Tri-Motor airplane was also tested, and it was found that the noise level was around 100 db. This particular ship has its engine mounted by means of rubber shock absorbers. While no particular scientific attempt was made in soundproofing, the noise is less than in the Ford Tri-Motor, which is attributable to the discontinuity of surface between the engine and the fuselage. Owing to the very flexible engine mounts, the propellers precessed and created highly unpleasant beats, some of them lasting as long as 4 sec. A later type of the same airplane shows a certain improvement. The noise level was about 96 db and the pitch between 400 and 600 cycles, and in this airplane conversation was possible by shouting.

A very modern high-speed low-wing all-metal monoplane with a single-geared two-blade propeller was tested. This airplane was one of the first commercial high-speed airplanes in which an attempt was made to sound-insulate the ship scientifically. The highest noise level at a speed of 175 mph was about 96 db, the lowest was 86 db, and the noise distribution was unusually uniform. This ship at the original test had double windows. When these were removed, no appreciable decrease in noise level was experienced.

One of the authors was fortunate enough to fly from London to Paris in the new 44-passenger Handley-Paige Heracles. This is a four-engine biplane of a speed of about 90 mph, in which the cabin is divided into two compartments—one far ahead of the plane of the propellers and the other far behind. The space of the greatest acoustical disturbance is reserved for mail, for cargo, and the steward's compartment. This airplane uses four-blade wooden propellers, and was the quietest airplane flying up to 1933, as far as we know. The noise level was found to be around 78 db. Of course, when saying that it was the quietest airplane, we must make reservation that this type of airplane would be entirely unsuitable for American air transportation, where speeds of 90 mph are considered obsolete. The design of this ship is of great interest in showing how much may be accomplished by merely removing the passengers as far as possible from the plane of the engines and propellers.

A pair of wooden propellers were built and tested on the E.A.T. Condor No. 4. First, we tested it with the standard three-blade and then with the four-blade. The average noise with the three-blade propellers was 96 db, and with the four-blade 92.5 db, a reduction which our calculations predicted.[8] With the wooden propeller, we found that beats and momentary vibrations due to sudden changes of azimuth lasted but 2 to 3 sec, while with the metal propeller we could hear and feel them as long as 6 to 8 sec. This is again natural, if one considers the fact that the damping constant[9] for wood is 0.022, while that for aluminum is 0.0034 and for steel 0.0023. Hence, any vibration will be damped out quicker in a wooden propeller than in a metal one.

Numerous other airplanes and special tests have been made by the authors during the past year, but the foregoing examples have all been selected to show certain factors of the problem and to make the following list of principles more easily understandable. From these tests, we can conclude certain basic features to be included in the design which will make an airplane inherently more quiet.

(1) The engines should be mounted as far ahead of and as far away from the cabin as practical, consistent with good aerodynamic efficiency

(2) Three- or four-blade slow-moving propellers are better acoustically than two-blade fast-moving propellers

(3) Great care must be exercised to prevent the transmission of vibration to the cabin by solid conduction; thus, vibration control through the use of properly designed vibration insulators is imperative

(4) Double windows are at present of no value in an airplane cabin

(5) The cargo compartment should be located preferably in such a position as to form a barrier to the source of noise

(6) Exhaust manifolds should be located at a point as far removed from the cabin as possible, and either wing or nacelle should be used as a sound shadow between exhaust and the cabin

(7) The cabin should be as tight as possible, as small leaks will off-set the value of elaborate acoustical treatment

(8) Ventilating and heating systems should be made quiet by filtering out the noise before admitting the air to the cabin

(9) The inside of the cabin should be covered with a material which will absorb as much noise as possible; the material should be particularly

effective in the absorption of the predominant frequency of that particular airplane

(10) All accessories, such as lamp shades, ash trays, curtain rods, etc., should be fastened very securely to the structure to prevent acting as localized sound sources

(11) Avoid the use of large unsupported panels of thin section, as they act as loud-speaker diaphragms

(12) Blanket materials which will pack are not suitable and should be avoided.

While all of the foregoing points are self-evident, the last one needs amplification. Some sales people have products which are excellent when used under different conditions than on the airplane. Not knowing what vibration and constant changes of acceleration will do to this material, they convince the airplane manufacturer to use it.

Recently, in rebuilding one of Eastern Air Transport's "Condors" (1928 model), it was interesting to discover what time and vibration had done to the old sound-insulating mattress when the fabric was removed. We found that about 75 per cent of the blanket was pulverized and packed on the sides and bottom of the airplane, leaving whole areas without treatment. Therefore, only materials which will not pack should be used on airplane soundproofing work.

And so Condor No. 3 was rebuilt, and we made the cabin a tight shell, using a newly developed material called Onazote,[10] an extremely light exploded and molded rubber compound (weight 4.75 lb per cu ft, impact strength 75 lb per sq in.). Further, we endeavored to improve the reverberation control by the use of certain fabrics and a fluffy backing of same. The ventilating system was also changed, and all doors were gasketed.

[10] The inventor of this material is Mr. C. L. Marshall, Montclair, N. J. The material, while excellent in many respects, is not available commercially as yet.

[8] Davis, in his paper on "Noise," *Jl. Royal Aero. Soc.*, vol. 36, no. 224, claims that a decrease of 100 fps in tip speed will give a reduction of 10 db of the noise generated by the propeller. The N.A.C.A. has worked out the relation $db_{prop.} = 40$ to $60 (\log_{10} V_2/V_1)$, where V_1 and V_2 are the tip velocities.

[9] See Kimball, "Vibration Prevention in Engineering," J. Wiley, p. 133; S. J. Zand, "Vibration of Instrument Boards and Airplane Structures," *S.A.E. Journal*, November, 1932.

Fig. 7 View of the Cabin Before Interior Was Put In

FIG. 8   FINISHED CABIN

Being an old ship, it was impossible to change the structure, and naturally the results were far from perfect. Nevertheless, the noise level was brought down from 97 to 85 db in the front of the cabin, the center of the cabin showing an even greater improvement. This was the first ship in which it was possible to talk without much difficulty.

Since that time, other Condors have been rebuilt, using similar methods but different materials, with very gratifying results. On most of the ships it was possible to reduce the noise level to about 85 to 88 db.

When plans were made for the new Condors, Model XT-32, we were retained for the acoustical work at the beginning of the project. Thus, it was possible to incorporate in the basic design many features contributing to quietness. The location of the propeller was chosen so that its plane passes between the cockpit and the cabin proper, intersecting the radio compartment. This room is separated from the cabin by a bulkhead and fully gasketed door of generous dimensions. This bulkhead was treated with an acoustical material which would be particularly suited to absorb frequencies generated by the propeller. The engines were mounted on rubber shock absorbers of such characteristics

and so located as to give the power-plant assembly the lowest natural frequency far removed from the operating range. We used shear-type rubber shock absorbers, which give the maximum vibratory isolation with minimum weight (see Fig. 6). The whole cabin structure is insulated from the main structure by means of specially designed felt shock absorbers (see Fig. 7). The cabin was built of a panel made of a combination of two materials giving a transmission loss of about 25 db at 300 cycles, and weighing less than 0.4 lb per sq ft. The windows are floating, which was accomplished by the use of a radically different construction in mounting the pane to the frame. Further, we used a special non-shatterable glass which was made in accordance with our specifications. This glass is lighter, acoustically as efficient as any standard glass, and its vibration-damping coefficient is ten times higher than ordinary safety glass. The result is that these windows do not "sing."

The floor was not specially insulated because the baggage and mail compartment was located underneath it, and this has such considerable mass that it acts automatically as a sound barrier. The cabin framework was insulated from the main structure by means of a felt shock absorber (see Fig. 7).

The ventilating systems, both intake and exhaust, were provided with acoustical filters of such characteristics as to attenuate the predominant frequencies.

The reverberation control included such refinements as the making of the parcel rack of a blanket material which acts as an efficient sound absorber (see Fig. 8).

The exhaust manifolds, tuned to unison, were located under the lower wing, their clamps being insulated from the structure by asbestos gaskets.

Attention was paid to all details, and the result was a noise level of about 78 db at cruising speed of 145 to 150 mph with four passengers and no baggage. Fully loaded, the noise level dropped to about 76 db (at 150 mph), or no more than in the average railroad car (at 60 mph). Conversation is fully possible, and radio loud speakers installed in the cabin can be distinctly heard and enjoyed by the passengers.

For quick reference, all the ships tested have been listed in Table 1, which gives also certain basic features of these ships.

TABLE 1   DIFFERENT AIRPLANES COMPARED

(Noise level above 1 mb)

No.	Airplane	Power plant	Speed, mhp	Average noise level, db	Soundproofing material	Conversation level	Remarks
1	Condor No. 2	Conqueror 2-geared 3-blade propeller	110	98.5	Dry zero blanket	Possible with difficulty	Test before overhaul
2	Condor No. 2	Same	110	86.0	Seapack-plywood and Sperry soundproofing	Possible with normal voice up to 3 to 4 ft	After overhaul. The 86 db is with ventilators closed; when open, 92 db
3	Condor No. 3	Same	110	96.5	Dry zero blanket	Possible with certain effort	Before overhaul
4	Condor No. 3	Same	110	85.0	Onazote and Sperry soundproofing	Possible with normal voice up to 3 to 4 ft	After overhaul; added weight 285 lb
5	Condor No. 5	Same	110	94.5	Dry zero blanket	Possible with certain effort	Test before overhaul
6	Kingbird	2 Wright J-6, direct 2-blade	105	99.0	Dry zero blanket	Possible with great difficulty	Test after overhaul
7	Kingbird	Same	105	100.0	Dry zero blanket	Possible with great difficulty	Test before overhaul
8	Condor No. 4	2 Conquerors, 3-blade geared propeller	112	95.0	Dry zero blanket	Possible with difficulty	....
9	Condor No. 4	2 Conquerors 4-blade wooden propeller	107	92.0	Dry zero blanket	Possible with difficulty	Decidedly different pitch: less vibration
10	Ford	Wasp 3 direct, 2-blade	112	110.0	None; no reverberation control	Impossible, not even shouting	Highly unpleasant, vibration excessive
11	Ford	Same	115	106.0	None; no reverberation control	Impossible, not even shouting	Highly unpleasant; vibration excessive
12	A tri-motor	3 radials, direct, 2-blade	105	100.0	Unknown; no reverberation control	Possible with great difficulty	....
13	A tri-motor	Same	110	98.5	Seapack; poor reverberation control	Possible with effort	....
14	A high-speed low-wing monoplane	Radial geared 2-blade propeller	165	88.0	Downy material; poor reverberation control	Possible with little effort	Spectrum of noise in the neighborhood of human speech; ventilating system noisy
15	A high-speed low-wing monoplane	Same	190	90.0	Same	Possible with some effort	With improved ventilating and cabin interior this airplane could be made about 10 db more quiet
16	Curtiss Condor XT-32	2 geared Cyclones, 3-blade propeller	150	76.0	Seapack, insulite, special cement, Sperry soundproofing	Possible in normal tones; with raised voice, full length of the cabin	The most quiet ship flying at 150 mph

7

Excerpted from National Advisory Committee for Aeronautics Technical Memorandum 1195, October, 1948, 21p.

ON THE SOUND FIELD OF
A ROTATING PROPELLER*

L. Gutin

The sound field of a rotating propeller is treated theoretically on the basis of aerodynamic principles. For the lower harmonics, the directional characteristics and the radiated sound energy are determined and are in conformity with existing experimental results.

1. INTRODUCTION

A rotating propeller produces periodic disturbances of the medium which cause a sound of low frequency. The fundamental tone of this sound equals the product of the number of blades and the number of revolutions per unit time, since each spatial configuration of the propeller is repeated with exactly the same frequency. The present work aims at a theoretical investigation of the sound field produced by the propeller. First a short report on earlier investigations in this field will be given.

The directional properties of propeller sound were first investigated theoretically by Lynam and Webb.[1] They suggested two hypotheses on the acoustic action of the propeller. Both hypotheses were based on actually arbitrary assumptions and led to directional characteristics (cf. figs. 1 and 2) which are not in agreement with the observed ones (cf. figs. 3 and 4).

---

* "Uber das Schallfeld eimer rotierenden Luftschraube" Physikalische Zeitschrift der Sowjetunion, Band 9, Heft 1, 1936, pp. 57-71. English translation printed in National Advisory Committee for Aeronautics, Technical Memorandum No. 1195, October, 1948.

1. Lynam, E.J. and Webb, H.A., The Emission of Sound by Airscrews, R. & M. No. 624, 1919. (Two theories of Lynam and Webb are known to us only through the treatment of Paris and Kemp since the original report was unfortunately not accessible.)

Hart[2] attempted to develop a theory which was supposed to be free of arbitrary assumptions; however, he tacitly made the same assumption as Lynam and Webb in their second hypothesis; namely, that each disturbed element in the propeller plane acts as a simple nondirectional source; accordingly, he obtained the same result.

Experimental investigations were undertaken by Paris[3] and Kemp.[4] Both experimented with a two-blade propeller; its diameter was 4.5 meters, the number of revolutions 13.9 per second. It is true that a noticeable discrepancy exists between their results in quantitative respect; essentially, however, they agree. Both directional characteristics show an asymmetry with respect to the rotational plane, a principal maximum at about $\psi = 115°$, and a secondary maximum at $\psi = 40°$. This peculiar asymmetry contradicts both hypotheses by Lynam and Webb.

Paris obtained, by a combination of the two hypotheses, a directional characteristic (cf. fig. 6) which reproduces the main outlines of the experimental curves; however, his method lacks physical foundation, and the number of arbitrary assumptions is too large.

The hypotheses mentioned above did not yield any information on the sound output of the propeller. This problem was experimentally examined by Kemp. He obtained the values 17.8 watts for the first harmonic and 7.4 watts for the second.[5]

In the following, a theory based on aerodynamic principles shall be developed, which determines quantitatively the sound field of a propeller.

2. SOME AERODYNAMIC RESULTS

A few facts concerning the action of the propeller, which are known from aerodynamics, shall be briefly quoted.[6]

The cross section of a propeller blade looks very similar to that of a wing (of fig. 7); the action of the propeller is based on this fact. Motion of a wing element relative to the medium causes, exactly as for a wing, a pressure increase on the concave side and a pressure

2. Hart, M.D., The Aeroplane as a Source of Sound, R. & M. No. 1310, 1929.

3. Paris, E.T., Phil. Mag. vol. 13, no. 99, 1932.

4. Kemp, S.F., Proc. Physical Soc. (London), vol. 44, pt. 2, p. 151.

5. In the report by Kemp values half as large as these are given by mistake. Compare Paris, E.T., Phil. Mag. vol. 16, no. 60, 1933.

6. Compare with this paragraph Mises, R.V., The Theory of Flight, pp. 116-121.

decrease on the convex side. The resultant air force may be separated into two components of which one has the direction of the propeller axis and indeed is a thrust in the direction of the forward motion; the other one is opposed to the rotation and represents a drag force. The thrust forces of single elements add up to the total thrust force of the propeller, the drag forces result in a torque opposed to the torque of the motor.

A few important quantitative results which will be used subsequently are derived from the so-called momentum theory of the propeller. The propeller is assumed to be at rest and the air moving towards it from the front, opposed to the flight direction, with the velocity V. The effect of the propeller then consists in an acceleration of the approaching mass of air which finally assumes at a certain distance behind the propeller the velocity V + w. Let ρ be the density of the medium, P the magnitude of the thrust force, S the area described by the propeller, W the power supplied to the propeller, $W\eta$ the part of the power converted into translational energy. Then the following relations hold:

$$P = \rho S \left[V + \frac{w}{2} \right] w$$

$$W\eta = \rho S \left[V + \frac{w}{2} \right]^2 w$$

By eliminating w from both equations one obtains an expression for the dependence of the thrust on the power. For static conditions (V = 0) there results:

$$P = \sqrt[3]{2\rho S W^2 \eta^2} \tag{1}$$

The obvious relation

$$M = \frac{W}{\alpha} \tag{2}$$

shall be mentioned. (M = torque of the motor, α = angular velocity of the rotation.)

3. DISTURBANCE FORCES IN THE PROPELLER PLANE

Two forces act on each element of the propeller: a thrust and a drag force. From the well-known theorem of mechanics, it follows that each element exerts forces of equal magnitude and in the opposite direction upon the medium. The points of application of these forces may be imagined concentrated in one plane since the axial dimensions of the propeller are very small in comparison with the wave lengths of the first harmonics; this plane is subsequently denoted as the plane of rotation.

A propeller element will be considered, the distance of which
from the axis equals R; let dR be its radial length, a its width,
measured in the projection on the rotational plane. Let the forces
exerted upon the medium by the element be A(R) dR (in the direction of
the axis, opposed to the flight direction) and B(R) dR (in the direction of the rotation).

It is clear that

$$n \int_0^{R_0} A(R) \, dR = \int_0^{R_0} dP = P$$

$$n \int_0^{R_0} B(R)R \, dR = \int_0^{R_0} dM = M$$

n representing the number of blades, R_0 the length of the blade.

Let it first be assumed that the forces are uniformly distributed
over the width of the element.

In the element R dR dθ of the rotational plane, forces
$A(R) \, dR\frac{R \, d\theta}{a}$ and $B(R) \, dR\frac{R \, d\theta}{a}$ act on the medium during the time interval
in which this element is covered by the projection of the propeller
element. If the overlapping starts at the time t = 0, it will be
ended at time $t = \tau = \frac{a}{R\alpha}$, to return after $t = T = \frac{2\pi}{n\alpha}$. These periodical forces may be developed in Fourier series

$$F_1(t) = \begin{cases} A(R)\dfrac{R \, d\theta}{a}dR & (0 < t < \tau) \\ 0 \quad . \; . & (\tau < t < T) \end{cases} = \sum_{m=1}^{\infty} A_m \cos(mn\alpha t - \epsilon_m) + A_0$$

$$F_2(t) = \begin{cases} B(R)\dfrac{R \, d\theta}{a}dR & (0 < t < \tau) \\ 0 \quad . \; . & (\tau < t < T) \end{cases} = \sum_{m=1}^{\infty} B_m \cos(mn\alpha t - \eta_m) + B_0$$

One obtains

$$A_m = \frac{2}{m\pi} A(R) \frac{R}{a} \sin\left[m\pi\frac{\tau}{T}\right] dR\, d\theta$$

$$B_m = \frac{2}{m\pi} B(R) \frac{R}{a} \sin\left[m\pi\frac{\tau}{T}\right] dR\, d\theta$$

(3)

In a second area element $R\, dR\, d\theta$, shifted with respect to the first by the angle θ in the rotational direction, there act periodical forces of the same magnitude but retarded by the time $\frac{\theta}{\alpha}$. The corresponding Fourier developments are

$$F_1\left[t - \frac{\theta}{\alpha}\right] = \sum_{m=1}^{\infty} A_m \cos(mn\alpha t - mn\theta - \epsilon_m) + A_0$$

$$F_2\left[t - \frac{\theta}{\alpha}\right] = \sum_{m=1}^{\infty} B_m \cos(mn\alpha t - mn\theta - \eta_m) + B_0$$

For the first harmonics $\frac{m\pi\tau}{T} = \frac{mna}{2R}$ is usually small (the blade parts lying close to the center where R is small are eliminated since they make almost no contribution to the air forces), and one may equate

$$\sin\left[\frac{m\pi\tau}{T}\right] = \left[\frac{m\pi\tau}{T}\right]$$

One then obtains

$$A_m = \frac{n}{\pi} A(R)\, dR\, d\theta$$

$$B_m = \frac{n}{\pi} B(R)\, dR\, d\theta$$

(4)

It shall be shown now that these expressions are valid also when the air forces are not uniformly distributed over the blade width. Using for instance for the distribution of thrust the relation

$$A(R)\, dR = \int_0^a A(R)\, dR\, f(s)\, \frac{ds}{a}$$

with

$$\int_0^a f(s)\, ds = a$$

116

(s is counted from the leading edge of the blade), one obtains for the Fourier coefficients the expression

$$A_m = \frac{2}{T} A(R) \frac{R}{a} \, dR \, d\theta \left| \int_0^\tau f\left[\frac{at}{\tau}\right] e^{-i2m\pi t/T} \, dt \right|$$

For the first harmonic

$$\left| \int_0^\tau f\left[\frac{at}{\tau}\right] e^{-i2m\pi t/T} \, dt \right| \cong \left| \int_0^\tau f\left[\frac{at}{\tau}\right] \, dt \right| \,^7$$

From $\int_0^a f(s) \, ds$, it follows that

$$\int_0^\tau f\left[\frac{at}{\tau}\right] \, dt = \tau$$

and hence

$$A_m = \frac{n}{\pi} A(R) \, dR \, d\theta$$

For ϵ_m one obtains

$$\epsilon_m = \arctan \frac{\int_0^\tau f\left[\frac{at}{\tau}\right] \sin 2m\pi\frac{t}{T} dt}{\int_0^\tau f\left[\frac{at}{\tau}\right] \cos 2m\pi\frac{t}{T} dt} < \arctan \frac{\sin 2m\pi\frac{\tau}{T}}{\cos 2m\pi\frac{\tau}{T}} = 2m\pi\frac{\tau}{T}$$

4. THE SOUND FIELD

The point of origin of coordinates is assumed at the propeller center, the y-axis and z-axis in the rotational plane (cf. fig. 8). The x-axis is assumed to coincide with the flight direction. In the rotational plane polar coordinates are introduced, with the polar angle being counted from the y-axis, in the direction of the rotation. Let it first be assumed that the rotation, observed from the flight

---

7 A more accurate estimation gives (for $\frac{2m\pi\tau}{T} < \frac{\pi}{2}$)

$$\int_0^\tau f\left[\frac{at}{\tau}\right] \, dt \sqrt{\cos^2 2m\pi\frac{t_1}{T} + \sin^2 2m\pi\frac{t_2}{T}}$$

with t_1 and t_2, representing certain mean values. One can name distributions for which $t_1 \cong 0$ and $t_2 \cong \tau$, but for the distributions occurring in practice the difference between t_1 and t_2 is much smaller.

direction, is counter clockwise. The forces acting on the medium in an element R dR dθ are (for the first harmonics)

$$X = - \frac{n}{\pi}A(R)e^{i(kct-mn\theta-\epsilon_m)} dR\ d\theta$$

$$Y = -\frac{n}{\pi}B(R)\ \sin\theta\ e^{i(kct-mn\theta-\eta_m)} dR\ d\theta \qquad (5)$$

$$Z = \frac{n}{\pi}B(R)\ \cos\theta\ e^{i(kct-mn\theta-\eta_m)} dR\ d\theta$$

If one assumes, for reasons of simplification, that at the time t = 0 the center line of one of the blades coincides with the positive y-axis, ϵ_m and η_m will be at any rate smaller than $\frac{2m\pi\tau}{T}$, therefore small, at least for the lower harmonics.

The velocity potential produced by a concentrated force with the components X, Y, Z is[8]

$$\Phi = - \frac{i}{4\pi k\rho c} \left[X\frac{\partial}{\partial x} + Y\frac{\partial}{\partial y} + Z\frac{\partial}{\partial z}\right] \frac{e^{-ikr}}{r} \qquad (6)$$

If one inserts the expressions for the force components and notes that

$$\frac{\partial}{\partial x}\left[\frac{e^{-ikr}}{r}\right] = \frac{\partial}{\partial r}\left[\frac{e^{-ikr}}{r}\right] \cos\psi$$

$$\frac{\partial}{\partial y}\left[\frac{e^{-ikr}}{r}\right] = \frac{\partial}{\partial r}\left[\frac{e^{-ikr}}{r}\right] \cos\chi$$

$$\frac{\partial}{\partial z}\left[\frac{e^{-ikr}}{r}\right] = \frac{\partial}{\partial r}\left[\frac{e^{-ikr}}{r}\right] \cos\nu$$

(ψ, χ, ν are the angles that the radius vector r makes with the axis), one obtains for the total velocity potential

$$\Phi = \frac{in}{4\pi^2\rho ck} \int_0^{R_0} \int_0^{2\pi} \left[A(R)e^{i(kct-mn\theta-\epsilon_m)} \cos\psi \right.$$

$$\left. + B(R)e^{i(kct-mn\theta-\eta_m)}(\cos\chi\ \sin\theta - \cos\nu\ \cos\theta)\right] \frac{\partial}{\partial r} \frac{e^{-ikr}}{r} dR\ d\theta \quad (7)$$

8 Compare Lamb, H., Textbook of Hydrodynamics, Teubner 1931, p. 567. A concentrated force Fe^{ikct} is therefore equivalent to an acoustic double source of the strength $\frac{iF}{kc\rho}$.

118

The point at which the force acts is assumed to lie in the xy-plane which can always be attained by a suitable choice of the system of coordinates y,z in the rotational plane; let it further be assumed that r is large. Then it follows that:

$$\cos \chi = \sin \psi; \quad \cos \nu = 0$$

$$r = r_1 - R \cos \theta \sin \psi$$

$$\frac{\partial}{\partial r} \left[\frac{e^{-ikr}}{r} \right] \approx -ik\frac{e^{-ikr}}{r} \approx -ik\frac{e^{-ikr_1}}{r_1} e^{ikR \sin \psi \cos \theta}$$

After substituting into equation (7) and carrying out the integration with respect to θ one obtains

$$\Phi = \frac{i^{mn}n \, e^{ik(ct - r_1)}}{2\pi\rho c \, r_1} \int_0^{R_0} dR \left[A(R)e^{-i\epsilon_m} \cos \psi \, J_{mn}(kR \sin \psi) \right.$$

$$\left. - B(R)e^{-i\eta_m}\sin \psi \, \frac{J_{mn-1}(kR \sin \psi) + J_{mn+1}(kR \sin \psi)}{2} \right]$$

$$= - \frac{i^{mn}n \, e^{ik(ct - r_1)}}{2\pi\rho c \, r_1} \int_0^{R_0} \left[-A(R)e^{-i\epsilon_m} \cos \psi \right.$$

$$\left. + \frac{mn}{kR}B(R)e^{-i\eta_m} \right] J_{mn}(kR \sin \psi) \, dR \qquad (8)$$

Note: $\int_0^{2\pi} e^{iz \cos \Phi - im\Phi} \, d\Phi = 2\pi i^m J_m(z)$; where J_m = Bessel function of the first kind of the mth order.

One can easily satisfy oneself that the expression for Φ remains unchanged in the case of opposite direction of rotation.

119

For the magnitude of the sound pressure amplitude there results

$$P = \rho \left| \frac{\partial \Phi}{\partial t} \right| = \frac{kn}{2\pi r} \left| \int_0^{R_0} \left[-A(R) e^{-i\epsilon_m} \cos \psi \right. \right.$$

$$\left. \left. + \frac{mn}{kR} B(R) e^{-i\eta_m} \right] J_{mn}(kR \sin \psi) \, dR \right|$$

If one inserts

$$A(R) \, dR = \frac{dP}{n}$$

$$B(R) \, dR = \frac{dM}{nR}$$

$$k = \frac{m\omega_1}{c}$$

(ω_1 = circular frequency of the fundamental tone), P is

$$P = \frac{m\omega_1}{2\pi cr} \left| \int_0^{R_0} \left[-\frac{dP}{dR} e^{-i\epsilon_m} \cos \psi + \frac{nc}{\omega_1 R^2} \frac{dM}{dR} e^{-i\eta_m} \right] J_{mn}(kR \sin \psi) \, dR \right| \quad (9)$$

The sound power may be calculated according to the formula

$$W = \int_0^{\pi} \frac{p^2}{2\rho c} 2\pi r^2 \sin \psi \, d\psi \quad (10)$$

In order to execute in equations (8) and (9), respectively, the integration with respect to R, one would have to know the distribution of the thrust and drag forces along the blade; it can be determined for instance from aerodynamic model tests. However, it is still possible to approximately evaluate the integrals if the aerodynamic data are less detailed. Since ϵ_m and η_m are small and the Bessel functions occurring in the inegrand may be regarded for the first harmonics as monotonically increasing with R it follows on the basis of the mean value theorem that:

$$P = \frac{m\omega_1}{2\pi cr} \left[-P \cos \psi \, J_{mn}(kR_1 \sin \psi) + \frac{ncM}{\omega_1 R_2^2} J_{mn}(kR_2 \sin\psi) \right] \quad (11)$$

R_1 and R_2 are certain mean values.

If the number of blades is small, one can, for the first harmonics, put R_1 and R_2 about equal to the radius R_c of the circumference on which runs the point of application of the resultant thrust force of a single blade. In most cases it equals about $0.7 - 0.75 \ R_0$.

$$P = \frac{m\omega_1}{2\pi cr} \left[-P \cos \Psi + \frac{ncM}{\omega_1 R^2} \right] J_{mn}(kR \sin \Psi) \quad (R \cong R_c) \quad (12)$$

This contention shall be discussed for the case of a two-blade propeller.

In the integral

$$\int_0^{R_0} \frac{dP}{dR} e^{-i\epsilon_1} J_2(kR \sin \Psi) \, dR$$

one may equate

$$J_2(kR \sin \Psi) = \frac{1}{8} k^2 R^2 \sin^2 \Psi$$

One then obtains

$$\frac{k^2 \sin^2 \Psi}{8} \int_0^{R_0} \frac{dP}{dR} e^{-i\epsilon_1} R^2 \, dR = \frac{k^2 \sin^2 \Psi}{8} R_1^2 \int_0^{R_0} \frac{dP}{dR} e^{-i\epsilon_1} \, dR$$

$$\cong P \frac{k^2 R_1^2 \sin^2 \Psi}{8}$$

$$\cong P J_2(kR_1 \sin \Psi)$$

A comparison of the relations

$$\int_0^{R_0} \frac{dP}{dR} R \, dR = PR_c$$

and

$$\int_0^{R_0} \frac{dP}{dR} R^2 \, dR = PR_1^2$$

shows that

$$R_c < R_1 < \sqrt{R_0 R_c}$$

The upper limit is reached when the forces are distributed with the maximum of nonuniformity, namely, concentrated at the ends of the blade. For the distributions occurring in practice, there is only little difference between R_1 and R_c.

In the integral

$$\int_0^{R_0} \frac{dM}{R^2 dR} e^{-i\eta_1} J_2(kR \sin \psi)\, dR$$

the factor

$$\frac{J_2(kR \sin \psi)}{R^2}$$

is in first approximation fully independent of R and one can, therefore, substitute $R = R_1$; then

$$\int_0^{R_0} \frac{dM}{R^2 dR} e^{-i\eta_1} J_2(kR \sin \psi)\, dR = \frac{J_2(kR_1 \sin \psi)}{R_1^2} \int_0^{R_0} \frac{dM}{dR} e^{-i\eta_1}\, dR$$

$$\cong \frac{M}{R_1^2} J_2(kR_1 \sin \psi)$$

Thus one obtains for the fundamental tone of a two-blade propeller the following expression for the sound-pressure amplitude

$$P = \frac{\omega_1}{2\pi cr} \left[-P \cos \psi + \frac{2cM}{\omega_1 R^2} \right] J_2(kR \sin \psi) \tag{13}$$

(with R being slightly larger than R_c).

For the second harmonic generally $R_1 \neq R_2$; however, approximately on may assume also in this case

$$P = \frac{2\omega_1}{2\pi cr} \left[-P \cos \psi + \frac{2cM}{\omega_1 R^2} \right] J_4(kR \sin \psi) \tag{14}$$

(In this case, the difference between R and R_c is larger.)

5. COMPARISON WITH THE EXPERIMENTAL RESULTS

The theoretical results are now to be compared with the experimental data of Paris and Kemp. For this purpose, the values P and M for the propeller used by these investigators must be known; unfortunately, they are not given directly; however, they may be calculated approximately on the basis of other given data. Both investigators experimented with a two-blade propeller of diameter of 4.5 meters and a number of revolutions of 13.9 per second. The propeller operated under static conditions. If one assumes that the motor developed under static conditions its full power (600 hp), one obtains according to the formula (1)

$$P = 1690 \text{ kg}$$

(under the assumption of $\eta = 0.75$, a quite probable value).

For the torque one obtains

$$M = 515 \text{ kg m}$$

Using these numerical values, two directional characteristics were calculated for the fundamental according to the formula (13), corresponding to the two values $R = 0.7R_0$ and $R = 0.75R_0$ (cf. figs. 9 and 10).

For the second harmonic a directional characteristic was calculated according to the formula (14) for the value $R = 0.75R_0$) cf. fig. 11)

A comparison with the experimental directional characteristics (cf. figs. 3 and 4) shows that the main characteristics of the experimental curves for the fundamental tone are well borne out by the theory. The agreement with Paris' results is particularly good. For the second harmonic the agreement with Kemp's curve (fig. 5) is less satisfactory.

For the sound power results, according to the formula (10) by means of graphical and numerical integration, the values 34 W and 30 W for the fundamental corresponding to the values $R = 0.7R_0$ and $R = 0.75R_0$, and 6.5 W($R = 0.8R_0$) and 4.7 W($R = 0.75R_0$) for the second harmonic.

The agreement with Kemp's experimental values (17.8 W and 7.4 W) may be regarded as satisfactory if one takes into consideration, on the one hand, the approximate character of the present calculation of P and M, on the other, the possible experimental errors. (In this connection, the noticeable quantitative discrepancy between Kemp's and Paris' curves for the fundamental should be pointed out.)

Let it be noted that the formula (6) which forms the basis for the calculation of the sound field was derived from the equations of motion for small motions. This circumstance may lead to an underestimation of the higher harmonics. Perhaps this is the reason for the increasing discrepancy between the theory and Kemp's experimental values for the higher harmonics. However, a final decision of this problem requires further and more accurate experimental data.

Translated by Mary L. Mahler
National Advisory Committee
for Aeronautics

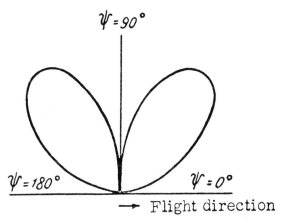

Figure 1. Calculated directional characteristic for the fundamental tone of a two-blade propeller according to Lynam and Webb (hypothesis 1).

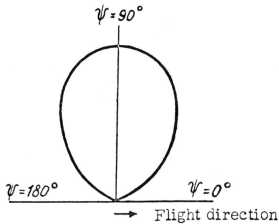

Figure 2. Calculated directional characteristic for the fundamental tone of a two-blade propeller according to Lynam and Webb (hypothesis 2).

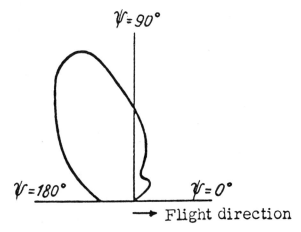

Figure 3. Experimental directional characteristic for the fundamental tone according to Paris.

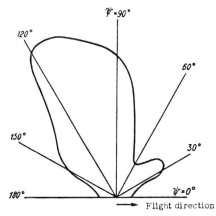

Figure 4. Experimental directional characteristic for the fundamental tone according to Kemp.

125

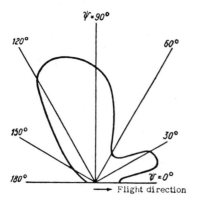

Figure 5. Experimental directional characteristic for the second harmonic according to Kemp.

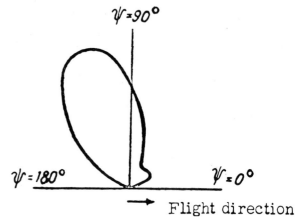

Figure 6. Calculated directional characteristic for the fundamental tone according to Paris (combination of the two hypotheses by Lynam and Webb).

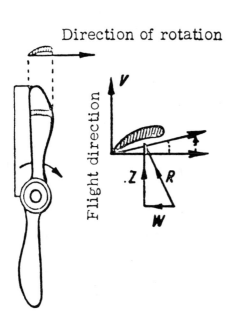

Figure 7.

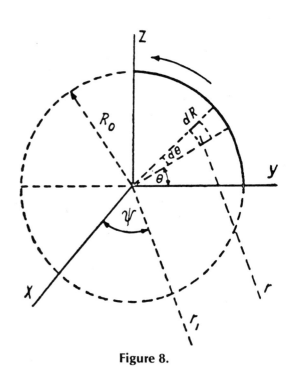

Figure 8.

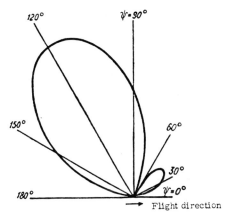

Figure 9. Calculated directional characteristics for the fundamental tone ($R = 0.7R_0$).

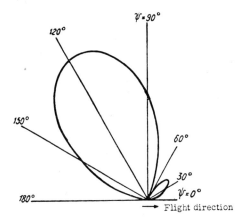

Figure 10. Calculated directional characteristic for the fundamental tone ($R = 0.75_0$).

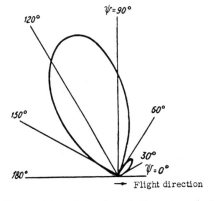

Figure 11. Calculated directional characteristic for the second harmonic.

Reprinted from *Acoust. Soc. Am. J.* **25**:367–378 (1953)

Physical Characteristics of Aircraft Noise Sources*

H. E. von Gierke

Aero Medical Laboratory, Wright Air Development Center, Wright-Patterson Air Force Base, Ohio

Available basic characteristics of different aircraft noise sources under the condition of zero forward speed are summarized: Total acoustic power; acoustic mechanical efficiency; directivity and frequency characteristics are given for the rotation and vortex noise of propellers, for the exhaust noise of reciprocating engines, and for different types of jet engines. The physical mechanisms underlying the different noise sources are discussed and the influence of changes in the parameters such as tip speed or pitch of the propeller and diameter or velocity of the gas jet are shown. In cases where measurements on the actual propulsion systems are incomplete, the basic physical conclusions are drawn from experiments on model airscrews and small air jets. The changes in the characteristics of the noise generators during flight are discussed briefly.

I. INTRODUCTION

THE physics of the generation of aircraft noise is a field in which acoustics must concede that it is actually a child of aerodynamics. This is true, not only because the aircraft thrust is a source of aircraft noise and therefore an aerodynamic one, but also because the noise generation mechanism is closely connected with aerodynamic phenomena. If one desires to study the noise, an unwanted by-product, one should consider it in relation to its cause, the propulsion, whereas for studying the aerodynamic effect, the simultaneous radiation of acoustic power is usually not considered important and is therefore neglected. On the other hand, as can be shown today for propulsion by propellers and as will probably be shown sometime for the pulse jet, one gains a more complete understanding of propulsion if one considers not only static forces, but also the influence of the acoustic radiation field on aerodynamic flow. One is still far from the goal of knowing the exact relation between the aerodynamic and the acoustic effect of the propulsion units or, more practically speaking, between sound level and thrust. But today one sees this goal clearly and has many valuable approaches. In the following short summary of present day knowledge pertaining to aircraft noise sources, an attempt is made to point out as clearly as possible how the characteristics of these noise sources can be deduced from the aerodynamic phenomena connected with propulsion. The possibilities for noise reduction at the source (the practical importance of which increases daily) as well as their limits are made clearer by such an approach. It is not the purpose of this survey to give complete quantitative data on all noise sources. Such data can be found in the cited references where they are discussed at length. It is our purpose to compare the basic physical characteristic of these noise generators and to show where our physical knowledge is still limited and where further work has to be done. The noise sources considered are: propellers; reciprocating engine exhausts; turbojet, ram jet, pulse

jet and rocket engines. In addition, the aerodynamic noise generated in the boundary layer surrounding the moving airplane, a source which is common to all types of airplanes in flight, is discussed.

II. THE CHARACTERISTICS OF AIRCRAFT NOISE SOURCES ON THE GROUND

Today's aircraft noise problems,[1-3] of primary importance, are related to static or low forward speed operation of the aircraft. Nearly all investigations on the noise produced by various propulsive units are made, therefore, during static operation on the ground. In order to characterize completely the noise field produced by any given noise source, one must know:

(a) How the total acoustic power radiated into the air is related to the mechanical and operational characteristics of the engine.
(b) How the total acoustic power output of an engine is distributed over the frequency spectrum.
(c) How the power of the different spectral bands is distributed in space; i.e., what is the directivity pattern as a function of frequency.

What is desired is the ability to calculate the complete sound field for a given engine, when the mechanical characteristics are available.

Usually, such a complete set of data is not considered necessary. Instead of (b) only a characteristic frequency spectrum is selected and given. This is then assumed to be approximately valid for all angles around the engine. Instead of (c) the directivity pattern for a few bands radiating major portions of the power are presented.

These characteristics are discussed briefly below and typical examples are given. Unfortunately, our knowledge is not complete and very often not all data, (a) through (c), are published for one and the same noise source. The graphs present, therefore, the general functions and orders of magnitude and do not represent a complete set of data, (a) through (c), obtained under the same conditions. For some of the noise

---

[1] R. O. Fehr, J. Acoust. Soc. Am. **24**, 772–775 (1952).
[2] Leo L. Beranek, J. Acoust. Soc. Am. **24**, 769–772 (1952).
[3] Harvey H. Hubbard, "A Survey of the Aircraft Noise Problem with Special Reference to its Physical Aspects," National Advisory Committee for Aeronautics, TN2701 (May, 1952).

---

*Presented by invitation during the Aircraft Noise Symposium which was part of the San Diego meeting of the Acoustical Society of America, November, 1952.

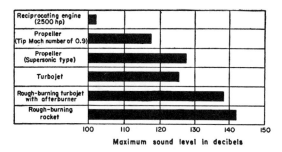

FIG. 1. Comparison of the maximum over-all sound levels generated by various propulsive units. (Data obtained from references 3 are adjusted to correspond to a thrust of 5000 lb and a distance of 300 feet.)

sources, the respective quantities have not been published.[4] A rough, general comparison of the maximum over-all sound levels generated by various propulsive units producing equal thrust is presented in Fig. 1.

(a) Total Acoustic Power

Approximate data on the acoustic power produced by the two most common noise sources are given in Figs. 2 and 3. The curves are calculated from measurements in the distant field near the ground, on various types of propellers and jet engines in use today.[5] The actual power radiated by the sound source might be somewhat higher than the one measured in the far field since high amplitude effects result in increased absorption in the near field. Figure 2 shows how the total acoustic output of a propeller increases with the Mach number of the propeller tip and with the power loading per blade. The maximum acoustic power is around 20 kw, which is about 10 percent of the total mechanical power delivered to the propeller. The acoustic power output becomes less and less sensitive to changes in horsepower and tip speed when these

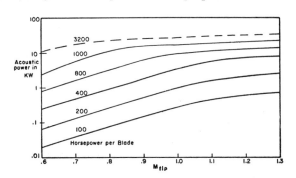

FIG. 2. Acoustic power output of propellers as a function of tip Mach number and horsepower per blade. (Read acoustic power directly for a two-blade propeller. Use the following factors to correct for other numbers of blades: 3 blades, times 1.5; 4 blades, times 2; 5 blades, times 2.5; 6 blades, times 3. Multiply the power obtained by 2 if the blade tips are blunt) (data from reference 5).

[4] See, however, pertinent new data in other papers presented in this Aircraft Noise Symposium.
[5] The staff, Bolt Beranek and Newman, *Handbook of Acoustic Noise Control*, Vol. I, "Physical Acoustics," Chapters 4–5 (WADC Technical Report 52–204) (to be published).

two quantities become very large. Figure 3 shows how the total acoustic output of ram jets and turbojets increases with their fuel consumption. The data show the best available approximation obtained by measurements on jets of different make and size. The maximum powers measured, to date, are in excess of 10 kw and are higher than those of the propeller. In the case of the jet, at full power, the acoustic power is only about 1 percent of the kinetic energy per unit time delivered by the gas jet and is about 0.2 percent of the available fuel power. Curves such as these can serve as starting points for all calculations of propeller and jet noise to obtain the space average sound levels that one needs for solving test cell or comfort level problems. Measurements on individual propellers and engines are given in Figs. 4 and 5. Here the acoustic-mechanical efficiency (i.e., the ratio of the acoustic power to the expended mechanical power) is plotted for different noise sources as a function of the operating conditions. It is very

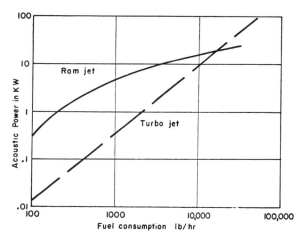

FIG. 3. Acoustic power output of turbojet and ram jet engines as a function of the fuel consumption at full power operation (data from reference 5).

instructive to see what percentage of the total mechanical energy produced or transmitted by the propulsion unit goes into noise and how this efficiency can be expected to increase as we go to engines of still higher power. Figure 4 shows the rapid increase with tip Mach number that was measured for the rotational sound of a model airscrew.[6] The maximum value observed was 30 percent! Fortunately the curves level off at supersonic tip speed and seem to approach a constant percentage (compare with Fig. 2). From tip speed of 0.9 Mach on, the fourth harmonic was more powerful than the fundamental. In Fig. 5 is shown how the efficiency for the vortex noise produced by a model airscrew having a National Advisory Committee for Aeronautic's profile increases with tip speed.[7] This

[6] W. Ernsthausen, "Sound and Vibration in Aircraft," Vol. II, Chap. VII-A, in *German Aviation Medicine, World War II* (Department of the Air Force, U. S. Government Printing Office, 1950).
[7] W. W. von Wittern, "The Relation between Vortex Noise and Wind Resistance," Tech. Data Digest **19**, 20 (1951).

vortex power is integrated over the frequency range, up to at least 100 kc and, therefore, a large part of the power would be attenuated near the airscrew. That is why it does not contribute very much to the practical noise problem. For an aerodynamically good profile the efficiency η for the vortex noise is nearly constant between 10^{-2} to 10^{-3} over a wide range of tip speeds. Only when the latter approaches Mach number 1 does the efficiency increase rapidly. The thrust producing profile with 4° blade angle produces, naturally, a smaller percentage vortex noise (lower curve). For large airscrews, different efficiency functions would be obtained because the Reynolds numbers are different. Nevertheless at high velocities efficiencies of the order of a few percent are to be expected. Only a few measurements of the vortex noise of large air screws in flight at Mach 0.8 are reported.[6] On the propeller axis the vortex noise produced sound pressures at 10 kc of about 100 dynes/cm². Figure 6 presents the efficiency for the exhaust noise of reciprocating engines. It decreases relatively with increasing rpm of the engine. Therefore, the total exhaust noise increases more slowly as a

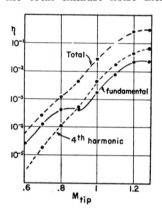

FIG. 4. Acoustic-mechanical efficiency η of the propeller rotation noise of a model airscrew as a function of tip Mach number. (Overall noise, fundamental and fourth harmonic are shown) (data from reference 6).

function of the rotational speed of the engine than does the rotational sound of the propeller.[8] Hence, the propeller noise power at full engine power is usually greater than the exhaust noise power. In addition, when the noise power radiated is the same, the propeller produces a higher sound pressure at a given distance than does the engine exhaust because of the directivity of radiation. The efficiency for the jet noise given in the two top curves of Fig. 7 is defined here as the ratio of acoustic power output to the rate of delivery of kinetic energy or the power in the jet stream. Here the most powerful jets were measured to have an efficiency of 1 percent also[9,10] and the acoustic energy

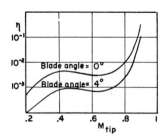

FIG. 5. Acoustic-mechanical efficiency η of the vortex noise from a model airscrew as a function of tip Mach number. (Parameter: blade angle) (data from reference 6).

is not proportional to the kinetic energy, but increases at least with its square; i.e., with the sixth power of the jet exit velocity. (Here as well as in the discussion of the noise of a turbojet engine only the noise produced by the actual jet stream is considered. The compressor and turbine noise is negligible at operational power of the engine and their sound generating mechanisms are fairly well known.[5,9])

(b) The Frequency Spectrum

The next question is: How is the total power given in Figs. 2–7 distributed over the frequency spectrum? In the case of the propeller the rotational noise, the predominant noise of the propeller, has a discrete frequency spectrum harmonically related to the blade passage frequency. At small subsonic tip speeds the intensity of the higher harmonics is smaller than that of the fundamental. With increasing tip speed the harmonics become more significant and at supersonic tip speeds some of the higher harmonics become stronger than the fundamental frequency.[3,6,11] In either case, the main part of the energy is at frequencies below 1500 cps. The pulse jet engine, designed so that its jet issues from the nozzle in periodic bursts of around

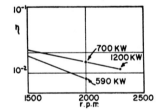

FIG. 6. Acoustic-mechanical efficiency η of the exhaust noise of different reciprocating engines as a function of the rpm of the engine (data from reference 6).

fluenced significantly by this change. The total calculated power at the highest engine rpm is now 75 kw instead of 69 kw. The directivity patterns of the lower octave bands (Fig. 2, a-c) are only slightly changed in shape. For the two lowest bands the maximal sound level was still found close to the jet boundary. With increasing frequency this maximum shifted outwards, as observed before, at the higher frequencies. For the 150–300 cps band the maximum was at 165°. This seems to indicate, as discussed in Part III(b) of this paper, that for those low frequencies the measurements made are still in the near field of the noise source. (Distance 50 feet=32 nozzle diameters). At 200 feet distance (128 nozzle diameters) the maximum of the lowest band (20–75 cps) is shifted from the jet boundary to 155°. The larger the distance (100, 200, 400 feet) the more the directivity factor has increased, but the maximum was always about at the same angle (140°–150°).

[8] H. Wayne Rudmose and Leo L. Beranek, J. Aeronautical Sciences 14, 79–96 (1947).

[9] von Gierke, Parrack, Gannon, and Hansen, J. Acoust. Soc. Am. 24, 169–174 (1952). The measurements described in this paper have been repeated at angular intervals of 5.6° around the engine. The directivity patterns for the new overall (Fig. 1 of the cited paper) at the two higher power settings of the engine show sharp maxima around 150° and 210° and a minimum on the jet axis. The original curves which were interpolated between the 22.5° increments of angle, should be corrected accordingly. The total over-all power output of the engine and the power in the bands is not in-

[10] L. W. Lassiter and H. H. Hubbard, "Experimental Studies of Noise from Subsonic Jets in Still Air," National Advisory Committee for Aeronautics TN2757 (August, 1952).

[11] H. H. Hubbard and A. A. Regier, "Sound from a two blade propeller at supersonic tip speeds," National Advisory Committee for Aeronautics RM L51C27 (1951).

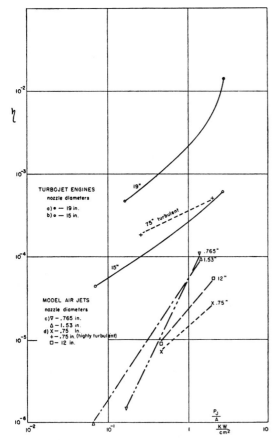

FIG. 7. Acoustic-mechanical efficiency η of the noise of turbojet engines and subsonic, cold-model air jets as a function of the power passing through the unit area of the jet stream. Figures on the lines indicate diameter of the circular nozzle. (Data obtained from the following: (a) reference 9; (b) reference 10; (c) reference 24; (d) reference 10.

100 per second naturally has this predominant frequency.[3,12,13] There is also some small harmonic content and some aerodynamic mixing noise. A typical octave band analysis of the noise of a small pulse jet is given in Fig. 8. The exhaust noise of a reciprocating engine is principally of the same type. The most intense noise component corresponds to the firing frequency of the engine. In contrast to these noise sources the noise of turbojet and ram jet engines and the noise of rockets show a continuous frequency spectrum extending from the subaudible to the ultrasonic range. This noise, generated mainly through the turbulent mixing of the issuing gas with the resting air, has a random character. Average spectra of the sound pressure level in the audiofrequency range are given in Fig. 8. The peaks of the spectra shift to higher frequencies with decreasing nozzle diameter. Above the peak frequency the total

[12] Leslie W. Lassiter, "Noise from Intermittent Jet Engines and Steady Flow Jet Engines with Rough Burning," National Advisory Committee for Aeronautics TN2756 (August, 1952).

[13] Paul S. Veneklasen, "Measurements and Analysis of the Physical Characteristics of Pulse-Jet Engine Noise," American Helicopter Company, Inc., Report No. 163-K-1 (September, 1951).

energy per cycle decreases[9] about as $1/f^2$, where f is the frequency. It must be recalled that sometimes one observes rough burning in these "steady flow" engines, especially on turbojets with afterburner. Then, these engines produce some discrete frequency components with a fundamental of around 100 cps and their spectrum becomes somewhat similar to that of a pulse jet.[12]

(c) The Directivity Pattern

In Fig. 9 is shown how the noise of the different sources is distributed in space when measured near the ground at zero speed. Always only one-half of the rotational symmetrical directivity patterns, which were adjusted to show equal maximum values, is presented. The propeller noise has its maximum at an angle of about 30° behind the plane of rotation. It is there about 5 db above its space average. With increasing tip speed this maximum shifts forward and for supersonic tip speeds it is in the plane of rotation. The exhaust noise is only slightly directional. The noise of turbojets and rockets has a sharp maximum around 150° near the jet axis. (0° is in front of the engine.) The over-all noise of the turbojet engine has for the angle of this maximum a directivity index of about 6 db. The noise energy radiated towards the hemisphere behind the engine, in the direction of the jet, is about 13 db greater than the noise radiated in the front hemisphere. The higher the frequency of the noise the more the maximum shifts forward, towards an axis perpendicular to the jet. Certainly this is qualitatively true at distances of 30 to 60 nozzle diameters. It is not yet established that this holds also in the asymptotic distant field. At least the maxima for the different frequency ranges probably are not separated so much there (compare footnote 9 and Part IIIb). The pulse jet radiation is much less directional. Most sound is radiated into the backward hemisphere with pronounced maximum around 150°.

III. THE MECHANISMS OF NOISE GENERATION

The characteristics presented in Part II for power, frequency distribution, and directivity give most information about the noise sources that one needs if

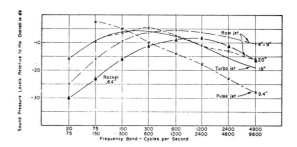

FIG. 8. Sound pressure level in octave bands relative to the overall noise level of turbojet, ram jet, pulse jet, and rocket. The figures on the lines indicate the nozzle diameters in inches. (Data from references 5 and 13 and measurements by the author.)

one is only interested in the effects of the noise generators, that is, in their outside fields. They give the basic information from which one is able to estimate the noise levels under certain operating conditions and in cetain setups such as test cells. When one is interested in more than these noise levels as such and when one is not willing to accept the high acoustic power outputs of engines as unchangeable, then one must look more closely at the noise sources in order to learn why and how the sound is generated. Only when much more detailed knowledge is gained can one judge whether aircraft noise may be reduced at its source or, in other words, one requires the additional information to judge whether the design of equivalent propulsion units with less disturbing noise output is possible. In the following, the mechanisms of noise generation underlying our major noise sources, the propeller and the jet, are discussed in more detail.

(a) The Propeller Noise

The propeller rotation noise is probably the noise source which we understand best of all aircraft noises. We have a complete theory on it which is, at least approximately, still valid at tip speeds higher than sound velcoity.[6,14—17] The theory is in good agreement with careful model experiments in anechoic rooms and with measurements on actual airscrews. The physical meaning of the theory is that the rotating blade periodically excites each point of the circular area of its path by a twofold effect. The first one is a symmetrical displacement of the air due to the finite thickness of the blade (torque effect). The second is a concentrated force due to the circulation flow around the propeller with pitch (thrust effect). Because of the first effect, the spatial points thus excited radiate in the manner of nondirectional sound sources. Because of the second effect, they radiate like dipoles with their axes perpendicular to the plane of rotation. The sound field results from interferences among the radiations from all points of the area. The sound field itself in turn affects the source, thus changing the structure of the force on the blade. The mass forces of the sound field determine the thrust coefficient of the propeller, changing it from the constant value it has in an incompressible fluid (i.e., at slow velocities). The radiation forces determine the additional resistance. The mass effect, the thrust, is necessarily connected with the radiation losses, the sound. Therefore, the sound effect is of necessity incorporated in the aerodynamic principle of the airscrew, expecially at higher velocities where the radiation resistance becomes much larger than the mass reactance. Finally at very high velocities nearly all energy delivered to the rotating airscrew

[14] L. Gutin, Physik. Z. Sowjetunion 9, 57 (1936).
[15] A. F. Deming, J. Acoust. Soc. Am. 12, 173 (1940).
[16] W. Ernsthausen and W. Willms, Jahrb. deut. Luftfahrtforschung I, 410 (1938).
[17] W. Ernsthausen, Akust. Z. 6, 245 (1941).

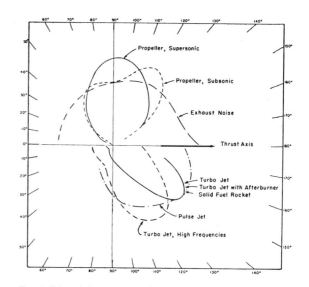

FIG. 9. Directivity patterns of the over-all sound pressure amplitude around various aircraft noise sources. The patterns are adjusted to show equal maximum value. Only one half of the rotationally symmetrical patterns is shown. 0° is in front of the engines.

goes into acoustic radiation. It is therefore evident from the acoustical point of view that the propulsive efficiency of a propeller must decrease with increasing speed and that the conventional propeller is inherently a low speed device. The propeller theory in its most general form[18] can therefore calculate the aerodynamic properties as well as the acoustic radiation field of the airscrew. A rotating propeller without any pitch, described by a distribution of simple sources alone, has the maxima of all harmonics of its sound field in the plane of rotation. The propeller with geometric pitch has, due to the additional dipole radiation (influence of blade angle), the maximum of its directivity pattern at an angle of about 30° behind the plane of rotation. In Fig. 10 the measured radiation pattern of a particular airscrew at tip Mach number 0.9 (upper half) is compared with the calculated pattern for the same airscrew (lower half). The maximum of the higher harmonics shifts towards the axis of rotation. At supersonic tip speeds the influence of the dipole radiation becomes negligible compared to the simple source radiation and directivity pattern as well as power function become at supersonic speed similar to the one for the profile without blade angle.[17] The maximum of the directivity pattern is in the plane of rotation. An increase in tip speed at supersonic tip speeds results in no further shift of the maximum but in a decrease of the directivity factor. At supersonic speeds, where a shock wave is formed at the leading edge of the propeller, the theory is not rigorously valid but still gives qualitatively useful results. A comparison between the calculated and measured power of the different harmonics radiated from a model

[18] W. Ernsthausen, Z. angew. Math. und Mech. 31, Heft 1/2 (January-February, 1951).

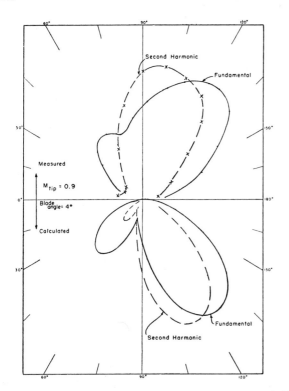

FIG. 10. Measured (upper half) and calculated (lower half) sound field of a model airscrew (blade angle 4°) at tip Mach number 0.9. The sound pressure is plotted in an arbitrary linear scale (data from reference 17).

airscrew as a function of tip speed is shown in Fig. 11. Up to Mach 0.9 the theory is even quantitatively correct. If one stays in this graph on the vertical lines, one sees how fast the total power of the fundamental or of the harmonics increases with tip speed. If one stays on the horizontal lines, which have the Mach number as parameter, one has the power as a function of the order of harmonic and the number of blades. In other words, one obtains the power spectrum of the noise. At low tip speed the power decreases with increasing harmonic number. At Mach 0.9 the spectrum is flat and at still higher Mach numbers the spectrum increases, reaches a maximum and drops off again as is observed for the supersonic propeller.[11] The graph explains also how the power decreases with increasing number of blades. The theory for the propeller rotation sound is quite complete. One understands that most of the noise is necessarily connected with the thrust, and knows how to design more quiet propellers.[19] (It might be mentioned that contrarotating airscrews as well as obstacles like the fusilage in the flow field of a propeller increase the sound output considerably.[6,8])

The vortex noise of the airscrew, which is not so important but still appreciable, especially when we take into consideration the ultrasonic part of the

spectrum, is satisfactorily explained too.[20,21,7] It has a continuous spectrum over a certain preferred band width. With increasing tip speed this band shifts to higher frequencies as is demonstrated in Fig. 12. This noise is caused by the vortex separation behind the propeller profile. Acoustic energy is radiated if the vortex circulation varies with time at a point at rest relative to the body. At low velocities we observe discrete vortices which leave the profile periodically. It is a kind of feedback mechanism which maintains the periodic vortex separation. At higher velocities we observe more and more irregular vortex separation and turbulence, but there is no fundamental difference in the sound generating mechanism. The vortex separation is connected with impulses at the edge of separation and we have there a sound source equivalent to a dipole. At low velocities the maxima of radiation are at right angles to the direction of flow. If the propeller blade moves, we have a moving sound source and, as theoretically shown,[22] more and more energy is radiated in the direction of motion. The maxima of the dipole characteristic turn towards the axis of motion as shown on Fig. 13. But this characteristic is now fixed to the rotating blade; it rotates with the blade. The vortex noise is therefore always greatest in front of the blade in the direction of its rotation and disappears

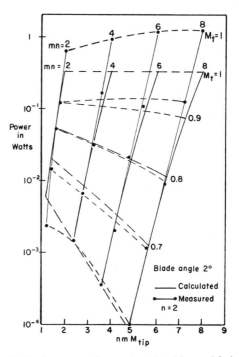

FIG. 11. Total power per harmonic radiated by model airscrews (blade angle 2°) as a function of the tip Mach number and the number of blades. n = number of blades, m = number of harmonic. Measured values for $n=2$ (data from reference 17).

[19] A. A. Regier and H. H. Hubbard, "Factors affecting the design of quiet propellers," National Advisory Committee for Aeronautics RM L7H05 (1947).

[20] E. Z. Stowell and A. F. Deming, "Noise from rotating cylindrical rods," National Advisory Committee for Aeronautics TN519 (February, 1935).
[21] M. F. Dowell, Gen. Elec. Rev. 42, 210–217 (1939).
[22] H. L. Oestreicher, Tech. Data Digest 19, 16 (1951).

behind it. At a fixed point in space where we record this noise, it is therefore always modulated at the blade passage frequency. The acoustic power of the vortex noise from propellers increases on the average, with the sixth power of the Mach number as does the power from most other vortex sounds.[15,20,23] Due to the motion of the source, the sound power may increase still faster with tip speed. Whereas the sound power of aerodynamically poor profiles (cylindrical rod) increases continually with the mentioned sixth or still higher power, the acoustic power radiated by a aerodynamically good profile as used in actual propellers does this only in the rough average and shows a typical course as a function of tip speed; the power remains almost constant in a certain range of velocity (compare the efficiency in Fig. 5). This brings out the interesting fact, that the vortex energy is directly connected with the aerodynamic flow resistance of the propeller profile.[7]

As already mentioned, only very few vortex noise measurements on full scale propellers appear to have been published.

(b) The Noise Produced by Gas Jets

The mechanism of noise generation by aircraft jet engines has not been completely investigated. Most of the existing publications are limited to sound pressure measurements in the distant field and there has been no systematic attempt to separate, under well controlled conditions, the different possible mechanisms for noise generation by an actual engine. We are not able to decide what part of the total noise is generated by combustion irregularities (i.e., by fluctuations of the

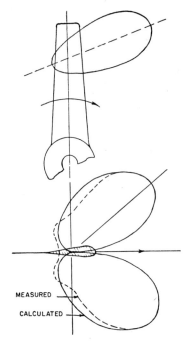

MEASURED

CALCULATED

FIG. 13. Measured and calculated directivity pattern of the vortex noise of a rotating profile. The sound pressure is plotted in an arbitrary linear scale. Tip Mach number 0.4 (data from reference 7).

exhaust momentum), what part is generated by turbulence of the jet radiating in the mixing region right behind the nozzle, and what part at supersonic velocities of the jet is caused by shock waves forming at the nozzle exit. Naturally these three possibilities are probably never completely separable since each of them has a reaction on the other. Non-uniform fuel burning rates not only modulate the exhaust velocity, but change the turbulence of the jet and its mixing zone and also change the shock wave pattern at the nozzle exit. On the other hand, the vortex separation of the mixing zone may have a reaction on the flame. Such flames have a tendency to operate periodically especially when they are coupled with a selective feedback, for instance a cavity or a tube. The extensive literature on sound sensitive or singing flames shows that flames may respond to the density and velocity variations of a sound wave, and may synchronize with the sound and amplify it. Rough burning in afterburners is a typical example of the acoustic reaction of a cavity on the combustion.

Noise generation by pulsatile flow, where the jet is already modulated before leaving the nozzle, is inherent in the principle of operation of a pulse jet, where the flow is 100 percent modulated. The noise generating mechanism is in principle the same for steady flow engines with rough burning[12] where the pulsations are smaller. The exhaust noise of a reciprocating engine is of the same type. The sound power can be approximately calculated when the pressure fluctuations in the exit nozzle or the exhaust duct are known.[6,12] The

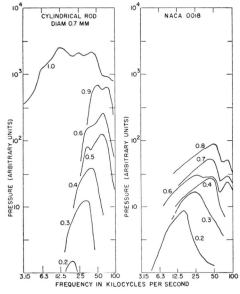

FIG. 12. Frequency spectra of the vortex noise of rotating cylindrical rods and model airscrews. The parameter shown is tip Mach number (data from reference 7).

[23] W. Holle, Akust. Z. 3, 321 (1938).

fact that the expelled exhaust gases have a high velocity in one direction is not considered in any of the theoretical approaches and therefore the slight directional effect towards the exhaust is not obtained. No theoretical relation between thrust and sound is available. Evaluation of the few available data on pulse jets indicate that the total sound power of one engine increases approximately as the fourth power of the thrust,[12] and as the fourth power of the fuel consumption.[13] The pulsating flow sound source generates, in addition to the operating frequency and its harmonics, a continuous noise from turbulent mixing which should be modulated by the fundamental frequency and is probably the main cause of the slight maximum around 150°. This part of the noise should be similar to that from continuous flow engines.

The noise source of gas jets issuing continuously from a chamber into the atmosphere is located in the mixing region behind the nozzle. Here, where the issuing gas mixes with the stationary surrounding air mass, disturbances are produced which are propagated as sound. The main part of the noise of a turbojet engine with smooth flow seems to come from this source and therefore studies on jet noise have concentrated, until now, on the sound production by smooth flowing, cold gas jets. The flow irregularities, introduced by combustion are eliminated in such experiments, and the jet has only the natural turbulence connected with the particular velocity and geometrical configuration. The similarity between the noise characteristics of such model jets and actual turbojet engines operating under high thrust conditions has been shown.[10] It may be interesting, therefore, to summarize here the results of studies on model air jets, to show the well established facts as well as the unsolved problems.

Figure 7 summarizes the data on the total acoustic power output of air jets. The only measurements selected[10,24] are those where the total sound output was either directly measured in a reverberation room or could be integrated from the directivity pattern. The ratio of the acoustic power output P_{ac} to the power in the jet stream $P_j = \frac{1}{2}\rho U^3 A$ (A = area of the jet, U = exit velocity, ρ = density of gas) is given as a function of the power per unit area in the jet. It is shown that the efficiency η is about the same for all the jets investigated (diameters from 0.75 inch to 12 inches). The agreement is satisfactorily good when one realizes that the several nozzles and jet tail pipes were not at all geometrically similar and the upstream flow conditions were not the same. The limits within which this efficiency can vary for the same jet and the same mean velocity, as a result of changes in its state of turbulence, are illustrated by an additional experiment on the 0.75 inch diameter jet. This jet was made highly turbulent by the addition of two 90° pipe bends external

to the nozzle. The efficiency became about 30 times higher and shifted to the upper curve, 0.75 inch diameter "turbulent." Since, in the turbojet engine, the jet is extremely turbulent it is not surprising to find that the efficiency of a 15 inch turbojet is in the same range and that the efficiency of a still higher power turbojet (19 inch) is still higher. For comparison and for supporting a general theory the efficiency of edge tones on jets 1 mm wide shall be mentioned too: it increases from $\eta = 10^{-6}$ to $\eta = 10^{-4}$ when P_j/A increases from 10^{-5} to 10^{-3} kw/cm². (Here, owing to the feedback mechanism of the edge, the stream noise becomes a pure tone and the stream dissolves completely into discrete vortices.) All the efficiencies presented increase on the average in proportion to the jet power: i.e., the acoustic power increases with the sixth power of the velocity, as it is reported also for other types of aerodynamic sound.[20,23] Sometimes a slightly higher power coefficient, up to the eighth,[24] is observed. In another quite extensive investigation,[25] (the only one which includes velocities up to Mach 1.4) the acoustic power increased with the sixth power of the velocity, on the average, for all subsonic exit velocities, but increased with powers of the velocity up to the 15th in the supersonic range. The relatively higher power output at supersonic velocity appears to be connected with the shock pattern in the jet, since by suitable noise reduction devices (wire gauze extension) the shock train can be eliminated and the increase of the acoustic power with the sixth power of velocity can be extended to the case of the supersonic jet. In Fig. 7 all velocities, even in the case of the turbojet, were subsonic. The acoustic power of the turbulent mixing noise is therefore about proportional to $\rho^2 A U^6$. If a minimum of noise is desirable for a given thrust an increase in nozzle diameter and a corresponding decrease in velocity should be the best solution. The dependence of the acoustic power on ρ is substantiated only by a few measurements with helium and Freon at room temperature. In these investigations the temperature is not believed to have a direct effect on the noise generated except in so far as it changes density and turbulence. Nevertheless, not enough measurements are available to give P_{ac} accurately as a function of density and temperature over a wide range and therefore to find the general relation between η and Mach number.

The frequency spectra of model jet noise also showed close similarity to the spectra found for turbojet noise. The larger the nozzle diameter the more prominent were the low frequency bands. It is not possible, now, to give a general formula for this shift of the spectrum since on this point there is some disagreement among the findings of the different authors. Generally they find only a very slight shift of the peak of the sound

[24] H. M. Fitzpatrick and R. Lee, "Measurements of Noise Radiated by Subsonic Air Jets," David W. Taylor Model Basin Rept. 835 (November, 1952).

[25] R. Westley and G. M. Lilley, "An Investigation on the Noise Field from a Small Jet and Methods for Its Reduction," The College of Aeronautics, Cranfield, England, Report No. 53 (January, 1952).

spectra to higher frequencies when the exit velocity increases; only in the case of the jet with high turbulence[10] does this peak frequency increase about in proportion to the velocity. Considering the difficulty of keeping the jet turbulence and mixing free from any influence of resonating cavities or tubes (upstream or downstream of the nozzle), as reported at lower velocities,[26] it is not astonishing that no clear picture was obtained. Further work is required here.

The directivity patterns of the model jets were found generally to be in agreement with the patterns given for the turbojets, Fig. 9. At Mach 0.3 the over-all noise has its maximum around 135° (0° upstream to the nozzle, center at the nozzle). In the experiments so far reported this maximum shifts with increasing Mach number towards the jet axis. For Mach 0.9 the maximum of the over-all noise and of the lower frequencies is around 165°. At supersonic jet exit velocities, the maximum over-all noise lies at angles inside the jet boundaries, at least for the distance of 108 nozzle diameters, where the reported data were measured. The maxima for the higher frequencies always turn away from the jet axis, as was observed for the turbojet engine, and are for the highest frequencies around 135°. A shift farther outwards was never observed. (The helium jet showed at Mach 0.9 its over-all maximum at 135°.) An interesting study of the near noise field of a one-inch jet revealed some clue to the behavior of the noise radiators of the jet. The noise sources of the different frequency bands were found to extend over a certain distance along the jet beginning at the nozzle exit. This distance is approximately proportional to the jet speed and increases as the frequency decreases. The strength of the radiators seems to increase to a maximum along the downstream jet axis shortly before they disappear. As an example in Fig. 14 the measured near noise field of one quadrant is shown. For the frequency band for which this figure is valid, at a distance of 108 nozzle diameters, the maximum intensity was measured around 135° as is shown in the figure. The figure demonstrates that the smaller is the distance from the nozzle exit, the closer to the jet axis the maximum of the "directivity pattern" shifts. For the low frequencies, up to about 800 cps, the distribution of assumed sound radiators did not come to an end over the investigated downstream distance of 30 nozzle diameters (Mach number around 1) which means that up to this distance the maximal pressure was still observed along the jet boundary. Considering these findings one must say that practically all reported directivity measurements were made too close to the jet exit. (Compare footnote 9.) Most measurements on turbojets, from which the data in Part II were summarized, were made at distances less than 50 times the nozzle diameter. For the low frequencies this distance should not be less than 500 jet diameters. The directivity patterns of turbojet engines as well

[26] H. E. von Gierke, Z. angew. Physik 2, 97 (1950).

as those of model jets should be redetermined at distances that will permit observation of any changes in directivity pattern that occur with changes in jet velocity. Likewise, the change in directivity with frequency would be verified under really asymptotic distant field conditions.

In the case of the model jets the turbulence in the jets (that means only the axial velocity fluctuations were measured) was in the order of 1 percent of the mean velocity.[10] For the high turbulence configuration it was up to 10 percent. A rough estimation shows that the ratio of the energy stored in the turbulent motion to the energy supplied to the jet is of the order of 10^{-4}. Since η in Fig. 7 is in the range of 10^{-5} one cannot say that turbulence as such is an inefficient sound radiator in this case. Only the conversion of direct power to alternating power is inefficient. A closer correlation between total turbulence and acoustic power output as

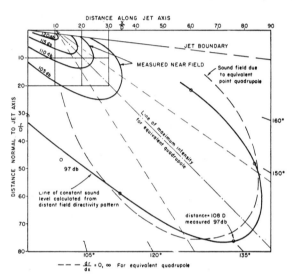

FIG. 14. Lines of equal sound level (frequency band 6400–12800 cps) for the near noise field and the far field of a model jet (one quadrant only). Jet exit velocity $M=1$. Jet diameter $D=1''$. x=distance along jet axis, r=distance normal to jet axis. The sound field resulting from a lateral quadrupole at rest at $x/D=10.5$ and its line of maximum intensity is also indicated. Angles are measured with respect to nozzle exit (based on data of reference 25).

a function of jet velocity has not been worked out. Measurements of the turbulence in the jet of a turbojet engine have not been reported.

The only theoretical approach to the jet noise problem is given by the general theory of Lighthill,[27] "on sound generated aerodynamically." This theory fills a long felt need in theoretical acoustics. It gives general equations for the sound radiated by a fluid flow with regular turbulence. Such a flow radiates, in the most general case, like a static distribution of acoustic quadrupoles. It is predicted that the power radiated by such a quadrupole field is $P_{ac}=K\rho(U^8/a^5)l^2$

[27] M. J. Lighthill, Proc. Royal Soc. (London) A211, 564–587 (1952).

($a =$ sound velocity, $l =$ linear dimension, here jet diameter) where the "acoustic power coefficient" K is expected to be only slightly dependent on Mach number and Reynolds number. This means that the acoustic-mechanical efficiency η should be $\eta \propto KU^5/a^5 = KM^5$. As can be seen in Fig. 7 and as was previously discussed η was experimentally found to increase with the first to fifth power of M or U, but most measurements indicate the third power of U. According to Lighthill η will increase still faster with U when the quadrupoles are in motion with respect to the surrounding air. Thus far there are an insufficient number of accurate measurements available to decide what would be the most accurate application of the Lighthill theory to the jet noise of the mixing region. The lateral quadrupoles, which he derives as the equivalent sound radiators of the turbulent shear flow of the jet, would have a directivity pattern with maxima along lines at 45° to the jet axis. A similar pattern was found for the high frequencies of the jet noise and for noise from jets of low velocity. If maxima closer to the jet axis are really found for the low frequencies and the overall at higher speeds, this could indicate that the quadrupoles are in relative motion to the surrounding air, whereby the radiation patterns are shifted toward the direction of motion; i.e., closer to the jet axis. Unfortunately, as pointed out before, one cannot be too sure that the existing directivity patterns were obtained at a sufficiently great distance to use them to check the application of Lighthill's theory. The pattern around such lateral quadrupoles at rest has some similarity with the field distribution given in Fig. 14. The field of such an equivalent static quadrupole is there indicated. The point radiator would have to be located, in this example, about 10 jet diameters downstream of the nozzle. The jet's axial flow velocity shows its major decrease in the region of 4 to 25 nozzle diameters downstream. This is the same region where the major distribution of sound sources appear. This is in agreement with theoretical

results that the sound intensity of the radiators is proportional to the square of the velocity gradient. The aforementioned findings led to development of noise reducing devices[25] designed to reduce the rate of shear. Teeth extensions to the nozzle proved to be quite promising. The teeth extensions corrugated the jet boundaries and diffused the shock pattern at supersonic velocities. At subsonic velocities a noise reduction of 5 db was obtained, whereas with certain teeth configurations at supersonic velocities noise reductions up to 12 db were observed. The theory has not been extended explicitly to a jet having properties other than the surrounding air (i.e. the case of the heated jet and a jet consisting of other gases). If high thermal power is conducted with the jet stream, there should be a point where it influences the conversion of mechanical power into acoustical power.[28] The model studies should be extended to include these cases. Generally it can be said that these model studies have increased considerably our knowledge of the basic physical mechanism of jet noise generation and in spite of the many unanswered or half-answered questions one may use them to explain roughly the observed physical characteristics—until a complete theory is developed which combines the different observations and explains them rigorously.

IV. THE AIRCRAFT NOISE SOURCES IN FLIGHT

So far, all data and discussions have been concerned only with the noise of the airplane on the ground. That nearly all of the characteristics change in flight is often not taken into consideration. Until recently the noise produced in flight was only a serious problem for occupants of the aircraft. For this purpose information about the sound levels in the aircraft in flight is easily obtained and has been published for propeller driven engines[8,29] and jet engines.[6,29] The new important source of noise is the aerodynamic noise generation in the boundary layer of the airplane as it moves through the air. It results from turbulence, vortex motion, and wakes. In high speed jet aircraft the aerodynamic noise may contribute 95 percent of the noise energy and might become the practical limiting factor rather than the engine noise. It is not radiated to any appreciable extent to the surrounding air and since it occurs only at high speed, it is of no concern to the observer on the ground. For him the moderate and low forward speed conditions as in take-off, climb, and landing are of great and daily increasing interest.[30] This problem, as to how the characteristics of the propulsion units are changed in flight has not been attacked in a way which

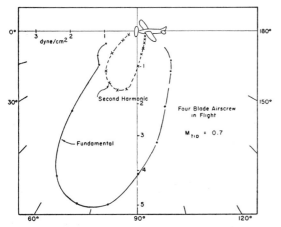

FIG. 15. Directivity pattern of the rotational sound (fundamental and second harmonic) of the airscrew of a slow aircraft in flight. The pattern, valid with respect to the moving aircraft, was measured at a distance of 650 feet (data from reference 6).

[28] Osman K. Mawardi and Ira Dyer, J. Acoust. Soc. Am. 25 389 (1953).

[29] S. C. Ghose, J. Royal Aeronaut. Soc. 54, 697 (1950).

[30] Field, Edwards, Kangas, and Pigman, "Measurement of Sound Levels Associated with Aircraft, Highway and Railroad Traffic." U. S. Department of Commerce, Civil Aeronautics Administration, Washington D. C., Tech. Development Report No. 68 (July, 1947).

produces practical data. Whereas, at least some observations are reported for the propeller driven aircraft,[16] basic information on jet engines is completely missing. Too often without justification the static data are applied to the airplane in flight The characteristics of the sound sources on the ground will be changed in flight by two effects:

(1) The effect of motion on the sound radiation when the source itself radiates otherwise with the same velocity amplitude as on the ground.
(2) A change in the sound generator itself.

The first effect, theoretically investigated for the simple source, dipole[22] and quadrupole,[27] turns the directivity pattern toward the direction of motion and the radiated power increases with the velocity. At Mach number 0.9 dipoles radiate about 20 db and quadruples up to 40 db more power than at rest. So we may expect for higher flight speeds a different directivity pattern and power output of the aircraft than that on the ground. In addition, for an observer at rest, the frequency spectrum will be modified according to the Doppler effect. The frequency is increased for sound emitted forward and decreased for sound emitted backwards. The second effect of moving the sound source is that the source strength and the frequency spectrum of the generator itself change.

For the case of the flying jet engine both effects are influencing the sound radiation. The source itself is naturally changed since the sound generation depends on the velocity of the issuing gas jet relative to that of the surrounding air. This velocity decreases with increasing flight speed, the radial velocity gradient in the jet boundary which gives, according to Lighthill, the strength of the quadrupoles, becomes smaller. Owing to this effect the noise in flight should become relatively less than on the ground. As another factor (in addition to the two mentioned effects due to motion) altitude might have some hard to predict effect on the jet noise. The turbulence at the jet might increase; the smaller ρ of the medium of propagation might decrease the sound levels. At high altitude, the jet noise is no real problem either for the occupant or for the observers on the ground. At present we have only little information how in flight, at low altitude, the acoustic power output of a jet engine changes with respect to static condition. A few recent measurements on only one engine indicate that the decrease of the power output because of the changed source strength is more significant than the increase because of the motion effect. Around the jet engine in flight at Mach number 0.7 considerably less sound power was measured than under static condition. But owing to the moving sound source effect the directivity pattern was shifted forward and had its maximum in front of the airplane.

The situation is similar for propeller noise. Here in flight, the aerodynamic angle of attack becomes about zero, which means that the dipoles of the equivalent sound source disappear. One has less sound power

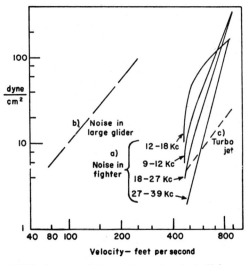

FIG. 16. The increase of the aerodynamic noise in flight as a function of the airplane's velocity. (a) Sound pressures in different frequency bands, in a fighter cockpit (reference 6). (b) Total aerodynamic noise in a large glider (reference 30). (c) Total noise of turbojet engine on the ground at 20 feet distance (90°). The velocity in this case is the jet exit velocity.

radiated, the maximum of the directivity pattern shifts forward and the spectrum changes. Therefore if one wants to correlate sound characteristics of a propeller in flight with the data measured on the ground, one should at least operate the airscrew on the ground with a blade angle which corresponds to the angle of attack at the considered flight speed. In addition to the forward shift of the directivity pattern in flight due to the disappearing dipole radiation of the airscrew, the moving sound source effect shifts the pattern still farther towards the direction of motion so that the maximum of the pattern might be in front of the plane of rotation. Such an effect where the maximum of radiation shifted to an angle of 15° in front of the plane of rotation was already observed at relatively slow speed.[16] In Fig. 15 such an inflight characteristic with respect to the moving airplane is shown.

The boundary layer noise and general aerodynamic noise inside the airplane in flight have been investigated by several authors for special planes and flight conditions.[3,6,8,31] In all these cases frequency spectrum and energy were influenced by the sound transmission characteristics of the fuselage wall. General physical characteristics can therefore not yet be given even for the simplest case of the noise on plane boundaries. The peak energy in the spectrum apparently shifts to higher frequencies as the velocity increases. High energy levels are at least present[6,29] from 500 cps up to 70 kc in the boundary layer. They may go up to 150 db but are easily attenuated by the fuselage wall. The latter will, to a large extent, determine the spectrum measured

[31] O. R. Rogers, "Noise level and its variation with position and air speed in the XCG-4 glider," No. 29618 MR No. EXP-M-51/VF2 (Addendum 13) Materiel Command, Army Air Forces (June 23, 1942).

on its inner side. The measured sound pressure of the aerodynamic noise in the cabin was found to increase with the 2.3 power of the velocity[30] for relatively slow velocities and with up to the eighth power at higher velocities. This looks reasonable when compared with the increase of the jet noise, Fig. 7. Some of the measurements are summarized for comparison in Fig. 16. In the cockpit of certain fighter airplanes sound pressure levels around 130 db were recorded at an airspeed of 500 miles per hour.

At higher altitude, we have generally to expect lower aerodynamic noise levels for the same air speed. The ratio of the acoustic energy (transmitted to the wall) to the turbulence in the boundary layer has not been closely investigated in spite of the many studies on this subject from a purely aerodynamic point of view.

It can be said in summary that one has for nearly all aircraft noise sources data and design curves which are useful as guides for the calculation of sound levels to be expected from aircraft operating under certain specified conditions. On the other hand, despite the many studies which have been carried out, one is still far from a complete understanding of the physics of many noise sources. A further development and application of the theory of sound produced aerodynamically with its two major parts, the boundary layer noise and the jet or mixing noise, and study on aircraft noise sources in flight will help most to fill the gaps in our present day knowledge.

9

Experimental Studies of Jet Noise*

H. H. HUBBARD AND L. W. LASSITER

Vibration and Flutter Branch, National Advisory Committee for Aeronautics, Langley Field, Virginia

The mixing region of a jet is observed to be a complex noise generator. The noise produced is highly directional and is affected by various geometric and flow parameters as well as by conditions in the settling chamber upstream of the nozzle. Noise measurements for a family of circular model air jets ranging in diameter from ¾ to 12 inches are consistent with available data for a turbojet engine. The intensity of the fluctuating pressure field near the jet is greatest at an axial distance of approximately two diameters downstream from the nozzle exit and decrease generally with increasing distance. The frequency spectrums recorded near the jet boundary are usually peaked, the peak frequencies being higher near the jet exit than at points farther downstream. These noise frequencies generally increase with increasing jet fluid velocity and decrease with increasing jet size. Hot wire surveys of turbulence (axial velocity fluctuation) in the jet stream indicated spectrums which were very similar in quality to the noise spectrums recorded just outside the jet boundary and at the same axial stations.

INTRODUCTION

THERE is considerable interest in fundamental jet noise studies since it is generally believed that an understanding of the nature and sources of the noise is necessary before any great progress may be expected in controlling it. The NACA is currently engaged in experimental studies aimed at an understanding of some of the basic laws of jet noise generation. These studies which have made use of model jets have led to an evaluation of the significance of some of the geometric and flow parameters involved. The present paper will discuss some of the effects on the observed noise of such variables as jet velocity and the observer's azimuth and distance. Some hot wire studies of spectra of turbulence in the jet stream will also be described and the results will be correlated with noise measurements.

SOME SIGNIFICANT PARAMETERS

Some of the parameters significant in jet noise generation are given in Fig. 1. The noise observed is a function of several geometric and flow parameters as listed in the figure. For given flow conditions the noise is affected by the quantities β, Z, and D which are defined as the observer's azimuth angle, and distance, and the jet diameter, respectively. In order to compare the model jet data with those for larger ones it was necessary to study some of the effects of scaling. This was done by means of tests with a family of air jets ranging in size from ¾ to 12 inches in diameter. It was found that at a given azimuth angle β, the significant parameter in regard to over-all pressure magnitudes was the nondimensional ratio Z/D. At equal Z/D values approximately equal over-all sound pressures are measured for different sizes of jets.[1]

At any given point O, additional effects are observed due to variations in the jet fluid parameters, such as density, velocity, Mach number, sound speed and turbulence, and the conditions upstream of the nozzle. In the schematic diagram of Fig. 1, where the fluid issues from a settling chamber to the atmosphere, the experimenter has some control over the noise generated. The mixing region of the jet may be the predominant source, particularly at the higher jet velocities, whereas the geometry and flow conditions in the settling chamber may have a large influence on the noise at the lower jet velocities. In the interest of brevity the present discussions will deal only with the currently more widely used subsonic or unchoked jets and, unless otherwise noted, will be confined to the noise from the mixing region.

COMPARISON OF MODEL AND ENGINE DATA

Model jets have been used extensively in noise studies, and the results of these tests are consistent with available measurements[1] for turbojet engines. The next two figures will illustrate some of the findings of the model tests and will allow a comparison with engine data. Figure 2 is a polar diagram which illustrates some

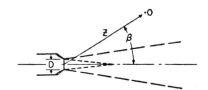

PARAMETERS AFFECTING NOISE AT OBSERVER
(a) GEOMETRIC
    OBSERVER'S DISTANCE (Z)
    OBSERVER'S AZIMUTH (β)
    JET DIAMETER (D)

(b) JET FLUID
    DENSITY
    VELOCITY
    MACH NUMBER
    TURBULENCE
    SOUND SPEED

(c) CONDITIONS IN SETTLING CHAMBER

FIG. 1. Parameters affecting jet noise at the observer's position.

---

*Presented during the Aircraft Noise Symposium which was part of the San Diego meeting of the Acoustical Society of America, November 13–15, 1952.

[1] L. W. Lassiter, and H. H. Hubbard, "Experimental Studies of Noise from Subsonic Jets in Still Air," NACA TN 2757, August, 1952.

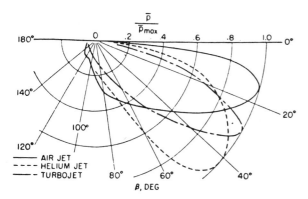

FIG. 2. Comparison of over-all sound pressure radiation patterns for turbojet exhaust and model jets.

of the directional properties of the over-all noise (20 cps–38 000 cps) from three different jets at a constant exit Mach number. For convenience the sound pressures in the direction of maximum radiation have been adjusted to equality. Let us first note the solid curve. This is a typical radiation pattern for low temperature air jets, and it is seen that the noise is a maximum at an azimuth angle of about 15°. As a matter of interest, a helium jet was tested to determine the effects of greatly increased velocity and sound speed of the jet fluid, and those data are shown by the curve of short dashes. The most noticeable result is that the direction of maximum radiation is now at approximately 45°. It was also found that data for a turbojet engine fitted in between these two extreme cases in a manner shown by the curve of long dashes. This result would be expected on the basis that the corresponding values of velocity and sound speed are intermediate between those of the two model jets. Thus the engine and model data are found to be closely related in regard to their directional properties.[1]

Experiments have shown also that the jet noise increases at a rapid rate with an increase of the jet fluid velocity. This effect is illustrated in Fig. 3 where sound pressures in dynes/cm² are plotted as a function of the jet velocity. Data were obtained at a distance of 16 jet diameters and at an azimuth angle of 90°. Model tests showing that the sound pressure varies directly with the jet fluid density have made it possible to compare data from jets of different densities. Thus the data of the model jets of Fig. 3 have been adjusted to a density comparable to that of the engine data. The curve of small dashes is for an air jet with smooth flow in the jet inlet. The curve of long dashes is for an air jet with the flow upstream of the nozzle made turbulent by passage through two 90° pipe bends. It will be noted that there is a general increase in the noise level when the flow is made more turbulent. For a large number of tests the data obtained with model jets indicated slopes varying from 2.1 to 3.7 for straight line plots of the type shown in Fig. 3.[1] Sound pressures for a turbojet exhaust also increased rapidly as the velocity increased as shown by the solid curve which has a maximum slope

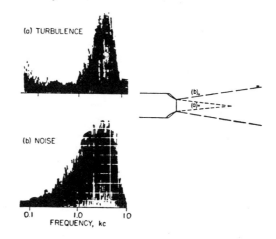

FIG. 4. Comparison of the noise spectrum near the boundary of a model jet with the spectrum of turbulence on the jet axis.

of approximately 3.0. These slopes compare with a value of 4.0 obtained theoretically by Lighthill.[2]

Of special interest in regard to the engine data of Fig. 4 is the observation that the slope of the curve is greatest at the higher velocities. A study of the spectra indicated that there are important discrete-frequency components present at the low engine speeds whereas at the higher speeds they are not noticeable. Apparently the noise from the mixing region of the jet is of sufficient intensity at the higher jet velocities to override any discrete frequency noise due to the turbine, resonances, etc.

CORRELATION OF NOISE AND TURBULENCE DATA

So far the discussions have dealt with some of the parameters which affect the noise and have indicated that the noise from a jet may be predicted if the various

[2] M. J. Lighthill, Proc. Roy. Soc. (London) A211, 564–587 (1952).

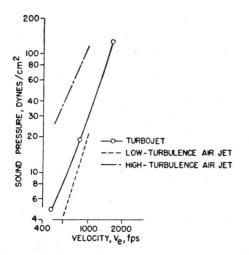

FIG. 3. Effect of exit velocity on over-all sound pressures from turbojet and model jets.

geometric and flow parameters are known. The remaining discussions will deal with studies of a fundamental nature which describe the noise field close to the jet and will attempt to correlate these findings with the results of hot wire surveys of axial velocity fluctuations in the jet stream. Figure 4 shows some of the hot wire data obtained on the jet axis at point (a) and allows a comparison with noise data recorded just outside the boundary of a one-inch jet at point (b) and at the same axial station. These photographs of the viewing screen of a Panoramic Sonic Analyzer Model AP-1, show directly the relative magnitudes on the vertical logarithmic scale as a function of frequency on the horizontal logarithmic scale.

Jet noise spectra at large distances are usually random in nature and contain a wide range of frequencies. Recent tests have shown however that at locations close to the jet boundary the noise spectra, while still random in nature, become peaked as shown in Fig. 4(b). As at greater distances, these measured noise spectra are affected by microphone position and the size and velocity of the jet and hence may vary widely. They were found to be very similar in quality to the turbulence spectra as detected by means of a hot wire inside the jet and as illustrated by Fig. 4(a). For purposes of correlating noise and turbulence data the frequencies at which these spectra tend to peak have been determined from the records and are used in the preparation of Figs. 5 and 6.

Figure 5 allows a comparison of frequencies as a function of jet velocity for noise measured near the jet boundary and for turbulence (axial velocity fluctuations) measured on the center line of the jet and at the same axial station. The frequencies were determined from records such as those described in Fig. 4. It may be seen that the predominant frequency of turbulence increases directly as the velocity increases, as shown by the dashed line. The predominant noise frequency also increases generally as the velocity increases in a

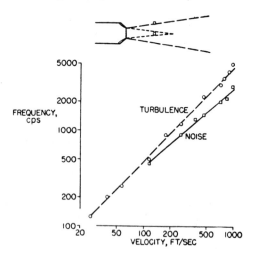

FIG. 5. Comparison of noise and turbulence frequencies as a function of exit velocity for a model jet.

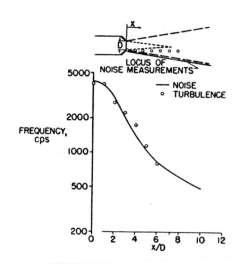

FIG. 6. Comparisons of noise and turbulence frequencies at various distances downstream from the exit of a model jet.

manner similar to the frequency variation of turbulence. Comparable noise data for a model jet twice as large indicated that the noise frequencies were approximately one-half those for the one-inch jet of Fig. 4. Jet noise frequencies appear to be proportional to the exit velocity and inversely proportional to the diameter, and thus conform approximately to the Strouhal relation which describes the shedding of vortices from a rotating rod.

It was of interest also to make similar comparisons at several points near each other and along the fringe of the jet for a constant exit velocity. These latter results are given in Fig. 6 where frequency data are compared at several distances downstream of the jet exit. The noise spectra were measured just outside the jet stream, whereas the turbulence spectra were measured at points one jet radius from the center line of the jet and for a velocity of approximately 700 feet per second. Here again the trends are similar and the agreement is good. It may thus be concluded tentatively from the data of Figs. 5 and 6 that the noise is associated in some way with the axial velocity fluctuations in the jet stream.

PRESSURE SURVEYS CLOSE TO THE JET

In addition to the physiological effects of jet noise, serious structural problems have become associated with jet engine operation. Failures of the secondary structures of wings and fuselages in some configurations have resulted from accelerated fatigue due to the intense fluctuating pressure field near the jet orifices. Surveys close to the boundary of a one-inch diameter model jet have yielded some interesting results which may explain some of the recent difficulties encountered in jet airplanes. Over-all values of fluctuating pressure amplitude (20 cps–38 000 cps) are plotted in Fig. 7 for varying values of x/D and y/D where x and y are defined as axial and radial distance, respectively.

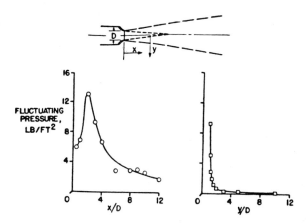

FIG. 7. Variations of over-all fluctuating pressures as a function of axial and radial distances from model jets for a constant jet exit velocity.

When the measurements are made axially along the jet and just outside of the boundary, the pressures increase to a maximum at a distance of about two diameters and then decrease somewhat at greater distances. Pressures along a line of increasing radial distance decrease rapidly with distance as shown in the left-hand figure at an x/D value of 3, much as though the source was near the fringe of the jet.

In vibrations the frequency of the exciting force may be of greater importance than its amplitude. Hence it is of interest to determine the manner in which the frequencies vary along the jet boundary. Figure 8 presents the relative frequency content and comparable amplitudes of the pressures at three points along the jet boundary. Point A corresponds to the point at which the maximum amplitudes were measured, and it is seen that the spectrum is rather sharply peaked in the vicinity of 3000 cps. At points farther from the orifice,

the pressures are somewhat lower and the energy seems to have shifted rapidly to the lower frequencies until at point C the frequency band centers near 400 cps. As the size of the jet increases there is a tendency for the noise frequencies to decrease, and thus for larger jets frequencies may be in the critical structural range.

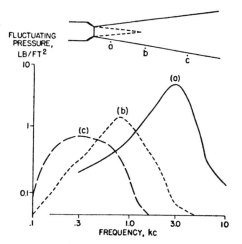

FIG. 8. Comparison of fluctuating pressure spectra measured at three points near the boundary of a model jet.

CONCLUSION

The basic noise studies which are briefly discussed in the present paper have described some effects of scaling and have evaluated some of the significant parameters in jet noise generation. Model jet noise studies have provided a useful background of experience for the interpretation of full-scale measurements and in addition have provided results of direct application to physiological and structural problems.

10

Reprinted from *Acoust. Soc. Am. J.* **25**:385–389 (1953)

The Noise of Choked Jets*†

ALAN POWELL

Department of Aeronautical Engineering, University of Southampton, England

The noise of a jet changes character after the pressure ratio exceeds the critical value appropriate to sonic exit velocity, the general roar being dominated by a loud "whistling" or "screeching." Schlieren photographs show that sound waves of ultrasonic frequency are caused by the transition of the initially laminar boundary layer to turbulence and also by this turbulence interacting with the shock waves of the flow. Larger disturbances have also been noted, involving both the jet stream and some of the air external to the jet, and these also give rise to sound waves which have been photographed: it is these which are held responsible for the audible effects. A two-dimensional study has shown the latter phenomenon to be enhanced, and it is shown how the system of disturbances is self-maintained by virtue of sound waves creating initially small disturbances at the jet exit. The directionality of the sound field has been predicted and found in agreement with experiment, and the dimensions of the motion are compatible with the suggested mechanism. The relation to edge tones is pointed out and the mechanism indicated, a photograph of this phenomenon also being shown. Finally mention is made of how the characteristic noise of jets working above the critical pressure might be reduced, the suggested methods having been found successful in practice.

I. INTRODUCTION

THE problem of aircraft noise has been steadily increasing, and the introduction of jet propulsion has accentuated the position. It is the jet efflux itself which is probably the greatest single source of noise of such aircraft. The engine powers are all the while being increased, as are the jet pressure ratios, particularly for rocket assisted or propelled machines. The need for information as to the mechanism by which such noise is generated is great, both from the point of view of understanding its characteristics and also of assessing the possibilities of a reduction of the noise output at source.

Whilst the noise arising from the general turbulence of a jet issuing at a subsonic velocity increases, even from a very low speed, with a power of the dynamic head to a power close to 3.6, after sonic exit speed has been reached and the pressure ratio further increased so as to "choke" the jet, the noise increases irregularly and much more rapidly, perhaps as the ninth power roughly, high peaks possibly appearing.[1] The character of the noise also changes from the general roar typical of subsonic jets having single flat maxima to spectra dominated by a loud harsh and confused "whistling" or "screeching." It is the latter aspect of the research programme at the University of Southampton which is the subject of this paper.

II. AN OPTICAL STUDY OF CIRCULAR JETS

A study of the flow of air jets was made using a two-mirror Toepler schlieren method[2] with a few extra refinements,[1] the jet exit diameters being one inch.

When a sonic jet exhausts with a pressure greater than the ambient value, the initial widening of the stream is followed by a contraction to conditions approximating to those at the exit, the process then recommencing. This results in a characteristic repetitive pattern which we shall call "cellular" and present in all the photographs displayed in this discourse.

This cellular pattern is evident in Fig. 1 where it is the most prominent feature, the turbulence of the stream also being clearly shown. The jet stream leaves the edge of the orifice with a laminar boundary layer, but small disturbances can be distinguished, and after a short distance, about one-third of the exit diameter, the layer breaks up and becomes turbulent, producing a ragged edge to the stream. Sound waves proceeding at angle of about 45° to the direction of the stream can be seen to be emanating from this region where turbulence of the boundary is setting in. In other cases,

* Presented in behalf of the author by Osman K. Mawardi during the Aircraft Noise Symposium which was part of the San Diego meeting of the Acoustical Society of America, November 13–15, 1952.
† Communicated by Professor E. J. Richards.
[1] A. Powell, "A 'Schlieren' Study of Small Scale Air Jets and Some Noise Measurements on Two-Inch Diameter Air Jets," University of Southampton, December, 1951.
[2] N. F. Barnes, and S. L. Bellinger, J. Opt. Soc. Am. **35**, 497 (1945).

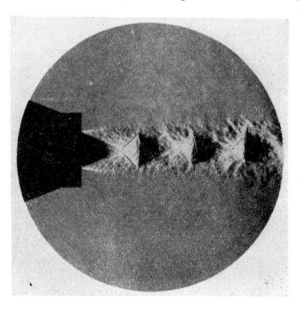

FIG. 1. Cellular nature of air jet flow.

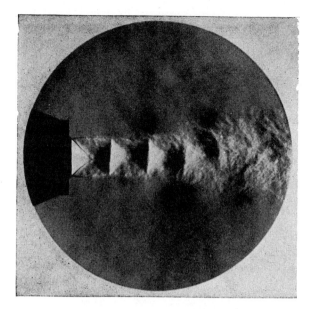

FIG. 2. Large disturbances associated with jet stream.

similar sound waves appear to originate over a much greater length of the stream and probably result from such turbulence interacting with shockwaves. This latter type of mechanism might be responsible for a portion of the sound field of Fig. 1, where the shocks can be distinguished as being of conical form, with the apex flattened and pointing upstream from the end of the cell. The frequency of this sound is of the order of 40 kc/sec. This phenomena was first reported[1] in 1950 where the corresponding frequency for a $\frac{1}{4}$-in. diameter jet was 80–100 kc/sec.

Another phenomenon then reported was the existence of larger disturbances, which involve not only the stream but also some of the external flow as may be seen in Fig. 2. These appear to take the form of incomplete torroidal vortices, sometimes resulting in a symmetric pattern but more often approaching an alternate arrangement as in the figure. Here the density gradients associated with the disturbances are easily seen, and it will be observed that the stream is undergoing local lateral displacements as these disturbances move along, the effect becoming greater towards the end of the cellular pattern. This pattern appears as a "negative" of the previous one, since the knife edge was active in the opposite direction.

In Fig. 3 these disturbances can also be seen, though rather less clearly since the higher pressure ratio has resulted in the end of the cellular pattern and those disturbances in that region, moving farther downstream and away from the observed field. In addition to the high frequency sound waves already mentioned, waves of much lower frequency are also visible on each side of the jet and moving in the upstream direction. Like the larger stream disturbances, they are in antiphase on each side of the stream, and it has been observed that they usually emanate from the region of the end of

the cellular pattern[1] where it is degenerating to general turbulence and where also the stream disturbances have been observed to be most vigorous. In this particular case the frequency of the sound is about 10 kc/sec which is similar to that of the audible "screeching" or "whistling."

III. THE TWO-DIMENSIONAL PHENOMENA

The cases we have just seen are awkward to deal with experimentally, since the disturbances have to lie in the right plane to be photographed and the dimensions are inconveniently large. A change to the two-dimensional case was therefore made, so that the larger disturbances and their associated sound field should become more prominent.

The jets used issued from a rectangular orifice 0.7 in. ×0.12 in. so that the flow might approach two-dimensional conditions, and Fig. 4 shows a typical photograph obtained. The stream disturbances are again alternate, being now closely packed and as before extending to the end of the cellular pattern, which is almost obliterated by the disturbance pattern and may be seen only with difficulty, the cells, however, undergoing appreciable lateral displacements.

Two forms of sound wave are also evident, corresponding to those of circular jets discussed earlier. The high frequency downstream radiation is clearly seen, and the upstream waves now take a clearly defined form. As before the waves are in antiphase on each side of the jet, and the phenomenon as a whole was much more stable than in the case of the circular jet. From a knowledge of the optical system and the observed sound field it is possible to estimate[3] the sound level to

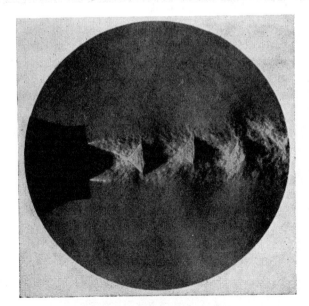

FIG. 3. Example of jet with low frequency sound waves moving upstream.

[3] A. Powell, "The Schlieren Method in Acoustics," University of Southampton (to be published).

be about 165 db at the edge of the photograph shown. It was found that the wavelength was uniquely dependent upon the pressure ratio, the relation closely following that for the spacing of the cellular pattern. Empirical rules obtained from a series of photographs or one particular nozzle design are

$$\text{cell spacing} = s = 1.89(R-R_c)^{\frac{1}{2}}\times\text{jet depth},$$

$$\text{upstream wavelength} = \lambda = 0.63(R-R_c)^{\frac{1}{2}},$$

where "jet depth" means the smaller of the dimensions of the orifice. R is the pressure ratio, R_c being the critical value for sonic exit velocity. The close similarity between these results suggests that they are in some way connected.

By using reflectors inclined at 45° to the stream (Fig. 5) or by rotating the critical knife edge of the optical system 90°, it was found that there was in addition radiated in a comparatively narrow beam normal to the jet, not detected in the other photographs by virtue of the insensitivity of the optical system to density gradients parallel to the critical knife edge. It was interesting to note that the frequency of this sound directed normally to the stream was precisely double that observed to be passing upstream.

The mechanism by which these stream disturbances are created, and the salient characteristics of the sound produced (i.e., the powerful radiations in the upstream direction and normal to the stream at double the frequency) have already received some attention.[1,4]

It was suggested that the sound waves passing the exit give rise to the stream disturbances. The position is rather similar to that of sensitive jets,[5] where the disturbances originate as a result of the distortion of

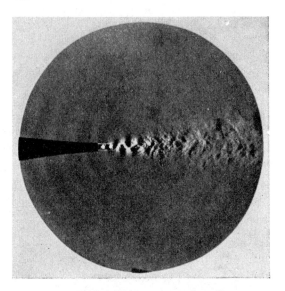

FIG. 4. Typical jet from rectangular orifice.

[4] A. Powell, "On the Noise Emanating From a Two-Dimensional Jet Above the Critical Pressure," University of Southampton, June, 1952 (to be published).

[5] E. N. daC. Andrade, Proc. Phys. Soc. (London) **53**, 329 (1941).

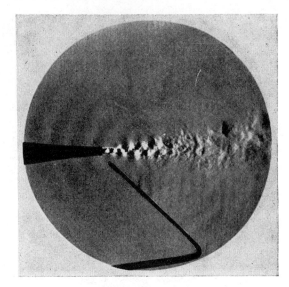

FIG. 5. Jet in presence of 45° reflector.

the stream caused by that part external to the orifice moving with the acoustic vibrations of the atmosphere, yet that about to leave being constrained from lateral movement. In the present case the fluctuating sound pressures at the orifice will result in the boundary taking up slightly different directions on leaving the orifice since the pressure ratio across it will be fluctuating. Since the fluctuations are of opposite sense on each side of the stream, the stream as a whole will be deflected very slightly from side to side and, together with instability, will give rise to amplification of the disturbances which after a certain distance become strong enough to be photographed.

It is supposed that these disturbances interact with the cellular pattern, most probably with the shock waves near the end of the cells and so give rise to pressure fluctuations at the boundary and this will in part be radiated as sound, since they will then be hydrodynamically unbalanced owing to the pressure changes arising from traversing the shocks. It seems significant that the characteristic noise of choked jets does not appear immediately when sonic speed is reached, but after the pressure ratio has been further slightly increased. This feature is also found in the formation of shock waves. Thus there will be a number of sources with spacings equal to those of the cellular pattern and whose relative phase is dependent upon the wavelength of the stream disturbance. Now it can be shown[4] that the directional properties for the fundamental and harmonic of such a combination on one side of the stream can be expressed as

$$D_f = \cos 2\alpha \cdot \cos \alpha,$$

$$D_h = \cos 4\alpha \cdot \cos 2\alpha,$$

being the factors to apply to the arithmetical sum of the strengths of the four equal sources considered to obtain the relative strength in the direction β to the

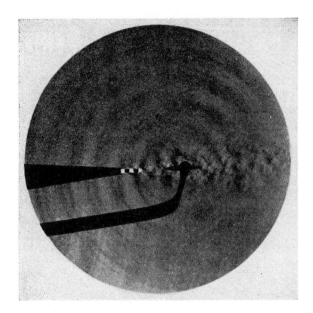

FIG. 6. Schieren photograph of jet when an
edge tone is formed.

stream direction, where

$$\alpha = (\pi s/\eta)(1 - M\cos\beta),$$

in which s is the spacing of the cells, η the wavelength of the stream disturbance which is moving at a speed of M times the speed of sound. Now, if it is assumed that the upstream radiation is at the fundamental frequency, the Mach number M can be obtained immediately from the disturbance wavelength relative to the radiated wavelength. By taking typical values of these parameters it is easily shown that the fundamental has a large relatively powerful lobe in the upstream direction, limited to within 45° or so of the axis and that the harmonic has a powerful but narrow beam normal to the axis, but there are no other lobes of comparable magnitude. The choice of four such sources is an approximation to the actual state of affairs, being taken since it was over that number of cells that the disturbances were most vigorous, but similar results are obtained for slightly different numbers of sources. Thus the characteristic directional properties can be accounted for, and it remains to show that the dimensions of the motion are consistent with such a mechanism.

This was done in the following manner. The sound can be assumed to originate at an effective source, which can be located fairly accurately at the center of the arcs formed by the sound waves. It is at once evident that there may be more than one complete cycle of the stream disturbance between the orifice and the effective source (distance d apart) and, indeed, more than one wavelength of sound wave. Nor will there be necessarily an integral number of cycles or wavelengths. A further complication is that the dis-

turbance speed is not necessarily constant nor negligible compared with the speed of sound. If a certain phase of sound wave is assumed to leave the effective source, a sound wave of similar phase, but l cycles ahead, will reach the orifice a certain time later during which time the disturbance originally at the source will have moved a certain distance beyond it. Let this disturbance which was originally at the orifice now be k disturbance wavelengths, i.e., $k\eta$ from the orifice. Then the distance between the orifice and the source can be written as

$$d = k\eta - Mc(d - l\lambda)/c,$$

and we can use the value $M = \eta/\lambda$. In this way we can find k by using data from any chosen photograph. It will be noted that η is the measured disturbance wavelength near the source, so that k is not synonymous with the number of cycles, but is an effective value. Different values of k were obtained for two different designs of nozzle, being attributable to different rates of instability and turbulence. The actual values found were 1.9 and 2.9, indicating that an additional cycle of the disturbance was present in the latter case. Having determined the value of k, which was found to be consistent with the photographic evidence, the frequency or other property may be found for different pressure ratios. Actually it was the disturbance spacing η which was compared with experiment. Substantial agreement was found: the coefficient of variation of 3 percent and mean fractional deviation of about 0.6 percent are satisfactory in view of the probable experimental error.

IV. GENERAL CONSIDERATION

It will perhaps have been noticed that this phenomenon has much in common with edge tones, since there also the sound effectively emanates from a certain point. An edge was placed in the stream, and an edge tone did ensue, a schlieren photograph of the motion being shown in Fig. 6. The curious double structure of the sound waves is an extreme case of a tendency noticed in many photographs, the waves in the lower half of the picture having been reflected off the local structure. A most important feature is the phase relationship of the sound waves on each side of the jet, so enabling stream disturbances to be created in the manner suggested earlier for the jet alone. This phase difference also exists in the normal low speed case, as can be seen by traversing a pair of microphones and comparing their outputs on an oscilloscope. This observation has made possible an explanation of the edge-tone phenomena, enabling a purely theoretical result to be obtained which is in good agreement with experiment.[6]

Now it can be shown that for such a process to be

[6] A. Powell, "Edge-tones and Associated Phenomena," University of Southampton, August, 1952 (to be published).

maintained, the total "gain" around the circuit must be unity and the same is evidently true of the jet whistling above choking, which as we have seen can be self-maintained and superimposed on the general turbulence. Thus if the efficiency of some part of this cycle could be reduced, the amplitude of the motion would decrease or cease altogether. There are clearly several ways of doing this: for example, ventilation of the nozzle to remove the excess pressure, so weakening the shock-wave formations, or the introduction of additional turbulence or thickening of the boundary layer to lower the rates of amplification.

The first might be achieved by a gauze cylinder forming an extension to the nozzle,[7] the other by a comparatively slight nicking of the exit edge of the nozzle or by introducing cambered radial vanes into the flow, so as to tend to set up steady axial vortices at the jet boundary and others internally. Such techniques have been tested in practice,[8] and an appreciable success achieved: the characteristic noise of a choked jet being much reduced or completely eliminated, in some cases with a decrease in noise level of as much as 20 decibels.

[7] Such a method of ventilation was suggested by the College of Aeronautics, England.

[8] A. Powell, "On the Mechanism and Reduction of Choked Jet Noise," University of Southampton (to be published).

AN INVESTIGATION OF THE NOISE FIELD
FROM A SMALL JET AND METHODS FOR
ITS REDUCTION

R. Westley and G. M. Lilley
Department of Aerodynamics
The College of Aeronautics,
Cranfield, England.

SUMMARY

Sound measurements have been made on the noise from the jet of a
one inch diameter convergent nozzle at atmospheric temperature and at
speeds above and below choking. The noise level and spectrum have
been investigated in both the near and distant fields. The results
agree in some measure with the predictions of the Lighthill theory,
that the elementary sound radiator is an acoustic quadrupole. The
agreement is more marked if attention is confined to the higher frequencies.

Simple empirical formulae are derived giving the overall sound
intensity and frequency spectrum in terms of the position relative to
the jet, the stagnation pressure excess over the atmospheric pressure,
and the frequency.

The results of tests on various noise reduction devices are discussed. These tests indicate promising lines of investigation. The
maximum reduction in total noise level was about 10 db.

INTRODUCTION

In this report the following definitions of boundary layer, jet
and aerodynamic noise are used.

Definitions

Boundary layer noise is the noise created in the boundary layer
of moving fluids adjacent to solid surfaces (due either to the turbulence or some other cause).

Jet noise is the noise created in the mixing region between a jet
of moving fluid and an external medium due to the presence of turbulence or some other vortex motion. This excludes mechanical noise,
in particular noise due to rotating machinery.

Aerodynamic noise is noise resulting from air in motion and
includes both boundary layer and jet noise. It also covers such noise
as that arising from the wakes of bluff obstacles and that created
inside ducts.

It is well known that air jets and boundary layers cause noise.
At present, however, very little is understood about how the noise is
caused. Interest in this subject has recently increased with the high
speeds and jet engines of modern aircraft. Noise is now a serious and
growing source of distress to passengers in aircraft and to persons on
the ground.

An allied problem which initially prompted the investigation of
aerodynamic noise arose in the interpretation of measurements made in
a low turbulence wind tunnel. In a low turbulence wind tunnel it is
difficult to differentiate between fluctuations in velocity due to
noise and turbulence. It was considered that measurements both of
noise and turbulence in an air jet would be a convenient means of making a preliminary attack on this problem.

The present series of experiments were planned

a. to make measurements of noise level and frequency around an air
    jet at speeds above and below choking so as to provide data ena-
    bling the problem of noise creation in the jet to be studied,

b. to check existing theories, notably the Lighthill theory, on jet
    noise,

c. to investigate devices for jet noise reduction.

The principal theoretical contribution to the noise created in
boundary layers and jet is that due to Lighthill[1]. The papers by
Billing[2] on moving sources of sound, the estimation of pressure
fluctuations in isotropic turbulence by Batchelor[3] and on stellar
dipoles and quadrupoles by Schwarzschild[4] represents the only additional theoretical work on this subject. A recent paper by Powell[5]
has reported on some interesting experimental work in connection with
the visualization of sound waves from a converging jet at speeds above
choking.

FURTHER DEFINITIONS AND PRESENTATION OF RESULTS

The following definitions of certain quantities will be used in this report.

Coordinates of the Microphone

The position of the microphone relative to the nozzle exit will be measured, in the distant field, in terms of spherical polar coordinates r, θ with the jet axis as datum.

In the near noise field it will be found more convenient to define the microphone position in terms of cylindrical coordinates r, x; r is measured normal to and x along the jet axis. The sound field is assumed to be axially symmetric.

The Nozzle Pressure Ratio

The jet exit speed (M_1) below choking can be found from the nozzle excess pressure ratio

$$\mathcal{P} = \frac{p_s - p_o}{p_o}$$

and

$$M_1^2 = \frac{2}{\gamma - 1} \left[(1 - \mathcal{P})^{\frac{\gamma}{\gamma-1}} - 1 \right].$$

Choking of the jet occurs at values of $\mathcal{P} \geqslant 0.893$.

Above choking the rate of mass flow (slugs per second)

$$m \approx \frac{0.01231 \, A_1 \, p_o (\mathcal{P} + 1)}{\sqrt{T_s}}$$

and

$$U_1 = 60.10 \sqrt{T_s}$$

where

A_1 = area of nozzle exit (sq. ft.)

p_o = atmospheric pressure (lb. per sq. ft.)

T_s = temperature at contraction entry (^{o}K)

$$U_1 = \text{speed at nozzle exit (f.p.s.)}$$

Reynolds Number

The Reynolds number, based on the nozzle exit diameter and the jet exit speed, of the tests described in this report, can be obtained from the relation

$$\frac{R}{M_1} = 0.59 \times 10^6$$

where

$$R \equiv \frac{\rho_1 U_1 d}{\mu_1}$$

M_1 = Mach number at the nozzle exit.

APPARATUS

Test Rig

The noise measurements were made with a horizontal air jet which emerged from a circular nozzle of one inch diameter. The nozzle was placed fifteen feet above ground level and about fifteen feet from the nearest building in order to reduce sound reflections. A view of the test rig is shown in Figure 1.

The air compressor, cooler and settling tank were situated in the main laboratory and were connected to the jet by about 30 feet of 6 inch diameter iron piping. The piping passed through two brick walls to the outside of the laboratory, through a gate control valve and then up over a supporting pylon to a contraction and the nozzle. The contraction had an area ratio of 36:1 and was designed by the method given in Reference 6 to give a monotonic increase in velocity along the boundary wall from entry to exit. A gauze screen was fitted at the contraction entry.

During each test the compressor was run continuously. Except at very low jet speeds, the noise from the compressor reaching the microphone placed in the jet field was small compared with the jet noise. The jet speed was maintained steady by adjusting the compressor inlet valve and the gate control valve. The pressure drop across the latter was kept as small as possible in order to avoid further noise. The air supplied to the jet was undried but condensed water vapour was drained off from the settling tank and the vertical pipe leading to the nozzle. No icing up of the jet nozzle occurred.

The static pressure at the contraction entry was measured on a mercury manometer.

Noise Measuring Apparatus

The jet noise was measured using a 74100-C objective noise meter together with a 74101-A octave filter, both manufactured by Standard Telephone Ltd. The octave filter has 16 ranges and noise levels in any of the following frequency bands may be measured.

Octave Number	Frequency Band (cycles/second)	Octave Number	Frequency Band (cycles/second)
11	37.5 - 75	21	50 - 100
12	75 - 150	22	100 - 200
13	150 - 300	23	200 - 400
14	300 - 600	24	400 - 800
15	600 -1200	25	800 - 1600
16	1200 -2400	26	1600 - 3200
17	2400 -4800	27	3200 - 6400
18	4800 -9600	28	6400 -12800

The microphone, of the moving coil type, was sensitive over the range 50 - 10,000 c/s. The sensitivity was not constant over the whole frequency range and a "microphone correction" had to be applied as follows.

Octave Number	Correction	Octave Number	Correction
28	Add + 5.5 db.	24	+ 1.0
18	+ 6.0	14	+ 1.0
27	+ 5.0	23	+ 0.5
17	+ 5.5	13	+ 0.0
26	+ 4.5		
16	+ 4.0		
25	+ 3.0		
15	+ 2.5		

An additional correction was necessary to the lower octave readings if high noise levels were present in the higher octaves. For this reason it has been necessary to discount a large number of readings obtained in the octaves below 13. The contribution to the total noise, from these lower frequencies, was however very small.

Distant noise field tests were made with the microphone pointing toward the jet nozzle. The microphone was set in the horizontal plane containing the jet axis. It was loosely suspended by rubber strips in a wire box frame 15 in. x 15 in. x 15 in. The frame was covered with muslin to prevent atmospheric winds blowing on the microphone. The frame was supported by a vertical pole 15 feet high. The position of the microphone was located with the aid of a theodolite.

Near noise field tests were made with the microphone in a vertical plane below the jet axis and pointing toward the axis. The microphone was supported on a movable beam and no wind shield was fitted.

Total head traverses were made across the jet at various stations downstream of the jet exit. The pitot tube was made of hyperdermic tubing of 1.0 mm external diameter. The pressures were measured on a mercury manometer.

Noise reduction devices. The nozzle end of the contraction (Figure 2) was constructed so as to permit the connection of various noise reducing devices (Figure 3). These devices were as follows:

a. A cylindrical wire gauze extension of 1 inch diameter and 1.5 inches long. The mesh of the gauze was 30 per inch.

b. A slotted extension (Figure 3b). This was 1.75 inches long and the slots, at 0.75 inches from the end, were alternately bent inwards and outwards.

c. Teeth extension (Figures 3 c.1, 2, 3, 4). The teeth were set at various angles γ^o (positive if inwards) to the jet axis.
    c.1. Six teeth   0.25 in x 0.25 in   .
    c.2. Twelve teeth   0.125 in x 0.125 in
    c.3. Twenty-four teeth 0.0625 in x 0.0625 in
    c.4. Six teeth ⎤   0.125 in x 0.375 in
         Six teeth ⎦   0.125 in x 0.25 in

d. Diffuser and slot (Figure 3d). A circumferential slot of variable width was situated at the throat of the diffuser.

DETAILS OF RESULTS

The tests may be grouped under the following headings:

A. The distant sound field.

B. The near sound field.

C. The velocity distribution in the jet.

D. Noise reduction devices.

The atmospheric pressure p_o, atmospheric temperature T_o, and the jet stagnation temperature T_s were recorded during each test. These quantities varied according to the daily atmospheric conditions. The noise level meter was not of sufficient accuracy to detect any changes of noise intensity due to these variations. Mean values of

p_o, T_o, and T_s were

$$p_o = 29.9 \pm 0.5 \text{ in. Hg}$$

$$T_o = 10^\circ \pm 10^\circ C$$

$$T_s = 17^\circ \pm 5^\circ C$$

The curves of noise level at each octave are labeled in each diagram where possible with the appropriate octave number. In order to distinguish the curves at various parts of the speed range, where intersection occurs, the following set of symbols will be used:

Octave Number	28	18	27	17	26	16
Frequency	12,800 -6,400	9,600 -4,800	6,400 -3,200	4,800 -2,400	3,200 -1,600	2,400 -1,200
Symbol	☉	●	X	↖	△	∨

Octave Number	25	15	24	14	23	13	22	12
Frequency	1600 -800	1200 -600	800 -400	600 -300	400 -200	300 -150	200 -100	150 -75
Symbol	∼	,	⊡	✳	↘	▽	+	Y

Octave Number	21	11	Total Noise
Frequency	100 -50	75 -37.5	
Symbol	⌢	◇	✳

A. The Distant Noise Field

The variation of noise level with $\bar\omega$. The noise level plotted against $\log(\bar\omega/d)$ shows that for $\bar\omega/d > 50$ the noise intensity follows an inverse square law

$$I \approx 1/\bar\omega^2$$

for both high and low frequencies and for total noise. The result is true for speeds above and below choking.

The variation of total noise level with θ. ($\mathcal{P} = 0.9$; $\bar\omega/d = 108$). The total noise level was obtained for values of θ up to 150°. It was found impossible to take accurate measurements at less than 20°. These results show that a maximum occurs at $\theta = 28°$ and the noise level then decreases as the jet is approached.

The variation of octave noise level with θ. ($\mathcal{P} = 0.9$; $\bar\omega/d = 108$). The octave noise level is plotted against θ in Figure 4. The results show a similar trend to those for total noise, but the values of θ for maximum noise intensity, θ_{max}, vary in each

155

frequency band as follows:

Freq. c/s	300 -600	400 -800	600 -1200	800 -1600	1200 -2400	1600 -3200	2400 -4800	3200 -6400	4800 -9600	6400 -12800
θ_{max}	<20°	<20°	<20°	20°	25°	25°	26°	31°	38°	45°

It is significant that the angle of maximum noise increases as the frequency increases. (A lower maximum at $\theta = 100°$ occurs at the high frequencies. This is probably due to experimental error but is worth further investigation.)

The octave noise level is plotted against octave number (or log n) in Figure 5. It is shown that the noise spectrum is roughly similar at all angles. The octave number for maximum noise at a given value of θ is as follows:

θ	150°	120°	90°	60°	30°	20°
frequency c/s at maximum noise	3200 -6400	3200 -6400	3200 -4800	3200 -4800	2400 -4800	1600 -2400
octave number	27	27	17-27	17-27	17	26-17

The <u>variation of noise level with</u> $\mathcal{P}$. ($\theta = 90°$; $\omega/d = 108$). The noise level is plotted against $\mathcal{P}$ in Figure 6 for $\theta = 90°$. In Figure 6 it can be seen that above choking ($\mathcal{P} \geqslant 0.893$) the results for octaves lower than 14 not only register equal noise levels but have slopes similar to those corresponding to the highest frequencies. It is suggested that the octave filter was leaking from the high frequency to the low frequency bands at the high noise levels. Although the readings in the lower octaves at low total noise levels appeared to be correct the results for all octaves below 14 have been discarded, except for measurements in the near sound field.

It is shown that a maximum occurs at about $\mathcal{P} = 1.1$ followed by a minimum at about $\mathcal{P} = 1.3$. A rapid rise in noise intensity then follows.

The background noise, with the jet turned off was as follows:

Octave number	13	23	14	24	15	25	16	26	17	27	18	28
Octave Noise (db)	60	62	60	55	<50	<50	<50	<50	<50	<50	<50	<50

The rate of increase of noise intensity with jet speed. (θ = 90°; $\bar{\omega}/d$ = 108). The results from Figure 6 have been replotted in Figure 6 using a logarithm scale for abscissa. Similar results are obtained at other values of θ. It is shown that when $\wp$ < 1 and > 1.2 the intensity, both total and octave, follows the law

$$I \approx \wp^b$$

where the index b is a function both of frequency and θ. The values of b, below choking, approach the value of 4 as θ tends to zero. Between θ = 30° and θ = 90° the mean value of b, below choking, is about 2.5 and above choking at the higher frequencies is about 8.5.

The variation of the noise spectrum with jet speed. (θ = 90°; $\bar{\omega}/d$ = 108). The effect of jet speed on the noise spectrum is shown in Figure 8 for θ=90°. It can be seen that the spectrum is similar for all speeds in the range 0.3 < $\wp$ < 1.3. The maximum noise intensity shifts slightly toward the higher frequencies as the jet speed is increased.

B. The Near Noise Field

The average of a large number of results has been plotted as sound contour lines (lines of equal sound intensity) in Figures 9, 10 and 11 for a value of $\wp$ = 0.9. The diagrams are drawn for both total noise contours and octave noise contours. The low frequency contour diagrams (apart from octave 11) have not been reproduced. The contour line diagrams for octave 11 shown in Figure 11 show marked differences from the higher frequency diagrams beyond a distance of about 20 diameters from the nozzle exit. Since the high frequency noise level in this region is considerably lower than in the regions near to the nozzle exit, it is considered that the portion of the lowest frequency diagrams beyond, say, 20 diameters may be reliable.

The various contour diagrams clearly show that the jet contains a distribution of sound sources the extent of which increases in the downstream direction as the frequency is decreased. The approximate extent of these sources along the jet axis at $\wp$ = 0.9 is as follows:

	Position of noise sources
Total noise	0 - 30 diameters
Octave 28	0 - 10 diameters
27	0 - 14 diameters
26	0 - 16 diameters
25	0 - 18 diameters
24	0 - 24 diameters
11	Commencing at about 25 diameters and extending to distances far downstream

It can be seen that although the total intensity contour lines do not show marked directional characteristics, the contour lines at high frequencies show maxima at 45° to the jet axis measured from an origin coincident with the downstream position of the source distribution in the appropriate frequency band.

With the origin of coordinates at the nozzle exit, the direction of maximum noise approaches the jet axis as the frequency is reduced. This result confirms the observations made in the distant noise field.

The contour diagrams above do not extend to the regions upstream of the nozzle. Measurements obtained in this region, at various frequencies, did show a second maximum of lower magnitude at the highest frequencies, but the increase in noise level was only about 2 db and in general the contours showed regular continuations of the lines in the downstream region.

C. The Velocity Distribution in the Jet

The distribution of total head at various stations along the jet axis at $\wp$ = 0.9 is shown in Figure 12.

The boundary layer thickness at the nozzle exit was 0.08 diameters.

D. Noise Reducing Devices

Lighthill[1] has shown that the sound intensity is proportional to the square of the velocity gradient. It was considered, therefore, that a reduction in the noise intensity would be obtained by the use of devices which would reduce the rate of shear. Only those devices which did not cause excess blockage were tested.

It was found, at a very early stage in the investigation, that a drop of 10 db in noise intensity could be obtained if a solid object projected into the jet boundary.

The wire gauze extension. Below $\wp$ = 1 no change in noise level was obtained. Above this value of excess pressure ratio a reduction of 5 db occurred. The effect of the gauze was to delay the rapid noise increase which was present above $\wp$ = 1 (this increase can be seen in Figure 5). An optical investigation showed that the train of shock waves, present above choking, were eliminated by the introduction of the gauze. In the latter case, an outflow took place through the gauze, the flow expanded inside the gauze, and the jet emerged as a supersonic stream free from expansion and shock waves. As the pressure increased still further the length of gauze was insufficient to allow a complete expansion down to atmospheric pressure. The shock pattern then began to reform with a corresponding rapid increase in noise level.

The effect of teeth extensions to the nozzle. The noise levels for thirteen different teeth configurations are compared with the unmodified nozzle results in Figure 13. The corresponding noise spectrum variations are shown, for $\mathcal{P}$ = 0.67 and $\mathcal{F}$ =2.0, in Figures 14 and 15 respectively. It will be seen that reductions in noise level of 5 db below choking and 12 db above choking can be obtained with six teeth about 0.25 diameters square; three set in line with the jet axis and three set inwards at 30° to the jet axis. A complete investigation of the corresponding reduction in mass flow was not made, but preliminary measurements indicated that the change was small. The maximum reduction in noise level was obtained with all the teeth pointing inwards. Some blockage did then occur and the results are therefore not quoted.

The optical investigation showed that the teeth deflected the flow inwards whilst the gaps between the teeth allowed an outflow. This resulted in a corrugation of the jet boundary. In an overchoked jet the corrugation of the jet boundary was still present but the shock pattern was diffused.

The effect of the diffuser and slot. (θ = 30°; ω/d = 108). The results of a large number of tests, including the variation in slot width and diffuser angle, are given in Figure 16. The corresponding noise level curve for the unmodified jet (square end see Figure 16) is also included. The results showed some scatter but below choking no noise reduction occurred.

Above choking a reduction in the noise level was observed but this reduction was also present when the diffuser was removed (h = ∞), i.e., with the edge of the nozzle exit sharp. It appeared, therefore, that part of the high noise level obtained above choking with the unmodified nozzle was due to the blunt end at exit. It appears that a reduction of 3 db approximately can be obtained with a sharp edged nozzle. The tests were, however, not conclusive and further investigation of this effect is necessary.

A large reduction in noise level was experienced when the nozzle end projected into the diffuser, (Figure 16; h = -0.685 in.). At $\mathcal{P}$ = 1.47, with increase of pressure, a reduction in noise level of 11 db was obtained but, with decrease of pressure, the corresponding increase in noise level did not occur until $\mathcal{P}$ = 1.26. The reduction in noise level corresponded to a reattachment of the flow in the diffuser due to the expansion of the flow at the nozzle exit above choking. The flow in the diffuser was then supersonic and full expansion to atmospheric pressure took place. Below and just above choking breakaway of the flow in the diffuser was present.

DISCUSSION OF RESULTS

The Distant Noise Field

It can be assumed that as ω/d approaches infinity the effect of the complete jet will be equivalent to that of a point sound radiator at the origin. The measurement in the distant field can, therefore, be applied to determine the characteristics of this point radiator and their variation with jet speed. It has been shown that the high frequency noise has a maximum intensity along lines at 45° to the jet axis. The lower frequencies do not exhibit a maximum along lines of 45° but since the measurements were made at only 108 diameters from the nozzle exit it is probable that asymptotic conditions were not reached.

The spectral density I_n, at high frequencies, follows, therefore, the asymptotic relation

$$I_n \sim \frac{x^2\, r^2}{(x^2 + r^2)^3} \, .$$

The total intensity contours do not show maxima along lines at 45° to the jet axis, but appear to be weighted by the low frequency spectrum. If the lateral quadrupoles in Lighthill's theory[1] are at rest relative to the jet, the total intensity contours would indicate maxima along lines at 45° to the jet axis, but if they are in motion the angle for maximum intensity, relative to the jet axis, decreases as the relative speed increases. It appears, therefore, from the results above, that the low and high frequency sound radiators are moving at different speeds relative to the jet axis.

On the basis of dimensional analysis the following law for the total intensity at a point in the noise field has been obtained.

$$I = \frac{h(\theta)}{\omega^2} \, \wp^{\,b(\theta)} \, \rho_o \, d^2 \, a_o^3 .$$

The index $b(\theta)$ varies for speeds just below choking from 4.0 inside the jet to 2.0 at about $\theta = 65^{\circ}$. Lighthill's theory suggests the value of 4.0 for all θ. At very low speeds $b(\theta)$ approaches the value of 4.0, but at high speeds above choking it attains a value of about 8.5. The change in $b(\theta)$ above choking appears to be connected with the shock pattern in the jet. When the shock train is eliminated (e.g. by the use of the gauze extension) it is possible to extend the unchoked pressure law to the case of the supersonic jet. It appears therefore that a value of b about 3.0 is associated with the noise produced by the turbulence in the mixing region.

160

Powell[5] has observed that ultrasonic waves* are emitted from an overchoked jet containing a train of shock waves, but there is evidence to show that this phenomenon is quite distinct from the creation of noise below choking.

It is possible, however, that the irregular results above $\wp$ = 1.3, and in particular the "whistle" obtained at θ = 30°, $\wp$ = 1.6 at 10 kc/sec may be due to the very high frequency waves of the type observed by Powell.

The Near Noise Field

A full discussion of these results is given in Appendix 1 and only the salient remarks will be summarized here.

Although future investigations may throw light on how far the sound radiators created in the air jet retain their identity in motion down the jet, it is only possible, from the present measurements, to infer the distribution and strengths of the equivalent stationary sound radiators.

The high frequency noise radiators are distributed along the jet, near to the nozzle exit, for a distance which increases as the frequency decreases and is approximately proportional to the jet speed. The strength of the radiators reaches a maximum near the downstream end of their distribution along the axis.

It appears that the radiators behave like the lateral quadrupoles of Lighthill's theory although for the low frequency radiators the relatively long length of the jet along which they are distributed masks any directional effect which individual radiators may have and the evidence is therefore not conclusive.

The Lighthill theory also specifies a second lower maximum intensity along lines in the upstream direction at angles to the jet axis, which vary from 135° for a lateral quadrupole at rest, to about 115° when the quadrupole is moving at a Mach number of 0.9. The high frequency contours do show a second lower maximum between 90° and 110°, but there is no evidence of a second maximum in the contours for the total intensity.

It is very important to notice that the total intensity appears to be more heavily weighted by the lower frequencies at distances greater than 20 diameters from the nozzle exit. It is for this reason that past measurements have failed to reveal the marked directional character of the sound radiators.

---

* The jet diameter in Powell's experiment was 0.375 in. diameter. If the law $n_{max} \sim a_o/d$ is correct, then the corresponding frequency in these tests would be 10,000 c/s approximately.

The Velocity Distribution in the Jet

It has been shown that the jet velocity in the direction of the axis has its major decrease in the region 4 to 25 diameters downstream of the nozzle exit. It has also been shown that the velocity gradient $(\partial U/\partial r)$ decreases rapidly with the distance downstream of the nozzle exit. These statements are very significant when it is remembered that the major distribution of the sound sources occurs in this same region.

The Noise Reduction Devices

The object of these devices was to reduce the noise caused by the jets, bearing in mind their ultimate application to the jet units of aircraft.* The model tests were made to simulate the conditions of both the unchoked and the overchoked jet. Since the jet nozzle in the tests was stationary, the results may need modification when applied to flight conditions.

Of the various devices tested, it was found that the small teeth extensions (as used on the contraction sections of low speed open jet wind tunnels) in which half the total number of teeth pointed inwards whilst the remainder pointed along the jet axis, were the most satisfactory over the complete range of jet speeds above and below choking. The reduction in noise level was about 12 db above and 5 db below choking. It is suggested that the teeth could be incorporated in a device to vary the jet exit area.

The effect of the wire gauze extension in removing the shock pattern above choking is noteworthy apart from its sound reducing properties. A similar result should be obtained if it were fitted to a supersonic jet which emerged with excess static pressure.

The hysteresis effect obtained with the diffuser and slot is interesting but in view of its size, its possibilities are limited.

ACKNOWLEDGEMENTS

The authors wish to thank Professor A.D. Young who instigated this research and gave valuable advice at all stages in the investigation.

They also wish to acknowledge the help of Mr. E.G. Parkins and Mr. A. Peduzzi who assisted in setting up the apparatus and in the amassing of the experimental data, and they thank Mr. S.H. Lilley who was responsible for the construction of much of the apparatus and

---

* It has been observed that the high frequency noise in a jet aircraft increases with altitude. If this effect is due to overchoking at the jet exit, the suggested noise reduction devices may be used to delay this increase.

whose unfailing help was greatly appreciated.

They also wish to thank Mr. N. Fleming of the N.P.L. for the valuable assistance and advice he gave on numerous occasions during the investigation, and on whose unrivalled experience of the technique of noise measurement they were always able to draw.

Their thanks are due to the Director of the Royal Aircraft Establishment who kindly loaned the sound measuring equipment used in this investigation.

CONCLUSIONS

1. Measurements of the noise level and spectrum in the near and distant sound fields of a one inch diameter convergent nozzle at jet speeds above and below choking have shown that the elementary sound radiator (at least at high frequency) is a lateral acoustic quadrupole. This result is in agreement with Lighthill's theory of jet noise.

2. The high frequency noise radiators are distributed along the jet, near to the nozzle exit, for a distance which increases as the frequency decreases and is approximately proportional to the jet speed. The strength of the radiator reaches a maximum near the downstream end of their distribution along the axis.

3. An empirical formula is presented for the determination of noise intensity at a given position from the nozzle exit outside the jet.

4. Various noise reduction devices have been tested. Small teeth extensions, which caused only very slight jet blockage, produced reductions in noise level of 5 db and 12 db below and above choking respectively.

REFERENCES

[1] Lighthill, M.J., On Sound Generated Aerodynamically, Part I – General Theory, A.R.C. 14,407, F.M. 1630 (1951), Later published in Proc Roy Soc (A) 211, pp. 564-587 (1952).

[2] Billing, H., Sources of Sound in Motion, A.V.A. Monograph R and T 960 (1947).

[3] Batchelor, G.K., Pressure Fluctuations in Isotropic Turbulence, Proc. Cam. Phil. Soc., Vol. 47 pt. 2, pp. 359-374 (1951).

[4] Schwarzschild, Zeeman Shifts for Stellar Dipoles and Quadrupoles with Inclined Axis, Astrophysics Journal, Vol 112, p. 222 (1950).

[5] Powell, A., Some Experimental Observations on the Behaviour of Free Air Jets, University College, Southampton Research Note (1951).

[6] Lilley, G.M., Some Theoretical Aspects of Nozzle Design, London University M.Sc. Thesis (1945).

[7] Townsend, A.A., Measurements in the Turbulent Wake of a Cylinder, Proc. Roy. Soc. A. 190, p. 551 (1947).

[8] Corrsin, S., Investigation of the Flow in an Axially Symmetrical Heated Jet of Air, NACA ACR 3L23 (1943).

[9] Townsend, A.A., Momentum and Energy Diffusion in the Turbulent Wake of a Cylinder, Proc. Roy. Soc. A. 197, p. 124 (1949).

[10] Goldstein, S. (editor), Modern Developments in Fluid Dynamics, Vol. I, p. 288 (1938).

APPENDIX 1

The Sound Field Due to a Quadrupole and Verification of its Existence in the Jet Both Above and Below Choking

Lighthill[1] has determined that the intensity at (x) due to a point quadrupole at rest situated at the origin, is given by

$$I \sim \frac{x^2 r^2}{(x^2 + r^2)^3}.$$

Our results indicate that stationary distributions of sound sources, not necessarily continuous, exist at various positions in the jet. These sound sources have the directional properties of a lateral acoustic quadrupole. It is very necessary to point out, however, that these tests have not shown in all cases a second lower maximum noise intensity in the upstream region from the nozzle exit, which is an essential requirement of the quadrupole theory. It has been suggested that the jet nozzle will prevent this by obstructing the upstream direction. This suggestion appears unlikely except for the noise produced very close to the nozzle exit. It is possible, however, that the cause of this disagreement lies in the high speed of the jet whose upstream and downstream sound radiation are respectively decreased and increased as compared with a low speed jet. This suggested explanation appears to be the more likely one, since measurements below choking show that large changes in the noise field occur as the jet speed is increased.

The measurements show that the extent of the sound source distribution is a function of the jet speed and frequency. It appears that the complete sound source distribution is close to the nozzle exit only at very low jet speeds and at the high frequency. Although

Figure 10 for the high frequency, shows that the sound radiators, at least at the downstream end of their distribution, have the properties of lateral acoustic quadrupoles at rest, there is no direct evidence to show that these properties are also possessed by the low frequency sound radiators (see Figure 11).

These measurements cannot be used to predict accurately the characteristics of moving sound radiators in the jet (if indeed these are present) since the microphone readings give the mean sound intensity only over a long period of time compared with the characteristic time of the eddies responsible, presumably, for the phenomenon of noise. These results show (see Figure 10) that, at high jet speeds, the high frequency sound radiators exist in the region between the nozzle exit and 15 diameters downstream. The strength of these radiators reaches a maximum near the downstream end of their distribution. The low frequency sound radiators (see Figure 11) on the other hand, appear to be distributed over much greater lengths of the jet, but they still exhibit this characteristic maximum radiation of noise near the downstream end of their distribution.

One important feature of the experiments should be noted which, if later verified, may prove to play a very important part in the explanation of this phenomenon. It was noticed that when noise measurements were made using a wave analyzer having a relatively small band width, the readings displayed an irregular intermittency of the type already recorded by Townsend[7] and Corrsin[8] in measurements on free turbulent shear flow in wakes and jets respectively. This intermittency has been explained by Townsend[9] as being due to the irregular boundary between the jet and the undisturbed flow outside, (see, for example, Plate 24b of reference 10), which is caused by the billows of turbulent fluid moving outwards from the turbulent mixing region. These jets of turbulent fluid are probably large eddies which carry smaller eddies in dynamic equilibrium with each other. From the measurements of the noise frequency spectrum, it is evident that the small eddies, having high intensity and producing the high frequency noise, are confined to the regions close to the nozzle exit. On the other hand, since the large eddies, having high intensity, exist over much greater lengths of the jet, it might be expected that the corresponding output of low frequency noise is not restricted to relatively small volumes of the jet.

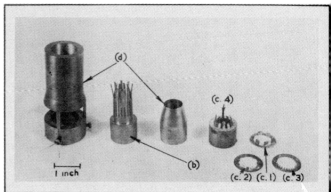

Figure 2. The contraction and nozzle exit. (Note: The Internal contour does not follow the external lines.)

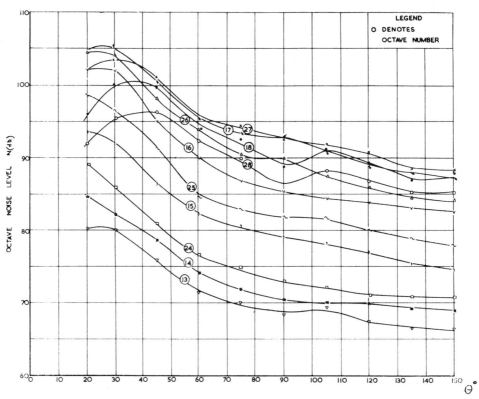

Figure 1. End view of the jet

Figure 3. The noise reduction devices

Figure 4. Variation of octave noise level with $\theta°$ (microphone correction applied)
$$\mathcal{P} = 0.9 \qquad \tilde{\omega}/d = 108$$

166

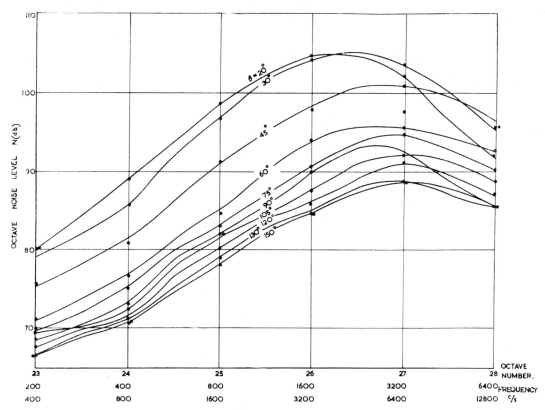

Figure 5. The variation of noise level with frequency (microphone correction applied) $\mathcal{P} = 0.9$ $\widetilde{\omega}/d = 108$

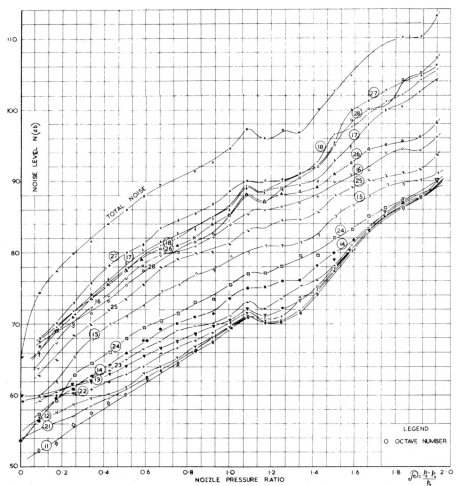

Figure 6. Variation of total and octave noise level with jet speed (microphone correction not applied) $\mathcal{P} = 90°$ $\omega/d = 108$

167

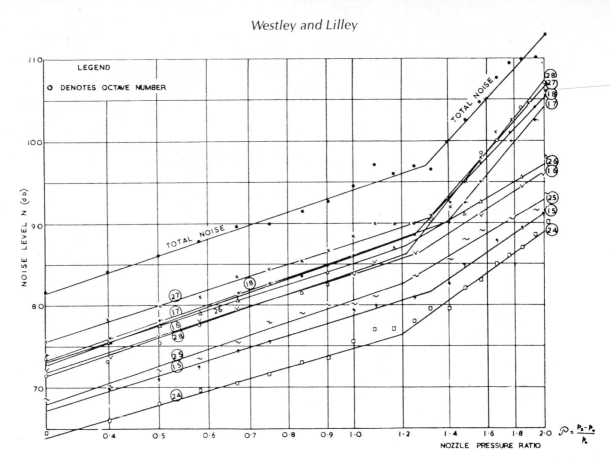

Figure 7. Rate of increase of noise level with jet speed (no microphone correction applied)
$\theta = 90°$ $\quad \widetilde{\omega}/d = 108$

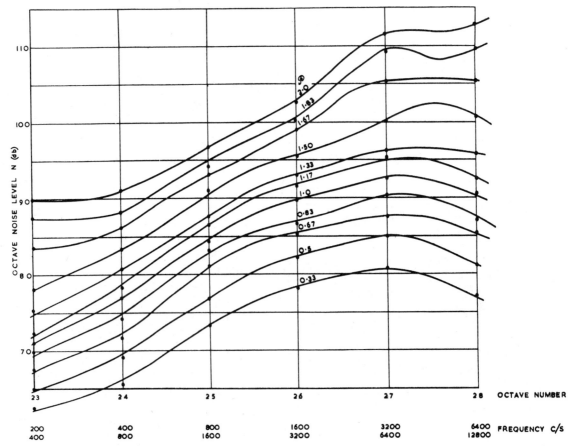

Figure 8. The noise level spectrum—effect of jet speed (microphone correction applied)
$\theta = 90°$ $\quad \widetilde{\omega}/d = 108$

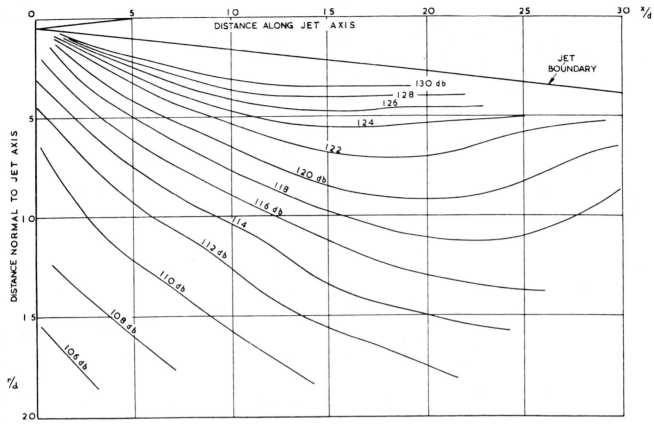

Figure 9. The near noise field (microphone correction not applied)
$\mathcal{P} = 0.9$ (total noise)

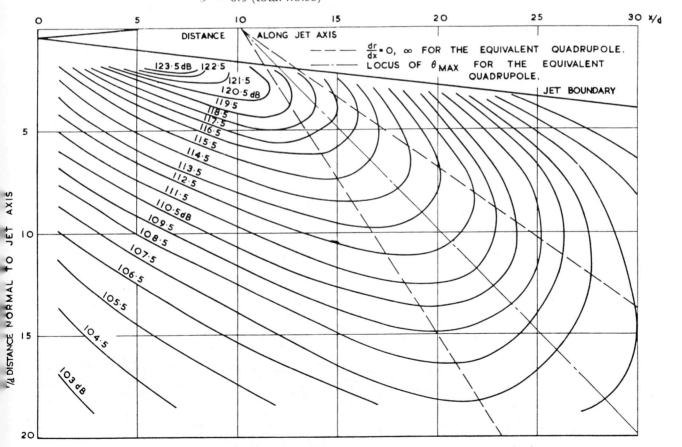

Figure 10. The near noise field (microphone correction not applied)
octave number = 28 $\mathcal{P} = 0.9$

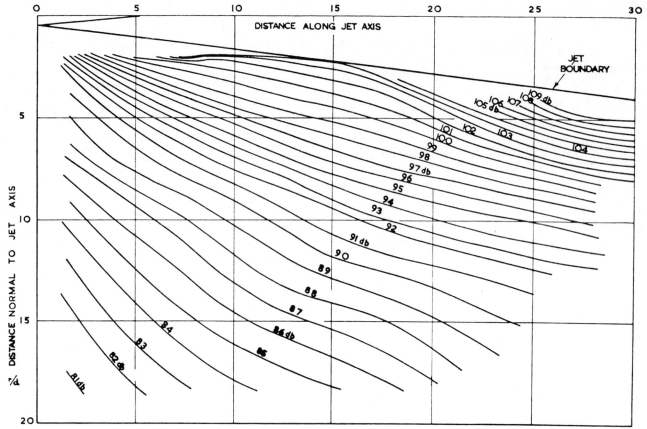

Figure 11. The near noise field (microphone correction not applied)   octave number =11 $\mathcal{P} = 0.9$

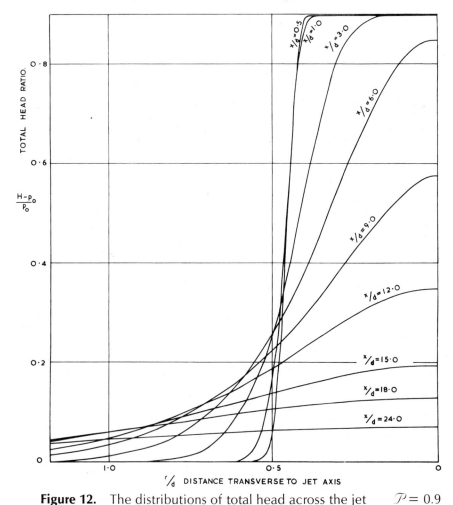

Figure 12. The distributions of total head across the jet $\mathcal{P} = 0.9$

170

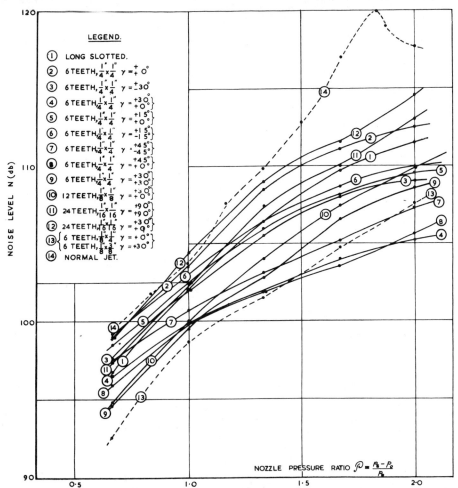

Figure 13. The effect of jet teeth on total noise—variation with jet speed
$\theta = 30°$ $\widetilde{\omega}/d = 108$

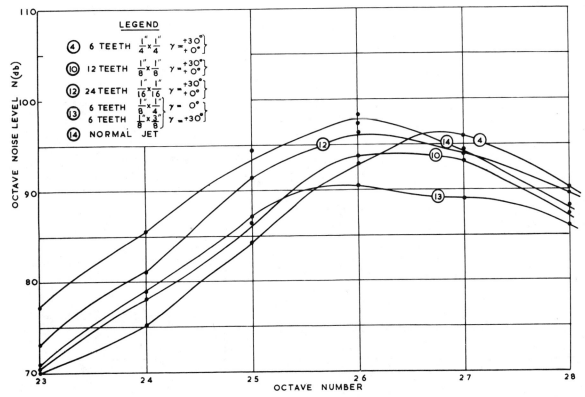

Figure 14. The effect of jet teeth on the noise spectrum
$\mathcal{P} = 0.67$ $\theta = 30°$ $\widetilde{\omega}/d = 108$

171

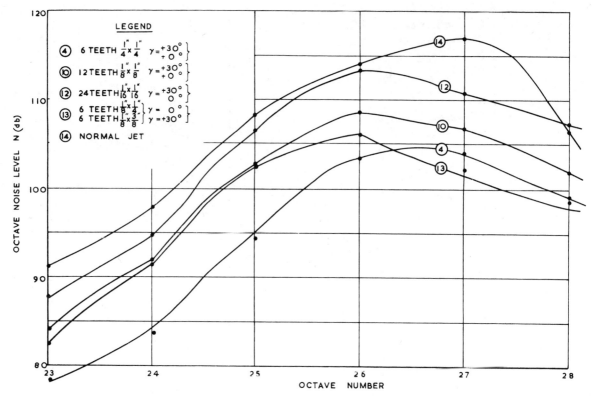

Figure 15. The effect of jet teeth on the noise spectrum
$$\mathcal{P} = 2.0 \qquad \theta = 30° \qquad \widetilde{\omega}/d = 108$$

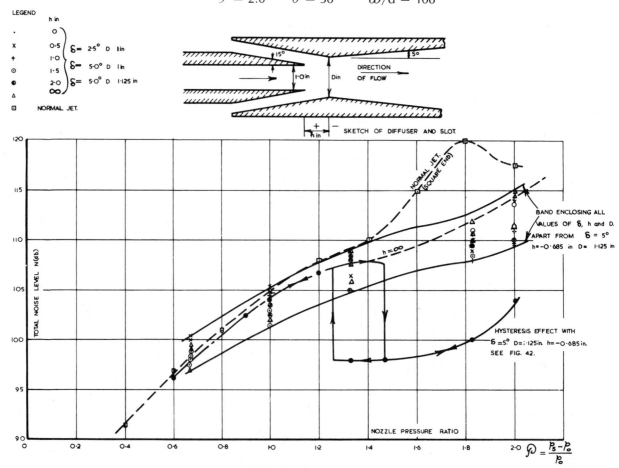

Figure 16. Noise reducing devices, the diffuser and slot effect of diffuser angle throat diameter and gap
$$\omega/d = 108 \qquad \theta = 30°$$

12

Reprinted from *Soc. Automot. Eng. Trans.* **70**:309–332 (1962)

Axial Flow Compressor Noise Studies

J. M. Tyler and T. G. Sofrin

Pratt & Whitney Aircraft Div., United Aircraft Corp.

This paper received the
1961 Manly Memorial Award

1. Introduction

During the development and initial operation of commercial jet transports, the aircraft industry was greatly concerned with the take-off exhaust noise problem. As commercial operations grew and incorporated procedures for reducing this take-off noise, the airport neighbors became increasingly more disturbed by the compressor inlet noise on approach. Meanwhile, Pratt & Whitney Aircraft was developing fan engines which gave an exhaust noise reduction of about 10 db. However, the quality and intensity of the compressor noise remained essentially unchanged so that the exhaust noise reduction, while obviously beneficial for the take-off condition, had the effect on approach of lowering the relatively moderate background against which the compressor was heard. From the standpoint of quality, the discrete-frequency compressor whine is more objectionable than broad band exhaust noise. Therefore, reduction of exhaust noise alone does not improve matters for the listener as the airplane approaches - in fact, it results in a stronger response.

Thus, the spotlight was focussed on the compressor noise problem more or less simultaneously by two factors:

1. The airport neighbors' reaction to the landing approach of the first jet transports.

2. The further unmasking of the compressor noise by the quieter exhaust of the fan engines.

Recognizing the increasing importance of compressor noise as exhaust noise is reduced, Pratt & Whitney Aircraft embarked on a major research and development program to obtain a better understanding of this whole problem.

Extensive facilities were developed for use in investigating the mechanisms of noise generation and transmission within the compressor. Several outside stands were used for measuring the noise radiated from full scale engines with and without sound-absorbing inlet and discharge ducts. Also, full scale compressor rigs were used for studies of many special aerodynamic and acoustic problems related to compressor noise. Many interesting papers could be written about the discoveries made in these studies. We have chosen to write about a portion of this whole area of study - the transmission of a pressure field from a compressor to the listener's ear.

In one sense we regret that so much space has been taken to cover but a small portion of the entire subject. However, the inclusion of relevant background material is intended to help make the picture clear to engineers who may not be familiar with the language and mathematics of acoustics.

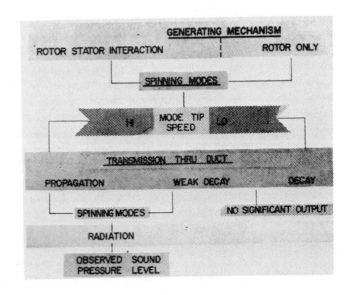

Fig. 1 — Organization of compressor noise study

2. Organization of Study

As a guide to the organization of the compressor noise study reported here, Fig. 1 has been prepared. Three general areas of study will be noted:

Generating Mechanisms - The nature of the sources of sound in an axial flow compressor is treated in this area. It is natural to consider first the case of a rotor alone, and to then take up the sources due to rotor-stator interaction. In both cases it is found that the sources produce rotating pressure patterns called spinning modes.

Duct Transmission - The behavior of the spinning modes as they are transmitted in a cylindrical duct enclosing the rotor is examined in this area. Depending upon the speed with which the generated spinning mode sweeps the duct wall, the magnitude of the associated pressure fluctuations, as a function of distance from the rotor, varies in one of two distinctly different ways. For sufficiently low tip speeds the intensity of the pressure pattern falls exponentially with distance, giving rise to the term "exponential decay," or, simply, "decay." In many cases the rate of decay is large enough to lower the intensity to a negligible value within a short length of duct. When the tip speed of the generated pattern is sufficiently high, however, the pressure fluctuations propagate in the duct as a wave of undiminished intensity. It then becomes important to consider topics included in the third area of study.

Radiation - In the case of propagating modes there will be a strong fluctuating pressure field at the face of the duct. The manner in which this field radiates sound into free space and the resulting angular directivity pattern is studied here as a function of operating parameters.

While the foregoing areas adequately partition the study into logical categories, a complete treatment of each in turn before proceeding to the next is probably not the best way to present material involving terms and concepts that may be familiar only to specialists. A more direct route to understanding the basic points of the study was considered to be a sequential selection of topics from each area, illustrated by reference to experimental techniques and results. Because the basic physical problems are intrinsically technical (one, for example, is the propagation of a polarized field in a wave guide), it is not possible to avoid mathematical description. Wherever possible, however, such material has been relegated to Appendixes so as not to appear to stand in the way of those seeking only a brief look at the subject. For convenience, symbols introduced in the text are collected and defined in Appendix A. During the course of the following presentation, an occasional reference to Fig. 1 may be helpful for orientation.

3. Generating Mechanisms - Rotor Only

This study is concerned with line spectrum type of compressor noise having components that are multiples of blade-passage frequency, rather than with broad-band noise of aerodynamic origin. Among the sources of this noise, of

greatest importance are effects due to steady aerodynamic blade loading and to blade thickness. Of less immediate concern are effects associated with vortex shedding, transient aerodynamic loading variations due to turbulence, and blade vibration.

The nature of the pressure field associated with steady aerodynamic and thickness effects is now considered in some detail. Fig. 2A is a developed view showing typical pressure contours around a rotor blade assembly at an arbitrary radius. If there are B blades in the rotor, they are spaced $2\pi/B$ radians apart, and the associated pressure contours also repeat in this interval. Fig. 2B is a plot of the pressure distribution as a function of angular position at a fixed distance forward of the blades. This distribution may vary in shape depending on details of the blade pressure contours, but two requirements must always be met:

1. Since the pressure pattern is associated with the rotor, it spins with rotor angular velocity $\Omega = 2\pi N$, when described in a fixed coordinate system. Therefore, the pressure must depend on a specific combination of the angular and time coordinates, namely, $(\theta - \Omega t)$. The form $p(\theta, t) = p(\theta - \Omega t)$ is standard for a traveling wave and says only that the pressure at a particular place and time arrives at another location θ radians forward at a time later by θ/Ω seconds.

2. Because the blades and associated pressure contours are spaced $2\pi/B$ radians apart, the pressure distribution must also repeat with this period. Recalling that a periodic function can be represented by a Fourier series, it is seen that at a fixed radius and axial location the pressure as a function of angle and time is given by

$$p(\theta, t) = \sum_{n=0}^{\infty} a_n \cos[nB(\theta - \Omega t) + \phi_n] \quad (3.1)$$

where a_n and ϕ_n are amplitude and phase parameters required to synthesize the particular pressure wave.

By means of this equation the pressure field in front of the rotor can be visualized in a manner that is both simple to recall and useful in the course of experimental and analytical work. Suppose a hypothetical photograph be taken at

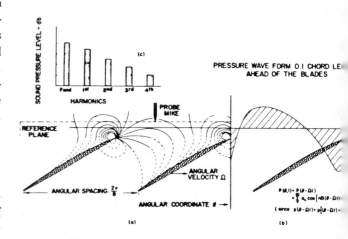

Fig. 2 — Rotor pressure field characteristics

some instant of time, say when t = 0. Then $p(\theta) = \Sigma\, a_n \cos$ (nB $\theta + \phi_n$) can be considered formed by a superposition of lobed patterns, each rotating at shaft speed. This concept is shown in the left hand diagrams of Fig. 3, illustrated for a (B = 4) blade rotor. Notice that the number of lobes in each pattern is an integer multiple, n, of the number of blades.

The complementary representation of the field is obtained by examining its time variation at a fixed angular position. Here $p(\theta, t)$ reduces to $p(t) = \Sigma\, a_n \cos$ (nB $\Omega t - \phi_n$). The right hand diagrams in Fig. 3 show the pressure-time wave form and its resolution into pure tones of harmonically-related frequencies BN cps (fundamental), 2BN (first harmonic), and so forth.

As suggested in this figure, the pressure-time dependence is readily obtained experimentally by means of a probe microphone and band-pass filter system. Fig. 4 shows some of the scale model compressor rigs from which extensive information has been obtained. These rigs are equipped with probe traversing capability in a cylindrical coordinate system which, together with suitable electronic gear, permits the pressure field to be completely determined as a function of three space coordinates and time.

As an illustration of the way this equipment is used to map the pressure field, suppose it is desired to obtain or verify experimentally the information shown in Fig. 3, concerning the pressure variation as a function of angular position and time at a constant radius from the axis in a plane forward of a 4-blade rotor. The probe microphone is traversed to the desired radial and axial location and set to a convenient angular reference position. Its output, an elec-

trical signal proportional to p(t), has the waveform shown in Fig. 5A. Harmonic analysis of this signal is conveniently made by means of an adjustable narrow band-pass filter. As the filter tuned frequency is slowly swept, its output rises sharply at frequencies observed to be multiples of blade-passage frequency. The filter output at these successive peaks corresponds to the pressure amplitude coefficients a_1, a_2 and so forth. A plot of these coefficients as a func-

Fig. 4A — Front 3/4 view of entire cantilevered rig

Fig. 4B — Side view of cantilevered rig

Fig. 3A — Pressure distribution

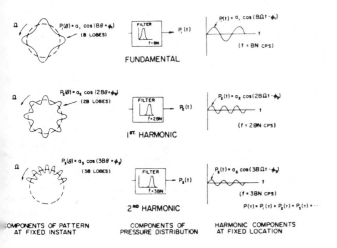

Fig. 3B — Pressure distribution

Fig. 4C — Inlet bellmouth, cylinder, and rotor

tion of frequency, called the spectrum, is conveniently made by an automatic spectrum analyzer, and is shown in Fig. 5B.

Having thus found the spectral composition of the pressure field, it remains to determine the manner in which each of the harmonic components is spatially distributed. Fig. 5C illustrates one of several possible experimental schemes. The pressure probe output is passed through a filter tuned to the desired harmonic of blade-passage frequency. The pure tone filter output is displayed on an oscilloscope, in which the sweep circuit is triggered by a rotor driveshaft contactor so as to fire every time the rotor completes a revolution. With the probe set at a reference position, the trace is as shown in the first of the series of photographs in Fig. 5D. If the probe is slowly turned along a constant radius arc in the direction of rotor spin, the pressure which

originally was being repetitively sensed and displayed at particular value of time after the shaft contactor triggere the sweep will now arrive at progressively later instants c time and will be displayed farther to the right of the screen Thus, the wave pattern is seen to drift slowly to the right a the probe is gradually rotated.

Fig. 5D shows photographs of the oscilloscope displa taken at equal intervals as the probe was rotated 90 deg From these it is clear that the trace shifts one complet cycle during a quarter turn of the probe and that the spatia distribution of the fundamental frequency component of press sure is indeed a 4-lobe pattern rotating at shaft speed When the experiment is repeated with the filter set to pass the $2BN = 8N$ frequency, it is observed that the wave shift one complete period to the right as the probe is moved $1/$

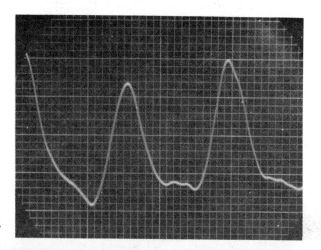

A

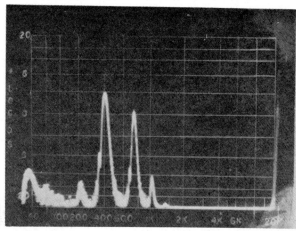

B

C

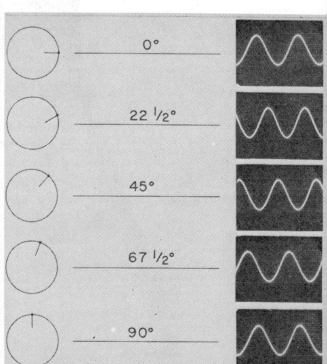

D

Fig. 5 — Displays of pressure fields and phase relationship

of a revolution, thus identifying an 8-lobe pattern.

From similar tests on several rotors it has been verified that the rotor pressure field consists of a superposition of lobed patterns all turning with rotor speed, N rps. The fundamental blade-passage frequency, BN, cps, is associated with a B-lobe pattern, and the harmonics, 2BN, 3BN cps, and so forth are associated with patterns having 2, 3 . . . and so forth times this number of lobes.

The foregoing description of the circumferential structure of the rotor field may well appear repetitious and unnecessarily detailed. However, since a sharp picture of these particular features is crucial to an understanding of all that follows, it is believed that the loss of brevity is here worth the price. In compensation, this section is concluded with only brief mention of two remaining points.

The radial pressure distribution from hub to tip may have a variety of forms depending on details of individual compressor design and operating conditions. This variation may be incorporated into the mathematical description of the rotor field by allowing the Fourier coefficients a_n and ϕ_n of Eq. 3.1 to depend on radius, r. The fluctuating pressure field can now be completely specified in some reference plane near the rotor by:

$$p(r, \theta, t) = \sum_{n=1}^{\infty} a_n(r) \cos[nB(\theta - \Omega t) + \phi_n(r)] \quad (3.2)$$

where the radial distribution functions $a_n(r)$ and $\phi_n(r)$ can be experimentally obtained by traversing the field radially.

A final comment may be of special interest to those concerned with the numerical evaluation of these coefficients in terms of compressor and operating characteristics. For reasons to be disclosed in the following sections, it turns out that further specification of the steady-state rotor field is unnecessary at this time. Independently of this fact, other important conclusions are more clearly derived and highlighted if the rotor field is specified in the simplest, most general way.

4. Duct Transmission

4.1 Duct Transmission - General - From a practical viewpoint, the pressure at the face of the compressor is of most significance to the extent that it affects the noise received by an observer on the ground. Unlike the case of a propeller, the pressure fluctuations must travel through an inlet duct before sound can be radiated into free space. It is with the features of duct transmission that this section is concerned.

The problem can be concisely formulated: Given that in fluid medium bounded by two concentric cylinders, the pressure is specified on a reference plane normal to the axis in the form of a spinning lobed pattern; find the pressure distribution throughout the cylinder so that the radiation from the end can in turn be evaluated.

The shortest route to an understanding of the results by those with limited experience in this area is probably by means of digression.

In order to introduce some essential concepts, we first consider a simpler related problem - transmission of sound in thin rectangular duct. There next follows a discussion of a special case of the cylindrical problem - a narrow annular duct. Results of this case can be interpreted in a straightforward way and provide a natural bridge to the general cylindrical duct, the last item in this section.

4.2 Duct Transmission - Thin Rectangular Duct - By thin, or two-dimensional, is meant a duct in which the thickness cross-section dimension is negligibly small compared with the width and thus allows representation as shown in Fig. 6A. Here the width extends in the y-axis direction, the x axis is aligned lengthwise, and the coordinate origin is conveniently taken at the intersection of the y axis with the bottom wall. The pressure at each point in the reference plane x = 0 oscillates at a frequency f cps with an intensity that is a function of location. Thus, in the reference plane the pressure is given by $p(y, t) = p(y) \cos \omega t$. $(\omega = 2\pi f)$. It may be imagined that this is accomplished by a membrane vibrating in an arbitrary way or by means of several loudspeakers driven at a common frequency but with arbitrarily different amplitudes. Fig. 6A shows such a distribution. The problem is to find the pressure at any point in a typical plane distant x from the reference. While it may be guessed (correctly) that this pressure varies in time with the same frequency, f cps, as the input, it is clearly outside the province of conjecture to determine the associated y-distribution function.

There is a straightforward 3-step procedure for arriving at this pressure distribution in cross-sections removed from the reference plane:

1. The distribution p(y) at x = 0 is resolved into a properly weighted sum of simple modes or characteristic functions.

2. The manner in which these modes are modified in amplitude and phase by travel in the duct is analyzed.

3. The modified modes are reassembled with the original weighting to give the resultant field.

The terms "modes" and characteristic "functions" refer respectively to the simplest ways in which the pressure can be distributed and their corresponding mathematical expressions. Any physically realizable pressure distribution can be expressed as a weighted sum of these modes. They are

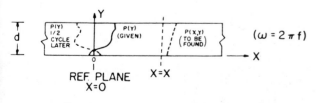

$$P(Y, t) = \left[P_0(Y) + P_1(Y) + P_2(Y) + \cdots \right] \cos \omega t$$

$$= \left[\sum_{q=0}^{\infty} a_q \cos q \frac{\pi}{d} Y \right] \cos \omega t$$

Fig. 6A — Input pressure distribution

called simple primarily because the necessary mathematical manipulation of their corresponding characteristic functions is less involved than would otherwise be the case. Fig. 6B shows the lower modes for a rectangular duct. It will be noted that, apart from the plane wave, they have the form of cosine waves with an integral number of half cycles: $p_q(y) = a_q \cos\left(\dfrac{q\pi}{d}y\right)$. It is essential to observe that, including the plane wave, the tangent to each curve at the top and bottom of the section is perpendicular to the wall surface. This means that the normal pressure gradient at the wall is zero, a condition necessary to insure the physical requirement of no airflow through the walls. For this reason, (called a boundary condition) the equally simple mathematical expressions involving $\sin(q\pi/d)y$ cannot possibly be used here.

The x-axis pressure dependence is now mathematically studied for a single representative mode: $a_q \cos(q\pi/d)y$. It turns out that this dependence takes one or the other of two completely different forms depending on whether the driving frequency is below or above a critical value, called cutoff. Cutoff frequency for a given mode, q, can be simply understood by noticing in Fig. 6B that the corresponding y-distribution has the appearance of an exact integer number, q, of half-waves standing between the walls. It becomes reasonable, then, to introduce the concept of wavelength in the y-direction, λ_y, where for the qth mode, clearly $\lambda_y = 2\,d/q$. If this wave were supposed free to travel in free space with natural velocity c in the y direction instead of being trapped between the walls, it would generate a frequency $f_q = c/\lambda_y$. This frequency is called cutoff for the qth mode. It is the frequency at which q natural half-wavelengths would exactly fit between the bounding walls. In reality, however, the q half waves in the duct are being cycled at an independent frequency, f. When the driving frequency, f, is less than cutoff frequency, f_q, of the qth mode, the transmission properties of the field in the x-direction are entirely different than when $f > f_q$.

The idea of a radical change in transmission properties as the driving frequency is gradually swept through cutoff

becomes more credible when it is pointed out that the following parameter relating f and f_q appears in the analysis:

$$k_x = \frac{2\pi}{c}\sqrt{f^2 - f_q^2}, \left(f_q = \frac{c}{\lambda_y} = \frac{c}{2\dfrac{d}{q}} = q\,\frac{c}{2d}\right)$$

Above cutoff, $f > f_q$ and k_x is a real number, whereas below cutoff k_x becomes imaginary. It would be surprising if the results did not have essential difference in the two cases. The complete pressure field is given by:

<u>case a - decay - $f < f_q$</u>

$$p(x, y, t) = a_q \cos q\,\frac{\pi}{d}\,y\; e^{-k_x x}\cos\omega t \qquad (4.2.1)$$

where:

$$\omega = 2\pi f$$

$$k_x = \frac{2\pi}{c}\sqrt{f_q^2 - f^2}$$

<u>case b - propagation - $f > f_q$</u>

$$p(x, y, t) = a_q \cos q\,\frac{\pi}{d}\,y \cos(k_x x - \omega t) \qquad (4.2.)$$

where:

$$k_x = \frac{2\pi}{c}\sqrt{f^2 - f_q^2} = 2\pi\sqrt{\frac{1}{\lambda^2} - \frac{1}{\lambda_y^2}}$$

$$\lambda = \frac{c}{f}$$

$$\lambda_y = \frac{c}{f_q}$$

By examination of these equations and Fig. 7 it can be seen that the following remarks apply:

1. In each case, the y-distribution is a cosine wave of the same shape as the input with a plus or minus maximum at the walls.

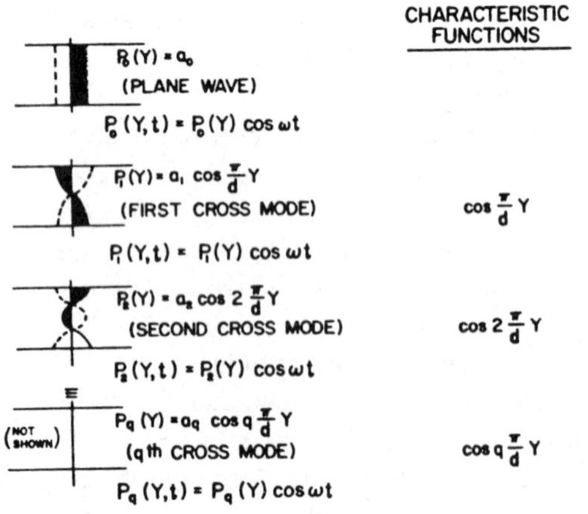

Fig. 6B — Components of input pressure distribution

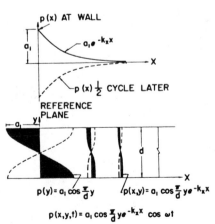

Fig. 7A — First cross mode - decaying field: $f < f_1$

2. The variation of pressure with x-distance and time takes different forms depending on the ratio of driving frequency to cutoff frequency for the y-distribution mode.

3. For driving frequencies lower than cutoff, the intensity or amplitude of the pressure fluctuations falls off exponentially with x-distance according to the factor $\exp - k_x x$. At all points located a fixed distance from the bounding walls, the pressure fluctuations are exactly in phase. The decay is greatest for frequencies far below cutoff and becomes negligible only when cutoff is closely approached. In this neighborhood the pressure intensity parallel to the walls is essentially the same throughout the length of the duct.

4. When cutoff frequency is exceeded, true wave motion propagates in the duct, as indicated by the case b factor, $\cos(k_x x - \omega t)$. Just above cutoff, the x-wavelength is appreciably greater than would be the case in free-field and the phase velocity in the x-direction is correspondingly high, as required to satisfy the relation $f = c/\lambda = c_x/\lambda_x = \text{const.}$ Farther above cutoff λ_x and c_x rapidly approach their free-space values.

Two additional comments are needed. First, no reference to the value of the length of the duct has been made. This intentional omission was based on the desire to defer unessential complications until the basic ground had been covered. That the duct length is indeed irrelevant, except possibly for a narrow range of frequency, can certainly be seen in the case of decay. If the driving frequency is reasonably far below cutoff, the decay parameter:

$$k_x = \frac{2\pi}{c}\sqrt{f_q^2 - f^2} = \frac{2\pi f_q}{c}\sqrt{1 - (f/f_q)^2}$$

is well approximated by $2\pi f_q/c = 2\pi/\lambda_y$. The x-dependent factor is then $p(x) = \exp - 2\pi x/\lambda_y$. This implies that in a length of duct equal to one y-wavelength the pressure will drop by an amount equal to $20 \log(\exp 2\pi)$ or roughly 50 db. Thus, only in a duct of extremely short length is the pressure arriving at the end sufficiently high to justify considering its backward reflection from the end termination. Although it is not equally obvious, it turns out that at frequencies reasonably higher than cutoff backward reflections from the (open) end of the duct can also be ignored. Consideration

of the effect of end boundary conditions in the neighborhood of cutoff has been deferred.

The concluding remarks for the rectangular duct concern an exceptional case: the plane wave, $q = 0$. This is the only case of waves in tubes usually covered in an introductory course in acoustics, but it is truly exceptional in a study of axial-flow compressor noise since the mechanisms for exciting it are either lacking or can be eliminated. It may also be considered exceptional in having a cutoff frequency of zero. Thus, a plane wave will always propagate. The velocity of propagation in the duct and the wavelength are the same as in free-field, and, in addition, the length of the duct and its manner of termination are extremely important considerations.

4.3 Duct Transmission - Narrow Annular Duct - The advantages of considering transmission of sound in a narrow annulus of the form shown in Fig. 8 are:

1. The problem involves but two space coordinates, as with the thin rectangular duct, and employs essentially the same characteristic functions for specification of the input pressure distribution.

2. A simple, pictorial, and easily recallable method of describing the complete pressure field is available in terms of velocity vectors and mach number.

3. Most important, the essential properties of the general problem are closely simulated by the annular case, and can be described as modifications of the special case results.

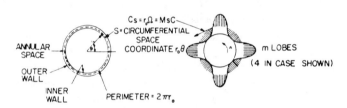

Narrow annular duct cross-section   Input spinning mode

Fig. 8A

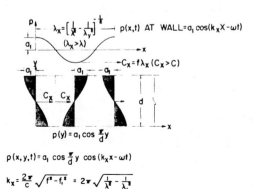

Fig. 7B — First cross mode - propagating field: $f > f_1$

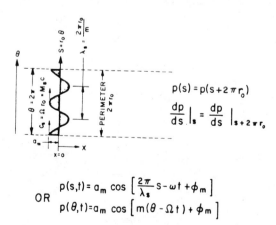

Fig. 8B — Developed view of annular duct, periodic boundary conditions

Fig. 8 shows the duct cross-section, in which the annular gap is negligibly small compared to the perimeter, $2\pi r_0$. The reference plane excitation is a lobed pattern spinning at Ω radians/sec, in which m full pressure wavelengths extend around the perimeter. This pressure wave, which may be generated by a rotor, for example, sweeps by the duct walls at velocity $c_s = \Omega r_0$. At any fixed location the effect is a harmonic pressure fluctuation at circular frequency $\omega = 2\pi f = m\Omega$.

For clarity of representation, the thin annular duct is shown as a developed view in Fig. 8C. Because the relative curvature is small, the wave field has the same behavior as it does in a thin rectangular duct, with the obvious exception that the boundary conditions in the s-direction are different from those existing when hard walls are present. Here, the condition on the s distribution is that at both ends of an arc $2\pi r_0$ long, located anywhere in the s-direction, the pressure and pressure gradient must be the same. The pressure distribution in the s-direction is accordingly given by

$$p(s, t) = a_m \cos\left(\frac{2\pi}{\lambda_s} s - \omega t + \phi_m\right).$$

In the above representation, a_m and ϕ_m are amplitude and phase coefficients, and λ_s, the wavelength in the circumferential direction, is evidently equal to $\frac{2\pi r_0}{m}$. Since arc length and angle are related by $s = r_0 \theta$, this last expression, when substituted in the p(s, t) equation gives the useful representations: $p(\theta, t) = a_m \cos(m\theta - \omega t + \phi_m)$, or $p(\theta, t) = a_m \cos[m(\theta - \Omega t) + \phi_m]$.

In this form, the reference plane pressure distribution can be naturally interpreted as an m-lobe pattern rotating at angular velocity Ω radians per second and generating at every point a fluctuating pressure with frequency $f = m\Omega/2\pi = mN$ cps.

It will be useful to introduce the circumferential velocity at which the pattern sweeps the annulus walls, $c_s = r_0 \Omega$, in terms of circumferential Mach number, M_s, and the speed of sound in free space, c, thru the relation $c_s = M_s c$. Then, the circumferential wavelength, λ_s, and velocity, c_s, are related to frequency by $f = c_s/\lambda_s$. In free space this frequency would be associated with a wave of length λ; traveling at velocity c. Therefore, the relations, $f = c/\lambda = c_s/\lambda_s$, require

that if c_s is related to c by $c_s = M_s c$, so also λ_s is given by $\lambda_s = M_s \lambda$.

As in the case of the rectangular duct, the pressure distribution in the x-direction depends critically on whether the driving frequency is higher or lower than cutoff frequency for the driving mode. The same parameter, k_x, now given by $k_x = \frac{2\pi}{c}\sqrt{f^2 - f_m^2}$, is involved, and it will be useful to modify this expression to the extent of taking the absolute value $|f^2 - f_m^2|$, of the quantity inside the radical, with the understanding that the pressure equation involving k_x takes the form of a decaying exponential if $f < f_m$, and a harmonic relation when $f > f_m$. By introducing the various connections among wavelength, frequency, speed of sound, and Mach number, k_x can be expressed in the following ways:

$$k_x = \frac{2\pi}{c}\sqrt{\left|f^2 - f_m^2\right|} \qquad (4.3.1a)$$

$$= 2\pi \sqrt{\left|\frac{1}{\lambda^2} - \frac{1}{\lambda_s^2}\right|} \qquad (4.3.1b)$$

$$= \frac{2\pi}{\lambda_s} \sqrt{\left|\left(\frac{\lambda_s}{\lambda}\right)^2 - 1\right|} \qquad (4.3.1c)$$

$$= \frac{2\pi}{\lambda_s} \sqrt{\left|M_s^2 - 1\right|} \qquad (4.3.1d)$$

$$= \frac{m}{r_0} \sqrt{\left|M_s^2 - 1\right|} \qquad (4.3.1e)$$

From the last form it is seen that the criterion for propagation can be phrased very simply. In order that the pressure field of a spinning lobed pattern propagate in the duct, the circumferential Mach number, M_s, at which it sweeps the annulus walls, must equal or exceed unity.

A discussion of the two types of transmission follows:

<u>Case a. Decay $f < f_m$, $M_s < 1$</u> - The pressure field of the mth mode is given by

$$p(s, x, t) = a_m \cos\left[\frac{2\pi}{\lambda_s} s - \omega t + \phi_m\right] e^{-k_x X}$$

or

$$p(\theta, x, t) = a_m \cos[m(\theta - \Omega t) + \phi_m] e^{-k_x X} \qquad (4.3.2)$$

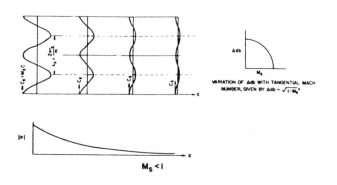

Fig. 9A — Properties of decaying field, $M_s < 1$

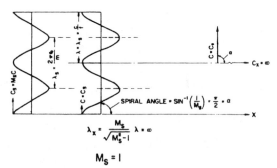

Fig. 9B — Properties of propagating field, $M_s = 1$

This field is illustrated in Fig. 9A. Lines of constant phase extend parallel to the x-axis and move purely in the circumferential direction at speed c_s. Along any such line the amplitude of the pressure decays according to the factor $e^{-k_x x}$. In a length Δx, the pressure level falls by $20 \log_{10} e^{k_x \Delta x}$. The decay rate, using Eq. 4.3.1d, can be expressed as:

$$\text{decay rate, } \Delta \text{ db} \left/ \frac{\Delta x}{\lambda_s} = 54.5 \sqrt{1 - M_s^2} \right. \quad (4.3.3)$$

(decibels per circumferential wavelength) or, using Eq. 4.3.1e:

$$\text{decay rate, } \Delta \text{ db} \left/ \frac{\Delta x}{r_0} = 8.69 \, m \sqrt{1 - M_s^2} \right. \quad (4.3.4)$$

(decibels per radius)

Starting at a maximum value for low circumferential velocity, the decay rate is seen to fall with increasing M_s, as shown in Fig. 9A. At any fixed Mach number the pressure drop from the reference plane to the duct exit varies directly with the number of lobes in the circumferential pattern. A tabulation of the pressure drop over a length equal to the radius in a thin annular duct made for selected values of m and M_s is given in Table 1.

At $M_s = 1$ the decay rate becomes zero, and a pressure field of constant intensity in the x-direction rotates in the duct as shown in Fig. 9B. Above this cutoff speed wave motion will propagate.

Case b.   Propagation $M_s > 1$ - The pressure field of the mth mode is given by

$$p(s, x, t) = a_m \cos \left[\frac{2\pi}{\lambda_s} s + k_x x - \omega t + \phi_m \right]$$

or

$$p(\theta, x, t) = a_m \cos \left[m\theta + k_x x - \omega t + \phi_m \right]$$
$$= a_m \cos \left[m(\theta - \Omega t) + k_x x + \phi_m \right] \quad (4.3.5)$$

Figs. 9B–9D portray the properties of the propagating field better than words. Starting at cutoff in Fig. 9B, the subsequent Figs. 9C and 9D, show the effect of increasing circumferential Mach number. In each case, the field consists of wave pattern moving in a spiral path in the annulus. In the developed views shown, the field appears as parallel wave train propagating in a direction inclined to the coordinate axis. Sections of the field are taken along the axes and in the propagation direction. Notice that in the propagation direction, the wave velocity and wavelength are those obtaining in free field. The wavelength decreases in the sequence, due to the successively higher frequencies being generated by the m-lobe pattern as its rotation speed is increased.

These considerations allow the annular duct field of an m-lobe rotating pressure pattern to be mapped without using any mathematics:

1. To a convenient scale lay out a series of parallel lines spaced a free-field wavelength apart. $\left(\lambda = \dfrac{c}{f} = \dfrac{c}{mN}, \text{ where} \right.$ $N = $ rps of rotating m-lobe pattern. $\Big)$

2. On the right hand edge of another sheet of paper place a series of marks spaced one circumferential wavelength apart $\left(\lambda_s = \dfrac{2\pi r_0}{m} \right)$.

Table 1 - Tabulation of Pressure Drop

Number of Lobes	Circumferential Mach Number, M_s			
m	0	0.25	0.5	0.75
1	9 db	8 db	7 db	6 db
2	17	16	15	13
4	35	33	30	27
8	69	67	60	53
16	139	134	120	106
32	278	269	233	206

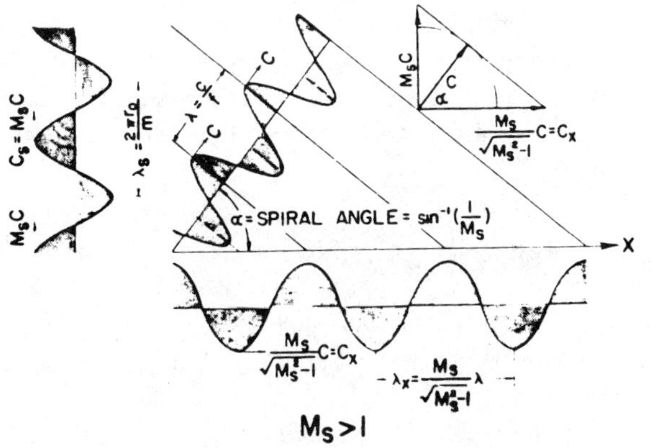

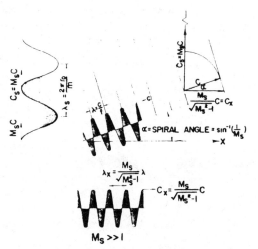

Fig. 9C — Properties of propagating field, $M_s > 1$

Fig. 9D — Properties of propagating field, $M_s \gg 1$

3. Orient the second sheet on the parallel lines so that the lines and marks coincide.

4. The line of marks is now in the correct developed position of the circumferential direction and the x-axis is perpendicular thereto.

This construction is possible, of course, only when $M_s \geq 1$. Viewed thus, as an inverse problem of fitting the duct to the field, the direct analytical approach may appear to have been unnecessary. However, in the case of decay, it is the most convenient method and, in the problem to be considered next, it is essential.

4.4 Duct Transmission - Concentric Cylindrical Duct - The air passageway in an axial flow compressor is bounded on the outside by the case wall and on the inside by the hub or spinner. Accordingly, this section considers the propagation of waves in a medium bounded by two concentric cylinders. (See Fig. 10.) As described in Section 3, the input pressure distribution in the reference plane has the form of a spinning lobed pattern. If the behavior of a single typical mode is determined as a function of the relevent parameters, the field of any input distribution can be obtained by superposition.

Consider an m-lobe pattern rotating at Ω_m radians per second and thus generating a sinusoidal pressure fluctuation with (circular) frequency $\omega = 2\pi f = m\,\Omega_m$. In the thin annular duct problem of 4.3 the pressure amplitude is sensibly constant across the annulus. When, as in the present case, the annular gap is not negligibly small, the radial variation of pressure must be taken into account. This variation can be represented by the following expression for the reference plane pressure:

$$p_m (r, \theta, t) = p_m (r) \cos [m(\theta - \Omega_m t) + \phi_m] \quad (4.4.1)$$

where $p_m (r)$ is the radial distribution function.

Generally, as in the case of the field of a rotor with circumferentially swept blades, for example, the phase angle, ϕ_m, is also a function of radius. It turns out that no essential differences result from supposing ϕ_m constant, and this will be done here for simplicity.

The only physical restriction on $p_m (r)$ is that its radial gradient must be zero at the bounding walls, in order to constrain airflow. A constant radial pressure will meet this condition, as will any of the cosine-waves which were introduced in the rectangular duct section (Fig. 6).

Therefore, a suitably weighted sum of such functions could certainly be selected to fit any physically realizable radial pressure distribution. If this were done, however, immediate technical difficulties would be encountered on attempting to determine the resulting field in the x-direction. There is a simple explanation of this difficulty. The expressions, $\cos q \dfrac{\pi}{d} y$, of 4.2 are called characteristic functions because they follow from the physical characteristics imposed on the medium by the bounding geometry. In a little more detailed terminology, the cosine function is the characteristic function. Its argument, or independent variable, or cosine table entry, is $q \dfrac{\pi}{d} y$.

The y-multiplier, $q \dfrac{\pi}{d}$, is called a characteristic number and may be considered as a scale factor which serves the purpose of fitting a cosine-wave between boundaries d units apart.

For a region bounded by concentric cylindrical walls, the corresponding characteristic functions turn out to be an infrequently used combination of Bessel functions. They are referred to here simply as E-functions. Fig. 11A summarizes their definition and Figs. 11B–11F illustrate their form. It will be noticed that their form depends on several parameters. Whereas, in the rectangular duct, a single function cosine did the job, here the very type of function depends on m, the circumferential lobe number, and on σ, the hub-tip ratio, or ratio of inner to outer cylinder radii. For any fixed pair of (m, σ) values there is a sequence of E-functions E_{m0}^{σ}, E_{m1}^{σ}, $E_{m\mu}^{\sigma}$ of increasing waviness which is roughly analogous to the successive cosine waves, $\cos \dfrac{\pi}{d} y$, $\cos \dfrac{2\pi}{d} y$, $\cos q \dfrac{\pi}{d} y$, of the rectangular case. It may be interesting and reassuring to notice that as the ratio, σ, is increased, so that the annular gap becomes narrower, the E functions increasingly take on the appearance of cosine waves. The value of the index, μ, gives the number of nodes or crossings of the zero pressure axis in the radial direction as does q in the rectangular case.

A complete analysis of this cylindrical duct problem is given in Appendix B. For the present, it is sufficient to summarize the reference plane input pressure distribution by means of Eq. 4.4.1, together with the radial variation function $p_m (r)$:

$$p_m (r) = \sum_{\mu=0}^{\infty} a_{m\mu}\, p_{\mu} (r) \quad \text{or} \quad a_m \sum_{\mu=0}^{\infty} a_{\mu}\, p_{\mu} (r) \quad (4.4.2)$$

where $p_{\mu} (r)$ are E-functions and the $a_{m\mu}$ or a_{μ} are weighting coefficients, by means of which an arbitrary distribution can be composed.

As in the previously considered duct cases, the nature of the x-variation of the pressure field is found to depend

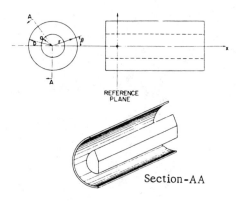

REFERENCE
PLANE

Section-AA

Fig. 10 — Annulus formed by two concentric cylinders

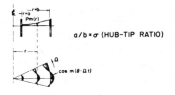

$$p_m(r) = \sum_{\mu=0}^{\infty} a_{m\mu}\, p_\mu(r) = a_m \sum_{\mu=0}^{\infty} a_\mu\, p_\mu(r) = \sum_{\mu=0}^{\infty} a_{m\mu}\, E_{m\mu}^{\sigma}(k_{m\mu}^{\sigma} r)$$

THE CHARACTERISTIC E- FUNCTIONS $E = E_{m\mu}^{\sigma} = E_{m\mu}^{\sigma}(k_{m\mu}^{\sigma} r) = E_{m\mu}^{\sigma}(b\,k_{m\mu}^{\sigma}\frac{r}{b}) = E_{m\mu}^{\sigma}(k_{m\mu}^{\sigma} r')$

ARE DEFINED IN THE RANGE $a \le r \le b$, $\sigma \le r' \le 1$, BY: $E_{m\mu}^{\sigma}( ) = J_m( ) + Q_{m\mu}^{\sigma}\, Y_m( )$,

WHERE J_m AND Y_m ARE BESSEL FUNCTIONS OF THE FIRST AND SECOND KINDS, OF ORDER m.

THE CHARACTERISTIC NUMBERS, $k_{m\mu}^{\sigma}$ AND $Q_{m\mu}^{\sigma}$ ($\mu = 0, 1, 2, \cdots$) ARE THE SUCCESSIVE PAIRED ROOTS OF THE SIMULTANEOUS EQUATIONS:

$$\left.\begin{array}{l} J_m'(k) + Q_m^{\sigma}\, Y_m'(k) = 0 \\ J_m'(\sigma k) + Q_m^{\sigma}\, Y_m'(\sigma k) = 0 \end{array}\right\} \quad \text{WHERE} \quad J'( ) = \frac{d}{d(\ )} J(\ ), \quad Y'( ) = \frac{d}{d(\ )} Y(\ )$$

Fig. 11A — Characteristic E functions for radial pressure distribution

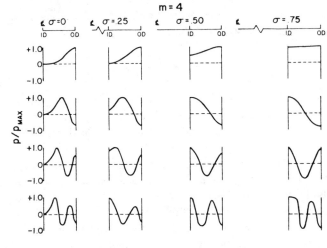

Fig. 11D — Characteristic E functions for radial pressure distribution, m = 4

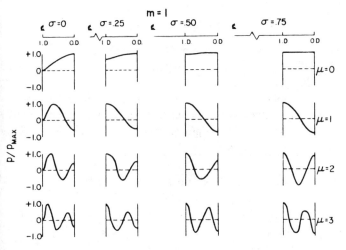

Fig. 11B — Characteristic E functions for radial pressure distribution, m = 1

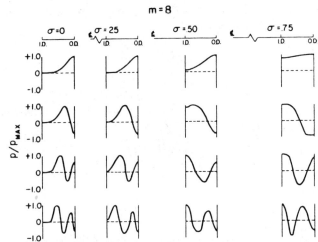

Fig. 11E — Characteristic E functions for radial pressure distribution, m = 8

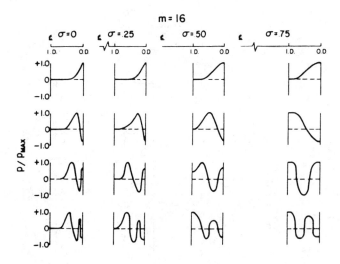

Fig. 11C — Characteristic E functions for radial pressure distribution, m = 2

Fig. 11F — Characteristic E functions for radial pressure distribution, m = 16

critically on whether the parameter, k_x, is real or imaginary. Here, k_x turns out to have the form:

$$k_{x\mu} = \frac{2\pi}{c}\sqrt{f^2 - f_{m\mu}^2}$$

where $f_{m\mu}$ is the cutoff frequency for the mode with an m-lobe circumferential pressure distribution and a single $p_\mu(r)$ characteristic E-function radial distribution having μ pressure nodes across the annulus.

Since $k_{x\mu}$ has different values for μ as well as m, the field properties upstream from the reference plane must be obtained by assembling the results for the transmission of each of the single modes which synthesize the input distribution. With this understanding, the following discussion applies to a single term of Eq. 4.4.2, denoted by $p_{m\mu}$.

Recalling that k_x is used in different ways, depending on whether f is smaller or larger than $f_{m\mu}$, and denoting by M_m the circumferential Mach number with which the m-lobe pattern sweeps the outer wall, $r = b$, the following forms can be evolved in the manner of 4.3.

$$k_{x\mu} = \frac{2\pi}{c}\sqrt{f^2 - f_{m\mu}^2} \qquad (4.4.3a)$$

$$= \sqrt{\left|\left(\frac{\omega}{c}\right)^2 - \left(\frac{2\pi f_{m\mu}}{c}\right)^2\right|} \qquad (4.4.3b)$$

$$= \sqrt{\left|\left(\frac{m\Omega}{c}\right)^2 - \left(k_{m\mu}^{(\sigma)}\right)^2\right|} \qquad (4.4.3c)$$

$$= \frac{m}{b}\sqrt{\left|M_m^2 - (k_{m\mu}^{'(\sigma)}/m)^2\right|} \qquad (4.4.3d)$$

$k_{m\mu}^{'\sigma}$ is a characteristic number, associated with the E-functions describing the radial pressure distribution, and, as the clutter of subscripts suggests, its value depends on hub-tip ratio, (σ), number of circumferential lobes, (m), and the number, (μ), of nodes or points of zero pressure across the annulus.

The transition from decay to propagation occurs when the circumferential tip Mach number, M_m, equals the value of $k_{m\mu}^{'\sigma}/m$. By comparing Eq. 4.3.1d with Eq. 4.4.3d it is seen that whereas, in the thin annular duct the criterion for propagation is simply $M_s \geq 1$, here the requirement, $M_m \geq k_{m\mu}^{'\sigma}/m$, must be further examined.

Consider a pair of fixed values for m and σ. Then, the successive values of $\mu = 0, 1, 2, \ldots$ represent more complicated radial pressure distributions. By recalling the results of the rectangular duct problem, it is to be expected that here also the higher cross-modes will have successively higher cutoff frequencies. Consequently, the tip speed at which an arbitrary radial distribution will just begin to propagate is governed entirely by the critical Mach number for the lowest radial mode in its makeup. Therefore, the transmission characteristics can be summed up as follows:

for decay	$M_m < M_m^*$
at cutoff	$M_m = M_m^*$
for propagation	$M_m > M_m^*$

The critical circumferential tip Mach number at the outer wall, $M_m^* = k_{m0}^{'\sigma}/m$, is tabulated in Table 2 for selected values of m and σ:

At cutoff, the pressure field can be visualized by referring to the corresponding pattern for the narrow annular duct, Fig. 9B. In the present case, similar developed views

Table 2 - Critical Tip Mach Number

m	$\sigma = 0$	$\sigma = 1/4$	$\sigma = 1/2$	$\sigma = 3/4$
1	1.84	1.64	1.36	1.15
2	1.53	1.50	1.34	1.15
4	1.33	1.33	1.29	1.14
8	1.21	1.21	1.20	1.14
16	1.13	1.13	1.13	1.12
32	1.08	1.08	1.08	1.08

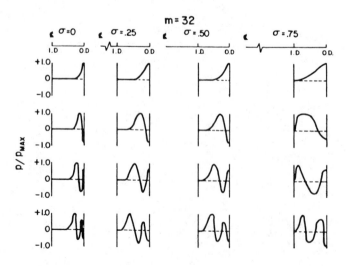

Fig. 11G — Characteristic E functions for radial pressure distribution, m = 32

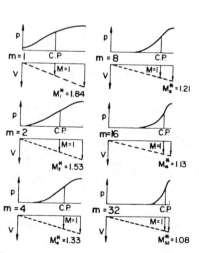

Fig. 12 — Critical Mach No. M_m^* in plain cylinder

apply at all radial locations with the essential difference that now the wave velocity varies with radial position, becoming $M_m^* c$ at the outer wall.

Attention should be directed to three interesting features:

1. The critical tip velocity $M_m^* c$ is always supersonic.

2. For a fixed hub-tip ratio, the critical speed approaches $M_m^* = 1$ as m increases.

3. For a fixed number of lobes, $M_m^* \longrightarrow 1$ as σ is increased.

The propagation requirement of supersonic tip speed may be made to appear less strange by discussing a few details. In the first place, since the circumferential velocity of the spinning pattern varies linearly with radius, the supersonic tip velocity does not mean that all portions of the pattern are supersonic. It would be strange indeed if exactly sonic tip velocity were required, for then virtually the entire pattern would be propagating at subsonic speed, a situation which would be in direct conflict with the information obtained from study of the thin annular duct.

For convenience, the portions of Fig. 11 giving the radial pressure distribution in the $\mu = 0$ mode for successive values of m in a plain ($\sigma = 0$) cylinder are assembled in Fig. 12. These distributions suggest, by analogy with the concept of center of pressure, the idea of introducing a similar point at which the total effect of the distributed pressure may be imagined to be concentrated for study of propagation. This point, called here the "center of propagation," is the point at which the circumferential Mach number at cutoff is exactly unity. Expecting this point to have some of the general properties of other such "effective" points (center of pressure, center of area) it may then appear more plausible that when m is small and the pressure is fairly spread out, the center of propagation is located somewhere inboard of the wall and, consequently, the wall Mach number is appreciably greater than unity. As m is increased, the pressure distribution and its center of propagation concentrate toward the bounding wall and the extreme tip Mach number need be only slightly supersonic.

The wallward drift of pressure with increasing m, and the attendent reduction of inboard pressure, is required to prevent the excessive tangential pressure gradients close to the axis that would result in the absence of this trend. Examined in the light of the preceeding comments, other features of the propagation criteria, including the effects of changing hub-tip ratio, may take on a more natural appearance.

As usual, the different field characteristics below and above cutoff require separate treatment. With the background information gathered from the discussion of the other duct problems, no elaboration will be needed now.

Case a: decay: $M_m < M_m^*$ - The pressure field for the (m, 0) mode is

$$p(r, \theta, x, t) = a_{mo} \, p_o(r) \cos[m(\theta - \Omega_m t) + \phi_m] \, e^{-k_x x}$$

where:

$$k_x = \frac{m}{b} \sqrt{(M_m^*)^2 - M_m^2} \qquad (4.4.5)$$

The field for the higher radial (m, μ) modes is

$$p(r, \theta, x, t) = a_{m\mu} \, p_\mu(r) \cos[m(\theta - \Omega_m t) + \phi_m] \, e^{-k_{x\mu} x}$$

where:

$$k_{x\mu} = \frac{m}{b} \sqrt{\left(\frac{k_{m\mu}^{'(\sigma)}}{m}\right)^2 - M_m^2} \qquad (4.4.6)$$

Decay rates and the superposition of decaying fields are discussed in Section 5.

Case b: propagation: $M_m > M_m^*$ - The pressure field for the (m, 0) mode is

$$p(r, \theta, x, t) = a_{mo} \, p_o(r) \cos[m(\theta - \Omega_m t) + k_x x + \phi_m]$$

where:

$$k_x = \frac{m}{b} \sqrt{M_m^2 - (M_m^*)^2} \qquad (4.4.7)$$

The field for the higher (m, μ) modes when they are above cutoff is

$$p(r, \theta, x, t) = a_{m\mu} \, p_\mu(r) \cos[m(\theta - \Omega_m t) + k_{x\mu} x + \phi_m]$$

where:

$$k_{x\mu} = \frac{m}{b} \sqrt{M_m^2 - \left(\frac{k_{m\mu}^{'(\sigma)}}{m}\right)^2} \qquad (4.4.8)$$

Mechanisms for generating propagating fields in a compressor, and the superposition of propagating fields are considered in Section 6.

5. Decay of Rotor Field

This section discusses some of the properties of the pressure field of a ducted rotor. The study of noise-generating mechanisms in Section 3 led to the resulting reference plane pressure specification:

$$p(r, \theta, t) = \sum_{n=1}^{\infty} a_n(r) \cos[nB(\theta - \Omega t) + \phi_n(r)] \qquad (3.2)$$

where:

B = Number of rotor blades

Ω = Rotor angular velocity

n = Harmonic number

$a_n(r)$ = Radial distribution of (nB) - lobe pattern rotating at Ω and generating a frequency $f = \dfrac{nB\Omega}{2\pi}$

In the following, attention is centered on a single frequency, so that the pressure of the nth harmonic is the nth term only, $a_n(r) \cos[nB(\theta - \Omega t) + \phi_n(r)]$. Using the information just discussed in 4.4, the input pressure at frequency nBN, assuming $\phi_n(r)$ constant, can be expressed as:

$$p(r, \theta, t) = a_m \sum_{\mu=0}^{\infty} a_\mu \, p_\mu(r) \cos[m(\theta - \Omega t) + \phi_m]$$

or

$$a_{nB} \sum_{\mu=0}^{\infty} a_\mu \, p_\mu (r) \cos [nB(\theta - \Omega t) + \phi_{nB}] \qquad (5.1)$$

It is essential to notice that the angular velocity of the (M = n B)-lobe pattern is exactly rotor speed, Ω. Since the tip Mach number for propagation, $M_m^* = M_{nB}^*$ decreases through the supersonic range as m = nB increases, it can be seen that all harmonics (n = 1, 2, 3 . . .) of blade-passage noise will propagate if the rotor tip speed is held above the supersonic critical tip speed for the fundamental (m = B)-lobe pattern. Conservatively - the field of a subsonic rotor will always decay. As discussed in 4.4, the speeds at which the higher radial modes, represented by $\mu = 1, 2, \ldots$ in Eq. 5.1, will begin to propagate are successively higher than that required for propagation of the lowest ($\mu = 0$) radial mode. Consequently, when the rotor speed is below critical, so that its $\mu = 0$ mode is decaying, the higher radial modes decay more rapidly and become insignificant. The rate of decay in the lowest ($\mu = 0$) radial mode, expressed as decibel drop per length of duct equal to the outer wall radius, is derived by the methods of Section 4 and can be written as

decay rate, $\dfrac{\Delta db}{\Delta x / b} = 8.69m \sqrt{M_m^{*2} - M_m^2}$

(decibels per outer wall radius)   (5.2)

Figs. 13a–13d present this information in graphical form for various values of σ and m (= nB). The high decay rates obtainable with a reasonably large number of blades are noteworthy.

Confirmation of these rates has been made on a large number of single stage compressor rigs by measuring the variation of pressure as a function of distance upstream of the rotor. Tests have been conducted with respect to the fundamental and several harmonics of blade-passage frequency over a wide range of speed, number of blades, hub-tip ratio, and scale. A sample of this information, presented in Fig. 14 was obtained on a 10-in. diameter, 8-blade rotor. The decay rates, obtained from the slope of the lines, agree with analysis near the rotor, but eventually, the data level out to a value called the floor. This floor level is set mainly by airstream turbulence effects in the noise generation process at the rotor.

In conclusion, it may be of interest to examine briefly how a flat radial pressure profile generated at the face of a

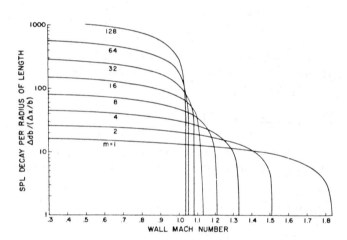

Fig. 13A — Cylindrical duct decay rates $\sigma = 0$

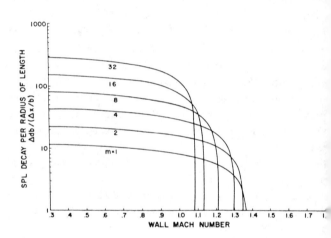

Fig. 13C — Cylindrical duct decay rates $\sigma = 0.50$

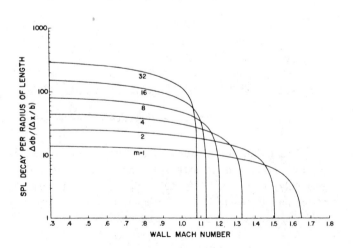

Fig. 13B — Cylindrical duct decay rates $\sigma = 0.25$

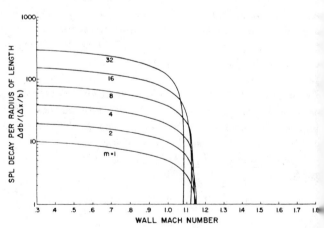

Fig. 13D — Cylindrical duct decay rates $\sigma = 0.75$

rotor, becomes transformed with distance in a decaying field. From Eqs. 4.4.6 and 5.1, and noticing the common time-angle dependence factor, the peak pressure, as a function of radius and upstream location is:

$$p_{max}(r, x) = a_m \sum_{\mu=0}^{\infty} a_\mu \, p_\mu(r) \, e^{-k_{x\mu}x} \qquad (5.3)$$

At the reference plane $x = 0$ a flat pressure profile requires that

$$p(r) = \sum_{\mu=0}^{\infty} a_\mu \, p_\mu(r) = 1 \quad \text{for} \quad a < r < b$$

Since the $p_\mu(r)$ are characteristic E-functions (4.4), the coefficients a_μ can be evaluated by a process similar to Fourier analysis. Then, Eq. 5.3 enables the pressure profile to be computed at selected distances from the rotor.

Fig. 15 shows the results of this calculation for several cases, using a 4-term approximation to the flat input profile. The peak levels have been adjusted to the same height in order to show better the change of shape - in reality of course, they are decaying. The rate at which the patterns come to resemble the corresponding lowest radial mode (Fig. 11) may be noticed.

6. Rotor-Stator Interaction

6.1 General - It will be shown in this section that rotor-stator interaction produces a plurality of spinning modes corresponding to a single frequency. Whereas the rotor field structure responsible for generating a frequency nBNcps consists of a single (nB)-lobe pattern rotating at shaft speed Ω, rotor-stator interaction noise at this frequency is generated by many patterns, each of which has a number of lobes and a speed of rotation which can be determined. If a particular interaction pattern has fewer lobes than the number of rotor blades, it must rotate faster than the shaft in order to generate the same frequency. Therefore, it becomes possible to produce propagating modes through interaction even when the rotor tip speed is subsonic.

6.2 Generating Mechanisms - The sources of interaction noise are:

1. Cutting of wakes of upstream stators by rotor blades.
2. Impingement of rotating blade wakes on downstream stators.

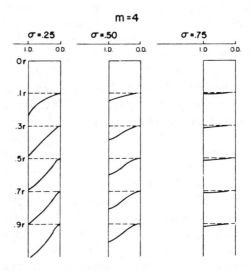

Fig. 15B — Higher mode degeneration with duct length m = 4

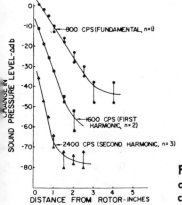

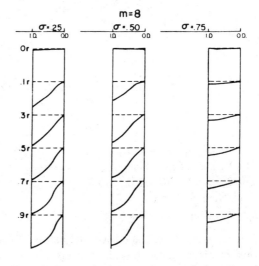

Fig. 14 — Experimental decay data from single stage compressor rig

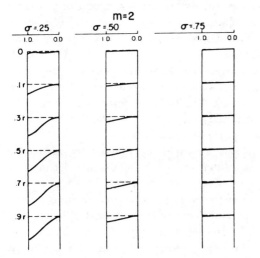

Fig. 15A — Higher mode degeneration with duct length m = 2

Fig. 15C — Higher mode degeneration with duct length m = 8

3. Interruption of the rotating periodic pressure field of the rotor by the proximity of reflecting objects, apart from wake effects.

As has been shown by the results of the previous sections, it is often unnecessary to specify the strength of a source and the influence of various factors on this strength. Provided that the essential structure of the source can be found, many important conclusions can be drawn. So here, it is not essential to assess the relative importance of these three sources of interaction noise, for, as can readily be shown, they all produce effects having an identical, common structure.

Consider a succession of rotating blades passing by a single upstream or downstream stator vane. In the neighborhood of this vane will be sensed a pressure fluctuation at blade-passage frequency that will, due to the physical presence of the vane, be separate and distinct from the fluctuation due to the periodic ripple of the rotor field itself. The distinction is immediately obvious if one takes the position of an observer rotating with the blades. If there is no physical irregularity near the rotor, the moving observer will be in a steady field and will sense no fluctuation. However, when a stationary vane is placed nearby, the observer will then sense a change in pressure, or pulse, every time he passes by the vane. To be sure, the magnitude of this fluctuation depends on the size, proximity, and aerodynamic influence of the obstacle, but it has the common property of recurring every time the blade completes a revolution.

In a stationary reference frame near the obstacle this effect is duplicated every time any other rotor blade passes by, and thus takes place at blade-passage frequency. It is convenient to refer to this phenomena as the "blade-vane event," or, more simply the "event." In a convenient reference plane in or near the interaction zone let an angular reference position ($\theta = 0$) be selected near the vane. At this point, the recurring events will be sensed as a periodic pressure fluctuation which can therefore be represented by:

$$p(t) = \sum_{n=1}^{\infty} a_n \cos (n\omega t + \phi_n),$$

or

$$= \sum_{n=1}^{\infty} a_n \cos (nB\Omega t + \phi_n), \qquad (6.2.1)$$

where:

Ω = Rotor angular velocity
B = Number of rotor blades
$\omega = 2\pi f = 2\pi BN = B\Omega$
n = Harmonic index
a_n, ϕ_n = Coefficients to fit any waveform

This fluctuation, described above at position $\theta = 0$, is clearly not confined to the immediate neighborhood of the vane, but rather is sensed in varying levels at every position around the duct. Accordingly, the pressure in the reference plane can be given in terms of position and time by

$$p(\theta, t) = \sum_{n=1}^{\infty} p_n(\theta) \cos [n\omega t + \phi_n(\theta)] \qquad (6.2.2)$$

This equation states that the pressure fluctuation at any position, θ, is a sum of harmonics, each having amplitude and phase that are perfectly general functions of position. Expanding the cosine term:

$$p(\theta, t) = \sum_{n=1}^{\infty} [p_n(\theta) \cos \phi_n(\theta)] \cos n\omega t -$$

$$\sum_{n=1}^{\infty} [p_n(\theta) \sin \phi_n(\theta)] \sin n\omega t \qquad (6.2.3)$$

Since the quantities in brackets are functions of position only, they may be expanded in Fourier series of the form:

$$\sum_{m=0}^{\infty} (A_m \cos m\theta + B_m \sin m\theta).$$

Making this substitution in Eq. 6.2.3, using the trigonometric product identities, and collecting terms, leads to the result:

$$p(\theta, t) = \sum_{n=1}^{\infty} \sum_{m=-\infty}^{\infty} p_{mn},$$

where

$$p_{mn} = a_{mn} \cos (m\theta - n\omega t + \phi_{mn}). \qquad (6.2.4)$$

p_{mn} may be thought of as the (m, n)th space-time component of the fluctuating pressure distribution due to the recurring blade interaction events at the single vane. The form given is for a circumferential traveling wave; if m is positive the wave moves in the same direction as the rotor, and if m is negative it rotates in the opposite sense. A particular m-lobe pattern rotates at angular velocity $n\omega/m = nB\Omega/m$ radians per second. Thus, any pressure distribution, $p(\theta, t)$ due to the interaction may be represented as a superposition of spinning patterns. The importance of this interpretation of the interaction will become clearer when the case of a vane array is considered. Description of the radial pressure distribution is intentionally avoided since it is not of primary interest at this stage, and would unnecessarily complicate the analysis.

6.3 Modes Produced by Rotor-Stator Interaction - In a multivane stator assembly, rotor-stator interaction events having the effect given by Eq. 6.2.4 occur at each vane. The time sequence of these events, or firing-order, may be worked out as a function of the number of blades and vanes, but, for the present, it is sufficient to point out that the effect of the rotor interacting with the entire vane assembly can be found by superposing, in the correct space and time sequence, the effect of a single event. The method for combining these events is now developed.

Consider an array of V identical vanes, equally spaced $\Delta\theta = 2\pi/V$ radians apart. If the rotor is turning at Ω radians per second, the time required for a particular blade to turn from one vane position to the next is given by $\Delta t = \Delta\theta/\Omega = 2\pi/V\Omega$ seconds. Let one of the rotor blades be identified by a mark and let a stopwatch be started, $(t = 0)$, at the instant the marked blade coincides with stator vane No. 1. Further, interpret Eq. 6.2.4 as the pressure time history at location θ due to the rotor interaction with vane No. 1. Specifically,

$$p_{mn}^1 = a_{mn} \cos(m\theta - n\omega t + \phi_{mn}) \qquad (6.3.1)$$

To find the time history at θ due to rotor interaction with vane No. 2, observe that, since the marked blade turns from vane No. 1 to vane No. 2, a distance of $\Delta\theta$, in a time $\Delta t = \Delta\theta/\Omega$, the pressure at location $(\theta + \Delta\theta)$ and time $(t + \Delta t)$ due to vane No. 2 events, is the same as the pressure at location θ, time t, due to No. 1 vane events. Therefore, the effect of the events due to vane No. 2 is obtained by simultaneously replacing θ by $\theta - \Delta\theta$ and t by $t - \Delta t$ in Eq. 6.3.1:

$$p_{mn}^2 = a_{mn} \cos[m(\theta - \Delta\theta) - n\omega(t - \Delta t) + \phi_{mn}]$$

Similarly, the effect due to a vane $q\Delta\theta$ radians away from vane No. 1 is:

$$p_{mn}^{1+q} = a_{mn} \cos[m(\theta - q\Delta\theta) - n\omega(t - q\Delta t) + \phi_{mn}]$$

Therefore, summing over all V vanes,

$$p_{mn} = \sum_{q=0}^{v-1} p_{mn}^{1+q} = \sum_{q=0}^{v-1} a_{mn} \cos[m(\theta - q\Delta\theta) -$$

$$n\omega(t - q\Delta t) + \phi_{mn}]$$

After some manipulation, the result can be expressed in a particularly revealing form: The pressure in the reference plane, due to rotor-stator interaction, is given by

$$\left[\begin{array}{c} p_{mn} = V\, a_{mn} \cos(m\theta - nB\Omega t + \phi_{mn}) \\ \text{where m is restricted to the following values:} \\ m = nB + kV, \quad k = \ldots -1,\, 0,\, 1,\, \ldots \end{array}\right] \quad (6.3.2)$$

This form differs markedly from the expression for the field of the rotor only. To make this clear, Eq. 6.3.2 can be written:

$$p_{mn} = V\, a_{mn} \cos\left[m\left(\theta - \frac{nB}{m}\Omega t\right) + \phi_{mn}\right] \qquad (6.3.3)$$

But

$$p_n = a_n \cos[nB(\theta - \Omega t) + \phi_n] \qquad (3.1)$$

is the pressure in the reference plane for the nth harmonic of the rotor only. It can be seen by comparing Eqs. 6.3.3 and 3.1 above, that the rotor field generating the nth harmonic of blade-passage frequency simply consists of an (nB)-lobe

pattern rotating at shaft speed. The interaction field, however, is more involved.

For a particular harmonic, n, of blade-passage frequency, the interaction field is a superposition of an infinite number of rotating patterns. The number of lobes in each pattern is given by the successive values of m, generated as an index, k, ranges over all positive and negative integers in the expression $m = nB + kV$. This amounts to adding and subtracting multiples of the number of vanes from the product nB. Each m-lobe pattern turns at a different speed, $nB\Omega/m$ radians per second, as is required to generate n times blade-passage frequency.

Notice that here, with an array of V vanes, the only resulting m-lobe spinning patterns comprising the nBN frequency field are those corresponding to those values of m given by the relation, $m = nB + kV$ $(k = \ldots -1,\, 0,\, 1,\, \ldots)$. In the case of a single vane, the interaction near field includes, generally, all positive and negative integral values of m, associated with amplitude and phase parameters, a_{mn} and ϕ_{mn}, the values of which depend on details of the particular vane-blade aerodynamics. When V of these interactions are superposed, only the specific $m(= nB + kV)$-lobe patterns survive in the near field, all others having phase relationships such that the sum over V yields a zero resultant. The term "near field," is used in the sense of being sufficiently close, axially, to the interaction zone such that many of these m-lobe patterns, which are spinning too slowly to propagate, have not travelled far enough to completely decay.

The existence of some of these patterns may be made more plausible by means of a simple demonstration involving reasonably few blades and vanes. Fig. 16A illustrates in diagramatic form the case of an 8-blade rotor interacting with a 6-vane stator. Eq. 6.3.2 predicts the existence of an $8 - 6 = 2$ lobe pattern, turning at $8/2 = 4$ times shaft speed for the 8N frequency. The rotor blades are represented by eight spokes, one of which is marked with a large dot for identification of rotor position. The six stator vanes are schematically shown as short radial segments extending inward from the duct wall. In the first diagram, the marked blade coincides with the right-hand stator vane, and a similar coincidence exists on the opposite side. The coincidences are identified by exterior arrows, and the positive portions of a 2-lobe pattern are suggested by the shaded areas. Successive pictures show intermediate positions of the rotor (identifying mark) and the coincidences (arrows) as the rotor makes 1/4 of a turn. Notice that in this time interval, the 2-lobe pattern has completed one whole revolution, and is therefore turning at a rate of 4 times the rotor speed.

Similarly, Fig. 16B represents the interaction of the 8-blade rotor with a 9-vane stator. Here, the expression for the smallest value of m gives $m = 8 - 9 = -1$. This result is interpreted as a 1-lobe pattern rotating in a direction opposite to that of the rotor. Its speed is eight times shaft speed. The substitution of -1 for m in Eqs. 6.3.2 or 6.3.3 results in the same sign for the θ and t terms, which is the standard form for a backward-traveling wave. A glance at the diagrams in Fig. 16B shows that a complete backward revolution of the 1-lobe pattern (identified by a single coincidence arrow) is accomplished in the time required for the rotor to make but 1/8 of a turn.

While these are convincing demonstrations of the plausibility of the phenomena predicted by Eq. 6.3.2, it has been considered essential to obtain actual experimental information on model compressor rigs incorporating stator vane assemblies. By carefully selecting the number of rotor blades, the number of stator vanes, and the operating rotor tip Mach number, it is possible to generate a multiplicity of patterns such that only one of these will propagate in the duct while the others simultaneously decay to negligible levels. The experimental procedures used for identifying a rotating lobe pattern have already been described in Section 3. It is sufficient to remark here that the results of the interaction analysis have been confirmed repeatedly on a large variety of blade-vane combinations.

A cautionary note may be appreciated by one interested in conducting similar experiments. Unless conditions are carefully arranged to specifically exclude the possibility, a duct traverse will usually be made in a pressure field having significant energy in more than one wave pattern. The superposition field of a number of patterns traveling at different speeds is complicated by having a signal level that varies with circumferential probe location, and a phase shift that may be a highly nonlinear function of location. Parenthetically, while procedures for resolving a superposition wave field have been developed and used in some experiments, they are to be avoided whenever possible.

With this understanding of the mechanism by which propagating modes can be generated in an axial-flow compressor, it is reasonable to pass directly on to topics in the third area of this study - the radiation problem.

7. Radiation

7.1 Nature of Problem - The third and final phase in the transmission of sound generated by the compressor to a

listener's ear is the radiation into free space of the fluctuating pressure field appearing at the open face of the duct. As the airplane flies by overhead, the reaction of an observer on the ground will depend largely on the intensity and the duration of exposure to the radiated sound. It is natural to inquire into the directional distribution of the radiated acoustic energy and to assess its characteristics in terms of compressor operating parameters. Accordingly, an analysis of the spatial distribution of the radiation field has been made, and is discussed in this section. Where possible, detailed mathematical formulation is minimized; however, a complete treatment of the problem is contained in Appendix C.

From a geometrical point of view, the sound radiated from a compressor can be considered as the sound radiated from the open end of a cylindrical duct. However, a preliminary analysis of the field of a narrow rectangular duct provides a convenient route to the actual development of the more realistic cylindrical case. This procedure is consistent with the treatment of duct transmission in Section 4.

7.2 Radiation of Sound from Thin Rectangular Duct - A thin rectangular duct transmitting higher order sound waves is illustrated in Fig. 17. The fluctuating pressure field at the open face of the duct can be described in terms of simple modes whose characteristic functions are of the form:

$$p_q(y, t) = a_q \sin\left(\frac{q\pi y}{2d}\right) \cos \omega t$$

where:

$$q = 1, 3, 5 \ldots$$

or:

$$p_q(y, t) = a_q \cos\left(\frac{q\pi y}{2d}\right) \cos \omega t \qquad (7.2.1)$$

where:

$$q = 0, 2, 4 \ldots$$

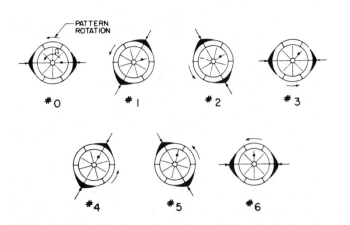

Fig. 16A — Demonstration of rotor-stator interaction patterns, example (A)

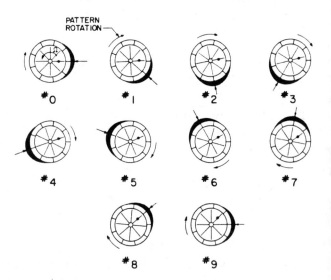

Fig. 16B — Demonstration of rotor-stator interaction patterns, example (B)

An element of area, (dy) located at position y in the face, can be considered as a simple source of sound; that is, a vibrating sphere with diameter small compared to the wavelength being radiated. The resulting pressure at a point in the far field, $dP_q(R, \psi, t)$, can be evaluated in terms of the source strength, and the total pressure at this point can then be expressed as the sum of the contributions of the elements in the duct face.

From the face pressure distribution given by Eq. 7.2.1 the acoustic particle velocity in the x-direction can be found, and the field pressure expressed as:

$$dP_q(R, \psi, t) = a_q k_x \cos\left[\frac{2\pi f}{c}(R-ct)\right] \sin\frac{q\pi y}{2d}$$

$$\cos\left[\frac{2\pi fy}{c}\sin\psi\right]dy \qquad (7.2.2)$$

In obtaining this expression it had been assumed, for mathematical simplicity, that the particle velocity at the duct extremities, $y = \pm d$, drops immediately to zero. While this assumption cannot be met physically, it allows a

preliminary study of the radiation field to proceed without delay. Integrating Eq. 7.2.2 over the face of the duct gives for the resultant field pressure:

$$P_q(R, \psi, t) =$$

$$\frac{A_q \cos\left[\frac{2\pi f}{c}(R-ct)\right]\sin\psi\sin\left(\frac{2\pi f}{c}d\sin\psi\right)}{R\left[\left(\frac{q\pi}{d}\right)^2 - \left(\frac{2\pi f}{c}\right)^2\sin^2\psi\right]} \qquad (7.2.3)$$

The directionality of the radiated field given by Eq. 7.2.3 can be clearly seen from Fig. 18 which shows in sequence the angular distribution at several frequencies for the q = 1 mode.

Notice that the peak of the radiation field tends to move toward the centerline as the frequency is increased and that the pressure field vanishes at the centerline.

In summary then, the field radiated from a rectangular duct possesses strong angular directivity which depends explicitly on frequency and mode number.

7.3 <u>Radiation of Sound from Cylindrical Duct</u> - The procedure for determining the sound field radiated by a cylindrical duct follows in essentially the same manner. The fluctuating pressure field at the open face of the duct is

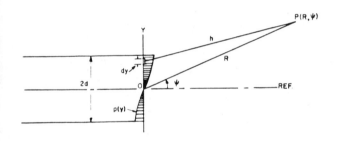

Fig. 17 — Rectangular duct showing coordinates and characteristic mode at duct face

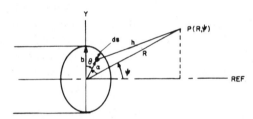

Fig. 19 — Cylindrical duct showing coordinates

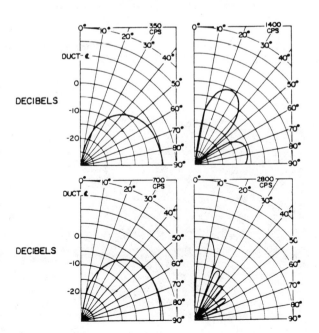

Fig. 18 — Computed radiation for thin rectangular duct 1 ft wide, q = 1 mode

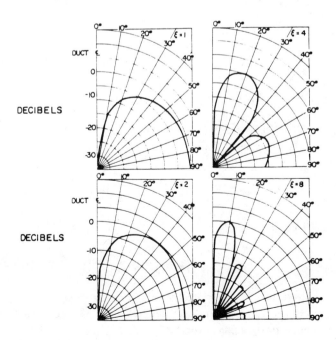

Fig. 20 — Computed radiation patterns for cylindrical duct for m = 1, μ = 0 mode

described best in terms of characteristic functions of the form:

$$p_{m\mu}(r, \theta, t) = a_{m\mu} J_m (k_{m\mu} r) \cos (m\theta - \omega t) \quad (7.3.1)$$

Where $J_m (k_{m\mu}r)$ is the special form taken by the characteristic E-functions of 4.4 in the case of no interior cylinder. Again, the pressure in the far field due to a simple source in the face of the duct is calculated. Fig. 19 shows the coordinate system used. The resultant radiated pressure $P_{m\mu}(R, \psi, t)$ is then obtained by summing the spatially weighted contributions of each simple source, or, more formally, integrating over the duct surface. Despite the appearance of cylindrical functions, that is, Bessel functions, in the description of the pressure at the duct face, the radiated field bears a strong similarity to the radiated field of a rectangular duct and can be written as follows:

$$P_{m\mu}(R, \psi, t) = \frac{A_{m\mu} \cos\left[\dfrac{2\pi f}{c}(R - ct)\right] \sin \psi \left[J_{m-1}\left(\dfrac{2\pi f}{c} b \sin \psi\right) - J_{m+1}\left(\dfrac{2\pi f}{c} b \sin \psi\right)\right]}{R\left[k_{m\mu}{}^2 - \left(\dfrac{2\pi f}{c}\right)^2 \sin^2 \psi\right]} \quad (7.3.2)$$

Before discussing the directivity implied by Eq. 7.3.2, it is convenient to recall some useful parameters for describing the pressure at the face of the duct. For simplicity, reference is made only to the fundamental blade passage noise $(f = BN)$. The concept of tip Mach number M_m for spinning modes has been discussed in Section 4 and the corresponding Mach number, M_m^*, at which a spinning mode transits from decay to propagation is called cutoff Mach number. The ratio of the Mach number of a spinning mode to its cutoff Mach number $\dfrac{M_m}{M_m^*}$ will be now labeled ξ_m and called cutoff ratio. That is:

$$\xi_m = \frac{M_m}{M_m^*} \quad (7.3.3)$$

where:

$M_m = BM_B/m$
B = Number of blades
M_B = Blade tip Mach number
m = Number of lobes

and:

$$M_m^* = \frac{k_{m,\mu} b}{m} = \frac{k_{m\mu}}{m} \quad (7.3.4)$$

Therefore:

$$\xi_m = \frac{BM_B}{k_{m\mu} b} = \frac{BM_B}{k'_{m\mu}} \quad (7.3.5)$$

Note that if $\xi_m > 1$ the m-lobe spinning mode propagates unattenuated down the duct and if $\xi_m < 1$ the mode decays exponentially as shown in 4.4.

The angular dependence of the radiation of an m-lobe mode contains only parameters associated with the com-

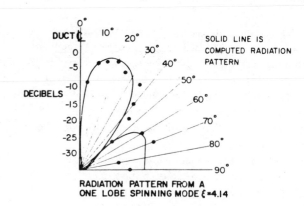

RADIATION PATTERN FROM A
ONE LOBE SPINNING MODE $\xi = 4.14$

Fig. 21A — Experimental radiation data taken on 10 in. diameter, single stage, 32-blade compressor

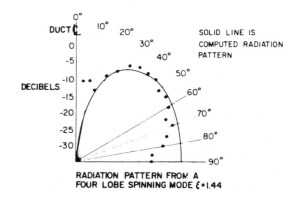

RADIATION PATTERN FROM A
FOUR LOBE SPINNING MODE $\xi = 1.44$

Fig. 21B — Experimental radiation data taken on 10 in. diameter, single stage, 32-blade compressor

Fig. 22 — Single stage rig equipped for radiation testing

pressor operation and the duct characteristic numbers $k_{m\mu} b$ which are listed in Appendix B. The strong directionality of the field can be seen by examining Fig. 20. Here, the angular dependence of the first radial mode ($\mu = 0$) is displayed in a sequence of diagrams for a one-lobe mode, computed for several cutoff ratios. It may be seen that as the cutoff ratio increases, the radiated field becomes focused toward the centerline. Since an increase in cutoff ratio is synonymous with higher frequency, it is clear that the directivity patterns of the cylinder and the thin rectangular duct have common features.

It may be remarked that for both ducts, the plane wave case (m = 0) exhibits the exceptional property of having the peak of the directivity pattern fall on the duct axis. All other modes have zero pressure in this direction.

Fig. 21 presents experimental radiation data taken on a 10-in. diameter, single stage, 32-blade compressor rig shown in Fig. 22. Two stator vane assemblies were run with the rotor operating at 100 rps. One assembly, containing 31 vanes, generated a single 1-lobe mode at a cutoff ratio of $\xi = 4.14$. The other produced a 4-lobe mode $\xi = 1.44$. Some departure from the theoretical curves is noticeable, particularly near the 90 deg position. This departure may be a consequence of the simplifying assumption made early in the radiation analysis with respect to the abrupt vanishing of the particle velocity at the edge of the duct.

The foregoing treatment affords a preliminary view of some of the radiation characteristics of axial flow compressor noise. In order to evaluate the noise received by ground observers as the airplane passes overhead, a more comprehensive study is required. Because the directivity patterns are much more complicated than those of jet exhaust noise and because compressor noise is in the high frequency range, the far field is much more dependent on atmospheric turbulence and on temperature and wind gradients. For this reason the constant altitude curves which may be calculated readily from the constant radius curves, are likely to differ appreciably from specific measured fly-by data. Although these factors are outside the scope of this paper it must be emphasized that final judgment of the noise from a jet engine compressor must be based on fly-by data rather than constant radius data.

3. Summary

The basic processes for the generation, transmission and radiation of noise in an axial-flow compressor have been explored by means of a combined analytical and experimental program. It has been demonstrated that the pressure field generated in the compressor consists of spinning mode patterns; that the number of lobes in each pattern, together with the frequency of the pressure fluctuations, gives the speed at which the pattern spins; and that there is a critical pattern speed which separates the subsequent behavior of the pattern into one of two different categories.

If the tip Mach number with which the pattern sweeps the duct wall is below critical, as is the case for a subsonic rotor, the intensity rapidly decays with distance from the rotor. If, however, the rotor is operating in combination with stator vanes, many patterns are generated, some of which may be spinning above their critical tip Mach numbers. These high-speed patterns propagate along the duct and radiate sound to the outside. An analytical process for determining the rotor-stator interaction patterns has been presented, and experimental procedures for confirming the presence of these patterns have been described. Finally, the angular distribution of sound radiated from the face of a compressor duct has been treated, so that the effects of the propagating patterns can be studied in free-field.

An understanding of these basic phenomona is requisite to intelligent analytical or experimental study of axial flow compressor noise. This paper is by no means exhaustive. Continued exploration in this area is required in order to answer important questions. However, the tools are now available to carry on in a productive manner.

APPENDIX A

LIST OF SYMBOLS

Symbol		Section First Used
B	= Number of rotor blades	3
N	= Rotor shaft speed, rps	3
Ω	= Rotor shaft speed, radians/sec	3
θ	= Angular coordinate, radians	3
t	= Time coordinate	3
p	= Pressure	3
n	= Harmonic number (n = 1 for fundamental)	3
a_n	= Amplitude coefficient	3
ϕ_n	= Phase angle	3
r	= Radial coordinate	3
f	= Frequency, cps	3
ω	= Circular frequency ($\omega = 2\pi f$)	4.2
x	= Coordinate in direction of duct axis	4.2
y	= Rectangular duct coordinate	4.2
d	= Rectangular duct width	4.2
q	= Index number	4.2
a_q	= Pressure amplitude in cross mode q	4.2
λ	= Wavelength	4.2
λ_y	= Wavelength in the y-direction	4.2
c	= Speed of sound in free-field	4.2
f_q	= Cutoff frequency for qth cross mode	4.2
k_x	= Wave number, Eq. 4.2.1	4.2
r_0	= Annular duct radius	4.3
Ω	= Angular velocity of spinning pattern	4.3
s	= Circumferential arc coordinate	4.3
c_s	= Wave velocity in s-direction	4.3
m	= Number of lobes or cycles of circumferential pressure variation	4.3
a_m	= Amplitude of m-th mode	4.3
ϕ_m	= Phase of m-th mode	4.3

M_s = Mach number of pattern in circumferential direction 4.3

f_m = Cutoff frequency for m-lobe pattern 4.3

k_x = Wave number, Eq. 4.3.1 .... 4.3

Δx = Change in x position 4.3

Δdb = Sound pressure level drop 4.3

λ_s = Wavelength in circumferential direction 4.3

Ω_m = Angular velocity of m-lobe spinning pattern 4.4

σ = Hub-tip ratio ($\sigma = a/b$) 4.4

a = Inner wall radius 4.4

b = Outer wall radius 4.4

μ = Radial mode index 4.4

$E_{m\mu}^{\sigma}$ = Characteristic E-function (see Fig. 11A) 4.4

$f_{m\mu}$ = Cutoff frequency of (m, μ) mode 4.4

$k_{x\mu}$ = Wave number 4.4

$k_{m\mu}^{'\sigma}$ = Characteristic number (see Fig. 11A) 4.4

M_m = Circumferential mach number for m-lobe pattern 4.4

M_m^* = Cutoff mach number for (m, o) mode 4.4

V = Number of stator vanes 6.3

q = Index (q = 1,2....V − 1) (used differently in 4.1) 6.3

k = Index (k = ... − 1, 0, 1, 2...) 6.3

R = Radial coordinate 7.2

ψ = Angle to observer 7.2

$P_{m,\mu}$ = Radiated pressure 7.3

$\xi_{m,\mu}$ = Cutoff ratio 7.3

APPENDIX B

Pressure Field In Concentric Cylindrical Duct

The pressure, $p = p(r, \theta, x, t)$, of an ideal fluid at rest in the annular region shown in Fig. 10 is found as a solution of the homogeneous wave equation,

$$\nabla^2 p - \frac{1}{c^2}\frac{\partial^2 p}{\partial t^2} = 0 \qquad (1)$$

satisfying the following boundary conditions:

1. p is a continuous function.

2. The radial component of the pressure gradient vanishes at the boundary walls; $\partial p/\partial r = 0$ at r = a and r = b.

3. At a reference plane, x = 0, normal to the cylinder axis, the pressure is specified as an arbitrary function of r, satisfying 1 and 2 and a periodic function of time.

The solution is found as a combination of the characteristic functions of 1 which satisfies the boundary conditions. The characteristic functions are obtained as solutions of the ordinary differential equations into which the wave equation separates on substitution of the form $p = f(r) g(\theta) h(x) T(t)$.

Characteristic Functions of θ - $g(\theta) = \cos(m\theta + \phi_m)$ or equivalent, where m = 0, 1, 2... is a characteristic number restricted to integer values by the requirement of continuity: $p(\theta) = p(\theta + 2\pi)$ and $p'(\theta) = p'(\theta + 2\pi)$.

Characteristic Functions of t - $T(t) = \cos(\omega t + \alpha)$ or equivalent form. ω is unrestricted, and the solution will consider a single arbitrary value of (ω) rather than a series.

Characteristic Functions of r - $f(r) = J_m(k_{m\mu}^{(\sigma)} r) + Q_{m\mu}^{(\sigma)} \times Y_m(k_{m\mu}^{(\sigma)} r) = E_{m\mu}^{(\sigma)}(k_{m\mu}^{(\sigma)} r)$. The argument, $(k_{m\mu}^{(\sigma)} r)$ may be written = $bk_{m\mu}^{(\sigma)}\frac{r}{b} = (k_{m\mu}^{'(\sigma)} r')$ where $a \le r \le b$, $\sigma \le r \le 1$, and $\sigma = a/b$.

The characteristic function $E_{m\mu}^{(\sigma)}$ is the sum of the Bessel function of the first kind and a weighted Bessel function of the second kind. The order of both Bessel functions, m, is the same characteristic number as used for the $g(\theta)$ dependence. This feature "couples" the r and θ solutions. μ is an index, taking on values 0, 1, 2 ... which orders the paired roots of the transcendental equations relating the characteristic numbers, $k_{m\mu}^{'(\sigma)}$ and $Q_{m\mu}^{(\sigma)}$. These equations are:

$$J_m'(k') + Q_{m\mu}^{(\sigma)} Y_m'(k') = 0$$

$$J_m'(\sigma k') + Q_{m\mu}^{(\sigma)} Y_m'(\sigma k') = 0$$

where primes denote differentiation with respect to the argument.

This combination of Bessel functions results from solution of the separated Bessel equation:

$$\frac{d^2 f}{dr^2} + \frac{1}{r}\frac{df}{dr} + \left(k^2 - \frac{m^2}{r^2}\right)f = 0$$

k first enters as an arbitrary separation constant, together with the previously considered m. It is later found to obey the simultaneous relations above, which are a consequence of boundary condition 2.

The limiting case of no interior cylinder ($\sigma = 0$) is simply $f(r) = J_m(k_{m\mu} r)$ since the presence of Y_m violates boundary condition.

Values of the first four paired roots $k_{m\mu}^{'(\sigma)}$ and $Q_{m\mu}^{(\sigma)}$ for $\sigma = 0$, 0.25, 0.5, 0.75 and m = 1, 2, 4, 8, 16, 32, were obtained by numerical methods and appear at the end of this appendix.

Characteristic Functions of x —

$$h(x) = \exp \pm i\sqrt{\left(\frac{\omega}{c}\right)^2 - (k_{m\mu}^{(\sigma)})^2} \times$$

x is the solution of the final separated equation. When $\omega/c > k_{m\mu}^{(\sigma)}$ this is equivalent to a cosine function. However, when $\omega/c < k_{m\mu}^{(\sigma)}$ the positive exponential is rejected on physical grounds giving

$$h(x) = \exp -\sqrt{(k_{m\mu}^{(\sigma)})^2 - (\omega/c)^2}\, x \quad (x \ge 0, \omega/c < k_{m\mu}^{\sigma})$$

The final solutions are assembled from these characteristic functions to fit boundary condition 3. The results for decay and propagation are given in 4.4.

Tabulation of Characteristic Numbers

Mode		$k_{m\mu}^{'(\sigma)}$ hub-tip ratio - σ			$Q_{m\mu}^{(\sigma)}$ hub-tip ratio - σ		
m μ	0	0.25	0.50	0.75	0.25	0.50	0.75
1 0	1.841	1.644	1.354	1.146	−0.129	−0.286	−0.357
1	5.331	5.004	6.564	12.659	−0.327	2.578	1.390
2	8.526	8.808	12.706	25.179	0.275	1.537	1.177
3	11.706	12.850	18.942	37.730	2.160	1.324	1.115
2 0	3.054	3.009	2.681	2.292	−0.029	−0.221	−0.391
1	6.706	6.357	7.062	12.819	−0.341	0.350	−0.361
2	9.970	9.623	12.949	25.258	−0.351	−0.222	−0.635
3	13.170	13.371	19.103	37.783	0.201	−0.431	−0.742
4 0	5.318	5.316	5.175	4.578	−0.001	−0.080	−0.363
1	9.282	9.240	8.836	13.443	−0.038	−0.418	0.881
2	12.682	12.437	13.892	25.575	−0.235	0.224	−0.014
3	15.964	15.501	19.735	37.994	−0.479	0.582	−0.276
8 0	9.648	9.647	9.638	9.109	0.000	−0.004	−0.245
1	14.116	14.115	13.796	15.713	0.000	−0.265	4.318
2	17.774	17.772	17.340	26.806	−0.001	−0.405	−1.684
3	21.229	21.212	22.136	38.830	−0.015	1.119	2.446
16 0	18.063	18.063	18.063	17.890	0.000	0.000	−0.070
1	23.264	23.264	23.259	22.790	0.000	−0.003	−0.351
2	27.347	27.347	27.234	31.272	0.000	−0,091	0.138
3	31.111	31.111	30.610	42.013	0.000	−0.456	0.355
32 0	34.588	34.588	34.588	34.581	0.000	0.000	−0.002
1	40.797	40.797	40.797	40.333	0.000	0.000	−0.291
2	45.510	45.510	45.510	45.209	0.000	0.000	−0.216
3	49.762	49.762	49.761		0.000	−0.001	

These characteristic numbers were calculated on an IBM 7090 at the UAC Computation Laboratory. A Bessel function routine, distributed through Share, and written at N.Y.U. Institute of Mathematical Sciences, was employed in finding the roots of the transcendental equations. The accuracy of the routine depends to some extent on the argument and order of the Bessel functions. However, of many values checked, all were accurate to at least the 7th significant figure. From these considerations, the $k_{m\mu}^{'(\sigma)}$ are believed accurate to one unit in the third decimal place. For reasons too detailed to recount here, the best that can be said at this time concerning the accuracy of the $Q_{m\mu}^{(\sigma)}$ is that they are believed accurate to a "few" units in the last significant figure.

APPENDIX C

Sound Field Radiated from a Cylindrical Duct

A cylindrical duct transmitting higher order sound waves is illustrated in Fig. 19. The fluctuating pressure field at the open face of the duct can be described in terms of simple modes whose characteristic functions are of the form:

$$p_{m\mu}(r, \theta, t) = A_{m\mu} J_m(k_{m\mu}r)\, e^{i(m\theta - \omega t)} \tag{1}$$

$$m = 0, 1, 2, \ldots.$$

$$\mu = 0, 1, 2, \ldots.$$

An element of area (ds) located at $p(r, \theta)$ can be considered as a simple source whose strength is weighted by the spatial distribution of acoustical particle velocity normal to the face of the duct. For simple harmonic motion the particle velocity $\overline{V}$ can be calculated from the pressure (p) as follows:

$$\overline{V} = \frac{1}{i\omega\rho}\, \text{grad } p \tag{2}$$

The component of particle velocity normal to the duct face v_x is then:

$$v_x = \frac{-A_m k_x}{\omega\rho} J_m(k_{m\mu}r)\, e^{i(m\theta - \omega t)} \tag{3}$$

Assume here for mathematical simplicity that the particle velocity vanishes at the duct extremity, r = b. Then, the strength of the simple source corresponding to (ds) is $d(Q) = 2v_x e^{i\omega t} ds$ where the factor (2) appears since the source can radiate only to the right. Accordingly, the pressure (p) in the far field due to a simple source is approximately $\frac{\rho}{4\pi h}$ $Q'(t - h/c)$.*

In terms of particle velocity, the pressure in the far field due to the simple source (ds) in the duct face is then:

$$dp = -i\omega\rho v_x\, e^{-i\omega(t - h/c)}\, ds \tag{4}$$

Combining Eqs. 3 and 4 gives the pressure at the point $P(R, \psi)$ due to (ds) as follows:

$$dp = \frac{A_{m\mu} i\omega k_x}{2\pi hc} J_m(k_{m\mu}r)\, e^{i(kh + m\theta - \omega t)}\, ds \tag{5}$$

where $k = \omega/c = 2\pi f/c$ and is called the wave number.

Here it should be remarked that (f) the frequency being referred to is the fundamental frequency or blade passing frequency. From the law of cosines

$$h^2 = R^2 + r^2 - 2rR\cos\alpha \quad \text{and} \quad \cos\alpha = \cos\theta\sin\psi$$

there follows:

$$h = R\left(1 - \frac{2r}{R}\cos\theta\sin\psi\right)^{1/2} \tag{6}$$

Expanding the radical in Eq. 6 gives the approximation

$$h = R - r\cos\theta\sin\psi \quad \text{for} \quad r \ll R \tag{7}$$

Combining Eq. 7 with Eq. 5 yields the pressure radiated

*See Morse, Vibration and Sound, p. 313.

from any simple source (ds) to the point in the far field $P(r, \psi)$. The integral over the surface of the duct face then gives the resultant field:

$$P_{mq}(R, \psi, t) = \frac{A_{m\mu}if}{cR} k_x e^{i(kR - \omega t)}$$

$$\iint r J_m(k_{m\mu}r) e^{i(m\theta - kr \sin \psi \cos \theta)} \, dr \, d\theta \quad (8)$$

$$(ds = r \, dr \, d\theta)$$

The integral I_θ from zero to 2π is exactly the integral representation of the cylindrical Bessel function of order

(m) argument $(kr \sin \psi)$, that is,

$$I_\theta = 2\pi i^m J_m(kr \sin \psi) \quad (9)$$

The integral over (r) now becomes:

$$I_r = \int r J_m(k_{m\mu}r) J_m(rk \sin \psi) dr$$

which can be evaluated directly:

$$I_r = \left[\frac{-bk \sin \psi}{(k_{m\mu}{}^2 - k^2 \sin^2 \psi)}\right] \left[J_m(k_{m\mu}b) J'_m(-kb \sin \psi)\right] \quad (10)$$

Simplifying Eqs. 9 and 10 in terms of well known recursive relations, combining constants, and substituting into Eq. 8 yields finally the resultant radiated field:

$$P_{m\mu}(R, \psi, t) = \frac{B_{m\mu} e^{i(kR - \omega t)} J_m(k'_{m\mu}) bk \sin \psi [J_{m-1}(kb \sin \psi) - J_{m+1}(kb \sin \psi)]}{R(k_{m\mu}{}^2 - k^2 \sin^2 \psi)} \quad (11)$$

In terms of cutoff ratio,

$$\xi_{m\mu} = \frac{k}{k_{m\mu}}$$

Eq. 11 can be written:

$$P_{m\mu}(R, \psi, t) = \frac{B_{m\mu} e^{i(kR - \omega t)} J_m(k'_{m\mu}) bk \sin \psi [J_{m-1}(\xi_{m\mu}k_{m\mu}b \sin \psi) - J_{m+1}(\xi_{m\mu}k_{m\mu}b \sin \psi)]}{Rk_{m\mu}{}^2 [1 - \xi_{m\mu}{}^2 \sin^2 \psi]} \quad (12)$$

For the exceptional case of the plane wave this reduces to:

$$P_{00}(R, \psi, t) = \frac{B_{00} e^{i(kR - \omega t)} kb J_1(kb \sin \psi)}{Rk \sin \psi} \quad (13)$$

13

Copyright © 1962 by the Acoustical Society of America
Reprinted from *Acoust. Soc. Am. J.* **34:**623-639 (1962)

Power Flow between Linearly Coupled Oscillators*

Richard H. Lyon and Gideon Maidanik

Bolt Beranek and Newman Inc., Cambridge 38, Massachusetts

The power flow between two independently and randomly excited harmonic oscillators is calculated assuming small linear coupling. It is found that for conservative coupling the power flow is proportional to the temperature (average modal energy) difference of the two oscillators. The constant of proportionality is symmetric in the parameters of the two modes and is positive definite although its magnitude depends on the relative sign of the inertial and stiffness coupling. An equivalent circuit for the energy flow between the modes is developed. Finally, we apply the formalism to obtain the motion transmitted through a two-stage vibration isolator. The method is then extended to the problem where two multimodal systems interact. A particular application is the power flow between several structural modes and a reverberant acoustic field. The reverberant sound field is considered as a temperature bath in which the structural modes are immersed. Using this model the steady-state partition of energy between the two systems and the parameters which govern this partition are computed. Analog circuits for this problem are also constructed. The radiation resistance for some of the normal modes of a beam are calculated and results of the formalism are illustrated experimentally.

PART I

I. INTRODUCTION

THE motion of systems with very many degrees of freedom which are excited by random sources is of central interest in the disciplines of room acoustics,[1] noise control,[2] kinetic theory,[3] and turbulence.[4] A feature

* Editor's Note: Part I was submitted for publication in this journal by the first author on August 24, 1960. It was subsequently withdrawn by the author and resubmitted in its present form as a portion of this paper.

[1] L. L. Beranek, *Acoustic Measurements* (John Wiley & Sons, Inc., New York, 1949), Chap. 18.
[2] I. Dyer, J. Acoust. Soc. Am. **31**, 922 (1949).
[3] I. Prigogine and R. Brout, contribution to *Transport Processes in Statistical Mechanics*, edited by I. Prigogine (Interscience Publishers, Inc., New York, 1958), p. 25.
[4] R. H. Kraichnan, Phys. Rev. **109**, 1407 (1958).

common to most of these studies is the presence of coupling between those "modes" of motion which one may regard as most convenient or natural in describing the response. This coupling may be strong or weak, linear or nonlinear, and its presence may be regarded as a small perturbation (room acoustics),[5] or as the essential controlling process of the phenomenon (turbulence).[4]

As an initial step in a broad attack on the types of problems mentioned here, we consider in detail possibly the simplest case imaginable. In this paper, we study the power flow between two randomly excited linear oscillators with small linear coupling between them. We find that even this simple problem yields results which find application in several instances of engineering interest.

The paper is divided into two parts. In Part I, we consider the analysis and give an example which is strictly concerned with a two-oscillator problem. We study the power flow between the oscillators by evaluating certain second moments of the response directly from the stochastic equations. After a discussion of these results, they are applied to a two-mode vibration isolation mount which has been treated previously. In Part II, we are interested in how the results are modified when two systems each containing many modes interact. A particular application is the power flow between several structural modes and a reverberant field which may be thought of as a dense collection of incoherent acoustic modes of a large room. We see how it is possible to combine modes in such a way that the simple two-mode model of Part I will have something useful to say about the multimode systems of Part II.

In Sec. II we set down the equations of motion and develop an expression for the power flow between modes in terms of second moments of these equations. Forming equations for these moments directly from the stochastic equations, we compute power flow and various moments correct to second order in the coupling parameter amplitudes.

In Sec. III, we find that for conservative coupling the power flow may be regarded as a heat flow in response to a temperature difference of the oscillators, the temperature defined in the Nyquist sense. In Sec. IV we make use of this idea to develop analog energy circuits to represent the system. In these circuits, power is the flow quantity and temperature is the drop. Energy leaks between the two systems through an energy conductance which is of second order in the coupling magnitudes.

In Sec. V, we apply these results to the motion transmitted through a two-stage vibration isolator. The results are immediate from our analysis and check well with previous calculations using another method. An introduction to the work in Part II may be found in Sec. VI.

---

[5] P. M. Morse and R. H. Bolt, Revs. Modern Phys. 16, 69 (1944).

II. STOCHASTIC EQUATIONS

We assume that the coupled oscillators satisfy the following set of equations:

(a) $\dot{u}_1 + \beta_1 u_1 + \omega_1^2 y_1 + A\dot{u}_2 + B_2 u_2 + Cy_2 = f_1,$

(b) $\dot{u}_2 + \beta_2 u_2 + \omega_2^2 y_2 + A\dot{u}_1 + B_1 u_1 + Cy_1 = f_2,$

(c) $\dot{y}_1 = u_1,$

(d) $\dot{y}_2 = u_2,$ (2.1)

where the coupling parameters A and C will cause inertial and stiffness coupling, respectively; the coupling parameters B_1 and B_2 will cause dissipative coupling when $B_1 = B_2$ and gyroscopic coupling when $B_1 = -B_2$. The sources f_1 and f_2 are assumed statistically independent and to have power spectra "flat" compared to the admittance spectra of the two oscillators. It is convenient to think of these sources as sequences of impulses of strength "a" for f_1 and "b" for f_2, of average frequency of occurrence ν_a and ν_b, respectively.[6] If $\sigma_a^2 = \langle a^2 \rangle$ and $\sigma_b^2 = \langle b^2 \rangle$, then the spectral densities of f_1 and f_2 will be $D_1 = \frac{1}{2}\nu_a\sigma_a^2$, $D_2 = \frac{1}{2}\nu_b\sigma_b^2$.

Suppose an impulse a occurs from source f_1 only. Then by Eqs. (2.1a) and (2.1b)

$$\Delta u_1 + A\Delta u_2 = a, \quad \Delta u_2 + A\Delta u_1 = 0.$$

Accordingly,

$$\Delta u_1 = a(1-A^2)^{-1}, \quad \Delta u_2 = -aA(1-A^2)^{-1}.$$

From the nature of f_1, subsequent increments of Δu_1 and Δu_2 (due to different impulses) are independent increments of velocity, and therefore in one second the amount of energy gained by mode 1 is

$$\tfrac{1}{2}\nu_a\sigma_a^2(1-A^2)^{-2},$$

while the energy gained by mode 2 is

$$\tfrac{1}{2}\nu_a\sigma_a^2 A^2(1-A^2)^{-2}.$$

Thus, the power from source 1 is

$$\langle f_1 u_1 \rangle = D_1(1+A^2)(1-A^2)^{-1},   (2.2a)$$

and similarly

$$\langle f_2 u_2 \rangle = D_2(1+A^2)(1-A^2)^{-1}.   (2.2b)$$

Also, since $\Delta u_2 = -A, \Delta u_1$ one has immediately

$$\langle f_1 u_2 \rangle = -D_1 A(1+A^2)(1-A^2)^{-2}   (2.2c)$$

and

$$\langle f_2 u_1 \rangle = -D_2 A(1+A^2)(1-A^2)^{-2}.   (2.2d)$$

Since an impulse produces no immediate increment of displacement,

$$\langle f_1 y_1 \rangle = \langle f_2 y_2 \rangle = \langle f_1 y_2 \rangle = \langle f_2 y_1 \rangle = 0.   (2.2e)$$

The above covariances between the sources and displacements and velocities define the statistical properties

---

[6] S. O. Rice, contribution to *Selected Papers on Noise and Stochastic Processes*, edited by Nelson Wax (Dover Publications Inc., New York, 1954), p. 153.

of the sources for our purposes. For sources which are not "purely random," the above results must be regarded as approximations which are appropriate as long as the previous conditions on bandwidth are observed.

The effective force which the motion of mode 1 causes to be produced on mode 2 is seen from Eq. (2.1) to be

$$f_{12} = -A\dot{u}_1 - B_1u_1 - Cy_1. \qquad (2.3)$$

Thus the average net power (time average equals stochastic average for stationary ergodic processes) delivered from mode 1 to mode 2 is $\langle f_{12}u_2 \rangle$. Similarly, the net power transferred from mode 2 to 1 is $\langle f_{21}u_1 \rangle$. Accordingly,

$$j_{12} = -A\langle \dot{u}_1 u_2 \rangle - B_1\langle u_1 u_2 \rangle - C\langle y_1 u_2 \rangle \qquad (2.4a)$$

and

$$j_{21} = -A\langle u_1\dot{u}_2 \rangle - B_2\langle u_1 u_2 \rangle - C\langle u_1 y_2 \rangle. \qquad (2.4b)$$

For a stationary process

$$\langle \dot{u}_1 u_2 \rangle = -\langle u_1\dot{u}_2 \rangle$$

and

$$\langle y_1 u_2 \rangle = -\langle u_1 y_2 \rangle,$$

and therefore we may write

$$j_{21} = A\langle \dot{u}_1 u_2 \rangle - B_2\langle u_1 u_2 \rangle + C\langle y_1 u_2 \rangle. \qquad (2.4c)$$

In order to evaluate the moments involved in Eq. (2.4), we return to the stochastic equation Eq. (2.1). Multiplying Eqs. (2.1a) and (2.1b) by u_1, y_1, u_2, and y_2 in turn and averaging we obtain the following set of equations

$$
\begin{bmatrix}
\beta_1 & 0 & 0 & 0 & 0 & B_2 & -C & -A \\
0 & \beta_2 & 0 & 0 & 0 & \beta_1 & C & A \\
0 & B_2 & 0 & 0 & 0 & \beta_1 & \omega_1^2 & 1 \\
B_1 & 0 & 0 & 0 & 0 & \beta_2 & -\omega_2^2 & -1 \\
-1 & 0 & \omega_1^2 & 0 & C & -A & B_2 & 0 \\
0 & -1 & 0 & \omega_2^2 & C & -A & -B_1 & 0 \\
0 & -A & 0 & C & \omega_1^2 & -1 & -\beta_1 & 0 \\
-A & 0 & C & 0 & \omega_2^2 & -1 & \beta_2 & 0
\end{bmatrix}
\begin{bmatrix}
\langle u_1^2 \rangle \\
\langle u_2^2 \rangle \\
\langle y_1^2 \rangle \\
\langle y_2^2 \rangle \\
\langle y_1 y_2 \rangle \\
\langle u_1 u_2 \rangle \\
\langle y_1 u_2 \rangle \\
\langle \dot{u}_1 u_2 \rangle
\end{bmatrix}
=
\begin{bmatrix}
\langle f_1 u_1 \rangle \\
\langle f_2 u_2 \rangle \\
\langle f_1 u_2 \rangle \\
\langle f_2 u_1 \rangle \\
0 \\
0 \\
0 \\
0
\end{bmatrix}. \qquad (2.5)
$$

The determinant of the matrix in Eq. (2.5) is evaluated under the assumption that the coupling constants are small ($|A| \ll 1$; $|C| \ll \omega_1^2, \omega_2^2$; and $|B| \ll \beta_1, \beta_2$). The result is

$$\Delta = -\beta_1\beta_2\omega_1^2\omega_2^2[(\omega_1^2-\omega_2^2)^2 + (\beta_1+\beta_2)(\beta_1\omega_2^2+\beta_2\omega_1^2)]. \qquad (2.6)$$

Solving Eq. (2.5) and keeping only terms linear in the coupling parameters we obtain

$$\langle \dot{u}_1 u_2 \rangle \Delta = \beta_1\beta_2\omega_1^2\omega_2^2$$
$$\times \{A[\beta_1\omega_2^4+\beta_2\omega_1^4+\beta_1\beta_2(\beta_1\omega_2^2+\beta_2\omega_1^2)](\theta_1-\theta_2)$$
$$-B_1(\beta_1^2\omega_2^2+\beta_1\beta_2\omega_1^2+\omega_1^4-\omega_1^2\omega_2^2)\theta_1$$
$$+B_2(\beta_2^2\omega_1^2+\beta_1\beta_2\omega_2^2+\omega_2^4-\omega_1^2\omega_2^2)\theta_2$$
$$-C(\beta_1\omega_2^2+\beta_2\omega_1^2)(\theta_1-\theta_2)\}, \qquad (2.7)$$

$$\langle u_1 u_2 \rangle \Delta = \beta_1\beta_2\omega_1^2\omega_2^2$$
$$\times \{A[(\beta_1^2\omega_2^2+\beta_1\beta_2\omega_1^2+\omega_1^4-\omega_1^2\omega_2^2)\theta_1$$
$$+(\beta_2^2\omega_1^2+\beta_1\beta_2\omega_2^2+\omega_2^4-\omega_1^2\omega_2^2)\theta_2]$$
$$+(\beta_1\omega_2^2+\beta_2\omega_1^2)(B_1\theta_1+B_2\theta_2)$$
$$-C(\omega_1^2-\omega_2^2)(\theta_1-\theta_2)\}, \qquad (2.8)$$

$$\langle y_1 u_2 \rangle \Delta = \beta_1\beta_2\omega_1^2\omega_2^2$$
$$\times [-A(\beta_1\omega_2^2+\beta_2\omega_1^2)(\theta_1-\theta_2)$$
$$+(\omega_1^2-\omega_2^2)(B_1\theta_1+B_2\theta_2)$$
$$+C(\beta_1+\beta_2)(\theta_1-\theta_2)], \qquad (2.9)$$

and

$$\langle y_1 y_2 \rangle \Delta = \beta_1\beta_2\omega_1^2\omega_2^2$$
$$\times \{A(\omega_1^2-\omega_2^2)(\theta_1-\theta_2)+(\beta_1+\beta_2)(B_1\theta_1+B_2\theta_2)$$
$$-C[(\omega_1^2-\omega_2^2)(\theta_1/\omega_1^2-\theta_2/\omega_2^2)$$
$$-(\beta_1+\beta_2)(\beta_1\theta_1/\omega_1^2+\beta_2\theta_2/\omega_2^2)]\}. \qquad (2.10)$$

The remaining four moments, namely: $\langle u_1^2 \rangle$, $\langle u_2^2 \rangle$, $\langle y_1^2 \rangle$, and $\langle y_2^2 \rangle$, can now be easily obtained from Eqs. (2.5), (2.6), (2.7), (2.8), (2.9), and (2.10).

In the above we have used the relations

$$\theta_1 = D_1/\beta_1 \quad \text{and} \quad \theta_2 = D_2/\beta_2, \qquad (2.11)$$

which are the effective temperatures of modes 1 and 2, respectively, in the absence of coupling (Boltzmann's constant equal to unity) by Nyquist's theorem.[7] We shall find it useful to think of the modes as having different temperatures and the coupling as a heat leak allowing power to travel from one mode to the other. Our results, as is seen, lend considerable support to this view. In the following sections we define the unperturbed temperature θ of the mode by Eq. (2.11) and by θ with a superscript (e.g., θ^s or θ^R) when we refer to the temperature of the mode as defined by the actual average energy in the mode.

III. CALCULATION OF POWER FLOW

The first two equations in Eq. (2.5) are

$$\beta_1\langle u_1^2 \rangle + B_2\langle u_1 u_2 \rangle - C\langle y_1 u_2 \rangle - A\langle \dot{u}_1 u_2 \rangle = \langle f_1 u_1 \rangle \qquad (3.1a)$$

and

$$\beta_2\langle u_2^2 \rangle + B_1\langle u_1 u_2 \rangle + C\langle y_1 u_2 \rangle + A\langle \dot{u}_1 u_2 \rangle = \langle f_2 u_2 \rangle. \qquad (3.1b)$$

Adding Eqs. (3.1a) and (3.1b) we obtain the power-balance equation

$$\beta_1\langle u_1^2 \rangle + \beta_2\langle u_2^2 \rangle + (B_1+B_2)\langle u_1 u_2 \rangle = \langle f_1 u_1 \rangle + \langle f_2 u_2 \rangle. \qquad (3.2)$$

[7] A. van der Ziel, *Noise* (Prentice Hall, Inc., Engelwood Cliffs, New Jersey, 1955), Chap. 2.

The left-hand side of Eq. (3.2) represents the power dissipated in the system as a whole and the right-hand side of Eq. (3.2) represents the power supplied by the sources.

Setting $B_1 = -B_2$ (gyroscopic coupling) we see from Eq. (3.2) that the dissipation in the system is inde-

pendent of the coupling and in this sense the coupling is conservative. In this case we note from Eq. (2.4) that $j_{12} = -j_{21}$. From Eqs. (2.4), (2.6), (2.7), (2.8), and (2.9) we obtain readily

$$j_{12} = -j_{21} = g_{12}(\theta_1 - \theta_2), \quad (3.3)$$

with

$$g_{12} = \frac{\{A^2[\beta_1\omega_2^4 + \beta_2\omega_1^4 + \beta_1\beta_2(\beta_1\omega_2^2 + \beta_2\omega_1^2)] + (B^2 - 2AC)(\beta_1\omega_2^2 + \beta_2\omega_1^2) + C^2(\beta_1 + \beta_2)\}}{(\omega_1^2 - \omega_2^2)^2 + (\beta_1 + \beta_2)(\beta_1\omega_2^2 + \beta_2\omega_1^2)}. \quad (3.4)$$

The constant g_{12} is symmetric in the parameters of the two modes involved and it is positive definite although its magnitude for a given A, B, and C depends on the *relative* sign of A and C. The power flow from mode 1 to mode 2 depends on the unperturbed temperatures of mode 1 and mode 2 only through the difference between them, $\theta_1 - \theta_2$, and the direction of flow is determined by the sign of this quantity. Thus the analog with heat flow is apparent and we may think of g_{12} as the energy conductance between the two systems.

When $B_1 = B_2$ (dissipative coupling) it is obvious from Eq. (2.4) that $j_{12} \neq -j_{21}$. In fact neither j_{12} nor j_{21} will be proportional simply to $(\theta_1 - \theta_2)$ but will contain, in addition to such a term, a term which is proportional to $\theta_1 + \theta_2$. This is not surprising since with dissipative coupling we do not have a strictly two-component system, but rather, a three-component system, the third being a reservoir at zero temperature which couples the other two components (modes 1 and 2) by its dissipative conductance of magnitude B. The system may be represented by the circuit given in Fig. 1. The power dissipated in the coupling is given by

$$j_D = B\langle(u_1 + u_2)^2\rangle, \quad (3.5)$$

which is, to first order of magnitude in B,

$$j_D = B(\theta_1 + \theta_2) = B(\theta_1 - 0) + B(\theta_2 - 0), \quad (3.6)$$

where the temperature of the coupling element is set to zero since there is no thermal noise generator associated with it. Thus, the effect of dissipative coupling is, to first order, merely to drain out energy from both modes. In this case too, the analogy with heat flow is clearly demonstrated.

IV. ENERGETICS OF MODAL INTERACTION AND THE POWER FLOW ANALOG CIRCUIT

Since we have looked at energy flow from a quasi-thermodynamic point of view in the preceding sections,

we now see how a development of this attitude leads us to reasonable ways of thinking about our problem.

From Eqs. (2.2a) and (2.2b) we see that the power output of the sources is independent of any mechanical parameters except A; and, to the order of approximation we have carried things out, it is even independent of this. Thus the random sources supply constant average power to the system. The stored energy in system 1 without coupling is of course

$$\langle E\rangle_{av} = \tfrac{1}{2}\langle u_1^2\rangle + \tfrac{1}{2}\omega_1^2\langle y_1^2\rangle = \langle u_1^2\rangle = D_1/\beta_1 \equiv \theta_1. \quad (4.1)$$

Thus, the ratio of stored energy to temperature is unity, which is the thermal capacity for a system with two deg of freedom. The rate of leakage of energy from system 1 is equal to the dissipation $\beta_1\langle u_1^2\rangle = \beta_1\theta_1$. Thus we may represent the energetics of the mode without coupling by the analog circuit shown in Fig. 2. Here the flow

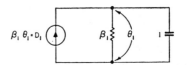

FIG. 2. Energy flow and storage in single mode.

quantity is power and the drop quantity is temperature. In the case of nondissipative coupling we may connect the two modes with an energy conductance g_{12} as shown in Fig. 3. Any dissipative coupling will place small conductances in shunt with β_1 and β_2 according to Eq. (3.6), but we are not entering these explicitly in this work.

It should be noted that it is slightly more elegant to use as a potential the reciprocal temperature θ^{-1} since then the drop times the flow quantity is entropy production which is the proper "dissipation function" for problems of this type.[8] This would have some theoretical advantages but makes no difference for the problem at hand (except that it is less convenient) so we do not adopt that convention here.

We also note that when coupling is present, the temperature in mode 1 is θ_1' and not $\theta_1 = D_1/\beta_1$. However, θ_1' will depart from θ_1 only slightly, in fact by the small term j_{12}/β_1 [see Eq. (3.1)]. In the case of two modes, it is satisfactory to ignore j_{12}/β_1 as compared with the temperatures of the mode, θ_1 and θ_1', to the order of accuracy maintained. In the event that systems 1 and 2

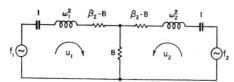

FIG. 1. Two oscillators coupled by a dissipative coupling.

[8] M. Lax, Revs. Modern Phys. 32, 25 (1960).

were attached to thermal reservoirs instead of random noise sources, then the generators of Fig. 3 would be of the constant drop variety. Again the change in drop across the conductance will be negligible to the order of accuracy we have maintained.

The circuit analog depicted in Fig. 3 makes not only the physical interpretation and possible extension easier to grasp, but transient solutions of the network can tell us the time history of the energy level in the two modes as a function of time. If the sources f_1 and f_2 are switched on at times t_1 and t_2 we can solve for the response on the basis of elementary R-C transient analysis. We do not bother to follow through such an analysis and we confine our attention to the steady-state conditions only.

V. RESPONSE OF TWO COUPLED RESONATORS TO RANDOM EXCITATION

We propose to analyze, as an example, the motion of the vibration isolator depicted in Fig. 4(a).[9] Its mechanical circuit is depicted in Fig. 4(b) and the equivalent mobility-type circuit in Fig. 4(c). Under the assumption of high Q resonances the mobility circuit may be transformed to the circuit given in Fig. 4(d) with the appropriate elements as indicated, where

$$\Omega_1^2 = k_1/m_1; \quad \Omega_2^2 = k_2/m_2; \quad Q_1 = \Omega_1 m_1/r_1;$$
$$\text{and} \quad Q_2 = \Omega_2 m_2/r_2. \quad (5.1)$$

The equations of motion of the system are

$$\dot{u}_1 + \beta_1 u_1 + \omega_1^2 y_1 + C y_2 = g, \quad (5.2a)$$
$$\dot{u}_2 + \beta_2 u_2 + \omega_2^2 y_2 + C y_1 = 0, \quad (5.2b)$$

where

$$u_i = (K_i)^{\frac{1}{2}} f_i; \quad y_i = (K_i)^{\frac{1}{2}} \int f_i dt; \quad \beta_i = R_i/K_i;$$

$$\omega_1^2 = 1/m_1 K_1; \quad \omega_2^2 = (m_1+m_2)/m_1 m_2 K_2; \quad g = (K_1)^{-\frac{1}{2}} v;$$

and

$$C^2 = 1/m_1^2 K_1 K_2.$$

For C to be considered small as required by our theory

$$|C| \ll \omega_1^2, \omega_2^2, \quad (5.3)$$

which is equivalent to requiring $m_2 \ll m_1$.

We note from Fig. 4(c) or (d) that

$$\langle v_2^2 \rangle = \left\langle \left(\int f_2 dt \right)^2 \right\rangle \Big/ m_2^2 = \langle y_2^2 \rangle / m_2^2 K_2, \quad (5.4)$$

and that the power spectrum of the source is

$$D_1 = \langle f_1 v \rangle = \langle g u_1 \rangle. \quad (5.5)$$

From Eqs. (2.5) and (5.2),

$$\langle y_2^2 \rangle = -(\langle y_1 u_2 \rangle / \beta_2 + \langle y_1 y_2 \rangle) C / \omega_2^2, \quad (5.6)$$

and using Eqs. (2.9), (2.10), and (5.4)

$$\langle v_2^2 \rangle \simeq K_1 D_1 \left\{ \frac{Q_1 \omega_1 (\omega_2^4 + \omega_2^2 \omega_1^2/Q_1^2) + Q_2 \omega_2 (\omega_1^4 + \omega_2^2 \omega_1^2/Q_2^2)}{(\omega_1^2 - \omega_2^2)^2 + \omega_1^2 \omega_2^2 [1/Q_1^2 + 1/Q_2^2 + (\omega_1/\omega_2 + \omega_2/\omega_1)/Q_1 Q_2]} \right\}. \quad (5.7)$$

The spectral density of the source velocity, $W(f)$ as defined by

$$\langle v^2 \rangle = \int_0^\infty W(f) df, \quad (5.8)$$

is related to the power spectrum, as defined in Eq. (5.5) by

$$K_1 D_1 = \frac{1}{4} W. \quad (5.9)$$

With Eq. (5.9) substituted in Eq. (5.7) the resulting equation is essentially in agreement with the result previously obtained.[9]

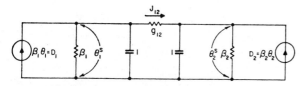

FIG. 3. Energy flow and storage in two coupled modes.

[9] This example has been treated previously in the literature [C. T. Morrow, B. A. Troesch, and H. Spence, J. Acoust. Soc. Am. 33, 46 (1961)]. It is our purpose here merely to illustrate the present methods in a case where the solution is known.

PART II

VI. INTRODUCTION

In Part I we deal exclusively with the interaction between two oscillators. We show how a thermodynamic or heat-flow analogy may be applied to these problems. We wish to extend this analogy to the coupling between an ensemble of modes of one system and a mode or an ensemble of modes of another system. Specifically, we concentrate on the coupling between a reverberant sound field and a structure.

We start by showing that the reverberant field may be thought of as constituting a temperature bath (Sec. VII) in which the structural modes (described in Sec. VIII) are immersed. Using this model we compute in Sec. IX the steady-state condition of the two interacting systems; i.e., we compute the partition of energy between the modes of the acoustic field and the modes of the structure and the parameters which governed this partition. The energy analog circuit which is introduced in Sec. IV is used in Sec. X as an alternate device to obtain expressions for the energy transferred between the structure and the field. In Sec. XI we calculate the radiation resistance for some of the normal

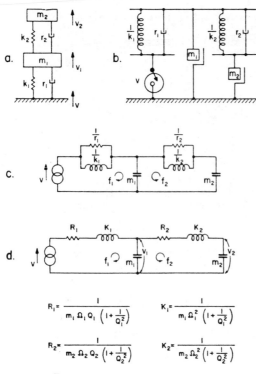

$$R_1 = \frac{1}{m_1 \, \Omega_1 \, Q_1 \left(1 + \frac{1}{Q_1^2}\right)} \qquad K_1 = \frac{1}{m_1 \, \Omega_1^2 \left(1 + \frac{1}{Q_1^2}\right)}$$

$$R_2 = \frac{1}{m_2 \, \Omega_2 \, Q_2 \left(1 + \frac{1}{Q_2^2}\right)} \qquad K_2 = \frac{1}{m_2 \, \Omega_2^2 \left(1 + \frac{1}{Q_2^2}\right)}$$

FIG. 4. Two-stage vibration isolator.

modes of a beam. In Sec. XII this radiation resistance is compared with that measured experimentally by allowing a supported beam to exchange energy with a reverberant acoustic field.

VII. REVERBERANT ACOUSTICAL FIELD AS A TEMPERATURE BATH

The main idea in thinking of a reverberant field as a thermal bath is that in the audiofrequency range one might represent the molecular agitation by a random superpostion of sound waves. Such a replacement is common in the statistical mechanics of crystalline solids.[10] If an oscillator were immersed in such a field, then that part of the thermal motion to which it would be sensitive would be in a narrow frequency range about its resonant frequency and if its resonance were in the audio range, then it would be our random sound waves which would account for its random motion.

Consider an oscillator in a diffuse sound field with a spectrum such as shown in Fig. 5(a), and suppose that the oscillator resonates at frequency ω_1. At this frequency the spectral density of the mean-square pressure is $S_p(\omega_1) = \langle p^2 \rangle / \Delta \omega$. Since this is the only frequency interval which the oscillator senses, we can turn the argument of the previous paragraph around and say that the sound field in this interval has a temperature related to the spectral density in a manner to be derived, and that the oscillator will come to steady-state condi-

---

[10] C. Kittell, *Introduction to Solid State Physics* (John Wiley & Sons, Inc., New York, 1956), 2nd ed., Chap. 6.

tion with the field with an average energy determined by that temperature.

Suppose the reverberant field is contained in a large room of volume V. The acoustic energy in a given interval of frequency is

$$E_R = \langle p^2 \rangle V / \rho c^2, \qquad (7.1)$$

where ρ is the ambient density and c is the corresponding speed of sound. This energy is shared by $n_R(\omega) \Delta \omega$ modes of the room where the mode density in the room is[5]

$$n_R(\omega) = \omega^2 V / 2\pi^2 c^3. \qquad (7.2)$$

The average energy per mode is just the ratio of these

$$\theta^R(\omega) = kT = 2\pi^2 c \langle p^2 \rangle / \rho \omega^2 \Delta \omega$$
$$= 2\pi^2 c S_p(\omega) / \rho \omega^2. \qquad (7.3)$$

In writing $\theta^R(\omega) = kT$ (k here is Boltzmann's constant) we are thinking of the modes of the room as an ensemble of oscillators each having two degrees of freedom. The oscillators have a given temperature which is determined by their location in frequency and their amplitudes are assumed statistically independent. It is evident from Eq. (7.3) that if all the modes of the room in all frequency intervals were to have the same temperature then the spectrum $S_p(\omega)$ would be fixed in the form $S_p(\omega) \propto \omega^2$ or a constant rise of 6 db/octave, or 9 db/octave on an octave-band basis as shown by the dashed lines in Fig. 5(a). Noise-producing devices do not normally have sound-level spectra of this form. For example, jet-noise spectra are typically like the "haystack" spectrum shown in Figs. 5(a) and (b). This means that various modes in different frequency ranges

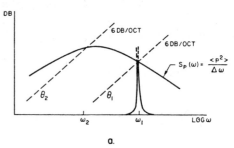

a.

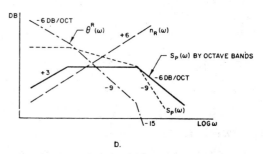

b.

FIG. 5. (a) Typical acoustic spectrum with constant temperature spectra superposed. (b) Derivation of temperature spectrum of diffuse field from sound spectrum in octave bands.

are at different temperatures, as for example in the frequency intervals near ω_1 and ω_2 as shown. The method of obtaining the acoustical temperature θ^R as a function of the frequency from the pressure spectral density is indicated in Fig. 5(b). It is clear that a mode at the lower frequencies will, in general, be at a higher temperature than its higher-frequency counterpart.

If a mechanical oscillator is placed in a steady sound field, it will interact with the oscillators representing the field, and power flow between them will resut. A steady-state condition will be reached between the reverberant field and the oscillator. It is important to realize that steady-state conditions can be established only with a "local" temperature (in frequency space) of the field if there can be no coupling between different frequencies, i.e., the oscillator must be linear. Assuming linearity, we may be guided by our previous consideration of the interaction between two oscillators and conclude that the interaction between the oscillator and the field oscillators will be confined mainly to those field oscillators that lie in a narrow frequency band centered at the natural frequency of the structural oscillator. We impose the restriction that the coupling parameters are not abnormally large in some region of the frequency removed from this band [see Eq. (3.4)]. Thus, if $S_p(\omega)/\omega^2$ is fairly smooth so that over the extent of the admittance spectrum of the oscillator the temperatures of the field oscillators are essentially the same, the oscillator will react with the field as if the field were a *constant-temperature bath* with a temperature corresponding to the temperature of these field oscillators. No conceptual extension is required to describe the situation where several linear mechanical oscillators are placed in a reverberant field. Each of these oscillators may see a different region of the field spectrum, each region with its own temperature.

We now turn to discuss the acoustic field in more-specific terms. We start by stating the equation of motion of the field

$$(\partial^2/\partial t^2)\psi(\mathbf{x}_q)+\alpha_q(\partial/\partial t)\psi(\mathbf{x}_q)+D_q\psi(\mathbf{x}_q)$$
$$= cg(\mathbf{x}_q)+c(\partial/\partial t)\phi(\mathbf{x})\delta(\mathbf{x}_q-\mathbf{x}), \quad (7.4)$$

where ψ is the velocity potential, D_q is the spatial operator $(=-c^2\nabla^2)$, $cg(\mathbf{x}_q)$ is the applied acoustic source density, $\mathbf{x}_q$ is the position vector in the field, $(\partial/\partial t)\phi(\mathbf{x})$ is the normal velocity of the structure surface (the properties of the structure are given in the next section), $\mathbf{x}$ is the position vector on the structure surface, and $\delta(\mathbf{x}_q-\mathbf{x})$ is defined by

$$\int_V \nu(\mathbf{x}_q)\chi(\mathbf{x})\delta(\mathbf{x}_q-\mathbf{x})d\mathbf{x}_q=\int_S \nu(\mathbf{x})\chi(\mathbf{x})dx, \quad (7.5)$$

with V being the volume containing the acoustic field, S the surface area of the structure, and $\nu(\mathbf{x}_q)$ and $\chi(\mathbf{x})$ any functions of the acoustic field and the vibratory motion of the structure, and finally α_q an *ad hoc* damping coefficient.

The eigenfunctions and eigenvalues of D_q which satisfy the appropriate boundary conditions are given by

$$D_q\psi_{(r)}(\mathbf{x}_q)=\omega_{(r)}^2\psi_{(r)}(\mathbf{x}_q), \quad (7.6)$$

where (r) stands for the three eigennumbers (one for each space component).

We assume that the $\psi_{(r)}$'s form a complete orthonormal set,

$$\int_V \psi_{(r)}(\mathbf{x}_q)\psi_{(k)}(\mathbf{x}_q)d\mathbf{x}_q=V\epsilon_{(r)}\delta(r,k), \quad (7.7)$$

where[5]

$$\epsilon_{(r)}=\begin{cases} \frac{1}{8} \text{ if all of the indices of } (r) \text{ nonzero} \\ \frac{1}{4} \text{ if two of the indices of } (r) \text{ nonzero} \\ \frac{1}{2} \text{ if one of the indices of } (r) \text{ nonzero} \\ 1 \text{ if all of the indices of } (r) \text{ zero} \end{cases}$$

and $\delta(r,k)$ is the Kronecker delta in (r) and (k).

We may then represent the solution to Eq. (7.4) in the form

$$\psi(\mathbf{x}_q)=\sum_{(r)} q'_{(r)}\psi_{(r)}(\mathbf{x}_q), \quad (7.8)$$

where $q_{(r)}'$ is the amplitude of the (r)th acoustic mode.

The damping of the acoustic modes arises from a combination of volume losses and boundary absorption. For the sake of convenience we assume lossless boundaries and have placed in our Eq. (7.4) the artificial damping α_q. Its introduction is merely a device to provide the modal damping $\beta_{(r)}$ which must arise mathematically from the integral

$$\int_V \psi_{(r)}\alpha_q\psi_{(k)}d\mathbf{x}_q=V\epsilon_{(r)}\beta_{(r)}\delta(r,k). \quad (7.9)$$

The pressure spectral density of the acoustic field is related to the modal amplitudes by

$$\sum_{(r)} \epsilon_{(r)}\langle \dot{q}_{(r)}'^2\rangle=\frac{S_p(\omega)\Delta\omega}{\rho^2c^2}, \quad (7.10)$$

where the summation is to be carried out only over the modes lying in the frequency band given by $\Delta\omega$ about ω. In stating Eq. (7.10) we imply that the acoustic field is reverberant in the sense that we have averaged the spectral density over the volume of the field and we are assuming $S_p(\omega)$ to be independent of position.

We shall have occasion to use a particular average over the acoustic modes

$$\Psi(\mathbf{x}_1,\mathbf{x}_2)=\langle \psi_{(r)}(\mathbf{x}_1)\psi_{(r)}(\mathbf{x}_2)\rangle_{(r)}, \quad (7.11)$$

where $\langle\ \rangle_r$ denotes an average over all the modes lying in a narrow frequency band where the mean wave number is k.

In order to simplify the analysis let us take for the $\psi_{(r)}(\mathbf{x})$'s the functions

$$\psi_{(r)}(\mathbf{x}_h)=\frac{1}{8}\prod_{j=1}^{3}\{\exp ik_j x_{hj}+\exp -ik_j x_{hj}\}, \quad (7.12)$$

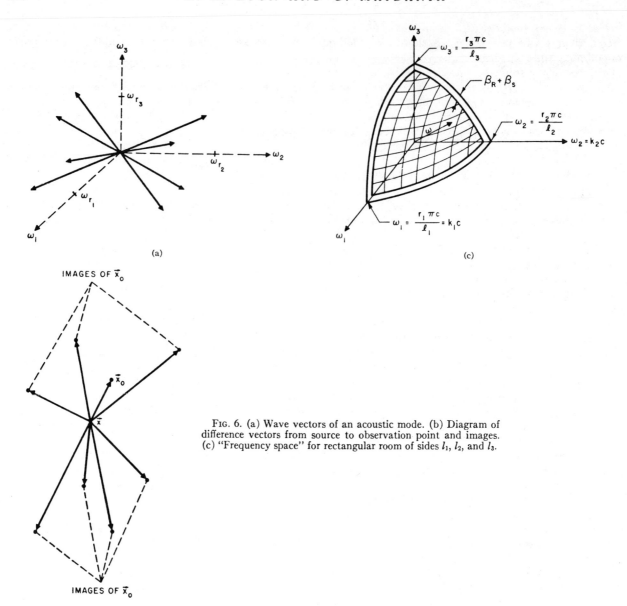

FIG. 6. (a) Wave vectors of an acoustic mode. (b) Diagram of difference vectors from source to observation point and images. (c) "Frequency space" for rectangular room of sides l_1, l_2, and l_3.

where $\mathbf{k}=\{k_j\}$ and $\mathbf{x}_h=\{x_{hj}\}$ with $j=1, 2, 3$ and $h=1, 2$. We have then

$$\psi_{(r)}(\mathbf{x}_1)\psi_{(r)}(\mathbf{x}_2)$$

$$=\frac{1}{64}\prod_{h=1}^{2}\{\prod_{j=1}^{3}[\exp(ik_jx_{hj})+\exp(-ik_jx_{hj})]\}. \quad (7.13)$$

There are 64 terms in this product, all of the form $\exp i\mathbf{k}\cdot\mathbf{x}$, where $\mathbf{k}$ is any of the 8 wave vectors of Fig. 6(a) and $\mathbf{x}$ is any of the position vectors of Fig. 6(b). As we sum over lattice points [see Fig. 6(c)], it is the coherent part of the terms which tend to contribute most to the sum. From Fig. 6(b) the terms with least phase shift will be those for which $|\mathbf{x}|$ is smallest. This is obviously $\mathbf{x}=\mathbf{x}_1-\mathbf{x}_2$. We may sum over the 8 wave

vectors by extending our sum (actually we replace it by an integral) to the entire sphere instead of the one octant of Fig. 6(c). If the field is diffuse (all directions equally likely) we have

$$\sum_{(r)}\psi_{(r)}(\mathbf{x}_1)\psi_{(r)}(\mathbf{x}_2)/N_R \rightarrow \frac{1}{4\pi}\int_0^{2\pi}d\phi\int_0^{\pi}d\theta$$

$$\times\sin\theta\,\cos\theta\,\exp(ikR), \quad (7.14)$$

where N_R is the number of modes of wave number k and

$$R=|\mathbf{x}_1-\mathbf{x}_2|. \quad (7.15)$$

Carrying out the integration,

$$\psi(\mathbf{x}_1,\mathbf{x}_2) \rightarrow \sin kR/kR. \quad (7.16)$$

It is useful to compare $\Psi(\mathbf{x}_1,\mathbf{x}_2)$ with the pressure cross correlation of a reverberant sound field[11]

$$P(\mathbf{x}_1,\mathbf{x}_2)=\langle p(\mathbf{x}_1,t)p(\mathbf{x}_2,t)\rangle/\langle p^2\rangle, \qquad (7.17)$$

and we are considering the pressure in a narrow frequency band with a mean wave number k. Substituting Eq. (7.8) in Eq. (7.17) we obtain

$$P(\mathbf{x}_1,\mathbf{x}_2)=\sum_{(r)(k)}\theta_{(k,r)}\psi_{(r)}(\mathbf{x}_1)\psi_{(k)}(\mathbf{x}_2)/\langle p^2\rangle, \quad (7.18)$$

where

$$\theta_{(r,k)}\propto\langle q_{(r)}'q_{(k)}'\rangle. \qquad (7.19)$$

For a reverberant field the modes are statistically independent and have equal mean energy. Thus,

$$\theta_{(r,k)}=\theta^R\delta(r,k); \quad \theta_{(r)}{}^R\simeq\theta_{(k)}{}^R\simeq\theta^R, \qquad (7.20)$$

where $\theta_{(r)}{}^R$ is the temperature of the (r)th mode. Substituting Eq. (7.20) in Eq. (7.18) we obtain

$$P(\mathbf{x}_1,\mathbf{x}_2)=\sum_{(r)}\psi_{(r)}(\mathbf{x}_1)\psi_{(r)}(\mathbf{x}_2)/N_R=\Psi(\mathbf{x}_1,\mathbf{x}_2). \quad (7.21)$$

VIII. STRUCTURE

In analogy with the acoustic field we may consider the modes of the structure as constituting an ensemble of oscillators. We assume they are statistically independent in the sense that their individual energies may be simply added to give the total energy of the system. This assumption requires the modes to be sufficiently separated in frequency space, have fairly high Q's, and be linear. On the assumption that the mode location along the frequency axis is a Poisson process of average interval $\delta\omega$ it can be shown that the condition that the probability that two or more modes overlap is less than 0.5 is $1.7Q\delta\omega/\omega < 1$.

We consider only one- or two-dimensional "structures" such as strings, beams, or panels. The equation of motion for the structure with acoustic loading is

$$m(\partial^2/\partial t^2)\phi(\mathbf{x})+\alpha_s(\partial/\partial t)\phi(\mathbf{x})+D_s\phi(\mathbf{x})$$
$$=f(\mathbf{x})-c\rho(\partial/\partial t)\psi(\mathbf{x}), \quad (8.1)$$

where $\phi(\mathbf{x})$ is the displacement of the structure at position $\mathbf{x}$, m is the mass of the structure per unit length or area, α_s is an *ad hoc* damping coefficient, D_s is the spatial operator appropriate to the structure, $f(\mathbf{x})$ is the applied-force density, $c\rho(\partial/\partial t)\psi(\mathbf{x})$ is the acoustic loading pressure.

The eigenvalues and the corresponding eigenfunctions of D_s are determined by

$$D_s\phi_{(m)}(\mathbf{x})=\omega_{(m)}{}^2\phi_{(m)}(\mathbf{x}), \qquad (8.2)$$

along with the appropriate boundary conditions and where (m) is the index of the set.

[11] R. K. Cook, R. V. Waterhouse, R. D. Berendt, S. E. Edelman, and M. C. Thompson, Jr., J. Acoust. Soc. Am. 27, 1072 (1955).

We assume that the $\phi_{(m)}$'s form a complete orthonormal set

$$\int_S \phi_{(m)}(\mathbf{x})m\phi_{(n)}(\mathbf{x})d\mathbf{x}=M_{(m)}\delta(m,n). \qquad (8.3)$$

The modal damping coefficients are determined by

$$\int_S \phi_{(m)}(\mathbf{x})\alpha_s\phi_{(n)}(\mathbf{x})dx=M_{(m)}\beta_{(m)}\delta(m,n), \quad (8.4)$$

where S is the surface of the structure and $M_{(m)}$ is the modal mass. We have assumed that the damping produces no coupling between the modes.

We expand the displacement

$$\phi(\mathbf{x})=\sum_{(m)} s_{(m)}'\phi_{(m)}(\mathbf{x}), \qquad (8.5)$$

where $s_{(m)}'$ is the amplitude of the (m)th structural mode.

The acceleration spectral density is related to the velocity of the structure by

$$\sum_{(m)} \epsilon_{(m)}\langle\dot{s}_{(m)}'^2\rangle=\frac{S_a(\omega)\Delta\omega}{\omega^2}, \qquad (8.6)$$

with

$$\epsilon_{(m)}=M_{(m)}/M, \qquad (8.7)$$

where the summation is to be carried out only over the modes lying in a narrow frequency band of width $\Delta\omega$ about ω. The total mass of the structure is M and we have averaged the velocity over the surface of the structure. For a simply supported beam $\epsilon_{(m)}=\frac{1}{2}$, for a simply supported panel $\epsilon_{(m)}=\frac{1}{4}$, and for other boundary conditions the values given above are good approximations, especially at high frequencies.

We confine our attention to two cases. In the first, we assume that in a narrow frequency band there is only one structural mode. In the second, we assume that there are many structural modes so that they form a reverberant field of structural vibrations. The cross correlation of structural vibration of a single mode is given by

$$\Phi(\mathbf{x}_1,\mathbf{x}_2)=\phi_{(m)}(\mathbf{x}_1)\phi_{(m)}(\mathbf{x}_2). \qquad (8.8)$$

On the other hand, using an argument similar to that given for Eq. (7.16), the cross correlation for a two-dimensional reverberant vibrational field may be shown to be[11]

$$\Phi(\mathbf{x}_1,\mathbf{x}_2)=\langle\phi_{(m)}(\mathbf{x}_1)\phi_{(m)}(\mathbf{x}_2)\rangle_{(m)}=J_0(k_bR), \quad (8.9)$$

where k_b is the mean wave number of the structural waves, and $\langle\ \rangle_{(m)}$ denotes an average over all the structural modes whose wave numbers are lying close to k_b.

Again in analogy with the reverberant acoustic field we assume for a reverberant vibrational field

$$\theta_{(m)}{}^s=\theta_{(n)}{}^s=\theta^s, \qquad (8.10)$$

where $\theta_{(m)}{}^s$ is the temperature of the (m)th structural mode. It is related to the spectral density by

$$\theta_{(\omega)}{}^s = \sum_{(m)} \epsilon_{(m)} \langle \dot{s}_{(m)}{}'^2 \rangle / N_s(\omega) = S_a(\omega)/\omega^2 n_s(\omega), \quad (8.11)$$

where $N_s(\omega)$ is the number of structural modes in the frequency interval $\Delta\omega$ and $n_s(\omega)$ is the modal density of the structure.

IX. COUPLING BETWEEN A REVERBERANT ACOUSTIC FIELD AND A STRUCTURE

Straightforward projection of the field equations of Secs. VII and VIII leads to the modal equations[12]

$$\ddot{s}_m + \beta_m \dot{s}_m + \omega_m{}^2 s_m + \sum_r B_{rm} \dot{q}_r = F_m, \quad (9.1a)$$

and

$$\ddot{q}_r + \beta_r \dot{q}_r + \omega_r{}^2 q_r - \sum_m B_{mr} \dot{s}_m = G_r, \quad (9.1b)$$

where

$$q_r = (\rho V \epsilon_r / M)^{\frac{1}{2}} q_r', \quad (9.2)$$

$$G_r = (c^2 \rho / M V \epsilon_r)^{\frac{1}{2}} \int_V g\psi_r dx_q, \quad (9.3)$$

$$s_m = (M_m/M)^{\frac{1}{2}} s_m', \quad (9.4)$$

$$F_m = (M_m M)^{-\frac{1}{2}} \int_S f\phi_m d\mathbf{x}, \quad (9.5)$$

$$B_{rm} = (c^2 \rho / V \epsilon_r M_m)^{\frac{1}{2}} \int_S \psi_r(\mathbf{x}) \phi_m(\mathbf{x}) dx. \quad (9.6)$$

We note from Eq. (9.1) that the coupling is gyroscopic. The conservation of energy equation is obtained from Eq. (9.1) by multiplying Eq. (9.1a) by $\dot{s}_m$ and summing over m, multiplying Eq. (9.1b) by $\dot{q}_r$ and summing over r, averaging the two equations and adding them. The result is

$$\sum_m \beta_m \langle \dot{s}_m{}^2 \rangle + \sum_r \beta_r \langle \dot{q}_r{}^2 \rangle = \sum_m \langle F_m \dot{s}_m \rangle + \sum_r \langle G_r \dot{q}_r \rangle. \quad (9.7)$$

The energy-balance equation is thus independent of the coupling and in this sense the coupling is conservative.

Because of the large number of equations of the form of Eq. (9.1) (there are many sound-field modes, in general, even in a narrow frequency band) an exact solution is difficult to obtain. Rather than attempt an exact solution of Eq. (9.1) we try to find an approximate solution which will provide most of the important effects of the interaction. We assume that F_m and G_r have the properties with respect to the mth and the rth modes as do f_1 and f_2 with respect to mode 1 and mode 2 discussed in Part I. Let us write Eq. (9.1) in the form

$$\ddot{s}_m + \beta_m \dot{s}_m + \omega_m{}^2 s_m + B_{rm} \dot{q}_r = F_m - \sum_{k \neq r} B_{km} \dot{q}_k \quad (9.8a)$$

[12] In the following we drop the parentheses on the subscripts noting that, in general, a set of indices is meant.

and

$$\ddot{q}_r + \beta_r \dot{q}_r + \omega_r{}^2 q_r - B_{mr} \dot{s}_m = G_r - \sum_{n \neq m} B_{nr} \dot{s}_n. \quad (9.8b)$$

We assume further that the q_r's and the s_m's are statistically independent, that $\sum_r B_{rm} \dot{q}_r$ has a flat spectrum with respect to the admittance spectrum of the mth mode, and that $\sum_m B_{rm} \dot{s}_m$ has a flat spectrum with respect to the admittance spectrum of the rth mode. We then set

$$F_m' = F_m - \sum_{k \neq r} B_{km} \dot{q}_k, \quad (9.9a)$$

$$G_r' = G_r + \sum_{n \neq m} B_{nr} \dot{s}_n. \quad (9.9b)$$

This means that F_m' and G_r' are also assumed independent and "white."

With these simplifying assumptions Eq. (9.1) is reduced to a set of two modes interacting and we may take over directly the formalism of Part I. Thus, the power flow from the mth mode to the rth mode is

$$j_{mr} = g_{mr}(\theta_m' - \theta_r'), \quad (9.10)$$

with

$$g_{mr} = \frac{B_{mr}{}^2(\beta_m \omega_r{}^2 + \beta_r \omega_m{}^2)}{(\omega_m{}^2 - \omega_r{}^2)^2 + (\beta_m + \beta_r)(\beta_m \omega_r{}^2 + \beta_r \omega_m{}^2)}, \quad (9.11)$$

where

$$\theta_m' = \langle F_m' \dot{s}_m \rangle / \beta_m \quad (9.12)$$

and

$$\theta_r' = \langle G_r' \dot{q}_r \rangle / \beta_r. \quad (9.13)$$

The energy-balance equations corresponding to the first two equations in Eq. (2.5) are

$$\beta_m \langle \dot{s}_m{}^2 \rangle + g_{mr}(\theta_m' - \theta_r') = \beta_m \theta_m' \quad (9.14a)$$

and

$$\beta_r \langle \dot{q}_r{}^2 \rangle - g_{mr}(\theta_m' - \theta_r') = \beta_r \theta_r', \quad (9.15a)$$

where we make use of Eq. (3.3).

We note that by multiplying Eq. (9.9a) by $\dot{s}_m$ and Eq. (9.9b) by $\dot{q}_r$ and averaging and once again applying Eq. (3.3) we obtain

$$\beta_m \theta_m' = \beta_m \theta_m - \sum_{k \neq r} g_{km}(\theta_m' - \theta_k') \quad (9.16a)$$

and

$$\beta_r \theta_r' = \beta_r \theta_r + \sum_{n \neq m} g_{rn}(\theta_n' - \theta_r'). \quad (9.16b)$$

We may thus write Eqs. (9.14a) and (9.15a) in the form

$$\beta_m \langle \dot{s}_m{}^2 \rangle + \sum_r g_{mr}(\theta_m' - \theta_r') = \beta_m \theta_m \quad (9.14b)$$

and

$$\beta_r \langle \dot{q}_r{}^2 \rangle - \sum_m g_{mr}(\theta_m' - \theta_r') = \beta_r \theta_r. \quad (9.15b)$$

We also note that Eq. (9.11) is consistent with the discussion of Sec. VII. We see that only the field modes which lie in the narrow band about ω_m will contribute appreciably to the power flow between the acoustic field and the mth structural mode. In fact we may set

$$g_{mr} \simeq \begin{cases} B_{mr}^2 \pi (\beta_r + \beta_m)^{-1} & |\omega_m^2 - \omega_r^2| < (\beta_m + \beta_r)^2 \\ 0 & |\omega_m^2 - \omega_r^2| > (\beta_m + \beta_r)^2. \end{cases} \quad (9.17)$$

It is important to realize that our equations are subjected to the limitation that $B_{rm}/\beta_r \ll 1$ and $B_{rm}/\beta_m \ll 1$. Thus, g_{rm} is second order in B/β. Consequently, the second terms on the left of Eqs. (9.14a) and (9.15a) are second order as compared with the leading terms. However, the second terms on the left of Eqs. (9.14b) and Eq. (9.15b) are not necessarily small because of the summations these terms may be quite large.

It can be appreciated that neither θ_m' nor θ_r' are necessarily equal to zero when F_m or G_r are equal to zero. We see from Eq. (9.8a) that when $F_m = 0$ the rth mode sees the structure as driven, the driving force being determined by the interaction of the structure with all the other acoustic modes, and thus as far as the rth mode is concerned the structure has an "unperturbed" temperature.

Since the interaction between the sound field and the structure is confined to modes lying in the same narrow frequency band it is sufficient to consider the interaction in each frequency band independently. In the following we relate all quantities to such a band without always stating it explicitly. Referring to Eq. (9.14b), we note that the first term on the left-hand side is proportional to the power dissipated in the structure, the second term on the left is proportional to the net power radiated by the structure, and the term on the right is proportional to the power supplied by the forces acting on the structure, the constant of proportionality being M. By definition then

$$\sum_m \epsilon_m \langle \dot{s}_m'^2 \rangle R_{\text{mech}} = M \sum_m \beta_m \langle \dot{s}_m^2 \rangle \quad (9.18)$$

and

$$\sum_m \epsilon_m \langle \dot{s}_m'^2 \rangle R_{\text{rad}} = M \sum_{m,r} g_{mr} (\theta_m' - \theta_r'), \quad (9.19)$$

where R_{mech} is the mechanical resistance of the structure, R_{rad} is the radiation resistance of the structure, and $\sum_m \epsilon_m \langle \dot{s}_m^2 \rangle$ is the mean-square surface velocity of the structure (averaged in time and space). It appears that in general R_{rad} and R_{mech} as defined by Eq. (9.19) will be dependent on the modal energy distribution which would present a severe limitation on their usefulness. However, if we confine our attention to one structural mode or to a reverberant structural vibration field, we see that R_{rad} and R_{mech} will achieve values independently of the energy distribution.

1. Single Structural Mode

Consider the case where the structure is driven by external random forces and the acoustic field is generated by the structure only. From Eqs. (9.9b) and (9.13)

we find that $\theta_r' = \theta_r = 0$ for all r, and from Eq. (9.14a) we have $\langle \dot{s}_m^2 \rangle \simeq \theta_m'$. Equations (9.18) and (9.19) reduce to

$$R_{\text{mech}} = \beta_m M \quad (9.20)$$

and

$$R_{\text{rad}} = M \sum_r g_{mr}, \quad (9.21)$$

respectively. From Eq. (9.15b) and using Eqs. (7.10) and (8.6)

$$S_p(\omega)/S_a(\omega) = (\rho/c)(R_{\text{rad}}/\beta_R)[2\pi^2 n_R(\omega)]^{-1}, \quad (9.22)$$

where we have set[5]

$$\beta_r = \beta_k = \beta_R. \quad (9.23)$$

Now consider the reciprocal case where the acoustic field is generated by external random forces, but no external forces are acting on the structure. Although $\theta_m = 0$, θ_m' is finite. Since there are very many acoustic modes even in a narrow frequency band we see from (9.16a) that

$$\beta_m \theta_m' \gg g_{rm}(\theta_r' - \theta_m'), \quad (9.24)$$

and thus we still have $\langle \dot{s}_m^2 \rangle \simeq \theta_m'$ [See Eqs. (9.14a) and (9.16a)]. In accordance with Sec. VII we may set $\theta_r^R \simeq \theta_k^R \simeq \theta^R$. With this approximation we obtain

$$S_a(\omega)/S_p(\omega) = \Gamma(\omega)\mu(\omega), \quad (9.25)$$

where

$$\mu(\omega) = R_{\text{rad}}(R_{\text{mech}} + R_{\text{rad}})^{-1} \quad (9.26)$$

and

$$\Gamma(\omega) = 2\pi^2 c/M\rho. \quad (9.27)$$

Equation (9.25) is in agreement with an expression derived previously by Smith.[13]

2. Reverberant Field of Structural Vibration

We assume that the properties of the vibrational field and the structure satisfy

$$\beta_m \simeq \beta_h \simeq \beta_s \quad (9.28)$$

and

$$\theta_m^s \simeq \theta_h^s \simeq \theta^s. \quad (9.29)$$

Since one usually has $n_s(\omega) \ll n_R(\omega)$, by a similar argument as for the single structural mode,

$$R_{\text{mech}} = \beta_s M \quad (9.30)$$

and

$$R_{\text{rad}} = M \sum_{r,m} g_{mr}/N_s(\omega). \quad (9.31)$$

With the approximations given by Eqs. (9.28) and (9.29) it follows that where the structure alone is directly excited

$$S_p(\omega)/S_a(\omega) = (\rho/c)(R_{\text{rad}}/\beta_R)[2\pi^2 n_R(\omega)]^{-1}, \quad (9.32)$$

[13] P. W. Smith, Jr., "Estimating structural response to noise," Paper M2, 59th Meeting of the Acoustical Society of America, June 1960, and I. Dyer, P. W. Smith, Jr., C. I. Malme, and C. M. Gogos, "Sonic fatigue resistance of structural designs," BBN Report No. 876, Contract AF33(616)-6340, March 1961, p. 27.

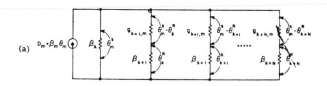

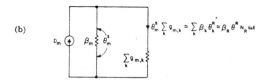

FIG. 7. (a) Energy interaction between one structural mode and several acoustic modes when the structure is driven. (b) Circuit contraction.

and when the acoustic field is excited

$$S_a(\omega)/S_p(\omega) = \Gamma(\omega)\mu(\omega),\qquad(9.33)$$

where

$$\Gamma(\omega) = 2\pi^2[n_s(\omega)/M](c/\rho).\qquad(9.34)$$

In panels at high frequencies

$$n_s(\omega) = S\sqrt{3}/2\pi t c_L,\qquad(9.35)$$

where S is the surface area, t the thickness, and c_L is the longitudinal velocity. From Eqs. (9.34) and (9.35) we have

$$\rho_s{}^2\Gamma(\omega) = \pi\sqrt{3}\rho_p c/\rho c_L,\qquad(9.36)$$

where $\rho_s = M/S$ and $\rho_p = M/St$.

We have thus established the steady-state relation between the modal temperature of the structure and the modal temperature of the acoustic field in terms of the physical parameters of the two systems. We note from Eqs. (9.25) and (9.33) that when $R_{rad} \gg R_{mech}$ the modal energy of the structure becomes equal to the modal energy of the sound field. This is to be expected if the reverberant field is behaving as a temperature bath in which the structural modes are "immersed." When the above inequality does not hold, then the temperature of the structural mode assumes a value a fraction $\mu(\omega)$ less than the equilibrium value.

X. ENERGY ANALOG CIRCUITS FOR ROOM-STRUCTURE INTERACTION

In Sec. IV we have demonstrated how one may represent the energetics of a single mode or two slightly coupled modes by analog circuits of energy conductance and storage elements. In the present section we wish to show how the simple two-mode interaction may be extended to the interaction of a single structural mode with several acoustic modes. In Sec. IX we have carried this out analytically, but here we manipulate the circuits directly.

It is possible and helpful to regard the present section as a diagrammatic presentation of the derivation in the preceding section, which it is. On the other hand, the circuit representation has its own logic and methods which can proceed independently of the analytical framework and which some may find more appealing. Accordingly, we develop our ideas in this section in terms of circuit notions and merely reference the analytical counterpart of our results as we proceed.

We first consider a directly excited structural mode which radiates by its coupling with the acoustic modes. The energy circuit for the interaction between the structural mode and a single acoustic mode is given by the equivalent circuit of Fig. 3 where the coupling-conductance value is given by Eq. (9.11). Suppose there are N acoustic modes in the resonance bandwidth of the structural mode. Then the energy circuit will be as shown in Fig. 7(a). Analytically, the flow conservation equation for this circuit is given in Eq. (9.14b). When $g_{rm}{}^{-1}$ greatly exceeds $\beta_r{}^{-1}$, then the circuit contraction to Fig. 7(b) is possible. The N conductances g_{rm} in parallel result in a total conductance of energy and a radiation resistance given by Eq. (9.21).

When the sound field is directly excited, the N independent acoustic modes which couple into the structural mode may be represented by a sequence of two-mode circuits. Since the temperature of the structural mode is the addition of the temperatures induced by all the acoustic modes coupled to it [see Eq. (9.12b) with $\theta_m = 0$], we merely combine these circuits to a common load as shown in Fig. 8(a). If the acoustic temperatures θ_r are equal as we have assumed, then the nodes A, B, C, etc., may be joined to form the contracted circuit shown in Fig. 8(b). The structural temperature is obviously given by an expression like Eq. (9.25) from observation of the circuit.

Circuit manipulation does not provide us with new answers concerning the transfer of energy between the structure and sound field but it does provide a simple graphical picture of the methods and assumptions which we have used. It may in addition have some utility as a computational technique for more-complicated interaction problems by analog circuits or computers.

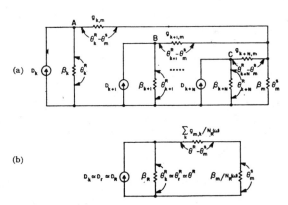

FIG. 8. (a) Energy interaction between one structural mode and several acoustic modes when the acoustic modes are driven. (b) Circuit contraction.

XI. COMPUTATION OF THE RADIATION RESISTANCE

In Sec. IX we showed the importance of R_{rad} in determining the coupling between a reverberant acoustic field and a structure. In this section we wish to compute this radiation resistance of a supported beam and in the following section to test these computations experimentally. In computing R_{rad}, we require the correlation of the acoustic field $\Psi(\mathbf{x}_1, \mathbf{x}_2)$ evaluated on the structure surface. We assume for simplicity that this has the same form as it does at points removed from the structure. This means that the structure is assumed to have small curvature and of large dimensions compared to an acoustic wavelength. We are therefore solving the problem of a baffled beam in this section.

From Eqs. (7.11), (8.7), (9.6), (9.17), and Eq. (9.21), or Eq. (9.31) we have

$$R_{\text{rad}} = (2\pi\epsilon_s\epsilon_R)^{-1}\rho c k^2 \iint\limits_S dx_1 dx_2 \Psi(\mathbf{x}_1, \mathbf{x}_2)\Phi(\mathbf{x}_1, \mathbf{x}_2), \quad (11.1)$$

where we set $\epsilon_r = \epsilon_k \simeq \epsilon_R$ and $\epsilon_m \simeq \epsilon_h \simeq \epsilon_s$. At the higher frequencies, $\epsilon_R = \frac{1}{8}$ and $\epsilon_s = \frac{1}{2}$ for one-dimensional structures, and $\epsilon_s = \frac{1}{4}$ for two-dimensional structures.

We consider the radiation resistance of a simply supported finite beam. The coordinate system and the dimensions of the beam are shown in Fig. 9. We further assume that the beam modes are well separated in frequency space so that in a narrow range of frequency only one structural mode is present. We have then

$$\Phi(\mathbf{x}_1, \mathbf{x}_2) = \frac{1}{2}[\cos k_b(x_1 - x_2) \pm \cos k_b(x_1 + x_2)], \quad (11.2)$$

where $k_b = \pi n/l$ and $-\frac{1}{2}l < x_1, x_2 < \frac{1}{2}l$. The $+$ sign is for odd values of n and the $-$ sign is for even values of n (n is an integer). The distance between $\mathbf{x}_1$ and $\mathbf{x}_2$ is

$$R = \{(x_1 - x_2)^2 + (z_1 - z_2)^2\}^{\frac{1}{2}}. \quad (11.3)$$

Using the identity[14]

$$\frac{\sin kR}{kR} = \int_0^1 \cos\beta k(x_1 - x_2) J_0\{\alpha k(z_1 - z_2)\}(\alpha/\beta)d\alpha, \quad (11.4)$$

where $\beta^2 = 1 - \alpha^2$, and standard Fourier's transforms one may readily compute the radiation resistance for such a beam.

1. $k_b/k < 1$ (above coincidence) and with $\frac{1}{2}kl > 1$

$$R_{\text{rad}} = \frac{1}{2}S\rho c(kw)j^2(a), \quad (11.5)$$

where

$$a = kw[1 - (k_b/k)^2]^{\frac{1}{2}}, \quad (11.6)$$

[14] E. T. Whittaker and G. N. Watson, *A Course of Modern Analysis* (Cambridge University Press, New York, 1958), 4th ed., p. 384.

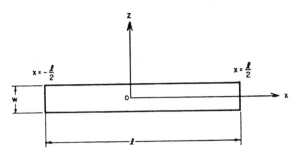

FIG. 9. Dimensions and coordinates system of the beam.

and[15]

$$j^2(a) = \frac{2}{\pi}\int_0^{\frac{1}{2}\pi} \frac{\sin^2[(\frac{1}{2}a)\sin\theta]}{[(\frac{1}{2}a)\sin\theta]^2}d\theta. \quad (11.7)$$

Thus for $\frac{1}{2}a > 1$ we obtain

$$R_{\text{rad}} \approx S\rho c[1 - (k_b/k)^2]^{-\frac{1}{2}}, \quad (11.8)$$

while for $\frac{1}{2}a < 1$ we obtain

$$R_{\text{rad}} \approx \frac{1}{2}S\rho ckw, \quad (11.9)$$

the well-known uniform strip radiation resistance.

We note that, in this region of wave numbers, for values of $\frac{1}{2}a > 1$ the radiation efficiency (defined by $R_{\text{rad}}/S\rho c$) is independent of w, whereas for $\frac{1}{2}a < 1$ it is directly proportional to w.

2. $k_b/k > 1$ (below coincidence) and with $\frac{1}{2}kl > 1$

(a) $\frac{1}{2}kw \gg 1$. In this case the radiation resistance is given by

$$R_{\text{rad}} = S\rho c(k/k_b)^2(1/kl) \times \{[2 - (k/k_b)^2]/[1 - (k/k_b)^2]^{\frac{1}{2}}\}. \quad (11.10)$$

Both Eqs. (11.8) and (11.10) are in agreement with the expressions derived by Gösele[16] who used a somewhat different approach.

(b) $\frac{1}{2}kw < 1$.

$$R_{\text{rad}} = \rho c(k/k_b)^2(w/\pi l)\{1/[1 - (k/k_b)^2] + (k_b/2k)\log(k_b + k)/(k_b - k)\}. \quad (11.11)$$

Again we see the direct proportionality between R_{rad} and w for small values of the latter.

(c) $\frac{1}{2}kw > 1$. Because the expression for this region is rather lengthy we do not include it here. It is sufficient to state that R_{rad} decreases monotonically from its value for $\frac{1}{2}kw \gg 1$ to that for $\frac{1}{2}kw < 1$. The difference between the values of R_{rad} for $\frac{1}{2}kw \gg 1$ and that for $\frac{1}{2}kw \approx 1$ is negligible and one may assume for all practical purposes that Eq. (11.10) holds good for $\frac{1}{2}kw > 1$.

[15] A. Powell, contribution to *Random Vibration*, edited by S. H. Crandall (Technology Press, Cambridge, Massachusetts, 1958), Chap. 8, pp. 8–18.
[16] K. Gösele, Acustica **3**, 243 (1953).

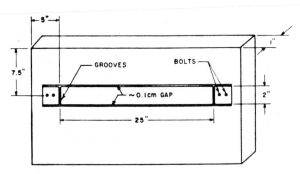

FIG. 10. Sketch of test beam and baffle.

3. $k_b/k \approx 1$ (at and about coincidence) and with $\frac{1}{2}kl > 1$

(a) $\frac{1}{2}kw \gg 1$. The radiation resistance at coincidence is given by

$$R_{\text{rad}} = S\rho c (kl)^{\frac{1}{2}}/3\sqrt{\pi}. \qquad (11.12)$$

(b) $\frac{1}{2}kw < 1$. In this case the radiation resistance is

$$R_{\text{rad}} = \frac{1}{4}S\rho ckw. \qquad (11.13)$$

It is evident from Eqs. (11.8), (11.9), (11.12), and (11.13) that, while for $\frac{1}{2}kw \gg 1$ one would expect a hump to appear at coincidence on a plot of the radiation resistances as a function of frequency, no such hump is expected for $\frac{1}{2}kw < 1$.

4. $\frac{1}{2}kl < 1$

In this region the odd (n odd) and the even (n even) modes will have a somewhat different value for the radiation resistance. This is to be expected since the odd modes cannot achieve as good a destructive inter-

ference as the even modes. The difference between the radiation resistance of the odd and even modes is most pronounced of course when the wavelength in the beam is long and the unpaired wave occupies sizeable portions of the length of the beam. In particular when $\frac{1}{2}kl \ll 1$ and $\frac{1}{2}kw < 1$ we have

$$R_{\text{rad}}(\text{odd}) = S\rho c(k/k_b)^2(wk/2\pi lk)\{1/[1-(k/k_b)^2] + (k_b/2k)\log(k_b+k)/(k_b-k)\}, \qquad (11.14)$$

and

$$R_{\text{rad}}(\text{even}) = S\rho c(klkw/2\pi)\{1/[1-(k/k_b)^2] - (k_b/2k)\log(k_b+k)/(k_b-k)\}. \qquad (11.15)$$

XII. RADIATION RESISTANCE FOR SUPPORTED BEAM (EXPERIMENTAL)

The radiation and internal resistance of a baffled beam were examined experimentally as a test of some of the results of Sec. XI. An approximately simply supported aluminum beam, in a baffle, was placed in a reverberant room. Two cases were studied; the beam driven by random forces (via 8% filter and a magnet-coil driver), and the room driven by a random source (via 8% filter and loudspeaker).

A diagram of the beam and baffle is shown in Fig. 10. The total mass of the beam is 1105 g. The Q's of the beam modes were measured by half-width and reverberation-time measurements. The agreement was quite good. The average values of the Q's so measured over several experiments are given in Fig. 11. Two rooms were used in this experiment, one with a volume of $\simeq 2400$ ft³ $(18 \times 14 \times 9.5$ ft) and the other with a volume of $\simeq 10\,000$ ft³ $(30 \times 20 \times 16$ ft). Their respective reverberation times as a function of frequency are shown in

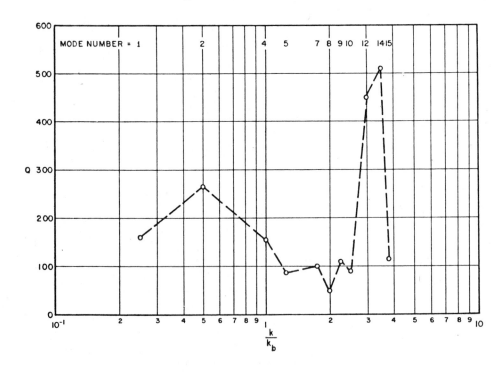

FIG. 11. Quality factor of the beam as a function of frequency.

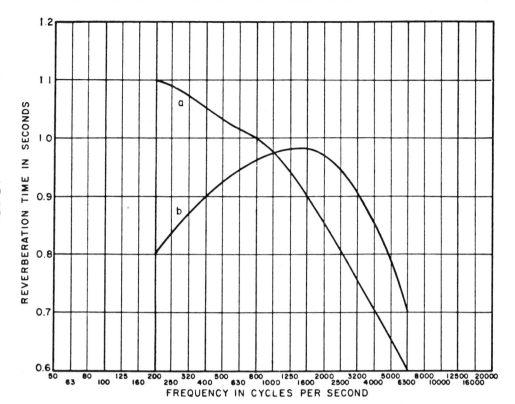

FIG. 12. Reverberation time of room (a) with dimensions (18×14 ×9.5 ft) and room (b) with dimensions (30×20 ×16 ft).

Fig. 12(a), (b). The curves given are average of several measurements.

A block diagram of the experimental arrangement is given in Fig. 13. With this arrangement it was possible to record the acceleration of the beam and the sound-level pressure in the room simultaneously, and the records could then be studied at leisure.

The experimental values of R_{rad} are given in Fig. 14. R_{rad} was determined using Eq. (9.22). The theoretical curve was computed using Eqs. (11.5), (11.11), and (11.13). The experimental values of $\mu(\omega)$ are given in Fig. 15. $\mu(\omega)$ was determined using Eq. (9.25) and also from the experimental values of R_{rad} and the Q's of Fig. 11 using Eq. (9.26). The Q's are related to $R_{mech}+R_{rad}$ by the equation

$$R_{mech}+R_{rad}=M\omega Q^{-1}. \qquad (12.1)$$

We consider that the agreement between theory and the experiment is satisfactory. That is, the values of R_{rad} as given by Eqs. (11.5) and (11.11) are in good accord with the experimental points plotted in Fig. 14. Also, there is a good agreement between the values of $\mu(\omega)$ as measured directly, using Eq. (9.25), and the values determined from R_{rad} and the Q's using Eq. (12.1). Unfortunately, the beam used in these experiments was such that it was difficult to make a highly critical test of the theory in the range $\frac{1}{2}kw<1$. Nevertheless the few points which are in the region of $\frac{1}{2}kw<1$ seem to support the theory (see Fig. 14). The considerable reduction in the radiation efficiency due to acoustical short circuiting across the gap between the bar and

baffle is clearly apparent. Because of the increase in damping when the gap was sealed off it was not possible to measure $\mu(\omega)$ with satisfactory degree of accuracy.

The wavelengths of the first and second modes are approximately 20 and 80 in., respectively. Thus, in this region the baffle is not sufficiently extended to avoid acoustical short circuiting between the two radiating surfaces of the beam. This leads to lower values of the radiation efficiency as compared with the values of a completely baffled beam as assumed in the theory.

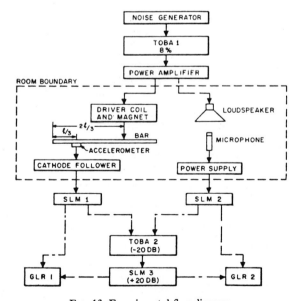

FIG. 13. Experimental flow diagram.

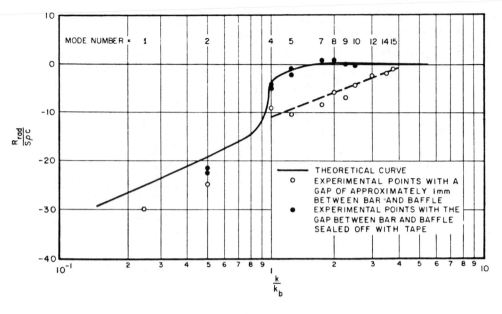

FIG. 14. The radiation efficiency of the beam (25×2×0.483 in.) in a baffle (35×15 in.).

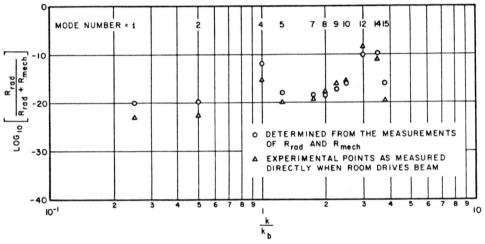

FIG. 15. Coupling factor $\mu(\omega) = R_{\text{rad}}(R_{\text{mech}} + R_{\text{rad}})^{-1}$ as a function of frequency.

ACKNOWLEDGMENTS

This work was begun at the University of Manchester, England, where Richard H. Lyon held a National Science Foundation Postdoctoral Fellowship. It was continued and completed at Bolt Beranek and Newman Inc. under the sponsorship of the Office of Naval Research and the Aeronautical Systems Division. The authors wish to acknowledge the interest and assistance of M. S. Bartlett at the University of Manchester, G. K. Batchelor and H. C. Longuet-Higgins at Cambridge University, and Ira Dyer, Manfred Heckl, and Preston Smith at Bolt Beranek and Newman Inc.

APPENDIX

Power Flow with Nonlinear Coupling

Suppose that instead of Eq. (2.1), one has a pair of coupled oscillators governed by

$$\dot{u}_1 + \beta_1 u_1 + \omega_1^2 y_1 + C y_2 + a_{11} y_1^3 + a_{12} y_1 y_2^2 = f_1, \quad (A1)$$
$$\dot{u}_2 + \beta_2 u_2 + \omega_2^2 y_2 + C y_1 + a_{22} y_2^3 + a_{12} y_2 y_1^2 = f_2.$$

The nonlinear coupling potential

$$V_c = \tfrac{1}{4}(a_{11} y_1^4 + 2 a_{12} y_1^2 y_2^2 + a_{22} y_2^4), \quad (A2)$$

occurs when the (instantaneous) stiffness of one mode is altered by the motion of the other. It arises in automotive suspensions[17] and beams with hinged or clamped ends.[18]

If $V_c = 0$, the problem reverts to that treated in the body of the paper. We are concerned here to see how small amounts of nonlinearity in the coupling affect the power flow. The technique we apply has come to be known as "equivalent linearization."[19]

We replace the Eqs. (A1) by the set

$$\dot{u}_1 + \beta_1 u_1 + \Omega_1^2 y_1 + \hat{C} y_2 + \epsilon_1 + \epsilon_{12} = f_1, \quad (A3)$$
$$\dot{u}_2 + \beta_2 u_2 + \Omega_2^2 y_2 + \hat{C} y_1 + \epsilon_2 + \epsilon_{21} = f_2.$$

If ϵ_1, ϵ_2, ϵ_{12}, and ϵ_{21} may be ignored, then the power flow

[17] R. F. Henry and S. A. Tobias, J. Mech. Eng. Sci. 1, 19 (1959).
[18] P. H. MacDonald, Jr., J. Appl. Mech. 22, 573 (1955).
[19] T. K. Caughey, J. Appl. Mech. 26, 341 (1959).

is given by [see Eqs. (3.3) and (3.4)]

$$j_{12} = [\hat{C}^2(\beta_1+\beta_2)/(\Omega_1{}^2-\Omega_2{}^2)^2](\theta_1-\theta_2), \quad (A4)$$

and we assume throughout that both $|\Omega_1-\Omega_2|$ and $|\omega_1-\omega_2| \gg (\beta_1+\beta_2)$. The error functions are

$$\epsilon_1 = \omega_1{}^2 y_1 + a_{11}y_1{}^3 - \Omega_1{}^2 y_1,$$
$$\epsilon_2 = \omega_2{}^2 y_2 + a_{22}y_2{}^3 - \Omega_2{}^2 y_2, \quad (A5)$$
$$\epsilon_{12} = Cy_2 + a_{12}y_1y_2{}^2 - \hat{C}y_2,$$
$$\epsilon_{21} = Cy_1 + a_{12}y_2y_1{}^2 - \hat{C}y_1.$$

One assumes that if $\langle\epsilon_1{}^2\rangle_{av}$, $\langle\epsilon_2{}^2\rangle_{av}$, etc., are minimized by the proper choices of $\Omega_1{}^2$, $\Omega_2{}^2$, $\hat{C}$, then the errors may in fact be neglected. Rather straightforwardly, the proper choices are

$$\hat{C} = C + 3a_{12}\langle y_1y_2\rangle,$$
$$\Omega_1{}^2 = \omega_1{}^2 + 3a_{11}\langle y_1{}^2\rangle, \quad (A6)$$
$$\Omega_2{}^2 = \omega_2{}^2 + 3a_{22}\langle y_2{}^2\rangle.$$

In the expressions for $\Omega_1{}^2$ and $\Omega_2{}^2$, we use the unperturbed values $\langle y_1{}^2\rangle = \theta_1/\omega_1{}^2$, $\langle y_2{}^2\rangle = \theta_2/\omega_2{}^2$, and for $\langle y_1y_2\rangle$, one has from Eq. (2.10),

$$\langle y_1y_2\rangle = [\omega_2{}^2\theta_1 - \omega_1{}^2\theta_2/\omega_1{}^2\omega_2{}^2(\omega_1{}^2-\omega_2{}^2)]\hat{C}. \quad (A7)$$

Although it is not necessary, let us evaluate the moments $\langle y_1{}^2\rangle$, $\langle y_2{}^2\rangle$, and $\langle y_1y_2\rangle$ at the average temperature $\theta_m = \frac{1}{2}(\theta_1+\theta_2)$, i.e., we set $\theta_1 = \theta_2 = \theta_m$. Then

$$\hat{C} = C - 3a_{12}\theta_m/\omega_1{}^2\omega_2{}^2,$$
$$\Omega_1{}^2 = \omega_1{}^2 + 3a_{11}\theta_m/\omega_1{}^2, \quad (A8)$$
$$\Omega_2{}^2 = \omega_2{}^2 + 3a_{22}\theta_m/\omega_2{}^2.$$

Putting Eq. (A8) in Eq. (A4) then, one has

$$j_{12} \simeq \frac{C^2(\beta_1+\beta_2)}{(\omega_1{}^2-\omega_2{}^2)^2}$$

$$\times \left[1 - \frac{6\theta_m}{\omega_1{}^2\omega_2{}^2}\left(a_{12} + \frac{a_{11}\omega_2{}^2 - a_{22}\omega_1{}^2}{\omega_1{}^2-\omega_2{}^2}\right)\right](\theta_1-\theta_2), \quad (A9)$$

which gives the correction to be applied to what one would have if nonlinearity were absent.

In the case of the supported-supported beam, if we measure y_1 and y_2 in units 2κ (twice the radius of gyration of the cross section), then one can show that $a_{11} = \omega_1{}^2$, $a_{22} = \omega_2{}^2$, $a_{12} = \omega_1\omega_2$. Thus

$$j_{12} = \frac{C^2(\beta_1+\beta_2)}{(\omega_1{}^2-\omega_2{}^2)^2}\left(1 - 6\frac{\theta_m}{\omega_1\omega_2}\right)(\theta_1-\theta_2), \quad (A10)$$

showing that the net effect of nonlinearity is to *reduce* the power flow between the modes.

If linear coupling is not present, then the efficiency of power transfer is very much reduced. Indeed, several arguments which one can construct lead to the conclusion that $j_{12} = 0$. The following argument gives a nonzero value for j_{12} and offers a result from which a number can be gained to guide one in experimental design.

A formal comparison of Eqs. (2.1) and (A.1) with $C=0$ in the latter leads one to set $\hat{C} = a_{12}y_1y_2$ (obviously not a constant). Noting that it is $\hat{C}^2$ which will occur in j_{12}, an average will give

$$\langle\hat{C}^2\rangle_{av} = a_{12}{}^2\langle y_1{}^2y_2{}^2\rangle_{av} \simeq \theta_1\theta_2 a_{12}{}^2/\omega_1{}^2\omega_2{}^2,$$

and substituting this in Eq. (3.4),

$$j_{12} = [a_{12}{}^2(\beta_1+\beta_2)/\omega_1{}^2\omega_2{}^2(\omega_1{}^2-\omega_2{}^2)^2]\theta_m{}^2(\theta_1-\theta_2). \quad (A11)$$

The form $\theta_m{}^2(\theta_1-\theta_2)$ suggests the interaction of two bodies which radiate to each other by a θ^3 law. The result Eq. (A11) represents a weaker effect than Eq. (A9) since it depends on a_{12}, but experiment or a better theory is needed to say how close Eq. (A11) comes to predicting any power flow which may occur. Again, for the supported-supported beam,

$$j_{12} = [\theta_m{}^2(\beta_1+\beta_2)/(\omega_1{}^2-\omega_2{}^2)^2](\theta_1-\theta_2), \quad (A12)$$

suggesting that the mean temperature may act as the coupling coefficient.

Part III

NOISE IN BUILDINGS

Editor's Comments on
Papers 14 and 15

14   SABINE
The Absorption of Noise in Ventilating Ducts

15   BERANEK, REYNOLDS, and WILSON
Apparatus and Procedures for Predicting Ventilation System Noise

There are three main noise problems in buildings: the insulation of air-borne noise; the insulation of structure-borne noise; and the acoustic design of heating, ventilating, and air-conditioning (HVAC) systems. Airborne noise may be caused by people's voices, human activities, or machinery. Structure-borne noise is created by structural vibrations carried through a building and then reradiated as sound. The structural vibrations may be caused first by airborne noise from internal and external sources such as people or vehicles and machines or by impacts caused by footsteps or other sources. Modern central-ized HVAC systems normally cause problems by distributing fan or blower noise throughout a building via the supply ductwork.

Airborne and structure-borne noise have been a problem in buildings for centuries. The problem seems to have become more acute in this century as the use of mechanical appliances, such as vacuum cleaners, dishwashers, clothes-washers, driers, elevators, pumps, compressors, and so forth, has increased in private and public buildings. The problem has been compounded in recent years by the construction of lightweight private homes, apartments, and other high-rise buildings. Lightweight materials such as plaster-board or gypsum board, unless used very carefully, do not insulate as well as does brick in older buildings against airborne sounds caused by people or vehicles and machinery. Papers on the theory and measurement of airborne and structure-borne insula-tion of partitions are included in the companion Benchmark volume on Architectural Acoustics [1]; the reader is directed to this volume for seminal papers on sound insulation. There are also several recent reviews of these topics in the form of book chapters, [2, 3].

Because of the overlap discussed with the Benchmark volume on Architec-tural Acoustics, papers reviewed here will be concerned with HVAC system noise only. HVAC system noise has become a major problem in buildings since the growth of the use of such systems beginning in the 1920s and 1930s. Prior to that time, ventilation was achieved naturally by opening windows. One problem with open windows in cities is that traffic noise and, near airports, aircraft noise are admitted. Ventilators can be designed that admit fresh air but prevent much of the sound from entering. An ingenious early type was described in 1929 in *Scientific American* [4]. However, in conditions where traffic or aircraft noise are intense or where the outside temperature is very low or high, opening windows for ventilation becomes impractical. In such cases, the use of HVAC systems becomes essential. Unfortunately, improperly designed HVAC systems can be unacceptably noisy. Most of the noise caused by such systems is produced by the fan or blower, although some noise can be generated by the air flow itself at sharp bends, grills, and so on.

Noise produced by a fan or blower can be attenuated by lining the ventilating duct with acoustic absorbing material. Although in the late 1930s there had been some theoretical attempts to predict the absorption of sound in a lined duct [5, 6], these theories had not then been applied to practical situations. Some empirical formulae for lined duct attenuation did exist, but they gave conflicting results. By using Sivian's analysis [5] and by assuming that the impedance of the wall materials used was purely resistive, Hale J. Sabine was able to derive a simple engineering formula for duct attenuation. Sabine published his results in 1940 (Paper 14). Sabine's assumption that the impedance of the duct lining material is purely resistive is now known to be grossly in error at low frequencies, although it is approximately true at high frequencies. However, Sabine's assumption enabled him to obtain a useful engineering formula for duct attenuation, which agreed fairly well with experiment. This formula has been used for many years and is still sometimes used today. Its advantage is that it relates the duct attenuation to the lining absorption coefficient $\bar{a}$, which is often available from the manufacturer, while complete impedance data are not.

The attenuation of lined ducts is also of interest in several other areas of noise control including jet engine inlets and silencer design. More recent theoretical studies have shown that the attenuation of a duct liner is quite complicated. Sivian [5] made the simplifying assumptions that: (1) the duct cross-section dimensions are much less than the acoustic wavelength (i.e., plane-waves only can exist); (2) the sound energy does not propagate in the absorbing material itself in a direction parallel to the duct axis (that is, the lining is *locally reacting*); and (3) the acoustic properties of the lining are described completely by the normal wall impedance (ratio of wall pressure to normal particle velocity). The theories of Morse, Willms, and Cremer [7, 8, 9] extended the theoretical model to include the case where the duct cross-section dimensions are greater than the sound wavelength (to include higher-order modes [cross modes]) so that sound can propagate in the duct in directions other than axial. However, these theoretical models still assume that sound energy cannot propagate in the lining in the axial direction. Scott [10] was the first to derive a theory to include axial propagation of sound in the lining, and Kurze and Ver [11] have derived a unified theory that reduces in extreme cases to Scott's homogeneous lining result and the locally reacting lining results of Morse and Cremer.

All the theories mentioned so far are rather complicated, and in practice noise control designers often resort to the use of either simplified design charts derived from the theories or experimentally measured results when designing acoustically absorbing lined duct systems. Many experimental data on lined ducts are available; however, some of the results are conflicting and questionable. For this reason Ver recently concluded a review for the American Society of Heating, Refrigeration, and Air Conditioning Engineers (ASHRAE) of the available experimental results and theories for the attenuation of sound in lined and unlined rectangular cross-section ducts [12]. Earlier reviews appeared in book chapters by Doelling [13], Embleton [14], and Leonard [15].

Leonard presents a design procedure based on Morse's theory. In his procedure a knowledge of the normal specific impedance of the lining material is required, and charts are then given so that the attenuation in dB/ft can be calculated. The attenuation of rectangular ducts with one absorbing wall, two absorbing walls of the same impedance, and four absorbing walls with different impedances on pairs of parallel walls can be predicted, as can the attenuation of a uniformly lined square duct. Leonard does not give any indication of the accuracy of this prediction scheme. Embleton presents a similar prediction scheme based on Ingard's calculations [16] using Cremer's theory.

So far, the attenuation of sound in a lined duct has been discussed without consideration of the sound source (fan noise) to be quieted or of the criterion for the sound pressure level in the receiving room. Beranek, Reynolds, and Wilson published a paper in 1953, reprinted here as Paper 15, which was probably the first attempt to treat ventilating system noise as a systems problem. They gave a procedure for measuring the sound *power* spectrum of the fan source and also presented a set of criterion curves so that acceptable levels could be determined for rooms of different types. With a knowledge of the source power and the acceptable levels in rooms, the amount of acoustic attenuation along the ductwork could be calculated instead of being determined more empirically. This systems approach suggested by Beranek and others in the 1950s has now become firmly adopted by ASHRAE. See, for example, Ref. 17 and earlier versions of the ASHRAE Handbook.

In Paper 15 the authors mention two difficulties with measurement of fan sound power in a duct: (1) radial standing waves exist in the duct above the duct cut-off frequency, and (2) the microphone placed in the duct to measure the fan sound also measures airstream noise. Beranek et al. suggested placing a single microphone with a windscreen at the duct centerline to estimate the fan sound power. However, they observed that the choice of location of the microphone is important and can introduce errors of as much as 5 dB in the fan sound power measurements. They suggested that further work needed to be done to determine the effect of the radial standing waves on sound power measured (p. 229) and to improve the windscreen design (p. 230). They also suggested measuring the sound power of fans on a spherical surface in a free field around a duct exit for fans that are not normally coupled to ducts.

Although the in-duct method of measuring fan sound power was suggested by Beranek et al. in 1953, it was not immediately adopted in the United States and has been little used in the United States or in other countries as a standard measurement method. One exception to this generalization is the in-duct fan sound power measurement standard [19] developed in Britain, BS 848.

In the 1940s and 1950s the National Association of Fan Manufacturers, Inc. (NAFM) recommended measuring the noise of fans by determining the average of seven sound pressure levels at the fan exit in a free field. This was an obviously unsatisfactory procedure. In the 1960s and 1970s most measurements of fan sound power have been made with a reverberant room method using a standard such as AMCA 300-67 [20]. Because this approach requires a special expensive facility, many workers have continued to try to overcome the problems inherent in the use of the in-duct fan sound power measurement method.

In the late 1950s radial sound level variations of as much as 10 dB in in-duct fan sound measurements were observed by several investigators [21–24]. This led to attempts to find an optimum radial single microphone position (rather than the duct center-line used by Beranek) in order to estimate fan sound power from sound pressure measurements [21, 22, 25–27]. Several workers concluded that the half-radius position was better than the duct center-line, and this has been used in BS 848. However in the early 1970s Bolleter and Crocker, by determining the sound power carried in different propagational duct modes, concluded that a position nearer the duct wall should be recommended for most fans [28, 29].

In parallel with this theoretical and experimental work on microphone location, several workers have attempted to design tubular microphone windscreens. These windscreens should ideally minimize the flow noise but not affect the fan sound measured. The flow noise may be considered to come from two sources: (1) the *self-noise* caused by the introduction of the microphone in the flow, and (2) the *initial turbulence* existing in the flow from the eddies cast off the fan. Conventional microphone nose cones and microphone windshields (such as the one mentioned by Beranek et al. in Paper 15) can reduce only self-noise. Tubular windscreens, sometimes called sampling tubes, can reduce both the self-noise and the initial turbulence. Several investigators, including Friedrich, Eichler, Bolleter, Crocker, Wang, Baade, Neise, and others, have investigated various kinds of sampling tubes [30–34]. Friedrich probably should be credited with first suggesting such a design [30]. However, the tubular windscreen now commercially available from Brüel and Kjaer is based on the more recent work of Wang and Crocker [33] and of Neise [34].

Research work on both optimum microphone location and tubular windscreen development has led to the recent development of ASHRAE and the International Organization for Standardization (ISO) in-duct fan power standards [36, 37]. This work has been summarized by Crocker and Wang [35]. It is expected that the in-duct fan-sound power method may eventually supplement or even supersede the reverberant room method [20] since it is less expensive and more convenient. Roland et al. have recently shown that for low flow speeds the sound power of a fan can be measured directly in a duct using the new two-microphone intensity technique [38, 39]. In this case a foam ball should be used to suppress flow noise on the microphone pair, and an anechoic termination should be fitted to the duct. Probably this new intensity approach for the measurement of fan sound power should be regarded as a laboratory procedure rather than as a simple method for use in a standard.

REFERENCES

1. Northwood, T. D., 1977, *Architectural Acoustics,* Benchmark Papers in Acoustics, Dowden, Hutchinson & Ross, Stroudsburg, Pa.
2. Harris, C. M., ed., 1979, *Handbook of Noise Control,* 2nd ed., McGraw-Hill, New York, pp. **22:**1–21.
3. Crocker, M. J., and F. M. Kessler, 1982, *Noise and Noise Control,* Vol. II, CRC Press, Boca Raton, Fla., pp. 51–135.
4. Anonymous, 1929, Noise Exclusion, Sound Waves from Noisy Streets Deflected by New Ventilator Which Admits Silent Fresh Air, *Sci. Am.* **104:**52.
5. Sivian, L. J., 1937, Sound Propagation in Ducts, *Acoust. Soc. Am. J.* **9:**135–140.
6. Rogers, R., 1940, The Attenuation of Sound in Tubes, *Acoust. Soc. Am. J.* **11:**480.
7. Morse, P. M., 1939, The Transmission of Sound Inside Pipes, *Acoust. Soc. Am. J.* **11:**205–210.
8. Willms, W., 1941, Sound Attenuation in Pipes with Internal Absorptive Lining, *Akust. Z.* **6:**150–165.
9. Cremer, L., 1953, Theory of Sound Attenuation in Lined Ducts of Rectangular Cross Section and the Resulting Maximum Attenuation, *Acustica* **3:**249–263.
10. Scott, R. A., 1946, The Propagation of Sound Between Walls of Porous Material, *Phys. Soc. London Proc.* **58:**358–368.
11. Kurze, U. J., and I. L. Ver, 1971, Sound Attenuation in Ducts Lined with Non-Isotropic Material, *J. Sound Vib.* **24:**177–187.
12. Ver, I. L., 1978, A Review of the Attenuation of Sound in Straight Lined and Unlined Ductwork of Rectangular Cross Section, *Am. Soc. Heat. Refrig. Air-conditioning Eng. Trans.* **78**(1):122–149.
13. Doelling, N., 1960, Dissipative Mufflers, in *Noise Reduction,* L. L. Beranek, ed., McGraw-Hill, New York, pp. 434–465.
14. Embleton, T. F. W., 1971, Mufflers, in *Noise and Vibration Control,* L. L. Beranek, ed., McGraw-Hill, New York.
15. Leonard, R. W., 1957, Heating and Ventilating System Noise, in *Handbook of Noise Control,* C. M. Harris, ed., McGraw-Hill, New York.
16. Bolt, Beranek and Newman, Inc., April, 1953, Physical Acoustics, vol. I (suppl. 1), Handbook of Acoustic Noise Control, *WADC Tech. Rept.* 52-204, pp. 217–240.
17. ASHRAE, 1976, ASHRAE Handbook and Product Directory, Systems Volume, Chapter 35, New York.
18. Wells, R. J., and R. C. Madison, 1957, Fan Noise, in *Handbook of Noise Control,* C. M. Harris, ed., McGraw-Hill, New York.
19. British Standards Institution, 1966, *Fan Noise Measurements,* British Standard BS. 848, Part 2.
20. Air Moving and Conditioning Association, Inc., 1967, *Test Code for Rating of Air Moving Devices,* AMCA Standard 300-67. Arlington Heights, Ill.
21. Van Niekerk, C. G., 1956, Measurement of the Noise of Ducted Fans, *Acoust. Soc. Am. J.* **28:**681–687.
22. Allen, C. H., 1957, Noise from Air Conditioning Fans, *Noise Control* **3:**28–34.

23. Kerka, W. F., 1957, An Evaluation of Four Methods for Determining Sound Power Output of a Fan, *Am. Soc. Heat. Refrig. Air-conditioning Eng. Trans.* **63:**367–388; Kerka, W. F., 1957, Heat, Piping, and Air Conditioning, *Am. Soc. Heat. Refrig. Air-conditioning Eng. Trans.* **29:**139–146.

24. Howes, F. S., and R. R. Real, 1958, Noise Origin, Power, and Spectra of Ducted Centrifugal Fans, *Acoust. Soc. Am. J.* **30**(8):714–720.

25. Dyer, I., 1958, Measurement of Noise Sources in Ducts, *Acoust. Soc. Am. J.* **30**(9):833–841.

26. Barrett, A. J., and W. C. Osborne, 1960, Noise Measurement in Cylindrical Fan Ducts, *Inst. Heat. Vent. Eng. J.* **28:**306–318; Osborne, W. C., 1964, *Am. Soc. Heat. Refrig. Air-conditioning Eng. Trans.* **70:**260–268.

27. Johnson, D. R., and D. M. Brown, January 1963, *Fan Noise Measurement Procedures,* Lab. Rep. No. 14, Heating and Ventilating Research Assoc., England.

28. Bolleter, U., and M. J. Crocker, 1970, Research Toward an In-Duct Fan Sound Power Measuring System. *Am. Soc. Heat. Refrig. Air-conditioning Eng. Trans.* **76** (pt. 2):110–119.

29. Bolleter, U., and M. J. Crocker, 1972, Modal Spectra in Hard Cylindrical Ducts, *Acoust. Soc. Am. J.* **51:**1439–1447.

30. Eichler, E., 1967, Reception and Self-Noise of Simple Lines and Perforated Pipes in Motion, *Acoust. Soc. Am. J.* **41A:**1615.

31. Friedrich, J., 1967, Ein quasischallunempfindliches Mikrophon für Geräuschmessungen in turbulenten Luftströmungen, *Tech. Mitt. RFZ,* Issue 1. Also translated by Strumpf and Kingsbury, Dept. of Architectural Engineering, Pennsylvania State University.

32. Bolleter, U., M. J. Crocker, and P. K. Baade, 1971, Tubular Microphone Windscreen for Microphones for In-Duct Fan Sound Power Measurements, *Acoust. Soc. Am. J.* **49A:**128.

33. Wang, J. S., 1973, *Optimization of Design of Sampling Tubes,* Ph. D. thesis, Purdue University; Wang J. S., and M. J. Crocker, 1974, Tubular Windscreen Design for Microphones for In-Duct Fan Sound Power Measurements, *Acoust. Soc. Am. J.* **55:**568–575.

34. Neise, W., 1973, Einflus der Mikrofunumströmung bei der Messung von Ventilatorgeräuschen im angeschlossenen Kanal, dissertation, Technical University, Berlin, published as DLR-FB.

35. Crocker, M. J., and J. S. Wang, 1974, In-Duct Fan-Sound Power Measurement Systems, *Am. Soc. Heat. Refrig. Air-conditioning Eng. Trans.* **80**(pt. 2):82–97.

36. American Society of Heating, Refrigerating, and Air-conditioning, Engineers, 1978, Method of Testing In-Duct Sound Power Measurement Procedure for Fans, *ASHRAE Standard 68-78.*

37. International Organization for Standardization (ISO), 1978, Acoustics—Determination of Sound Power Levels of Noise Sources—In-Duct Method, *ISO Draft International Standard 5136.*

38. Roland, J., M. J. Crocker, and M. Sandbakken, 1981, Measurement of Fan Sound Power in Ducts Using the Acoustic Intensity Technique, in *Proceedings of the International Congress on Recent Developments in Acoustic Intensity Measurement, Senlis, France,* pp. 237–243.

39. Roland, J., M. J. Crocker, and M. Sandbakken, 1982, Use of Acoustic Intensity Measurements to Evaluate the Sampling Tube Method of Measuring In-Duct Fan Sound Power, *Am. Soc. Heat. Refrig. Air-conditioning Eng. Trans.* **88**(pt. 2)

14

Reprinted from *Acoust. Soc. Am. J.* **12**:53–57 (1940)

The Absorption of Noise in Ventilating Ducts

Hale J. Sabine

The Celotex Corporation, Chicago, Illinois

Introduction

IN the 1940 edition of the *Heating, Ventilating, and Air Conditioning Guide*, the following statement is made: "At present there are no wholly rational or generally recognized methods of calculating the amount of duct lining necessary to accomplish a given reduction of noise level in the air traveling in a duct system; consequently some empirical method has been used." While this is perhaps a rather broad statement, it nevertheless reflects the fact that there are some inconsistencies between the various empirical rules which have been proposed, and that although the problem has been treated theoretically from at least two different approaches, the results have not yet been adapted to practical engineering use.

As a part of an experimental program to investigate the properties of a sound absorbing material designed specifically as a duct liner, it was desired particularly to establish if possible a means of calculating the amount of lining required to produce a given noise reduction by a method which would be both mathematically simple and reasonably accurate, and which at the same time would have a sound theoretical basis. More specifically, it was desired to determine experimentally the relation of the attenuation of sound in an absorbent lined duct to the length, size, and shape of the duct, to the frequency of the sound, and to the absorptive properties of the lining, and to check these experimental results against the predictions of the various theoretical and empirical formulae which have been developed.

Test Procedure

In attacking the problem from this standpoint, it was decided to set up an actual duct system, constructed in as close conformity as possible to actual practice, but also permitting accurate control of the variables under investigation. A diagram of the construction is shown in Fig. 1. The test duct was 30 ft. long and was constructed of sheet metal panels in such a way that it could be readily assembled or reassembled into various square or rectangular cross sections. Six sizes and shapes were tested, which included all combinations of 9, 12, and 18 inches. The absorbent lining used was a rigid rock wool sheet 1 inch thick, the entire length of the duct being lined throughout the tests. Fig. 2 shows the construction of the test duct, and Fig. 3 illustrates the fan, baffle chamber, and plenum chamber assembly.

The basic property of an absorbent lined duct which determines its noise reducing characteristics is the attenuation constant, or the drop in sound pressure level per unit length of duct. This quantity was measured directly in decibels per foot by means of a crystal microphone mounted on a carriage which was drawn through the duct by a pulley arrangement. In making these measurements the fan was shut off and a loudspeaker was placed in the tapered section immediately adjacent to the fan, as shown in Fig. 1, the baffles being removed. Warbled tones were used, at frequencies from 128 to 4096 cycles. Sound level readings, taken with a General Radio sound level meter, were made at several points inside the plenum chamber, inside the duct at intervals of 3 inches to 1 foot, and, when possible, at various points outside the exhaust end of the duct.

Results

Typical sets of data are shown in Figs. 4 and 5. The horizontal line at the left of the 0-foot ordinate represents the average sound level in the plenum chamber. These figures show that, as

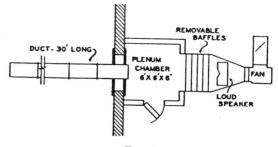

Fig. 1.

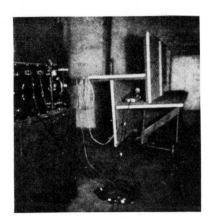

FIG. 2.

predicted by theory and already confirmed by other experimental work, the sound level drops off at a constant rate along the length of the duct within a region not too close to either end. Some random irregularity in the first few feet of duct appeared to a greater or less extent in all cases. Near the outlet end, clearly defined peaks and valleys due to end reflections were observed at the low frequencies, as also expected, with successive maxima or minima occurring one-half wave-length apart. Fig. 4 shows clearly why it is inadvisable to try to measure the over-all attenuation in an absorbent duct by taking readings only at the ends, particularly at low frequencies and with low absorption.

For all frequencies from 128 to 4096 and for all duct sizes tested, it was found that a straight line could be fitted, quite accurately in most cases, to the observed readings inside the duct, as shown in Figs. 4 and 5. The slope of this line in each case will hereafter be referred to as the attenuation constant, measured in db per foot.

INTERPRETATION OF RESULTS

The first step in interpreting the observed data was to relate the measured attenuation constants for each duct to the cross-sectional dimensions. The theories formulated both by Sivian[1] and by Rogers[2] state that the attenuation constant is directly proportional to the ratio of the perimeter, P, to the cross-sectional area, A, while

[1] L. J. Sivian, "Sound propagation in ducts," J. Acous. Soc. Am. 9, 135 (1937).
[2] R. Rogers, "The attenuation of sound in tubes," J. Acous. Soc. Am. 11, 480 (1940).

various empirical formulae which have been proposed state, respectively, that the attenuation constant is directly proportional to the 0.5 power of this ratio, inversely proportional to the average cross-sectional dimension, and inversely proportional to the smallest cross-sectional dimension. Fig. 6, representing the attenuation constants for a single frequency, shows that the data obtained in this investigation conforms more closely to the ratio P/A than to any other function of the cross-sectional dimensions.

In Fig. 7 the values of the attenuation constants measured for all six duct sizes are plotted against the ratio P/A for the frequencies 128 and 256 cycles. The highest value of P/A represents the 9×9-inch duct, while the lowest value refers to the 18×18 inch-size. The two points having the same value of P/A represent the 9×18- and the 12×12-inch sizes. Fig. 8 shows the same data for the frequencies 512 to 2048. At all of these frequencies, the attenuation is seen to be approximately proportional to P/A. At 4096 cycles, however, it was found that for the larger ducts the attenuation was much less

FIG. 3.

than would be expected on the basis of direct proportionality to P/A. This was interpreted to mean that the sound wave instead of filling the cross section of the duct uniformly over its entire length, tended to remain confined in a beam after the first few feet, and therefore to lose comparatively little energy at the absorbent walls.

The next step was to relate the measured attenuation constants to the frequency and the absorptive properties of the lining. This was done by checking the data against the theory developed by Sivian.[1] If it is assumed that the wave-length is greater than the largest cross-sectional dimension of the duct, and that the acoustic impedance of the duct lining is a pure resistance, R, then the attenuation constant, as derived by Sivian, can be shown to be independent of the frequency above 1000 cycles for values of less than 7 db per foot, and independent of the frequency above 100 cycles for values of less than 2 db per foot. Within this region, the value of the attenuation constant is shown by Sivian's analysis to be equal to

$$\text{db/ft} = (2195/R)(P/A) \quad (1)$$

where P and A are given in inches.

Assuming that the acoustic impedance of the lining was a pure resistance, R, which is approximately true of a material of the type used at all except the lowest frequencies, the value of R was calculated by the above formula at each frequency and for each duct size, from the measured attenuation constants and from the known values of P/A. This was permissible because of the fact that all of the measured values of the attenuation constant fell within the region of frequency independence. The average value of R for the six duct sizes at each frequency was substituted back in the formula and was thus used to determine the slope of each of the straight lines drawn through the observed points in Figs. 7 and 8.

A direct check of these calculated values of R could have been made by measuring the acoustic resistance and reactance of the lining material by a tube method. Since means for doing this were not available, the values of R were converted to corresponding reverberation chamber absorption coefficients, $\bar{\alpha}$, by the following formula, which is given by Morse[3] in different form,

$$\bar{\alpha} = 1 - \left(\frac{R - 2\rho c}{R + 2\rho c}\right)^2. \quad (2)$$

The values of $\bar{\alpha}$ thus obtained were checked by making actual reverberation chamber measurements, at Riverbank Laboratories, on the identical material used to line the duct.

In Fig. 9, the values of the absorption coefficient as measured in the reverberation chamber are compared with the values calculated from the measured attenuation constants by the above method. The vertical line through each point represents the range of the calculated coefficients for the six duct sizes at each frequency, the point representing the average. The points lie vertically in the order of increasing frequency. The lowest point, representing 128 cycles, is the farthest out of line. For the reverberation chamber tests, the material was mounted on sheet metal which was laid on 2×4 studs, in an effort to duplicate the mounting conditions in the duct, but evidently some discrepancy was introduced from this source. It was probable, also, that at this

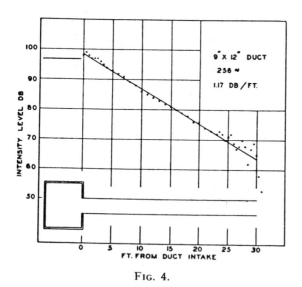

FIG. 4.

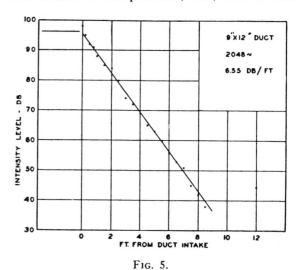

FIG. 5.

[3] P. M. Morse, *Vibration and Sound* (McGraw-Hill, 1936), p. 304.

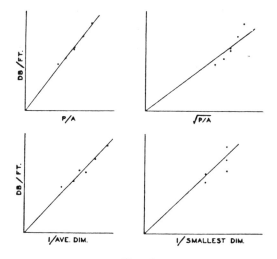

FIG. 6.

frequency the assumption of a purely resistive impedance was somewhat too far from the facts. It is intended to carry out direct impedance measurements of the material as a further check.

In general, however, the two sets of data show surprisingly good agreement in view of the number of assumptions and possible errors involved in arriving at both sets. It is therefore possible to reduce Sivian's analysis to a simple formula which will be sufficiently accurate for practical engineering work. It is believed that this formula should be stated in terms of the reverberation chamber coefficient of the lining material rather than its acoustic impedance, for the reason that the latter method of rating, while it may give more direct and complete information on the absorptive properties of the material, is not yet in common engineering use.

Equation (1) may be stated in terms of the reverberation chamber coefficient, $\bar{\alpha}$, by rearranging Eq. (2), but it may be further simplified by setting $1/R$ approximately equal to $0.00575\bar{\alpha}^{1.4}$. This relation is accurate within ± 10 percent for coefficients between 0.20 and 0.80. The approximate formula then reads

$$\text{db/ft} = 12.6\,\bar{\alpha}^{1.4}\,P/A. \qquad (3)$$

CONCLUSIONS

Formula (3) affords a simple means of calculating with reasonable engineering accuracy the decibel reduction in an absorbent lined duct for any single frequency, subject to certain limitations.

(1) The formula has been shown to hold for frequencies from 128 to 2048 cycles in ducts from 9 inches to 18 inches square. It was found that for ducts having the smallest dimension not over 9 inches, the formula was also valid for 4096 cycles. For larger ducts than any of those tested it would be expected that the upper frequency limit for which the formula would hold would be lower than 2048 cycles. However, as previously noted by Sivian, it was found that the analysis is valid for wave-lengths which are considerably smaller than the largest cross-sectional duct dimension.

(2) The formula is valid for duct shapes having dimension ratios as high as 2 to 1. Whether it will hold for shapes departing much farther from the square, will probably have to be determined by further experiment. Parkinson has observed that in these cases the attenuation is less than would be predicted on the basis of direct proportionality to P/A. In the tests reported here, however, the attenuation at all frequencies except 128 cycles was slightly higher for a rectangular duct than for a square duct of the same P/A ratio.

(3) The formula is based on the assumption that the acoustic impedance of the lining is mostly resistive. As noted above, this is probably true of most of the duct lining materials in

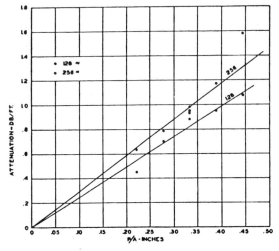

FIG. 7

general use, except possibly at low frequencies where diaphragmatic vibration of the duct wall might occur.

As stated above, the formula is applicable only to single frequencies or small frequency bands. In the case of a complex sound entering the duct, each frequency component will decrease along the duct at a rate determined by the absorption

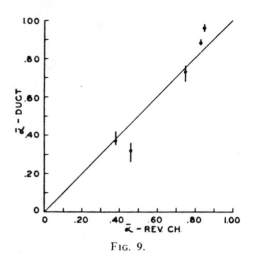

Fig. 9.

of the lining at that frequency. Thus, the over-all sound level will not, in general, drop off along a straight line for the entire length of the duct, but will follow a curved line whose shape is determined both by the frequency distribution of the original sound and by the absorption-frequency characteristic of the lining. If this information is known the shape of the attenuation curve for the over-all sound may be calculated by taking the logarithmic sum of the straight-line attenuation curves for each frequency component.

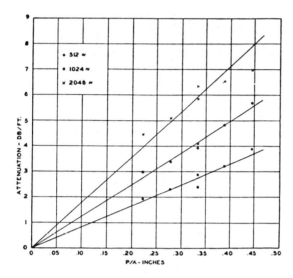

Fig. 8.

226

15

Reprinted from *Acoust. Soc. Am. J.* **25**:313-321 (1953)

Apparatus and Procedures for Predicting Ventilation System Noise*

L. L. Beranek, J. L. Reynolds,† and K. E. Wilson‡
Acoustics Laboratory, Massachusetts Institute of Technology, Cambridge, Massachusetts

A method has been developed for measuring the acoustic power output in watts and the frequency spectra of ventilating fans operating into ducts. Both the inlet and the exhaust noise are determined as a function of back-pressure. The fan is ideally connected to a circular section of duct about equal in diameter to the fan diameter and five to ten diameters in length. Three to six diameters from the input end of this duct an air straightener is located. The acoustic measurements are made with a condenser microphone enclosed in a wind screen located four to eight diameters from the inlet. The outlet end of the duct couples through an exponential connector to an anechoic wedge structure two feet square. This is backed by a plenum chamber containing an array of holes that can be progressively closed to produce increased back-pressure. Sound pressure level measurements are made in frequency bands and converted to acoustic watts per band. The inlet noise is measured over a hemispherical surface outdoors to obtain the directivity pattern and the acoustic power for each frequency band. Engineering formulas and charts for applying the data to prediction of the noise produced in rooms are presented, including the effects of room characteristics, duct losses, and the distance from the ventilating grille in the room. Criteria for permissible noise levels in various types of rooms and auditoriums are presented.

I. INTRODUCTION

THE ventilation design engineer has long been handicapped by his inability to predict quantitatively, prior to its installation and operation, the acoustical performance of a ventilation system. This difficulty is largely due to the lack of adequate knowledge of the acoustical characteristics of the various elements comprising the over-all system. Design engineers are becoming more aware of ventilating noises, but, at present, the standard acoustic tests as carried out by fan manufacturers[1] involve only over-all sound pressure level measurements around the outside casing of the fan. These tests are seldom conducted in free-field surroundings and give no information on the frequency spectrum of that fan.

This paper presents a method for determining the exhaust and intake frequency spectra of ventilating fans and gives procedures for using this information in predicting the performance of ventilating systems for rooms. The results of measurements made on five fans are presented in a companion paper.[2]

II. SOUND-POWER LEVEL

The sound-power level produced by a source of noise in a particular frequency band is logarithmically related to the sound power in watts produced by the source of noise in that band. By definition, the power level *PWL*

in decibels is

$$PWL = 10 \log_{10}\left(\frac{W}{10^{-13}}\right) \text{ db re } 10^{-13} \text{ watt}, \quad (1)$$

where W equals the acoustic power in watts. For example, if $W=1$, $PWL=130$ db. For each factor in power of 10 more or less, add or subtract 10 decibels. For example, if $W=100$ watts, $PWL=150$ db, and if $W=0.01$ watt, $PWL=110$ db.

The *total power level* is calculated from the total power in all bands.

The systems that we describe below for measuring fan noise are designed to determine the power levels in frequency bands and the total power level.

III. SYSTEM FOR MEASURING THE NOISE PRODUCED BY A FAN IN DUCTS

A. Initial System

General Description

The measuring system initially developed for determining fan noise is shown in Fig. 1. The main components are the fan, a measuring duct, and an anechoic termination.[3] If needed, the fan is coupled to the measuring section by a conical adapter. A heavy canvas coupling is used to prevent the structure-borne vibrations of the fan from entering the measuring section and giving false readings.

In the initial system, the microphone was placed at the center of the duct at a point about 8 feet from the fan. Since a United States Navy type A3 fan was the largest fan available for the tests, the inside of the duct was made to correspond to its $21\frac{1}{8}$-inch diameter. The measuring duct was 5 feet in length, of circular cross section, and constructed of $\frac{1}{16}$-inch galvanized steel coated with a $\frac{1}{4}$-inch thick layer of vibration damping

---

* This work was done at M.I.T. while two of the authors were engaged in postgraduate study. The statements contained herein are private opinions and assertions of the authors and do not necessarily reflect the views of the U. S. Navy.

† Lieutenant, U. S. Navy—now at Bremerton, Washington.

‡ Lieutenant (junior grade), U. S. Navy—now at San Diego, California.

[1] Bull. No. 110, National Association of Fan Manufacturers, 2159 Guardian Building, Detroit, Michigan (1950), first edition.

[2] C. F. Peistrup and J. E. Wesler, companion paper in this issue of the Journal, J. Acoust. Soc. Am. **25**, 322 (1953).

[3] L. L. Beranek and H. P. Sleeper, J. Acoust. Soc. Am. **18**, 140–150 (1946).

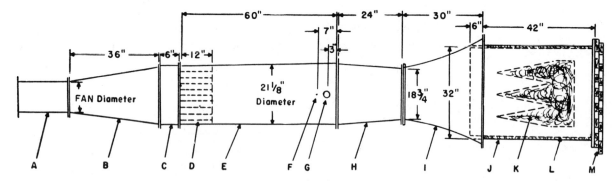

Fig. 1. Fan and duct system used in the experiments discussed in this paper. *A*—fan, *B*—conical adapter, *C*—canvas coupling, *D*—straightening vanes, *E*—measuring section, *F*—manometer fixture, *G*—microphone opening, *H*—adapter, *I*—exponential horn, *J*—acoustic termination, *K*—acoustic wedges 24 in.×24 in., *L*—fiberglas lining, *M*—back-pressure panels.

mastic. Conical adapter sections, 3 feet in length, were also constructed to adapt other fans to the measuring duct. Nine straightening vanes were inserted in the duct about 4 feet from the fan in order to reduce the turbulence of air flow. A manometer connection and microphone opening were placed at the top of the measuring duct as shown in Fig. 1.

An exponential horn, square in cross section for ease of construction and made of $\frac{1}{4}$-inch plywood, coupled the measuring duct to the anechoic termination. The square section of the horn made it necessary to insert a circular-to-square cross-sectional steel adapter between it and the measuring duct. Fitted to the large end of the horn was an anechoic termination. This consisted of a large square duct lined with a one-inch thickness of 6 lb/ft³ Fiberglas. In its center were suspended three 8×24-inch Fiberglas wedges, leaving an open space around the edges equal to the cross-sectional area of the duct. Incorporated in this termination was a method for quickly varying back-pressure. This was accomplished by covering the end with two contiguous perforated panels. The open area of these perforations was also equal to the cross-sectional area of the measuring duct. By sliding the outer panel with respect to the inner fixed panel, any desired area could be left open so as to determine the back pressure.

Instrumentation

Basically, the instrumentation for measuring the fan noise consisted of a microphone, a windscreen and an indicating device. An Altec-Lansing, model 21-B condenser microphone was used because of its small physical size and its relatively flat response over the desired frequency range. It was fitted with a windscreen to eliminate as much as possible the generation of noise by the airstream (see Fig. 2). The output signal was amplified by an Altec-Lansing line amplifier and measured on a Ballantine electronic voltmeter.

The microphone was calibrated to read absolute sound pressure level relative to 0.0002 microbar. This reference level was used throughout the work. The windscreen was calibrated, and the small corrections shown in Fig. 3 were found necessary at the higher

frequencies. The self-noise characteristics of this screen were measured at two wind speeds as a function of frequency and are shown in Fig. 4.[4]

In order to determine the noise levels in different frequency bands, one-third octave band filters were introduced in the system before the electronic voltmeter. Although that type of filter set is not used in the United States as widely as octave band filter sets, the larger number of frequency bands provides valuable information. Conversion from one-third octave to octave band readings is relatively easy. The filter set was calibrated and an average correction number for each band was established.

Back-pressures to the fans were measured directly by an Ellison inclined draft gauge, giving pressures in inches of water. Speed control of the dc current fan motors was obtained by varying armature current and, when necessary, field excitation.

Vibration Isolation

The canvas vibration break was found to be very necessary. Comparison of curves measured with and without it showed differences of as much as 10 db in some of the frequency bands. In well-designed ventila-

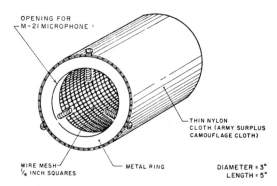

Fig. 2. Sketch of the windscreen used in the experiments. No covering was used over the open end, as it was not necessary for the purposes at hand.

[4] The authors are deeply indebted to Professor Robert W. Leonard, Physics Department, University of California at Los Angeles, for taking these data.

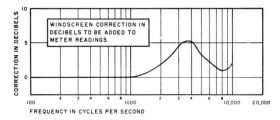

FIG. 3. Correction to the response of the sound measuring system due to the windscreen. Below 1000 cps no correction is necessary, but near 4000 cps the windscreen reduced the sensitivity of the system by 5 decibels.

tion systems, canvas breaks are commonly employed so that the use of them in the experimental set-up is believed desirable.

Longitudinal Standing Waves

The effectiveness of the anechoic termination was determined by measuring sound pressure levels along the axis in the horn itself and in the test section preceding it. Variations in levels of the order of ±1.5 db with position were found between 100 and 5000 cps. These variations included, however, not only reflections from the termination, but also nonuniformities in the sound field arising from the size and constructional details of the duct itself.

Transverse Standing Waves

The microphone was moved transversely across the duct in order to determine the effects of the radial modes of vibration in the tube. It was expected that lateral resonances would produce the principal inaccuracies in the measuring set up. The data from one of those tests are plotted as a function of frequency in Fig. 5. It is

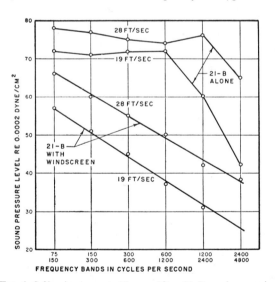

FIG. 4. Self-noise generated by an Altec 21-*B* condenser microphone with and without windscreen. The microphone was whirled in an anechoic chamber at a radius of 4.5 ft on a test stand carrying the preamplifier and batteries. These data were kindly taken for the authors by Professor Robert W. Leonard, Physics Department, University of California at Los Angeles.

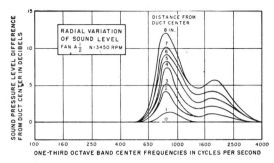

FIG. 5. Variation of sound pressure as a function of radial distance. A microphone position at the center of the duct was chosen because the data there were consistent with the power delivered to the termination.

seen that the principal resonances occurred at about 800 cps and at about double that frequency. The indicated differences of up to 10 db show that between 600 cps and 3000 cps errors will probably be introduced into the power level data amounting to as much as 5 db if a single microphone position is used. From a fan design engineer's standpoint this amount of error may be of some concern. However, inspection of the smooth curves obtained on a number of fans reveals that for practical application of the results in ventilation system design, the accuracy is adequate. A position at the center of the duct was ultimately chosen because the results obtained there correlated most closely with the total power delivered to the anechoic termination. Further research into the effect of these duct resonances on the total sound power is also indicated.

Effect of Reversing Fan

The change in power-level spectrum produced by reversing the fan, so that it sucked rather than blew air, was studied for one fan. The results are shown in Fig. 6. It is seen that the reversal produced only a small change in the amount of noise.

Determination of Power Level

The power level in any frequency band was determined by first measuring the sound-pressure level in decibels in that band at the center of the duct. The

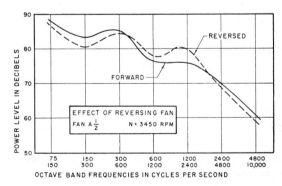

FIG. 6. Effect of reversing a small vaneaxial fan end-for-end.

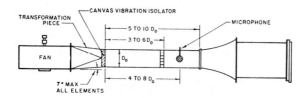

Fig. 7. Recommended test duct design.

SPL is defined as follows:

$$SPL = 20 \log_{10}\left(\frac{p}{0.0002}\right) \text{ db re 0.0002 microbar,} \quad (2)$$

where p is the sound pressure in dynes/cm² (microbars). This quantity is measured on a sound level meter with the "flat" or "C" network. If the sound intensity is uniform across the duct, the sound-pressure level in the duct is related to the power level by the equation,

$$PWL = SPL + 10 \log_{10}S - 0.5$$
$$+ 10 \log_{10}\left[\sqrt{\frac{K}{293}}\left(\frac{30}{B}\right)\right] \text{ db re } 10^{-13} \text{ watt,} \quad (3)$$

where S = cross-sectional area of the measuring section of the duct in square feet, K = absolute temperature of the air in degrees Kelvin, and B = total barometric pressure in inches of mercury. To the degree of accuracy usually necessary in ventilating system design, this formula can be approximated by

$$PWL \doteq SPL + 10 \log_{10}S. \quad (4)$$

B. Recommended System

General Description

It is recommended that future systems for measuring the noise characteristics of ventilating fans should be constructed more nearly like those described by the NAFM[1] for measuring fan performance (air volume, pressures, efficiencies, etc., at a stated density). In their Bulletin No. 110, they give duct and air straightener designs for 8 types of fans.

In brief, we believe that the fan is ideally connected to a circular section of duct about equal in diameter to the fan diameter and five to ten diameters in length (see Fig. 7). Three to six diameters from the input end of this duct an air straightener is located. The acoustic measurements are made with a condenser microphone enclosed in a wind screen located four to eight diameters from the input end. Measurements of temperature and barometric pressure should be made at all times. It is also believed that further studies should be undertaken on the best position of the microphone in the duct, as the data reported in Fig. 5 were taken for one fan only.

Instrumentation

The electronic measuring system should be similar to that used previously, except that in the United States an eight-band octave filter set can more readily be procured.

A superior type of windscreen, from the standpoint of low self-generated noise (due to the windstream) is believed to be spherical in shape with as large a diameter as is feasible in the duct being employed. The very thin nylon or silk cloth covering for the sphere can be stretched over a metal screen with 0.5-inch mesh squares. A particular calibration should be made to determine the effect of the windscreen on the frequency response of the microphone.

IV. MEASUREMENT OF NOISE IN FREE SPACE

Some fans do not couple to ducts. Even those that do often have one side or opening free. Measurements of the power levels produced by a free fan are accomplished by measuring the sound pressure levels in the different frequency bands at a number of positions on a spherical surface around the fan at a distance not less than four times the diameter of the fan. The acoustic power is obtained from the formula

$$W = (I_1 S_1 + I_2 S_2 = I_3 S_3 + \cdots I_n S_n + \cdots)10^{-7}, \quad (5)$$

where W = acoustic power in watts, $I_n = p_n^2/\rho_0 c$ = sound intensity in ergs/cm²/sec as measured at the nth position with a microphone, $\rho_0 c$ = characteristic impedance of air = 42.86 $(273/°K)^{\frac{1}{2}}(B/30)$ rayls, p_n = sound pressure in dynes/cm² measured at point n, S_n = area in cm² of the spherical surface for which the sound pressure equals p_n.

If the radiation pattern has an axis of symmetry, as is usually the case, measurements may be made at a number of points in a plane as shown in Fig. 8. In this case, the total acoustic power radiated is found from

$$W = \left(\frac{2\pi r^2}{\rho_0 c}\right) \sum_{n=1}^{N} p^2(\theta_n) \sin\theta_n \left(\frac{\Delta\theta}{57.3}\right) \times 10^{-7} \text{ watts.} \quad (6)$$

W = acoustic power in watts in a particular frequency

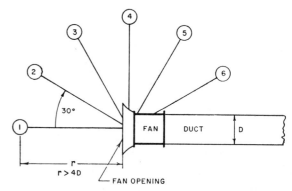

Fig. 8. Recommended positions for measuring the noise produced at the free end of a ventilating fan.

band, $r=$ distance from center of fan in centimeters $p(\theta_n)=$ sound pressure measured at a point n in dynes/ cm^2. $\rho_0 c=$ see definition above, $N=180°/\Delta\theta=$ number of measurements that were made in passing from a point directly in front of the source to one directly behind the source (0 to 180°). For this example, $N=6$. $\Delta\theta=$ separation in degrees of the successive points around the sound source at which measurement of $p(\theta_n)$ were made (see Fig. 8).

Usually a directivity pattern is plotted. A directivity pattern is the sound-pressure level in decibels plotted on polar graph paper as a function of the angle θ_n. The sound pressure p_n is found from the measured sound-pressure level with the aid of Eq. (2).

V. ACOUSTICAL DESIGN OF VENTILATING SYSTEMS

The acoustical design of ventilating systems for rooms can be broken into five parts: (a) source characteristics, (b) ducting characteristics, (c) grille characteristics, (d) room characteristics, (e) criteria for permissible sound levels.

A principal concern of this paper has been to describe a method for determining the noise characteristics of the source. In the companion paper,[2] data are presented on several specific fans and on typical unlined ducts with and without bends. The characteristics of lined ducts, lined bends, and grilles have been described previously in summary form in the Heating Ventilating Air Conditioning Guide[5] which should be required reading for ventilation design engineers. One summary graph will be presented here for convenience.

Let us now treat source and room characteristics and the criteria for permissible sound levels. Then we shall give an example of ventilation system design.

A. Source Characteristics

The noise characteristics of several vaneaxial and centrifugal fans were given in the accompanying paper[2]

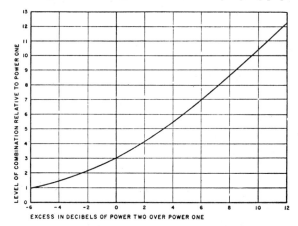

FIG. 9. Chart for combining two sound-pressure levels on an energy basis to obtain the sound-pressure level of the combination. Examples: 80 db plus 80 db yields 83 db; 80 db plus 86 db yields 87 db.

[5] *Heating Ventilating Air Conditioning Guide* (Am. Soc. Heating Ventilating Engrs. New York, 1952), Vol. 30, p. 871.

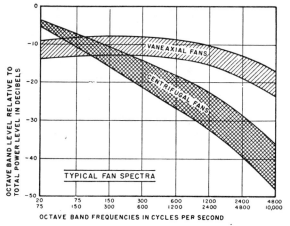

FIG. 10. Typical fan spectra for vaneaxial and centrifugal fans. The shaded areas give the expected variations due to various shapes of the fan blades and other design details. The data are taken from the companion paper by Peistrup and Wesler.

in terms of power levels in one-third octave bands plotted as a function of the mean frequencies of the bands. Usually, however, acoustic measurements of sound-pressure levels are made in octave bands. One-third octave band levels may be converted to octave bands by adding the powers contained in the three bands within an octave and converting to decibels again.

To assist in the summation of band levels, the chart of Fig. 9 is given. It permits easy determination of the level that would be obtained by a single band equal in width to two contiguous bands for which the levels are known. For example, suppose that the octave band level between 300 and 600 cps is desired when three one-third octave band levels are as follows: 80 db in the 284–360 cps band; 79 db in the 360–456 cps band; and 78 db in the 456–568 cps band. Using the chart, we find that the combination of the first two bands gives a level of 82.6 db. The combination of this level with the third band level gives a level of 83.9 db. The final level is that which would be measured by a filter with frequencies between 284 and 568 cps and is approximately equal to the desired level for a 300 to 600 cps octave band.

When the exact acoustic data on a fan are not available, Eq. (7) below and Fig. 10 may be used to yield estimated octave band levels. Equation (7) gives the total power level for all bands in decibels. Figure 10 gives the power levels in octave bands relative to the total power level.

Total power level (all bands)=

$$120.4+17.7 \log_{10}(HP/N)$$
$$+15 \log_{10}(N/6) \text{ db re} 10^{-13} \text{ watt,} \quad (7)$$

where $HP=$ total horse power delivered to the fan rotor by the motor, and $N=$ number of fan blades. Data taken on centrifugal fans by the NAFM procedure appear to lie about 40 decibels below the total power level.

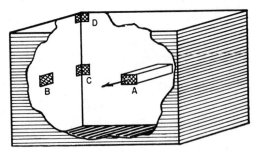

FIG. 11. Four positions of terminating a ventilating duct in a room.

B. Room Characteristics

In Fig. 11 we show, arbitrarily, four methods for terminating a ventilating duct in a room. These methods were chosen to correspond to four distinct acoustical situations. Other duct arrangements may be analyzed by interpolation among the four cases shown here.

In each of these cases, the noise power radiated into the room produces sound waves that are reflected from the walls many times to produce reverberant sound. Low frequency sounds radiate from the end of the duct equally in all directions. High frequency sounds are "beamed" in the direction the duct opening is facing. This beaming effect is described mathematically by the directivity factor Q.

The directivity factor Q is defined as the ratio of (a) the intensity produced in a stated direction at a given distance r from a source of sound to (b) the intensity produced at the same distance by a nondirectional source located in free space and radiating the same acoustic power. As an example, the directivity factor at low frequencies for a sound source located at the corner (see D of Fig. 11) of a room is equal to 8 because the source radiates the acoustic power into only one

quadrant of spherical space. Hence, 8 times as much power goes into the quadrant as would be the case if there were no corner.

The sound pressure level in the room at a distance r from the duct opening, as measured by a sound level meter using the "C" or "flat" network, is given approximately by the formula

$$SPL \doteq PWL + 10 \log_{10}\left[\frac{Q}{4\pi r^2} + \frac{4}{R}\right], \quad (8)$$

where PWL = power level of the source in a particular frequency band in db re 10^{-13} watt, Q = directivity factor (dimensionless), r = distance from the duct opening in feet, $R = \bar{\alpha}S/(1-\bar{\alpha})$ = room constant in square feet, $\bar{\alpha}$ = average absorption coefficient for the room at the mid-frequency of the band of noise being considered (dimensionless), S = area of the bounding surfaces of the room in square feet.

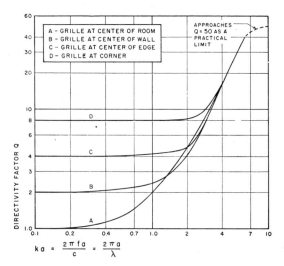

FIG. 13. Value of Q on a line directly in front of a ventilating grille as a function of the dimensionless constant ka.

A chart giving the room constant R as a function of the volumes of rooms of live, medium live, medium, and dead types is shown in Fig. 12. The values of $\bar{\alpha}$ for live rooms is assumed to be 0.05; for medium live rooms 0.15; for medium dead rooms 0.25; and for dead rooms 0.4. The relation between the surface area S and the volume V is assumed to be $S = 6V^{\frac{2}{3}}$.

Values of Q in front of the grille as a function of the dimensionless constant ka for the four entrance conditions shown in Fig. 11 are shown in Fig. 13. The quantity $k = 2\pi f/c$, where f is the mean frequency of the band of noise being considered and c is the speed of sound in ft/sec.§ The quantity a is the radius of grille in feet if it is circular, or equals $L/\pi^{\frac{1}{2}}$ if the duct is square with an area of L^2 ft². If the duct is rectangular

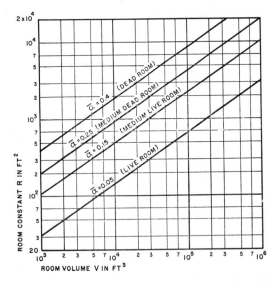

FIG. 12. Room constant R as a function of room volume V and average absorption coefficient $\bar{\alpha}$. It is assumed that the area of the surface of the room equals about six times the two-thirds power of the volume of the room.

_____

§ The speed of sound equals $(1052 + 1.11°F)$ ft/sec, where °F is the temperature in degrees Fahrenheit.

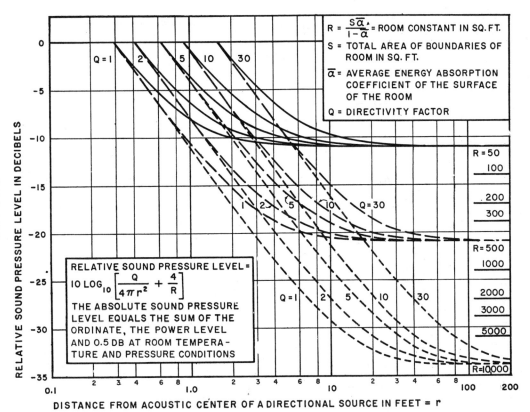

FIG. 14. Value of the second term on the right-hand side of Eq. (8) as a function of r, R and Q in English units.

with an area L_xL_y, the value of Q to be used lies between the Q's for two square ducts of areas L_x^2 and L_y^2.

For more accurate determinations of the sound pressure level, the quantity $\{0.5+10\log_{10}[(293/K)^{\frac{1}{2}} \times (B/30)]\}$ should be added to Eq. (8), where $K=$ absolute temperature in degrees Kelvin and B is the barometric pressure in inches of mercury.

The second term of Eq. (8) is plotted in Fig. 14, with SPL in db as ordinate, distance from the duct opening as abscissa, and the room constant R and the directivity factor Q as parameters.

It is seen from Eq. (8) and Fig. 14 that the noise from a ventilating duct is less at distances farther from it than nearby. Moreover, at large distances, the levels (averaged in space to eliminate standing waves) produced in the room depend only on the power level of the source and the room constant and not in the directivity factor and the distance r. The lines down the right-hand side of the graph give the relative sound pressure levels for large r.

C. Criteria

The criteria for the design of ventilating systems is usually based on hearing requirements for speech in the room. For good hearing conditions the arithmetic average (in decibels) of the long-time *RMS* speech levels in the three octave frequency bands between 600 and 4800 cps should be greater than the *speech interference level*. The speech interference level is defined as

the arithmetic average of the noise levels in those same three bands.[6] The assumption is made that the noise has a continuous spectrum.

Reference to Fig. 15 shows the complete criteria curves used by one consulting firm.[7] The *SC* numbers on the curves are equal to the speech interference level.

TABLE I. Tentative criteria for noise control
(Courtesy of Bolt Beranek and Newman).

Type of room	SC criterion
Concert halls	20
Broadcast studios	20
Legitimate theaters (500 seats, no amplification)	25
Music rooms	25
Assembly halls (amplification)	25
School rooms (no amplification)	25
Homes (sleeping areas)	25
Conference room for 50	25
Conference room for 20	30
Movie theaters	30
Hospitals	30
Churches	30
Courtrooms	30
Libraries	30
Small private office	40
Restaurants	45
Coliseums for sports only	50
Secretarial offices (typing)	55
Factories	40–65

[6] L. L. Beranek, Trans. Am. Soc. Mech. Engrs. **69**, 97 (1947). See also J. Acoust. Soc. Am. **19**, 257 (1947).
[7] Bolt Beranek and Newman, 16 Eliot Street, Cambridge 38 Massachusetts.

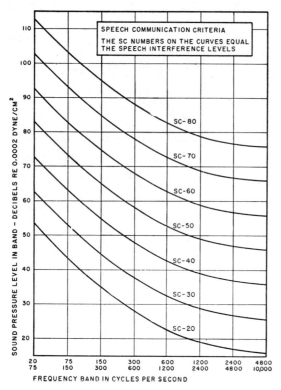

FIG. 15. Speech communication criteria curves
(courtesy of Bolt Beranek and Newman).

In Table I we show the *SC* value that we believe ought to be chosen for particular cases that the ventilation design engineer may encounter in practice.

D. Example of Design

Problem: As an example of the application of the procedure outlined above, let us assume that a room with a volume 100 000 cubic feet is to be ventilated by a centrifugal fan on which no noise data are available. The fan rpm is 300, the electrical horsepower is 4 hp, the airflow is 10 000 cubic feet per minute at 100°F, the fan blade-tip diameter is 36 inches and the number of blades is 60. Assume that the speech communication criterion for all parts of the room is *SC*-35 (see Fig. 15), and that the room is medium live. We are to determine the amount of attenuation in the duct system necessary for meeting the criterion and to suggest a suitable duct treatment.

Analysis: The total acoustic power level is found from Eq. (7) and equals

$$PWL = 120.4 + 17.7 \log(4/60) + 15 \log(60/6)$$
$$= 120.4 - 20.8 + 15 \doteq 115 \text{ db.}$$

The band power levels are found from Fig. 10, where we shall conservatively use the upper edge of the shaded area. The data are plotted in the second column of Table II. From Fig. 12 the room constant *R* equals about 2500 square feet.

For operation with an *SC*-35 criterion, which is not a very severe one, the grille velocity in feet per minute may be as high as 1000. Hence, the necessary grille opening is 10 square feet. This gives a value of *a* in Fig. 13,

$$a = (10/\pi)^{\frac{1}{2}} = 1.78 \text{ ft.}$$

The quantity $ka = 2\pi fa/c$. Because the temperature is 100°F, the speed of sound equals

$$c = 1052 + 1.1°F = 1162 \text{ ft/sec.}$$

Hence,

$$ka = 0.0096f.$$

From Fig. 13 we see that the directivity factor is lower when the grille is in the center of the wall than when it is at an edge or a corner. Also, we see that the grille will become fairly directive above 100 cps, i.e., when $ka > 1$.

Therefore, it is desirable to locate the grille high on the wall, if possible, in order to take full advantage of the directivity. Let us, however, assume the worst condition, where the grille must be at head level. Then

TABLE II. Detailed calculations for the example.

1 Band	2 Band PWL	3 Q	4 Fig. 14 DB values	5 Columns 2 +4	6 Cri- terion DB	7 Re- quired reduc- tion DB
20–75	111	2	−25	86	67	19
75–150	108	2.3	−25	83	58	25
150–300	105	4.2	−24	81	50	29
300–600	101	17	−18	83	43	40
600–1200	96	50	−13	83	38	45
1200–2400	93	50	−13	80	34	46
2400–4800	87	50	−13	74	31	43
4800–10 000	79	50	−13	66	31	35

Q will be as great as the values given by Fig. 13 (see column 3 of Table II). Of course, if the listeners are not directly in line with the grille, *Q* will not approach the high values assumed here and the noise reduction problem is simpler. In fact, at 45° to a perpendicular line the grille face, *Q* can be assumed to be 2 at all frequencies.

From Fig. 14, we find the numbers to be subtracted from the band sound-power level to obtain the band sound-pressure level. Let us assume that the nearest listener is 10 feet from the grille. Then from Fig. 14, we obtain the numbers given in column 4 of Table II.

Addition of the numbers in columns 2 and 4 gives the band sound levels in column 5. In column 6 we tabulate the *SC*-35 criterion numbers. The difference between column 5 and column 6 gives the required noise reduction between the fan and the room in decibels.

The reduction of the noise by 19 to 45 decibels will be possible by using a duct that is subdivided into cells. The number of the cells will depend on the length of the duct available for lining and on the thickness and type of acoustical absorbing material selected. Let us assume a typical rigid rock-wool or glass-fiber duct-lining board, with absorption coefficients in the eight octave bands as follows:

TABLE III. Absorption coefficients for a
1-inch thick typical duct-lining board.

Fre-quency band	20 75	75 150	150 300	300 600	600 1200	1200 2400	2400 4800	4800 10 000
α	0.05	0.1	0.3	0.7	0.8	0.8	0.8	0.8

Parkinson[8] has presented the empirical chart given in Fig. 16 for relating the attenuation in decibels per foot to the absorption coefficient and the length of a duct side. Let us assume that 60 feet of duct are available to line. Also, let us assume that the acoustic losses in the three lowest bands through the sidewalls themselves will amount to about 0.1 db per foot. Hence, we will pick up about 6 db by this process. We must pick up an additional 13 db in the 75–150 cps band. In this second band, we wish 0.3 db/ft loss. Reference to Fig. 16, shows that for $\alpha=0.1$, we need to divide the duct by means of splinters into sections that are no wider than (about) 4 inches. This would require, of course, sub-division into 10 compartments if the duct were 40 inches by 50 inches (i.e., ten 4-inch components plus the material). At higher frequencies there would be no trouble in obtaining the necessary noise reduction be-cause of the high values of the absorption coefficients.

[8] J. A. Parkinson, Heating and Ventilating, 23–26 (March, 1939).

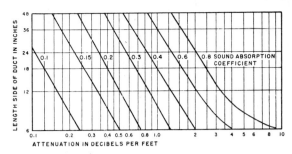

FIG. 16. Empirical chart for determining the size of a square duct lined on four sides as a function of the desired attenuation in decibels per foot with the sound absorption coefficient as pa-rameter. (After Parkinson, reference 8).

More efficient acoustical absorbing structures than parallel baffles in the duct can be constructed using bends and plenums. These structures usually introduce higher back-pressure into the system and must be specifically designed.

E. Field Data

Field data have been taken on several large ventila-tion systems. The data measured near the fan fell within the shaded region of Fig. 10. At distances of 30 to 100 feet from the fan, the noise in the two lowest frequency bands was 5 to 15 decibels lower than that predicted by Fig. 16. Presumably, sound passed through the thin sidewalls of the duct at low frequencies and was absorbed in the space provided for running the ducts.

Part IV

EVALUATION OF HUMAN
RESPONSE TO NOISE

Editor's Comments
on Papers 16 and 17

16   YOUNG
Don't Forget the Simple Sound-Level Meter

17   BERANEK
Criteria for Office Quieting Based on Questionnaire Rating Studies

Human response to noise is a very complicated subject. Noise has several effects on people, some well proved and documented, others less well proved, and some that remain controversial. Sufficiently intense noise (above about 140 or 150 dB) can cause immediate hearing damage. At much lower levels (between about 90 and 110 dB), as found in some industries, noise can cause permanent hearing damage if experienced several hours each day during a working life. Noise also interferes with speech and other related activities such as telephone use, listening to television and radio, and so on. This is a masking phenomenon. Noise can also attract our attention and interfere with activities such as sleep through an arousal or startle phenomenon. The effects of noise on people have been reviewed by several authors (see for example, Refs. 1, 2). Noise is thus seen to be the potential cause of hearing damage, masking of speech, and arousal or startle.

Masking and startle effects will obviously cause annoyance. However, many environmental and other factors unrelated to noise also annoy people. It is very difficult to separate the annoyance caused by noise from other contributing annoying factors. Sometimes it is claimed that noise can have other physiological effects on people, such as increased incidence of heart attacks, miscarriages, and tension resulting in high pulse rates, or social or other effects, such as increased absenteeism from work, more errors, and the like. However, whether noise can cause these additional physiological and social effects is controversial. Noise is definitely a stress agent, but some workers, such as Kryter, claim that unless the noise is intense enough to cause damage in the primary organ (the ear), it will not damage secondary organs (such as the heart). Kryter has also claimed that people usually become accustomed to noise so that such effects as sleep interference, productivity and errors at work, absenteeism, and so forth should disappear after a period of time [1]. More research is needed to resolve these controversies.

Because noise sources are varied in character and the responses of individuals and groups vary, several different measures have been devised to rank noise sources in terms of loudness, noisiness, and annoyance. It seems to be easiest to quantify the loudness of a noise. Attempts have also been made to quantify noisiness, although it is not at all clear that there is any real difference between the noisiness of a sound and its loudness [3]. As already observed, the annoyance of a noise is even harder to quantify, although, certainly, a louder noise is normally more annoying. Stevens has written an excellent review of the subject of annoyance and loudness [3].

Methods of measuring sound will be briefly reviewed in Part VI. Factors that seem to affect the loudness or annoyance of a noise include absolute level, frequency content, duration, and variability in level with time. Ideally, any noise measure should take into account some or all of these variables. Some measures have been designed to determine the different effects of noise on people, and some have been designed to account for the effect of different noise sources.

Young has written a comprehensive review of the many measures for rating noise that have been developed over the last fifty years [4]. Due to lack of space only a few of the more important measures will be mentioned here. Also because of space restrictions it will be possible to include in Part IV only two of the many important papers written on noise measures. In 1923 Fletcher suggested using a logarithmic scale for measurement of sound, not only because of the large range of values encountered, but also because of the Weber-Fechner relationship between sensation and stimulus [5]. By about 1930 the logarithmic measure had been adopted fairly universally, and the decibel had come into common use for measurement of sound and noise. In the 1920s and 1930s, work was conducted on the subjective loudness of sounds by several workers, including Fletcher and Steinberg in the United States and Barkhausen in Germany. In 1936 the British Standards Institution defined the unit of loudness level, the phon, in much the same way as it is still used today [6]. However, today the unit of loudness level, the phon, and the unit of loudness, the sone, have been standardized through the more recent work of Stevens and Zwicker [7, 8]. The loudness level in phons of a tone or narrow band of noise is the sound pressure level of a pure tone at 1000 Hz, which seems equally loud as the tone or band of noise. The sone, S, is a linear unit of loudness related to the loudness level, P, by $S = 2^{(P-40)/10}$. In the 1940s and 1950s the sone and phon were fairly widely used to rate loudness of noise sources. The application of this loudness scheme to broadband noise is complicated, since the total loudness must be obtained by a summation process over one-octave or one-third octave bands.

Although the sone loudness scheme was used by U.S. industry in the 1950s to rate the loudness of trucks [9], because of its complexity it has largely fallen into disfavor in the measurement of machinery noise. One exception to this is its use in rating the noise of small household fans [10].

The single-number ratings of noise obtained using the A-, B-, and C-networks on a sound level meter had been in use since the 1930s. However, during the 1950s the use of the A-, B-, and C-weighted sound levels was called into question. In the late 1950s, in a search for simpler noise measurements, engineers again suggested their use. The brief article by Young (reprinted as Paper 16) is an eloquent call for the reuse of the A-weighted sound level. It is interesting to note that the A-weighted sound level is now very much back in favor, and in 1984 many rules and regulations specify the use of the A-weighted sound level.

As discussed earlier, another very important effect of noise is its masking of speech. In investigations conducted around 1950, Beranek and Newman studied the noise levels in commercial offices to determine what levels people considered to be acceptable. Beranek later conducted a more extensive study at a large air force base, and the results of this study were published in 1956 (Paper 17). In this paper not only do we find the definition of *speech interference level* (SIL) that is still widely used, but also in Figure 25, the *room noise criteria* (early forms of the NC curves) that are also still in use. The speech interference level (SIL) of a noise is the average of the levels in the three octave bands: 600 to 1200, 1200 to 2400, and 2400 to 4800 Hz. SIL is no longer used; instead, the preferred speech interference level (PSIL), which is the average of the levels in octave bands centered at 500, 1000 and 2000 Hz, is used.

REFERENCES

1. Kryter, K. D., 1970, *Noise and Man,* Academic Press, New York.
2. Crocker, M. J., and A. J. Price, 1975, *Noise and Noise Control,* vol. I, CRC Press, Cleveland, Oh., pp. 42-98.
3. Stevens, S. S., 1970, On the Quantitative Evaluation of Noise, in *Transportation Noises,* J. D. Chalupnik, ed., University of Washington Press, Seattle, Wash., pp. 114-128.
4. Young, R. W., 1970, Measurement of Noise Level and Exposure, in *Transportation Noises,* J. D. Chapulnik, ed., University of Washington Press, Seattle, Wash., pp. 45-58.
5. Mackenzie, D., 1922, The Relative Sensitivity of the Ear at Different Levels of Loudness, *Phys. Rev.* **20**(ser. 2):331-348.
6. British Standards Institution, 1936, *British Standard B. S. 661.*
7. Stevens, S. S., 1961, Procedure for Calculating Loudness: Mark VI, *Acoust. Soc. Am. J.* **33:**1577.
8. Zwicker, E., 1960, Ein Verfahren zur Berechnung der Lautstarke (A Means for Calculating Loudness), *Acustica* **10:**304.
9. Apps, D. C., 1956, AMA - 125 - Sone - New Vehicle Noise Specifications, *Noise Control* **2**(3):13-17.
10. Clapp, D. E., and C. E. Neelley, 1978, A Program for Rating the Loudness of Consumer Fan Products, *Noise Control Eng.* **2**(1):12-17.

16

DON'T FORGET THE SIMPLE SOUND-LEVEL METER

Robert W. Young

U. S. Navy Electronics Laboratory, San Diego, California

FOR many purposes a "single-number" rating of a noise is needed. The likelihood of finding such a rating suitable for all situations seems small indeed, in view of the well recognized fact that the annoyance created by a given noise is related to the circumstances under which it is heard. One may still hope, nevertheless, that ratings can be found that will rank-order noises of similar character adequately for specific purposes. Two such possible ratings recently described in NOISE CONTROL are the NC-curves proposed by Beranek [1] and the loudness calculation devised by Stevens. [2] Also, the sound-level meter that has been available for many years affords a single-number rating; the A-sound level, for example, has often been used as a rating of fan noise. An opinion seems to have developed in recent years, however, that the "A and B networks [of the sound-level meter] are no longer considered authoritative in evaluating human reaction." [3] By contrast, the present article presents some evidence that the sound-level meter may be adequate—or at any rate almost as good as some other methods—when similar noises are to be compared in a given situation.

Judgments of loudness level were obtained by Quietzsch on thirty-seven widely different sounds. [4] From the measured spectra of these sounds he calculated the loudness level by an equivalent-tone method, making use of the particular procedure devised by Mintz and Tyzzer. [5] He also measured each sound with a broadcast program circuit noise meter and with a sound-level meter.

The German sound-level meters employed by Quietzsch contain weightings very similar to the A and B weightings prescribed in

American Standard Sound Level Meters for Measurement of Noise and Other Sounds, Z24.3-1944. These responses are pictured in Figure 1. The circuit noise meter he used was the Siemens & Halske Psophometer, Rel 3 U 311; the "ear-weighting" for this instrument, as provisionally specified in the CCIF Green Books (1954), Vol. IV, pp. 127–130, is also plotted in Figure 1. The Psophometer, Rel 3 U 311, affords a quasi-peak indication: the charging time constant is about 0.5 ms, the discharging time about 350 ms; the indicating meter has a natural period of about 1 second.

The three ratings obtained by Quietzsch (computed loudness level in phons, quasi-peak level, and sound level) are plotted against judged loudness level in Figure 2 for the thirty-seven different sounds. If there had been perfect correlation for any of the three ratings the respective points would have fallen on the 45° line. The average difference between a rating and the corresponding observed loudness level (i.e., the average departure from the 45° line) is listed in column *d* in the table.

For practical use in rank-ordering noises as to loudness it is only necessary that an increase in "rating" always correspond to an increase in loudness level; the functional relation need not be exactly linear. That is, the relation could be a smooth "trend" curve such as the one that has been arbitrarily sketched in Figure 2(c). [6]

It must be recognized that had a different group of sounds been selected (especially if some of the sounds were dominated by pure tones of frequency near 1000 cps) the scatter of points in Figure 2(c) might have been more pronounced. Nevertheless, the original

thirty-seven sounds do constitute a varied assortment, including sounds of music, motorcycles, electric bells, horns, machinery, typewriters. Those identified as 1 and 31 in Figure 2(c) are somewhat extreme: the first, having a steeply sloping spectrum, is "murmurs after passage through a wall" and the other is "hammer blows on an iron plate." One might expect much less scatter for sounds of similar character: witness, for example, the good correlation obtained by D. P. Loye for automotive vehicles, between measured A-sound level and computed loudness level. [7] Still other procedures that he employed are mentioned later.

A matter of prime importance is how closely individual "ratings" of a noise conform to the assumed relation between "rating" and judged loudness level. Since a constant error is of no effect in rank-ordering, the root-mean-square deviation (standard deviation) of individual ratings has been computed about the mean difference listed in the table. The results are listed there in column σ. There is also an entry (line 4) for the standard deviation about the "trend" curve. Obviously one prefers a method that results in a

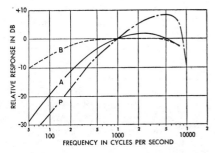

FIG. 1. Relative responses A and B for sound-level meters according to American Standard Z24.3-1944, and the response (P) for broadcast noise psophometer.

small scatter. It must be remembered that some of the scatter results from uncertainties in the original judgments of loudness level.

Professor Stevens has applied his promising method of loudness calculation to these same thirty-seven noises, and the results are evident in a scatter diagram already published.[2] He has kindly supplied me his numerical values of computed loudness level, from which I have obtained the statistics entered in line 5 of the table.

This brief article is not intended to be a comprehensive survey of methods of calculating loudness level. However, in lines 6, 7, and 8 of the table are included statistics for two further methods of calculation described by Quietzsch[4] and for a method recently devised by Munson.[8] The calculated loudness levels in the last instance were furnished me by Professor Harris.

It is evident from the table that, at least for data of the kind at hand, calculational procedures give more consistent results than does the simple sound-level meter. Whether the improvement is sufficient to justify the additional work will depend upon the situation at hand. The performance of the psophometer with its quasi-peak indicator is particularly encouraging. It may be remarked that a quasi-peak meter is incorporated in the General Radio Company Type 1556-A Impact Noise Analyzer and can readily be used to read the output of a sound-level meter.

Perhaps one reason the validity of the sound-level meter has been

FIG. 2. Loudness level (a) computed by equivalent-tone method (Mintz-Tyzzer procedure), measured quasi-peak level (b), and measured sound level (c), plotted against judged loudness level, all as obtained by Quietzsch for thirty-seven noises. Triangles in the bottom graph indicate that a Rohde and Schwarz sound-level meter was used, inverted triangles a Siemens & Halske meter; open symbols signify the B-weighting, solid symbols the A-weighting; (personal communication from G. Quietzsch).

questioned in the United States in recent years has been the frequent practice of measuring only the C-sound level; the necessity of reporting which weighting was used has been widely ignored. Notice that in the Quietzsch experiment, according to German practice, the sound levels were read with either the A or B weighting.

It may be remarked, incidentally, that the NC-curves[1] are similar in shape to the inverse of the A-weighting. Thus one might expect some correlation between the two methods of rating: for example, I have observed in a very limited set of ship noises that the A-sound level was 4 ± 2 db above the NC-rating.

The foregoing remarks are certainly not intended to disparage the calculation of loudness from a measured sound spectrum, nor to suggest that the present sound-level meter is adequate for all purposes. The spectrum measurement is often invaluable for engineering action, and some loudness calculations require little additional work. In certain situations a single-num-

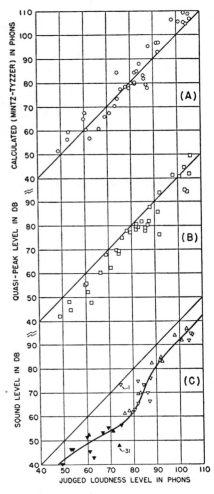

ber *measurement* is, however, the maximum complexity admissible. Therefore, in our search for simple ways to rate the noises of fans, automobiles, airplanes, and other everyday noises let us not ignore the possible virtues of weighting networks and quasi-peak indication. Indeed, let's see what we can get from the simple sound-level meter now available to us. ▲ ▲ ▲

Statistics on Different Methods of Rating Thirty-Seven Noises Studied by Quietzsch

The mean difference between a particular rating and the corresponding judged loudness level is d; the standard deviation of the differences about this mean is σ. (Exception: for the entries in line 4, d and σ are measured from the "trend" curve drawn on Fig. 2 c).

Method	d	σ
1. Equivalent-tone calculation	1.1	4.5
2. Quasi-peak meter	−4.5	4.0
3. Sound-level meter	−11.5	5.2
4. Sound-level trend	0.1	4.7
5. Stevens 1957	−1.5	4.1
6. Quietzsch, octave bands corrected	−0.4	2.7
7. Quietzsch, band division F	−0.6	3.1
8. Munson 1957	0.8	2.8

References

[1] L. L. Beranek, NOISE CONTROL 3, 1, pp. 19–27 (January 1957).

[2] S. S. Stevens, NOISE CONTROL 3, 5, pp. 11–22 (September 1957).

[3] H. C. Hardy, J. Acoust. Soc. Am. 29, 1330 (1957).

[4] G. Quietzsch, Acustica 5, 49–66 (1955).

[5] F. Mintz and F. G. Tyzzer, J. Acoust. Soc. Am. 24, 80–82 (1952).

[6] The present "trend" is reminiscent of relationships observed in similar tests by J. M. Barstow, J. Acoust. Soc. Am. 12, 150–166 (1940) and A. J. King et alii, J. Inst. Elec. Engrs. 88(II), 163–182 (1941).

[7] D. P. Loye, NOISE CONTROL 2, 4, pp. 56–60 (July 1956).

[8] *Handbook of Noise Control*, ed. Harris (McGraw-Hill Book Company, New York, 1957), p. 5–7.

17

Reprinted from *Acoust. Soc. Am. J.* **28**:833–852 (1956)

Criteria for Office Quieting Based on Questionnaire Rating Studies*

LEO L. BERANEK

Bolt Beranek and Newman Inc., 16, Eliot Street, Cambridge 38, Massachusetts

In order to discover what are the maximum noise levels that office personnel find acceptable and what their reactions are to noisy offices, a survey was carried out at a large air base. A questionnaire composed of 15 rating scales was administered to 190 people scattered over 17 different locations on the base. The rating scales allowed the workers to assess such things as the "noisiness" of their environment and to appraise the effect of noise on various aspects of their work, such as their ability to converse or to use the telephone.

The results obtained with the rating scales were compared with various physical measures of the noise. A high correlation was found between perceived noisiness and the measure called "speech interference level" SIL, which is the average of the sound pressure levels measured in the three octave bands between 600 and 4800 cycles per second. An even higher correlation was found between the ratings of noisiness and a computed measure of loudness level. More than two-thirds of those questioned stated that speech communication was an essential part of their activities and that

the more intense noises in their offices interfered with it. These correlations provide a basis for setting criteria for the maximum noise acceptable in terms of SIL and of loudness level. These results, plus those from a previous study, suggest that the maximum continuous noise levels acceptable for office spaces in which speech communication is important should not exceed an SIL of 40 db. Office employees are accustomed to a noise spectrum of a form that yields a loudness level of about 22 units higher than the SIL. Complaints are encountered when the loudness level exceeds the SIL by 30 or more units. Thus, with an SIL of 40 db the loudness level should preferably not exceed about 62 phons. Only under special conditions with this SIL would it be advisable to exceed a loudness level of 70 phons.

For intermittent noises, such as the noises produced by aircraft operations, evidence is presented to show that the mean (average) value of the SIL and loudness level should not exceed the numbers for steady noises if the work of office employees is not to be disturbed.

I. OBJECTIVES OF THE STUDY

THIS is a study of the noise conditions at a large Air Force base and of the reactions of office employees to the noise.

An early study of this kind by Professor R. B. Newman and the author[1–3] investigated commercial offices. A detailed presentation of the methods and findings of that program were withheld until a larger number of people could be questioned to see whether widely separated groups would respond similarly. Also, the earlier study suggested methods for improving the questionnaire and for administering the survey.

Since the problem of what constitutes acceptable noise levels in offices and other building spaces can be solved only by empirical methods, the aim of this study was to ascertain how people who work under different kinds and degrees of noisiness react to the noise. It was hoped that under actual field conditions, free from the artificialities of the laboratory, it might be possible to answer some of the questions that are basic to the design of measures for noise control.

Among these questions are the following:

1. Is unhampered speech communication the most significant desideratum for noise control? This question bears on the usefulness of rating noise in terms of

speech interference level (SIL),[3,4] which is defined as the simple average of the sound pressure levels in the three octave bands 600–1200, 1200–2400, and 2400–4800 cps.†

2. What factors other than speech interference are important, and what other aspects of the noise spectrum need to be controlled?

3. Do people agree with each other when they assess a noisy environment on a rating scale, and how do their ratings depend on the physical parameters of the noise?

4. Are the ratings made by office personnel at an air base consistent with those made by workers in commercial offices?[1] Or do different types of offices require different standards for noise control?

5. How must the standards developed for continuous noise be modified for intermittent noises such as those produced by aircraft operating on an air base?

II. PLAN OF THE STUDY

At least two approaches are possible in a study of this kind. One can try (a) to obtain objective behavioral evidence by observing directly the relation between "working efficiency" and noise, or (b) to sample attitudes and opinions through the use of rating scales on a questionnaire. Since behavioral evidence is notoriously difficult to obtain with reliability, the opinion survey was chosen as the more practical approach.

* This work was supported by the U. S. Air Force under Contract No. AF 33 (616)–2151.

[1] L. L. Beranek and R. B. Newman, paper entitled, "Speech interference levels as criteria for rating background noise in offices," presented before the Acoustical Society of America, Pennsylvania State College, June, 1950.

[2] L. L. Beranek, Transactions Bulletin No. 18, Industrial Hygiene Foundation, 4400 Fifth Avenue, Pittsburgh 13, Pennsylvania, pp. 26–33 (1950).

[3] L. L. Beranek, *Acoustics* (McGraw-Hill Book Company, Inc. New York, 1954), pp. 417–429.

[4] J. M. Pickett and K. D. Kryter, "Prediction of speech intelligibility in noise," Armed Services Technical Information Agency, Knott Building, Dayton 2, Ohio, Attn: DSC–SD. AFCRC–TR–55–4, June, 1955.

† Methods using five bands[4] and twenty bands[3] have been proposed. The three- and five-band methods give almost identical results for the spectra encountered in this study. The simplicity of the three-band method makes it especially attractive for engineering use.

A questionnaire employing a horizontal type of rating scale was decided upon. This is a simple device that can be used with psychologically naive individuals who have had no previous experience with rating. The scales used in this study were reasonably consistent with accepted practice,[5] although improvements in the questionnaire suggested themselves during the study and will be discussed in Appendix II.

An important decision that had to be made at the outset was whether a person questioned should be told that he was participating in a survey to determine the noise conditions in his building or whether he should be told that a general environmental survey was being made of attitude toward heat, light, noise, drafts, odors, and so on. For reasons of economy, questions were asked only about noise. It is possible, of course, that asking questions about noise only may have influenced the answers. But how much more the raters' answers were influenced by questions on noise alone than they would have been by a mixture of questions, is conjectural. The attitude taken here was that we wanted to learn how people interpret the adjectives along a rating scale when they are questioned directly about noise. We were further encouraged in undertaking this program by the fact that others have had success in establishing consistent scales by direct questioning.[6] Scales of this sort cannot be determined with high precision, but reasonable consistency is usually possible. A rating scale of the type used here is an *ordinal* scale,‡ that is to say, one interval along it is not necessarily equal in magnitude to other intervals along it.

Most of the items in the questionnaire asked for a rating of the noise along a scale of common adjectives. Typically, six points were located along the scale from left to right as follows: "very quiet," " quiet," " moderately noisy," "noisy," "very noisy," and "intolerably noisy." Other questions sought to find out which aspects of their work the office personnel thought were affected by the noise. From this information an attempt was made to relate the various adjectives to the personnel's reported ability to do their daily jobs and to the sound pressure levels measured in their working space.

Of course, many uncertainties lie in the way of determining what aspect of noise gives rise to a given judgment on the rating scale. Provided we are sufficiently modest in our expectations, however, the evidence collected allows us to make several generalizations. For example, speech communication appears to be one factor that is closely related to the ratings

of noise made by office personnel. Another correlate is loudness level, especially among personnel who do not communicate very often by voice, or who work where low-frequency noise is encountered. The prominence of these two factors does not mean that so-called annoying, disturbing, or fatiguing effects are not also of importance, because almost 40% of those questioned mentioned other effects. As we shall see, the plan and procedures followed in this survey lead to some instructive and useful correlations between subjective ratings on the one hand, and certain physical measurements on the other.

III. THE QUESTIONNAIRE

The questionnaire used in this survey is shown in its entirety on the following pages. It contains 17 items. The first three items provide scales on which the workers could rate the noisiness of their environment averaged over a period of "the past few months." Item 4 asks for a rating of the noise at the instant of filling out that item. While Item 4 was being answered, experimenters were measuring the noise levels in octave bands in the near vicinity. Items 5–8 concern the use of the telephone. Items 9 and 10 ask for an evaluation of the maximum noise that an individual feels he should be required to work in, and what aspect of his work would be most

[5] J. P. Guilford, *Psychometric Methods* (McGraw-Hill Book Company, Inc., New York, 1954), pp. 263–299.
[6] S. S. Stevens, J. Acoust. Soc. Am. 27, 815–829 (1955).
‡ A typical example of an *ordinal* scale is the rank ordering of a series of minerals according to hardness. The rank ordering is made on the basis of what minerals will scratch what other minerals. No information is obtained as to how much harder one is than another.

Sheet 1

SUBJECTIVE RATING OF NOISE CONDITIONS IN THIS ROOM

This questionnaire is part of a survey of noise conditions at Hill Air Force Base. You are asked to answer the questions below and to make ratings without consultation with your fellow employees. We want this to represent your own opinion. Your answers should apply to noise conditions that have existed during the past few months. But first,

Your age_____ Sex_____

Your job title_____

Room number or room name_____

Please mark your location in this room on a simple sketch below. All subsequent questions refer to this location.

How long have you worked in this room?_____

NOTE: In the questions that follow you are asked to rate the noisiness of this room on rating scales. An example of how to use a rating scale is given here:-

EXAMPLE

Suppose you are asked to rate the warmth of this room under average conditions on the scale drawn below. Assume you feel that on the average the room is slightly cooler than comfortable. Your rating would be an "X" located as follows:

Cold   Chilly   Comfortable   Warm   Hot

Note that you may put the mark anywhere along the scale, not just at the indicated names.

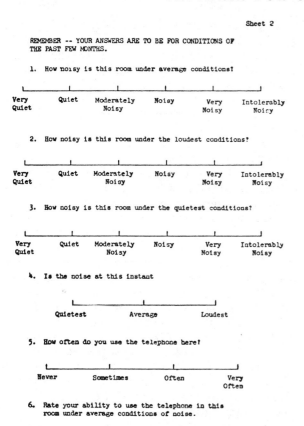

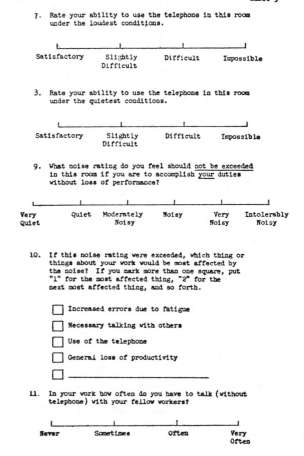

seriously affected if this maximum noise were to be exceeded. Items 11–13 deal with conversation between the subject and his fellow workers. Items 14 and 15 relate to the subject's evaluation of his hearing and of his sensitivity to noise. Items 16 and 17 seek to determine the source of the loudest noise in the worker's environment and the frequency with which the loudest noise disturbs his work.

IV. ADMINISTRATION OF THE SURVEY

Prior to the survey, a tour was made of the entire base to locate the important sources of external noise. On the basis of this tour, the base was divided into five areas as shown in Fig. 1. Area I included buildings that were exposed to external noise from aircraft in flight. Area II included buildings exposed to noise from aircraft in flight, and, of greater importance, to noise from aircraft warming up near the runways prior to takeoff. Area III included buildings next to the maintenance warmup area, where aircraft undergo extensive periods of operation to check out the engines and their controls after overhaul. Area IV included buildings adjacent to a group of six test cells (blocks) housing reciprocating engines (with propellers). These cells were in constant operation with from two to five engines running simultaneously. Area V included a building that was tightly

closed so that no external noise penetrated. Several buildings were chosen in each of the first four areas.

In administering the questionnaire, the experimenter entered the office space with a sound level meter, octave band analyzer, and some questionnaires. First he told the supervisor of the office that he was making a noise survey at the request of the Command and that he would like to hand out questionnaires to executive, supervisory, and stenographic personnel. The questionnaires were handed out to only a few people at a time so that the general noise and activities of the office could go on uninterrupted. The supervisor usually passed among the personnel in the office and informed them that they were to take time to fill out the questionnaires. The experimenter then proceeded to hand out one to four questionnaires at a time and to measure the noise in eight octave bands in the office at about the time the subjects were filling out Item 4. Usually the office routine was interrupted for only a few minutes while the first questionnaires were being handed out. As soon as the personnel saw that nothing very exciting was taking place, the noise of the office became normal and the experimenter could then pass from person to person without attracting attention. About a fourth of the office personnel in each building surveyed were questioned. This corresponded to about a sixth of the

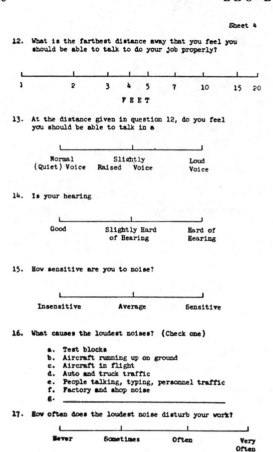

Sheet 4

total office personnel on the base. Only fully completed questionnaires were scored; eight were discarded. Of those scored, 5 were executives, 60 were engineers and analysts, 25 were foremen, and 7 were executive secretaries, or nurses with offices. This group of 97 employees is designated the "Executive Group." Also, there was a "Stenos Group" of 87 that included stenographers and clerks.

In addition to the noise measurements made concurrently with the answering of Item 4, noise level measurements were also made over a full day each at three selected locations. These measurements were made with a graphic level recorder which recorded through a 300–600 cps filter continuously from morning until evening. A third measurement consisted of using a magnetic tape recorder with which readings in eight octave bands could be obtained from time to time during the day to ascertain the spectra of the noises. By extrapolating from the continuous measurements between 300–600 cps and the sample measurements of full spectrum, it was possible to estimate the total spectrum and level of noise. An observer was present continuously at the graphic level recorder and every loud noise (airplane flying over, truck passing by, loud conversation, and so on), was identified on the record.

The "slow" meter switch on the General Radio Type 1550-A octave band analyzer was used for all the measurements not made with the graphic level recorder. Flyovers were of such infrequent occurrence and of such

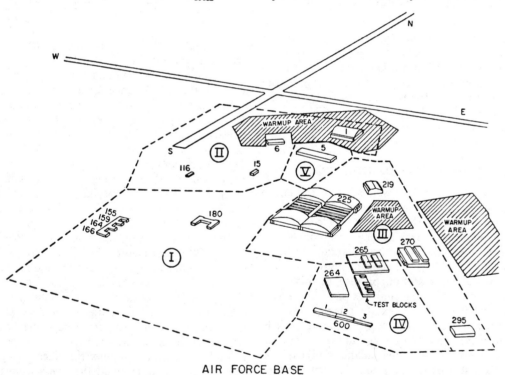

AIR FORCE BASE

FIG. 1. Hill Air Force Base shown divided into areas exposed to somewhat homogeneous external noise conditions. Of the several hundred buildings on the base, only those that were actually surveyed are shown. The categories are selected on the basis of open-window summertime conditions and are not applicable in all cases to the wintertime survey of this paper.

short duration that they could not be measured with the octave band analyzer. At no time was the microphone placed less than 10 ft from typewriters, accounting machines, or tabulating machines. Examples of noises from business machines measured at various distances from the microphone are given in Fig. 2. The distance of 10 ft was chosen because it was found that the people who used these machines usually turned them off when they had to converse. The nearest machine remaining in operation was generally 10 ft or more away. Some "silent" typewriters were used, and these registered lower levels than those indicated in Fig. 2.

The spectra of 14 locations in 12 buildings are plotted in Fig. 3 and of three locations in Building 600 in Fig. 4. They are grouped as follows: A, fairly low noise levels; B, somewhat higher; C, considerably higher. These groups will be discussed in greater detail in Sec. IX. The over-all and speech interference levels are given in Table I.

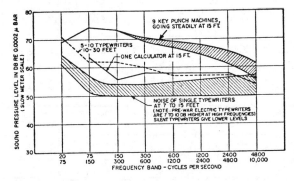

FIG. 2. Noise measurements of business machines operated continuously by skilled operators. The "slow" meter scale was used on the General Radio Type 1550–A octave band analyzer.

V. SCORING OF THE QUESTIONNAIRES

In order to facilitate the scoring of the questionnaires, each rating scale, regardless of its actual length in inches, was assigned a numerical length of 10. An exception was Item 4, which was used to derive a number between 0 and 10 corresponding to the rating scale of the three previous questions. For example, if the subject judged that the noise at "this instant" was halfway between "average" and "loudest," the number assigned to Item 4 was halfway between the numbers assigned to Items 1 and 2.

Since these rating scales are what are known as *ordinal* scales, it is proper to use the median as the measure of the "central tendency" of the subjects' responses.[7] Strictly speaking, it is not legitimate to use an arithmetic average. The median, which is determined by rank ordering the scores and finding the score that divides the distribution into two equal parts, has the additional advantage that its location is not unduly

[7] S. S. Stevens, Science 121, 113–116 (1955).

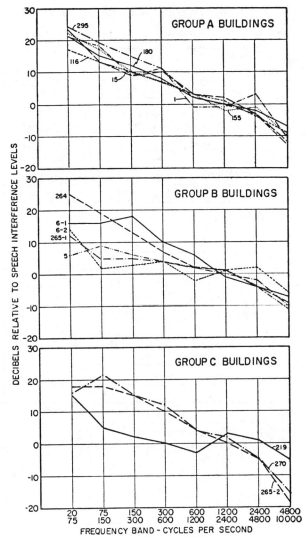

FIG. 3. Measured octave band spectra in decibels with respect to the speech interference levels. In several of the buildings, the noise levels varied in different rooms so that the buildings are split into divisions, viz. 6–1 and 6–2. The over-all levels and speech interference levels in decibels re 0.0002 microbar and the calculated loudness levels are given in Table I.

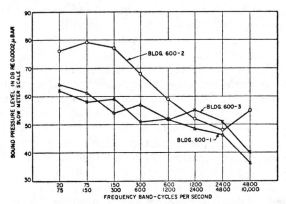

FIG. 4. Measured octave band spectra for building 600 in decibels re 0.0002 microbar. The over-all levels, the speech interference levels, and the calculated loudness levels are given in Table I.

TABLE I. Results of analysis of all valid questionnaires and of measurements of noise levels made as Item 4 was being answered.

Item	Description		Unit	Group A 1	15	116	155 159 164 166	180	295	6-2 / 6-1	Group B 264	265-1	5	Group C 270	265-2	219	Building 600 600-1	600-2	600-3
2	Median Noise Rating	Loudest	M.R.[a]	7.5	7.5	8.5	8.0	6.0	8.0	8.7	8.0	8.1	7.7	8.2	9.0	10.0	8.0	8.0	8.0
1		Average	M.R.	4.0	5.0	3.6	4.0	3.9	4.4	6.2	6.0	6.0	6.5	5.1	8.3	7.0	6.0	6.0	6.0
3		Quietest	M.R.	2.0	2.0	2.0	2.0	1.8	2.2	3.7	4.0	5.4	4.6	3.8	5.0	4.0	3.8	4.0	4.0
4		Present [b]	M.R.	4.0	4.5	3.6	4.0	3.4	4.2	6.2/5.2	5.9	6.6/6.6	6.7/6.7	6.8	8.6	5.5	4.0	7.0	5.0
	Measured SIL [b]		DB	40	44	37	39	38	35	56/46	52	60	63	60	65	53	49	53	52
	Measured Overall Level		DB	69	66	57	63	61	61	70/67	80	73	76	83	87	69	65	80	66
	Computed L.L.		Phons	64	65	56	61	59	58	76/71	76	79	81	86	92	72	69	85	70
5	Amt. Telephone Use		M.R.	10	9	6.7	7.0	7.2	7.0	9.5	7.0	6.8	3.3	8.0	9.9	8.0	9.5	7.0	5.5
7	Telephone Ability	Loudest	M.R.	6.7	6.2	6.4	6.7	3.5	3.8	9.0	7.5	7.0	6.0	6.7	7.8	10.0	6.6	6.0	6.0
6		Average	M.R.	3.0	3.3	1.8	3.3	0.8	0.3	6.6	4.0	6.0	3.7	5.3	5.3	5.0	3.3	3.3	3.6
8		Quietest	M.R.	0	0	0	0.2	0.1	0	0.5	1.0	2.9	0.5	1.0	2.0	0.5	0.5	1.5	1.5
9	Max. Desired Rating		M.R.	4.0	4.5	3.6	4.0	3.8	4.0	4.7	3.5	3.9	4.5	4.0	3.9	3.5	4.0	4.0	4.0
10	Number Reported Affected (First Res.)	Talking	M.R.	2	2	1	9	4	3	2	2	1	3	5	1	0	7	4	5
		Telephoning	M.R.	3	7	1	1	1	2	4	2	3	1	4	1	2	6	7	2
		General	M.R.	2	3	2	3	12	2	2	1	2	10	2	4	5	6	10	18
11	Amt. of Conversing		M.R.	6.7	6.7	7.0	6.7	7.2	6.7	6.8	7.0	6.9	5.0	6.7	8.3	6.7	7.0	6.6	6.0
12	Distance		Ft	10	8	8	10	10	10	10	5	8	4	8	6	10	4.0	4.0	5.0
13	Voice Level		M.R.	4.5	3.2	1.4	2.2	2.5	0.7	4.7	2.0	0.4	4.0	1.8	3.6	3.3	0.5	2.0	0
	Median of 5 and 11		M.R.	8.4	7.8	6.8	6.8	7.2	6.8	8.2	7.0	6.8	4.2	7.4	9.1	7.3	8.2	6.8	5.7
16	Source of Loudest Noise		-	A/C Runup	A/C Runup	A/C Runup	A/C Flight	Internal	A/C Runup	A/C Runup	Test Blocks	Factory shop	Internal	A/C Runup	Test Blocks	A/C Runup	Test Blocks	Test Blocks	Test Blocks
17	Frequency of Disturbance by loudest noise		M.R.	3.3	3.3	c	c	c	3.3	5.0	6.7	6.7	3.3	3.3	8.1	6.7	6.0	5.5	3.3
	Stenographers		No.	3	4	1	2	6	3	4	1	2	13	3	2	2	5	17	19
	Executives		No.	4	8	3	11	11	4	4	4	4	1	8	4	5	14	4	6

[a] Median rating.
[b] From Fig. 6.
[c] Note: Question 17 not included for these buildings.

affected by the score of a deviant rater who differs widely from the other subjects.

VI. ANALYSIS OF THE RESPONSES FOR ITEM 4

This section presents the correlation between the ratings of the noise "at this instant" and the sound level measurements in octave bands made at the same time.

The results for all items of the questionnaire are given in Table I. The hospital group (Buildings 155–166) was considered as one building. Since measured noise levels had substantially different values in two parts of Building 6, the data are reported separately as 6–1 and 6–2 for those two items only. Building 600 is divided into three areas, because the classification of the people involved and the stimuli in the three very large rooms contained in it are somewhat different. Since the number of people studied in each of the 13 buildings other than Building 600 was small, it was not deemed possible to obtain a curve relating the ratings on Item 4 to the noise levels in each of these buildings. Instead, the ratings of all the people in each building were plotted as a function of the measured speech interference levels for their respective buildings, as shown in Fig. 5. The median ratings on Item 4 and the median SIL's were determined and marked on the respective plots by cartwheels. The points so marked are tabulated in Table I as "Noise Rating—Present" and "Measured SIL."

The data from Fig. 5 are then plotted on Fig. 6 as the median noise rating for each building vs the median speech interference level in decibels for that building. As was found in earlier studies, there is a reasonably linear relation between the ratings and the SIL. On the other hand, ratings from five of the seventeen building spaces lie a significant distance from the heavy line shown. The rank-order correlation coefficient for the data in Fig. 6 is about 0.85.

Inspection of the measured octave-band noise levels shown in Figs. 4 and 5 indicates that variations in the shape of the noise spectra might account for part of the scatter on the curve. In particular, the levels in the low-frequency bands in buildings 600–2, 270, and 265–2 were high compared to the SIL's.

From previous experience with offices located near or

beneath printing presses we had learned that people object to low frequencies at high levels, even when conversation is easy. Hence, it seemed that some means should be sought for bringing the low frequencies into the picture. Because most of the spectra sloped downward, the bottom two bands determined the over-all level (OA). A rank-order correlation was computed for the median OA and median ratings for the 17 spaces. The coefficient of correlation was found to be 0.9 which seems to indicate excellent agreement. Since, as can be seen from Figs. 3 and 4, all the spectra have about the same shape, a rank-order correlation of median ratings and SIL would probably not be very different. More detailed product-moment correlations were computed using individual ratings rather than the medians of the ratings by buildings, and the results show a correlation

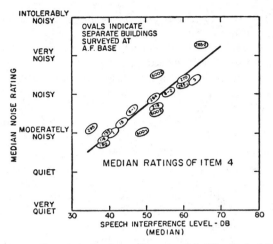

FIG. 6. Plot of the median ratings for the different buildings against the median speech interference levels measured as Item 4 was answered. The data are taken from Fig. 5. The rank-order correlation coefficient is about 0.85.

from Figs. 3, 4, and 5. The median SIL for each building was used to set the levels for the entire spectrum and the results are given in Table I as "Computed L.L." These data are plotted in Fig. 7 as median noise rating *vs* median computed loudness level in phons. The relation is quite linear over the range investigated and the scatter, except for building 600–1, is small. The rank-order correlation coefficients computed from this graph is better than 0.95, and is about 0.98 if building 600–1 is disregarded (reasons are given in Appendix I why this building might well be considered as a separate case).

Apparently the loudness level of a noise as computed by Stevens' method is a first-order determinant of how subjects categorize noises on a rating scale of the sort used in Item 4. This does not necessarily mean that the subjects are aware of loudness level as such. In fact

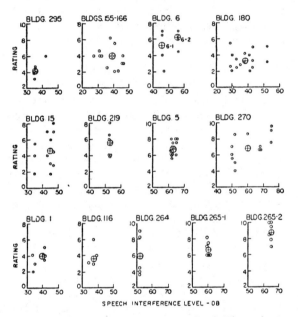

FIG. 5. Ratings *vs* speech interference level. The ratings are from Item 4 and the speech interference levels were measured at the time of rating. The cartwheels indicate the medians for both scales.

coefficient of 0.72 for SIL, compared to 0.52 for OA. The author appreciates, of course, that product-moment correlations are not applicable to *ordinal* scales, except that this one seems reasonably linear. In other words, there is a higher correlation between speech interference levels and individual ratings than between over-all levels and the individual ratings.

Recently, Professor S. S. Stevens of Harvard described a new technique for computing loudness level.[8] Professor Stevens has devised charts and formulas for converting from octave band levels to loudness in sones; for summing the loudness of the bands to obtain total loudness; and for converting from total loudness to loudness level in phons. According to Professor Stevens' method, loudness levels for the buildings were computed

[8] S. S. Stevens, J. Acoust. Soc. Am. 28, 807 (1956).

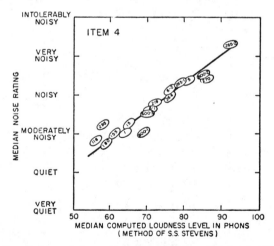

FIG. 7. Plot of the median ratings for the different buildings against the median computed loudness level in phons. The ordinate ratings are taken from Fig. 5. The abscissa was calculated for the spectra of Figs. 3 and 4, using the median SIL's of Fig. 5 to set the absolute levels. (See also Fig. 21.) The rank-order correlation coefficient is 0.95 with 600–1 or 0.98 without it.

they seem to be more conscious of speech interference and other noxious aspects of noise than they are of loudness level in the strict sense. All we can say is that the adjectives used on the rating scales to describe the various degrees of noisiness space themselves fairly uniformly along the scale of computed loudness level.

On the average, no appreciable difference was found, in the responses to Item 4, between the ratings of the executive group and the stenographers' group. This finding is substantially different from the results of the survey in commercial buildings.[1-3] In the earlier study, the executive group rated noise in the same way that the executive group rates the noise here, but the "commercial" stenographers rated a given noise lower by about one adjectival category than did the executive group. The small differences found here between the stenographers and the executive groups are due partly to the fact that the stenographers questioned were interspersed with the executive-engineering group, whereas the stenographers in the commercial companies were located by themselves in large rooms in which typewriting was almost the sole activity.

VII. RESPONSES TO ITEMS 5 AND 9–15

The medians of the answers to Item 5, "How often do you use the telephone," and Item 11, "How often do you have to converse," are given on Fig. 8. On the air base as a whole, people felt that they must talk directly (without telephone) to their fellow workers "often." Telephone conversation was more variable, ranging from "sometimes" to "very often." This result is consistent with the fact that on a repair and maintenance depot of this type, arrangements must frequently be made by telephone for materials, test schedules, and the movement of people and aircraft. Since the highly coordinated activities on the base are spread out physically, decisions and tasks cannot usually be completed within the confines of a single room.

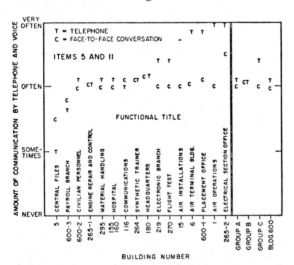

FIG. 8. Responses to Item 5 and Item 11 on amount of communication by telephone and by conversation.

TABLE II. First response to Item 10: What thing about your work would be most affected by noise ratings exceeding the desired rating?

Talking	43
Telephoning	44
General loss of productivity, increased errors due to fatigue, and "write-ins."	56
Total talking and telephoning—	
Listed as No. 1	87
Listed as No. 2	13

Item 9 asks the question, "What noise rating do you feel should not be exceeded in this room if you are to accomplish your duties without the loss of performance?" In most of the buildings, the answers fall into two categories. The first category includes those who make command (executive) decisions, and the second category, those who work as stenographers and clerks. The group that makes executive decisions indicated that the noise in their rooms should not exceed a rating of "moderately noisy" (4.0 on the 0 to 10 scale). Stenographers and clerks not required to make decisions seemed prepared to accept a condition that could be rated slightly higher than moderately noisy (4.3 on the 0 to 10 scale). A notable exception was in Building 600 where both groups voted for the same maximum noise ratings (4.0 in the 0 to 10 scale). These results also are different from those in the earlier study wherein the stenographers voted for a maximum noise rating of about 5.0 compared to 4.0 for the executives.

The results for Item 10 are given in Table II. Buildings 5 and 600–3 are omitted since, as Fig. 8 shows, speech communication is not very necessary in them. A large majority of the workers on the air base felt that noise levels above the desired maximum would, above all else, interfere with their ability to telephone and to talk. In fact, telephoning and talking were listed either first or second 100 times out of the 143 tabulated in Table II.

On the "write-in" blank in Item 10 a number of employees described disadvantages of excessive noise as "distraction that required them to re-add a column of figures," "general fatigue at the end of the day," "desire to escape to another building for a cup of coffee," "nervous irritability," "just want to get out of the noise for a while," and so forth. Occurrences of this sort, the workers felt, resulted in reduced ability to do their job.

The results for Items 12 and 13 concerning the necessary distances and maximum desired voice levels for speech communication are tabulated in Table III. In spaces where executive decisions are made, the workers felt that they should be able to converse at about 8 ft in a voice level between "normal" and "slightly raised." In Buildings 5 and 600, where filing,

TABLE III. Median answers to Item 12: What is the farthest distance over which you must talk?; and Item 13: At what voice levels should you be able to talk?

Buildings:	All but 600 and 5	600 and 5
Item 12	8 ft	4 ft
Item 13	Not more than slightly raised	

payroll, and personnel records constitute a large part of the activity, communication is necessary at 4 ft. It is probable that few of those answering the questionnaire could distinguish between a "normal (quiet) voice" and a "slightly raised voice" since over this range one adjusts his voice level according to his environment without knowing it. Hence, in Table III all scores below the middle of the scale have been lumped into one category.

The responses to Items 14 and 15 revealed that the civilian personnel on the base felt that they were neither unusually sensitive nor insensitive to noise, and the great majority assessed their hearing as "good."

VIII. RESPONSES TO ITEMS 1–3, 6–8, AND 16–17

This section deals with the responses to questions on the noise conditions of the past few months. Noise ratings and telephone effectiveness are correlated under average, loudest, and quietest conditions with the day-long noise level measurements. The causes of the loudest noises on the base were determined and the frequency with which these noises disturbed the work of the employees.

The noise in all the buildings fluctuated from minute to minute. The ratings given by the office personnel on the noise at the instant of questioning might have applied to the noise level at that instant or they might have applied to their memory of the noise levels during that morning or day. It was decided, therefore, for purposes of independent check, to make plots of the noise ratings as a function of both the speech interference levels and the loudness levels based on the day-long measurements of sound pressure level described in Sec. IV.

The graphic level recorder previously mentioned was located for a period of one full day in Buildings 166, 1, and 270, respectively. Building 166 was believed to be typical of those called "Group A" (see Fig. 3). In this group, the noise levels were on the average low, as can be seen from Table I. The noise became loudest when aircraft were flying over or warming up prior to takeoff at the end of a runway. Group B had higher noise levels due to internal factory operations, the use of business machines, or test block or aircraft operations around a freight depot (Building 6). The readings taken on the first floor of Building 1 with a window partly open are believed to be typical of Group B. Group C included areas affected by nearby aircraft runups following repair, or nearby test block opera-

tions. The readings taken in Building 270 are believed to be typical of this group. As can be seen from Table I, the median speech interference levels in Group C buildings at the time the questionnaire was administered were about 20 db higher than those in Group A, and about 5 db higher than those in Group B.

Even though Building 600 seems to belong in Group B, it has been singled out for particular study because approximately 65 questionnaires were handed out in this space alone. Within Building 600 there were, at the time of measurement, three different noise conditions that provide some interesting clues to the total noise problem. In the "past few months" averages there were no differences. The second floor of Building 1 was included in Group A because the noise conditions there were reasonably good in winter months when the windows were closed, and were not nearly so bad as those measured in the operations room on the first floor when the windows were open.

The results of the day-long records of the 300–600 band are shown in summary in Figs. 9–11. The shaded regions at the bottom of the graphs give the fluctuating variations from instant to instant in the offices, as determined during each half-hour period. The lines labeled "activity level" relate to the frequency of occurrence of the intermittent sounds during each half-hour. For example, on Fig. 9 between 1300 and 1330 hours, the recorded noise level in the 300 to 600 band was above 85 db twice; above 80 db four times; above 72 db eight times; and above 64 db sixteen times. The shapes of the contours have no meaning. Only the point located midway in the half-hour is significant.

The graphs show also the sources of the louder noises and the duration of those noises that persisted for a significant length of time.

It must be remembered that these noise levels were measured during the latter part of March in weather that ranged from 25°F to 50°F so that the windows in

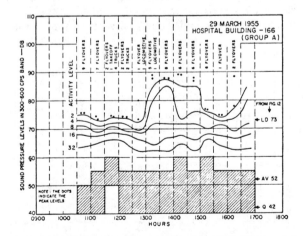

FIG. 9. Day-long sound pressure levels in 300–600 band in hospital building as taken from a graphic level record. The shaded region at the bottom indicates the general range of levels in the office. The contours are described in the text.

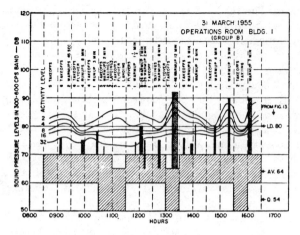

FIG. 10. Same as Fig. 9 except for operations room on first floor of Building 1. The cross-hatched regions indicate steady noises that came on for a time duration equal to the width of the vertical bars.

the buildings were usually closed. In summer months with the windows open, the levels of the noise originating outdoors in many of the buildings would be higher by perhaps 4 to 6 db in the lower three octave bands and 7 to 10 db in the upper five octave bands.

The original graphic level recordings were analyzed to yield a statistical description of the varying levels shown in Figs. 9–11. It was found that the noise levels could probably be divided into classes by 5-db steps, that is to say, 50 to 54.9, 55 to 59.9, etc. The sampling plan adopted consisted of obtaining momentary values of SPL in 5 db classes from the graphic level recordings. From a randomly selected starting point at the beginning of the first hour, values were drawn at regular two-minute intervals throughout a nine-hour period for each of the three buildings. This systematic sample seemed valid since the intermittent noises appeared to be randomly distributed in time and since the interval of the two minutes was long compared to the length of the short-time intermittent noises. Analysis of the thirty sample values for each hour showed that the samples

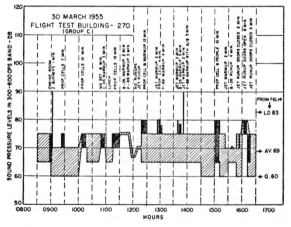

FIG. 11. Same as Figs. 9 and 10 except for flight test building.

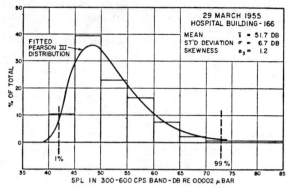

FIG. 12. Histogram and fitted Pearson III distribution curve for Group A noise.

for the entire nine-hour period in any one building could be considered to belong to a single population. The samples for all nine hours were then combined into a single day-long sample for each of the three buildings.

The histograms for the three day-long samples are shown in Figs. 12 to 14. These represent the actual

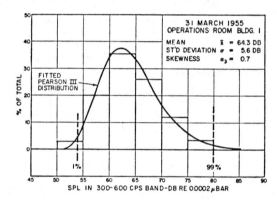

FIG. 13. Histogram and fitted Pearson III distribution curve for Group B noise.

frequency of occurrence of each of the 5 db classes of sound pressure levels on the sample drawn from the nine-hour recording.

A Pearson Type III distribution was found to provide a good empirical description of the distribution of

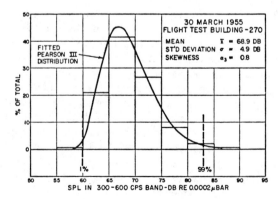

FIG. 14. Histogram and fitted Pearson III distribution curve for Group C noise.

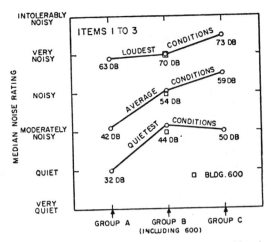

FIG. 15. Median noise ratings for Groups A–C. The decibel values are speech interference levels derived from day-long records.

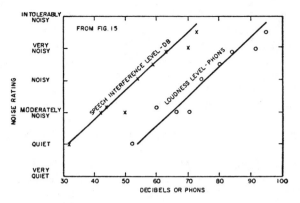

FIG. 17. Graphs of the medians of the responses to Items 1–3 plotted against the measured day-long speech interference levels and the calculated day-long loudness levels. The left-hand curve is obtained directly from Fig. 15. The loudness levels for the right-hand curve were obtained using the average spectrum shapes for Groups A–C given in Fig. 3.

sound pressure levels for these data.[9] This distribution uses three parameters, the *mean* (average), the *standard deviation* (root-mean-square value of the deviation from the mean), and the skewness (lack of symmetry between the upper and lower tails of the distribution). These parameters were computed for each of the day-long samples and are given on the figures. The solid-line distribution curves of Figs. 12 to 14 were plotted for these parameters using published tables for the Pearson Type III function.[10] From the published tables, it is possible to determine the abscissa (sound pressure level) corresponding to any percentage of the total area under the curve. Limits of 98% probability (1% and 99% are the limits) are shown by the dashed vertical lines on the curves of Figs. 12 to 14. To clarify further,

the 98% area limits are exceeded for 5.4 minutes in each nine-hour day at the upper end and for the same length of time at the lower end. It seems reasonable to assume that these limits roughly correspond to the loudest levels and the quietest levels the personnel had in mind when they answered Items 2 and 3 of the questionnaire. These values are shown as "Q," "AV," and "LD" on the right-hand sides of Figs. 9 to 11.

Representative spectra were determined from the samples recorded on magnetic tape of the background noise in these three buildings. If it is assumed that the SIL is distributed in a manner similar to the distribution found for the 300–600 cps band on the graphic level recordings, it is possible to convert to SIL by subtracting 10 db from the readings for the 300–600 cps band.

Figure 15 presents the median ratings of quietest, average, and loudest noise levels (Items 1–3) for the three groups of buildings (see Table I). The results for Building 600 are plotted as squares along with those of Group B. It is seen that for the buildings of Group A the noise ratings range from "quiet" to "very noisy" with the average conditions being rated "moderately noisy." For Groups B and C the conditions range from

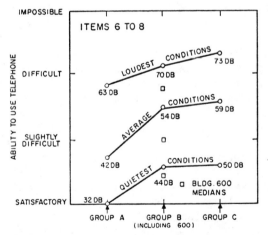

FIG. 16. Median ratings for ability to use the telephone for Groups A–C. The decibel values are speech interference levels derived from day-long records.

[9] M. G. Kendall, *The Advanced Theory of Statistics* (Hafner Publishing Company, New York, 1952), fifth edition, Vol. 1.
[10] *Statistical Tables* (Carver-Salvosa Tables of Pearson's Type III Frequency Function), edited by Harry C. Carver (Edwards Brothers, Inc., Ann Arbor, 1940). [Tables also appeared in Ann. Math. Statistics I (1930)].

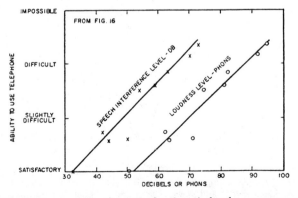

FIG. 18. Same as Fig. 17 except the above is for the responses to Items 6–8. The left-hand curve is taken directly from Fig. 16.

"moderately noisy" to more than "very noisy," with the average conditions being rated slightly more than "noisy." The SIL numbers are written beside the points.

A similar plot is shown in Fig. 16 for ability to use the telephone. The data for Fig. 16 were derived from the answers to Items 6, 7, and 8. In group A and Building 600, telephone conversation never becomes difficult, whereas in Groups B and C, telephone conversation becomes difficult when the noise is at its loudest.

The speech interference levels on Figs. 15 and 16 are derived by subtracting 10 db from the 300–600 cps "Ld," "Av," and "Q" levels in Figs. 9 to 11.

In order to derive a relation between noise rating and speech interference level from Fig. 15, the noise ratings for each of the nine points on Fig. 15 are plotted as a function of the mean of the nine SIL's given on the figure. This plot is shown by the nine crosses on Fig. 17.

A similar graph is presented in Fig. 17 (circles) for noise ratings plotted against loudness levels in phons. The loudness levels are calculated for spectra of the four average shapes given in Figs. 3 and 4. The absolute levels are determined by means of the SIL's on Fig. 15.

By the same means, the information on use of telephone (Fig. 16) is replotted as the crosses and circles on Fig. 18.

Comparisons are presented in Fig. 19 of: Fig. 17 (from Items 1 through 3 and the day-long measurements of noise level); Fig. 6 (from Item 4 and the immediate noise measurements); and the rating scale previously published (see reference 3, Fig. 13.24, p. 424). The agreement of the two air base curves with the results for the executive group in commercial offices is encouraging. Apparently, when asked to rate noises, office personnel in widely separated locations and in different types of organizations have in mind nearly the same noise levels when they make their noise ratings.

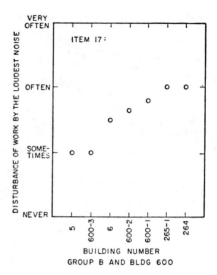

FIG. 20. Responses to Item 17, "How often does the loudest noise disturb your work?" Data for Group B and Building 600.

The answer to Item 17, "How often does the loudest noise disturb your work?" for the various buildings and and building groups is found in Table I. Building 600, as was found on other graphs, behaves as a member of Group B. The Group B and Building 600 data are tabulated in Fig. 20 for later reference.

IX. SPEECH INTERFERENCE LEVEL AND LOUDNESS LEVEL AS THE BASIS FOR NOISE CONTROL CRITERIA

In the previous sections we have chosen two variables against which to plot the ratings: The speech interference level (SIL), defined as the average of the sound pressure levels in the three octave bands, 600–1200, 1200–2400, and 2400–4800 cps; and the loudness level (LL) computed from octave band data by Stevens' method.

Which of these variables is the most valid measure for the specification of acceptable noise levels? This question is not easy to answer, for it is possible to justify either LL or SIL on the basis of the studies reported here. Because the noise spectra of this study had more or less the same shape (see Figs. 3 and 4), we cannot say how well the ratings would correlate with either of these measures in the presence of other types of noise spectra. Furthermore, there is no way to be certain of what each person has in mind when he places check marks on the rating scales of Items 1 to 4.

The chief justification for believing that loudness level forms at least a part of the basis for judgment is that the plots relating subjective ratings to loudness level are linear and the scatter of the points is small. This is an interesting empirical fact. Since these rating scales tend to show a linear relation to *loudness level*, they cannot, of course, show a linear relation to *loudness* as we measure it in sones—for loudness level in phons is roughly a logarithmic function of loudness in sones. The

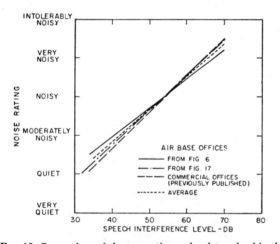

FIG. 19. Comparison of the two rating scales determined in this survey from Figs. 6 and 17 with the rating scale determined from the commercial office survey of references 1–3. The "average" line is used in later graphs.

nonlinearity between noise ratings and computed loudness in sones is shown in Fig. 21. Apparently the adjectives used on the rating scales correspond to points that are more nearly equidistant from one another in terms of phons than in terms of sones. But even this may be a coincidence, for it seems likely that the subjective ratings in terms of the adjectives used in the questionnaire are based on a complex judgmental process into which a number of factors enter. The judgments are probably compounded of reactions to the annoying and irritating aspects of the noise, especially to the interference produced on voice communication, and to the high levels in the remainder of the spectrum.

On the other hand, the fact that the ratings obtained by the questionnaire are related more linearly to loudness level (phons) than to loudness (sones) does not rule out the possibility that in rating the noisiness of their environments the subjects are essentially judging loudness as such. It appears that when they rate the noise in terms of descriptive categories the subjects tend to make judgments that are more relative than absolute. Thus, they place a moderate noise in the "moderate" category and they do not move their rating up to the next category until the loudness is increased by some given percentage. If this percentage is roughly constant from category to category, one can expect the ratings to show roughly a linear relation to loudness level as in Fig. 7, and a logarithmic relation to loudness as in Fig. 21. It has in fact been shown that rating scale judgments of such things as loudness, brightness, duration, heaviness, etc., are always nonlinear relative to the scale of subjective magnitude. Work on this problem by S. S. Stevens and E. H. Galanter is still in progress, but a preliminary report§ of it was made before the International Congress on Acoustics in June, 1956.

Speech interference level (SIL) is the other important variable against which one can plot the noise ratings. The SIL was chosen as the independent variable in Fig. 6 because the majority of the office workers questioned felt that the noise affects their ability to communicate by voice more than it affects anything else. This is evident from Table II and Fig. 8. The SIL is a measure of the interfering effect of noise on speech because calculated *articulation index*[3] is directly proportional to the difference between the decibel levels of the peaks of speech in the 300–6000 cps region and the decibel levels of the noise.

As a further indication that interference with voice communication is high in the minds of office personnel when they rate noises is the fact that those in Buildings 5 and 600–3 who do not communicate often by voice rate the noises below the average curve for the other buildings (see Fig. 6). Personnel in these two buildings

also stated that their work was disturbed substantially less often than did the other workers in Group B and 600–1 and 600–2. (Compare Fig. 20 with Fig. 8.)

Another significant indication that voice communication is important is that office personnel are consistent in (a) their choice of a noise level that should not be exceeded if their work is not to be affected by the noise, and (b) the distance and voice level at which they need to communicate. We learned in Sec. VII that the administrators, engineers, and secretaries desire SIL's that do not exceed about 40 db (Table I and Fig. 6). That level is about 10 db below the levels that will just barely permit person-to-person communication at 9 ft (see reference 3, p. 420). Eight feet is the distance over which office personnel believe they should be able to communicate in no more than a slightly raised voice. In other words, office workers seem to desire easy, reliable communication by voice at eight feet.

In order to check the plausibility of the responses to Items 12 and 13 concerning voice levels and distance and their relation to the responses to Item 9, which seeks the maximum desirable noise levels, a test was set up at a frequently used conference table in Cambridge, Massachusetts (hereafter called the Cambridge Test). The table was about 3 ft by 8 ft in size and could easily accommodate about six people. A loudspeaker was placed against the ceiling and a random noise spectrum was tailored to the same shape as the average of Group A (Fig. 3). At different times the SIL's near the heads of those around the table was varied over a range from 20 to 60 db. After a week of testing, those using the conference table (about 50 different people) concluded that, for no disturbance whatsoever, the SIL had to be held below 30 db. An SIL of 35 db gave only slight disturbance but was irritating. An SIL of 40 db was definitely the loudest that anyone believed permissible without causing loss of performance. Sizable disturbances occurred when the SIL reached 50 db. Sentence articulation at an SIL of 50 db was fairly high but listening was very difficult, demanding one's full attention. It was like trying to read very fine print in

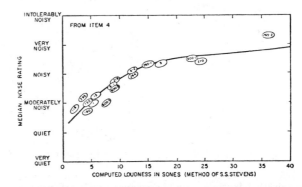

FIG. 21. Plot of the median ratings for the different buildings against the median computed loudness in sones. The ordinate and abscissa were obtained as for Fig. 7.

§ S. S. Stevens, "The calculation of the loudness of noise," Proc. Second Internatl. Congr. Acoustics, 1956 (to be published).

dim light. These results confirm the conclusion that an SIL above 35–40 db is to be avoided.

Another fact related to voice communication can be seen in the data for the three rooms of Building 600 (Table I). A comparison of Items 5 and 11 with Item 17 reveals that the noise is disturbing to the work of the people in Building 600 nearly in direct relation to the amount of voice communication that is necessary.

This evidence clearly indicates that office personnel who must communicate on the job are keenly aware that noise is deleterious to speech intelligibility. Since a large percentage of the personnel must communicate "often" to "very often," a prime requisite is that the SIL be held to a proper value. It should not exceed 40 db for continuous noises.

In addition to being able to communicate easily, office personnel seem also to desire a loudness level in phons that does not exceed the speech interference level in decibels by more than about 22 units. This statement is based on the evidence given in Figs. 6 and 7 for Buildings 600–2 and 265–2. For Building 600–2, (LL−SIL)=32 units and for Building 265–2, (LL−SIL)=27 units. In both buildings, for a given ease of speech intelligibility, the rating is higher than would be expected from the SIL alone by about one-

half a rating category, and, particularly in Building 265–2, complaints were vigorous.

The summary curves resulting from this study are given by the two solid lines in Fig. 22. The SIL line was taken arbitrarily as the mean of the three lines on Fig. 19. The loudness level line is drawn 22 db to the right of it. The relation between difficulty of telephoning and SIL is shown along the top of the graph (taken from Fig. 18).

Return now to the question of loudness level *vs* SIL as a basis for a criterion of acceptable noise in offices. It has been seen that both these measures work reasonably well for noises such as those encountered in this survey, but the question remains whether one or the other measure would prove the better basis for a criterion if other kinds of spectra were involved. This question cannot be answered in terms of direct empirical test, but a relevant aspect of the question can be explored by posing another question: How well do the two measures correlate? Obviously, if loudness level and SIL correlate perfectly, regardless of the noise spectrum being considered, it is a matter of indifference which is used.

One way in which the degree of this correlation can be tested is by means of the data obtained by

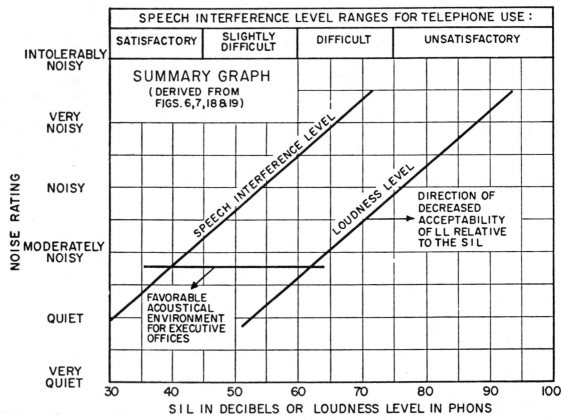

FIG. 22. Summary of findings from this survey for continuous noises. The solid lines relate the noise rating to the SIL and loudness level. A favorable acoustical environment for office personnel who must communicate often to very often by voice is one for which the SIL is below 40 db and the LL is below 62 phons. For a balanced spectra, the LL should preferably not exceed the SIL by more **than 22 units.** The SIL's yielding various qualities of telephone use are found by comparing the bottom scale with the scale above the graph.

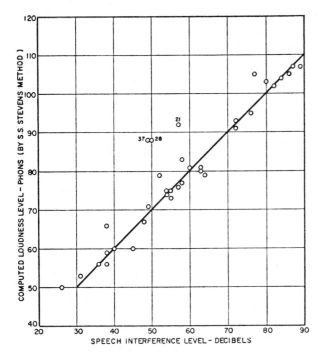

FIG. 23. Comparison of computed loudness level and SIL for 35 noises studied by Quietzsch (courtesy of S. S. Stevens).

Quietzsch,[11] who studied the loudness of 37 different noises representing a wide variety of spectra. For Quietzsch's noises, Professor Stevens computed the loudness levels. The author translated the levels in Quietzsch's bands (400–800, 800–1600, and 1600–3200 cps) into SIL bands (600–1200, 1200–2400, 2400–4800 cps). Two noises that Quietzsch obtained are omitted from this study because of incomplete data.

Figure 23 shows the relation between computed loudness level and SIL for the 35 different noises. For most of these noises the correlation between the two measures is obviously very high, the difference being about 20 units, but there are some marked discrepancies. Three of these points are labeled with the identifying numbers used by Quietzsch. The two points farthest from the line in Fig. 23 represent an electric bell (37) and the hiss of compressed air escaping from a small nozzle (28). Both noises have most of their energy in the high-frequency end of the spectrum. Noise No. 21 represents the exhaust noise of an airplane engine and has a high concentration of energy in the two lowest octave bands. Now the question is, if one were concerned with noises like these three (21, 28, and 37), should one specify a criterion for noise control in terms of loudness level or of SIL?

Evidence to answer this question is available in Bolt Beranek and Newman Inc. files. Figure 24 shows the noise situations in seven industrial offices and conference rooms that were judged unsatisfactory and in which

[11] G. Quietzsch, Akust. Beih. 1, 49–66 (1955). Bound in Acustica 5, 49–66 (1955).

the owners demanded corrective measures. The reduced noise levels that resulted from a change in the noise source are also shown on each graph. Comments of the users of these rooms, recorded by the engineers who handled their complaints, and observations by the engineers themselves, are given in Table IV. The results confirm those obtained during the air base survey. In those cases where the SIL was satisfactory [rooms (c), (d), and (e)] complaints were registered when the loudness level exceeded the SIL by 30 or more units. Even where the SIL was high and changed very little after the noise was altered [room (a)], comfort was increased significantly by reducing the low frequencies until LL minus SIL equaled only 23 units.

Coming back to the three deviant noises of Quietzsch mentioned above, it seems clear that if the criterion for speech were an SIL of 50 db, these noises would

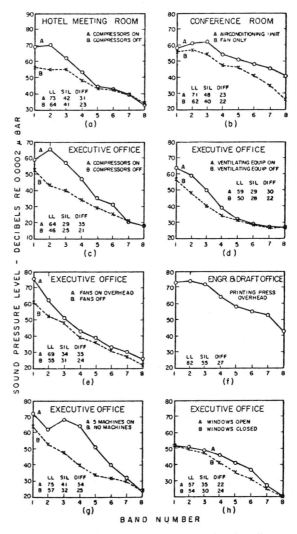

FIG. 24. Sound pressure levels vs band number for offices and rooms before (A) and after (B) changes in the noise. The eight bands, respectively, are 20–75, 75–150, 150–300, 300–600, 600–1200, 1200–2400, 2400–4800, and 4800–9600 cps. The calculated loudness levels and SIL's and their differences are given on the graphs.

TABLE IV. Comments of the users of rooms of Fig. 24 and the noise control engineers called in to correct the noise situation.

Room	Condition	"A" indicates the condition prior to correction. "B" indicates the condition after correction. Users' comments	Engineers' comments
(a)	A	Thought the noise intolerable.	Too noisy to be comfortable. Could hear reasonably well.
(a)	B	Background noise did not seem to be a problem.	Quite satisfactory. Should not be objectionable.
(b)	A	Conference of people seated 8 ft apart very difficult. Would rather turn unit off and be hot than not hear.	Difficult to hear at 10 ft. Not too hard to understand at 3 ft but noise omnipresent.
(b)	B	Just adequate for conference.	Noise situation better, but still too high for easy conference.
(c)	A	Noise levels absolutely unsatisfactory. Must be reduced.	Speech at 8 ft was not difficult. Low-frequency noise very annoying.
(c)	B	Satisfactory noise levels. Very pleased with improvement.	Speech situation not greatly improved, but the office now seemed pleasant.
(d)	A	Noise levels unsatisfactory. Desired correction.	Speech easy. Low-frequency noise annoying.
(d)	B	Noise levels satisfactory.	Speech just as easy as before. Noise seemed greatly reduced.
(e)	A	Noise levels very unsatisfactory.	Low-frequency noise very annoying.
(e)	B	Noise situation improved, but still not quite good enough.	
(f)		Noise situation terrible. Complained of fatigue and irritation.	Speech communication difficult. Noise sounded bad at all frequencies.
(g)	A	Don't like this noise. Very uncomfortable for conferences.	Difficult to confer at 10 ft. Low-frequency noise very annoying.
(g)	B	Excellent.	Low-frequency noise noticeable.
(h)	A	Conferences bothered sometimes.	Conferences bothered sometimes.
(h)	B	Conferences satisfactory.	Conferences satisfactory.

probably turn out to be objectionable, for their loudness levels would run about 40 units higher.

It must be concluded, therefore, that although SIL may be a necessary measure of satisfactory speech communication, there are some noises for which loudness level must also be considered. In the present state of knowledge, the best rule would appear to be to state the permissible noise in such a way that neither the SIL nor the loudness level exceeds its proper value. Thus two criteria are needed, neither of which should be exceeded. It should also be noted that there is little evidence from these studies as to the importance of the 4800–9600 cps band in setting he permissible loudness level, as all spectra except Nos. 37 and 28 of Quietzsch sloped downward with increasing frequency.

As an addendum, refer back to Fig. 2. It is seen that accounting machines and typewriters generally have SIL's between 50 and 60 db. It has just been stated that an SIL of more than 50 db is quite undesirable in executive offices, yet typewriters are often permitted in offices, even where conferences are held. The reason, as judged from direct observation, is that in many offices typewriter operation is intermittent and can be held up temporarily while other people are holding a critical conference or are trying to use the telephone.

The reader may have some trouble in matching exactly SIL numbers from Table I with graphs occurring after Fig. 19. This difficulty arises from the fact that the mean (see Fig. 19) of the three SIL *vs* rating lines was used on subsequent graphs.

X. CRITERION CURVES AND CHARTS

Using the information of Fig. 22 as a basis, we derive in Fig. 25 a set of spectrum curves (solid lines) for which the speech interference levels and loudness levels differ by about 22 units. The *SC* designations indicate the SIL's for the various curves. A spectrum lying in the region SC-30 to SC-40 is recommended as the permissible shape and level for noise in offices. It is believed that the spectra depicted by the solid lines correspond to the type of spectrum balance most often found in man-made noises (see Figs. 3, 4, 23, and 24).

The dashed "A" curves represent spectra for which the SIL and the loudness levels differ by about 30 units. Differences of more than 30 units often lead to undesirable noise environments. The region between the solid line and the dashed *A* portion of each curve encompasses a range of deviations that might be tolerated in situations where economic or other factors dictate a less rigid criterion.

The curves of Fig. 25 are suggested for trial as replacements for those given in reference 3, Fig. 13.29, p. 428.

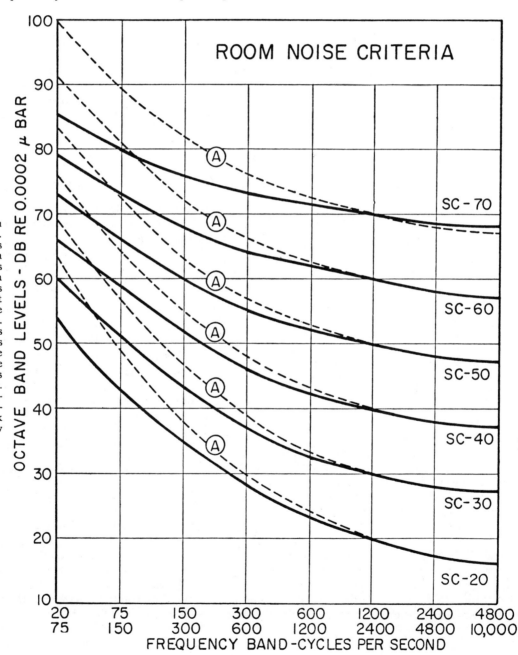

FIG. 25. Room noise criteria spectra. The SC numbers designate the speech interference levels for the curves. Each solid line curve has a loudness level that is 22 units above the SIL for that curve. The dashed A curves have a loudness level—SIL difference of 30 units. These curves are offered as replacements for Fig. 13.29, p. 428, Beranek, *Acoustics* (McGraw-Hill Book Company, Inc., New York, 1954).

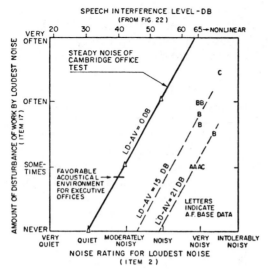

SPEECH INTERFERENCE LEVEL -DB
(FROM FIG 22)

FIG. 26. Summary of information obtained during this air base survey pertinent to the intermittent noise problem. This graph shows that for an intermittent noise to be equally as disturbing to office employees as a steady noise, it has the same average (in decibels) level determined from decibel readings samples every two minutes throughout the day.

Inherent in the derivation of the curves of Fig. 25 is the assumption that the noise is nearly steady. When the noise is intermittent and the loudest noises are of short duration, e.g., aircraft flyovers, the loudest noise may be permitted to exceed the sound pressure levels indicated by these curves without creating greater interference with the work in offices. Some evidence of the amount of deviation permissible is available from this study.

To explore this information, we refer first to Item 17 of the questionnaire where it was asked, "How often does the loudest noise disturb your work?" Also, under Item 2, it was asked, "How noisy is this room under the loudest conditions?" In Sec. VII of this paper it was decided to call the "loudest noise" that level that was exceeded 1% of the time (0.6 min of each hour). In a "steady" noise the loudest noise level and the average noise level differ very little. From the Cambridge office test described in Sec. IX, we had found the amount of disturbance of work in an executive conference room as related to the SIL of a steady noise. Finally from Fig. 22, a relation between SIL and noise ratings is available. From all of this information, Fig. 26 has been prepared.

The solid diagonal line on the left is from the Cambridge office test where the noise was steady (LD−AV =0 db). The two dashed lines are for Group A and Groups B−C, respectively, where the noise was intermittent. Referring to the scale along the upper edge of the graph, it is seen that the dashed lines are separated from the solid line by about the LD minus AV difference. From this observation, one concludes for an intermittent noise, e.g., aircraft flying overhead or warming up, that for equal disturbance the "average" SIL is approximately equal to the SIL of a steady noise.

TABLE V. Executive office noise criteria relating to Fig. 25.

SC curve of Fig. 25	Office condition
Below 25	Often judged too quiet because distant faint noises became distracting
25	Excellent
30–35	Good
40	Maximum permissible levels without interference with duties. Conferences around 7 ft table just satisfactory.
50–60	Telephone usage slightly difficult, particularly on long distance calls. Conferences at 3 ft just satisfactory. Satisfactory for secretarial typing pools where little voice communication is required.
60–75	Telephone use difficult. Bad environment.
Above 75	Telephone usage unsatisfactory.

Within the limits of our present knowledge, this statement applies to the loudness level in phons.

XI. CONCLUSIONS

Although considerable data have been amassed in this paper and during previous surveys, the author intends to continue studies of this type. In particular, the day-long noise measurements and the long-time noise ratings associated with them will be extended.

In executive office design, the criteria numbers shown in Table V are believed applicable. These criteria are given as SC numbers and relate to the complete noise spectra of Fig. 25. The levels shown on Fig. 25 apply to continuous noises. For intermittent noises, they are also believed to apply to the simple average of the decibel readings sampled every two minutes throughout the day.

In stenographic (typist) pools, it has been found that an SC criterion of 55 is judged satisfactory.[1-3] Present-day typewriters produce levels of this magnitude in offices with acoustically treated ceilings.

XII. ACKNOWLEDGMENTS

The author wishes to express his appreciation to K. N. Stevens and R. H. Bolt who helped administer the questionnaires; to D. Kyrazis, J. W. Richardson, and A. McConnell who obtained the day-long graphic level recordings; to W. Clark and S. Lukasik who calculated many of the statistical parameters; and to W. A. Rosenblith, K. N. Stevens, W. J. Galloway, A. C. Pietrasanta, and S. S. Stevens who read the manuscript critically. He is especially grateful to S. S. Stevens of Harvard University who supplied the information of Fig. 23 and the methods for calculating loudness level and to Geraldine Stone who lent an editorial hand to the preparation of the report. He also wishes to thank Dr. H. O. Parrack and Dr. H. E. von Gierke of the Aero Medical Laboratory, Wright Air Development Center, Wright-Patterson Air Force Base, Ohio, for encouraging the study and for helpful comments throughout.

APPENDIX I

Some Statistics from Building 600

The principal reason for passing out 65 questionnaires in one building, as compared to a maximum of 17 in any other, was to gather some valid information on the reliability of the data. Building 600 is called "Civilian Personnel." Here people are interviewed for jobs, payrolls are made up, and pay checks are written. All personnel problems of the civilian employees are handled in this building.

A further reason for choosing this building for study was its proximity to the test block for reciprocating engines. The cells in this test block were in nearly constant operation with from one to six engines running at a time.

Finally, Building 600 houses three types of activities. Most of the workers in Room 1 are men and their job is to recruit, interview, and hire personnel, Room 2 contains mostly women with executive and stenographic duties. Room 3 houses the payroll and bookkeeping group, nearly all women. Questionnaires were handed out to 19, 21, and 25 people, respectively, in the three rooms.

From Table I we see that for Building 600, the ratings of the noise on Items 1–3 and the rating of ability to use the telephone in Items 6–8 were substantially alike in the three rooms. The personnel in the three rooms agreed also on what they believed should be the maximum rating if their work was not to be affected. On most other matters, however, the three rooms differed in their ratings.

Histogram plots of the responses to Items 1–4, 9, and 17 of the questionnaire are given in Fig. 27. The medians and the upper and lower quartile divisions are marked on each. The character of the histograms is encouraging, as most will agree who have had experience with public opinion polls.

Items 4 and 17 are plotted separately for each of the three groups in Building 600. It was seen from Fig. 8 that the payroll group (600–3) uses the telephone least often and that the interviewers (600–1) use it most often. The need to converse is about the same in the three groups. Now refer to Item 17 of Fig. 27. The interviewers (600–1) who communicate by telephone "very often," state that their work is disturbed more often by the loudest noise than do the other two groups. This is especially true in comparison with the payroll group (600–3) who communicate by telephone much less often (see Fig. 8). The greatest spread between the

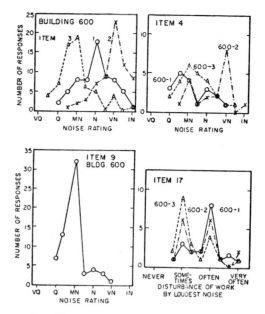

FIG. 27. Histograms for responses in building 600 to items 1–4, 9, and 17.

quartile points occurs for the data on Item 4, "Is the noise at this instant ···?" This variability was not unexpected for this building. In the first place, the noise levels were constantly changing as engines were stopped and started in the test block, so that the measurements were not necessarily made at the instant the person was considering Item 4. Second, some subjects may have interpreted the word "instant" as meaning "this morning" or "today." The noise that was measured at a particular time may not have been characteristic of the period the subject had in mind. Third, whereas voice communication is considered the most important thing to be affected by the noise, the subject has no opportunity to sample his communication ability at the time he is filling out the questionnaire.

The results for Room 1 of Building 600 require some explanation since the datum points for 600–1 stand out as mavericks on many of the graphs. Rooms 1 and 2 were usually exposed to nearly the same noise levels from the nearby test blocks. Near the end of the month, the number of engines operated per day is very large with the result that the high noise levels are almost continuous. Rooms 2 and 3 were surveyed on the last day of the month. The first day of the following month is usually a light day, and, in fact, activities had almost ceased in the test block when Room 1 was surveyed on April 1. The contrast with the noise of the previous days seemed to cause people in Room 1 to underrate the remaining noise in their offices. In fact, nearly every person who filled out a questionnaire on April 1 in this building remarked that the noise on that day was nothing and that the questionnaires should have been handed out earlier that week.

APPENDIX II

Suggested Revisions for the Questionnaire

The results themselves and observations made by those who have read a draft of this paper lead to the following suggestions for improvement of the questionnaire:

Item 4: Use the same rating scale as that used for Items 1 through 3 so as to avoid the possibility of nonlinearity in the transfer of ratings from the scale of Item 4 to the scale of Items 1 to 3.

Items 5, 11, and 17: Use a scale of the same length as that used for Item 1. The number of adjectival cues might be increased to include: *Never, Sometimes, Moderately Often, Often, Very Often, Continually.*

Item 13: Use a scale of the same length as that used for Item 1. The number of adjectival cues might be increased to include: *Whisper, Normal (Quiet) Voice, Slightly Raised Voice, Raised Voice, Moderately Loud Voice, Loud Voice, Very Loud Voice, Shouting.*

Item 10: Add more items such as: Necessity of redoing work because of distraction.

Part V

INDUSTRIAL MACHINERY NOISE
SOURCES AND CONTROL

Editor's Comments
on Papers 18, 19, and 20

18   BARR
*Enquiry into the Effects of Loud Sounds upon the Hearing of
Boilermakers and others who work amid noisy surroundings*

19   SABINE and WILSON
The Application of Sound Absorption to Factory Noise Problems

20   KARPLUS and BONVALLET
A Noise Survey of Manufacturing Industries

In discussing the topic of industrial and machinery noise it is useful to review briefly the use of machinery prior to the Industrial Revolution at the end of the eighteenth century and the rapid increase in use since.

The use of simple machinery dates back at least to the time of the Greeks and Romans [1, 2, 3, 4, 5]. In the Egyptian city of Alexandria in the third and second centuries B.C., the wedge, screw, wheel, and axle, and level and pulley were well known. However, because of the ready availability of slave labor in Greek and Roman societies, most of the power supply was manpower, and there was little incentive to use animal or water power [1, 2, 3, 4]. One can assume that the available machinery was not very noisy.

By the tenth century A.D. first water wheels and then windmills increasingly came into use. At first they were used only for grinding wheat, but by the eleventh and twelfth centuries A.D., water-driven forge-hammers and forge bellows came into use. In the thirteenth century, pulping mills (for making paper) and mills for sawing and for grinding cutlery were introduced and by the fourteenth, stamping mills for ore and for wire-drawing were in use [2]. Throughout the fifteenth, sixteenth, and seventeenth centuries increasing use was made of water mills as prime movers, some generating as much as 20 horsepower. In the Netherlands in the sixteenth and seventeenth centuries, windmills were used for many industrial purposes (some developing about 14 horsepower). However, we can only speculate on how noisy all these processes were, since records seem to be scanty. We can assume that some of these machines, particularly those involving metal forging, cutting, or grinding, could have been quite noisy. Blacksmiths were known to suffer from permanent hearing loss, as Fosbroke (a medical doctor) wrote in the Lancet in 1831:

> The blacksmiths' deafness is a consequence of their employment; it creeps on them gradually, in general at about forty or fifty years of age. At first the patient is insensible of weak impressions of sound; the deafness increases

with a ringing and noise in the ears, slight vertigo, and pain in the cranial bones, periodically or otherwise, and often violent. No wax is formed. It has been imputed to a paralytic state of the nerve, occasioned by the noise of forging, by certain modern writers, and by the old writers, to permanent over-tension of the membrane, which they compare to fixed dilation of the pupil [6].

A rapid increase in the use of steam-powered machinery occurred in Europe, particularly in Britain, in the late eighteenth century. In the 1770s and 1780s, the cotton industry in Lancashire and Yorkshire became mechanized. Production increased rapidly as cottage industry was replaced by factory manufacture. Cotton textiles accounted for only about 4 to 5 percent of British exports in 1802, while by 1815 they accounted for 40 percent and employed 100,000 workers in spinning factories and another 250,000 as weavers [8]. At first the machinery was powered by water, but after the invention of the steam engine by Watt in the 1770s, by steam engine.

Working conditions in the cotton industry then must have been very noisy, as they still are today. Conditions in Manchester, England in 1834 were described as follows: ". . . no man would like to work in a power-loom, they do not like it, there is such a clattering and noise it would make some men mad . . . " [7].

The British iron industry was revolutionized by the steam engine in 1775. It had an immense and immediate importance in the production of iron in the furnaces, hammers, and mills but also in the cheaper and better production of coal and ore that steam pumps in mines made accessible [8]. The steam hammer introduced by John Wilkinson in 1782 could strike 150 blows per minute. The availability of increasing quantities of cheaper iron and steel made possible all sorts of industrial production in Britain and later in the United States, France, and Germany. We have no reliable records of the noise conditions in such industries until late in the nineteenth century, but obviously the conditions were bad. Several nineteenth-century writers mention "boilermaker's deafness." In the United States, Holt discussed this problem in a paper read before the American Otological Society in 1882 [9]. Holt studied forty men from the steam boiler–making shops in Portland, Maine. In order to test their hearing he measured the distance at which each man could just hear the tick of his stopwatch and found that all the men suffered from some degree of deafness. He also found that the deafer ear was always on the side exposed to the noise and that the deafer men were those who had been employed longest in the construction work.

In 1886 Thomas Barr, surgeon to the Glasgow Hospital for Diseases of the Ear, presented a paper dealing with the effects of loud noise on hearing [Paper 18]. In 1890 he read another paper [10]. Barr's work was more extensive than Holt's. He studied 100 boilermakers, 100 iron-moulders, 100 letter carriers, and 100 men with normal hearing. He tested their hearing of a ticking watch, understanding of speech, and hearing of sounds of different pitch using an adjustable-pitch Galton's whistle. By using tuning forks and bone conduction tests in addition to the previous experiments, Barr correctly concluded that the noise damage was caused mostly at high frequencies and occurred in the inner ear (cochlea). It is surprising that with such simple equipment and procedures Barr came to the correct conclusions. One of Barr's concerns was that deafness would prevent workers from hearing sermons in church. It should be noted that Holt, using similar tuning-fork tests, incorrectly concluded earlier that noise-induced damage occurred in the middle ear. Barr recommended the use of rubber plugs as ear protectors. Because his paper appears to be the first important contribution in English on industrial noise deafness, it is reprinted here as Paper 18.

The United States lagged behind Britain as an industrial power in 1860. However, by 1890 it had assumed first place, and the value of its manufactured goods almost equaled the combined value of Britain, Germany, and France [11]. Between 1860 and 1914 manufacturing production in the USA was multiplied by twelve times [11]. This surge in industry was partly a result of the mainly American contributions of use of interchangeable parts and the production line. It is surprising that despite the noisy conditions that must have existed in U.S. industry in the early 1900s, more concern was not expressed about it. One finds brief mention (for example, Ref. 12), but much more discussion appears in the popular literature on community noise in cities. Even when mention is made it is often concerned with loss of efficiency rather than hearing damage: "A large part of the noise in a manufacturing plant may be translated into loss of power, unnecessarily rapid depreciation in equipment, and a reduced efficiency of employees resulting from the distraction which is created and from the indirect effect upon physical health . . . " [12].

The two remaining papers in this part discuss noise conditions in U.S. industry in 1943 and 1953. In Paper 19 Sabine and Wilson discuss noise measurements made in thirty-three separate industrial plants in the early 1940s. Unfortunately they report only a few representative results in Table I. They state that the highest and lowest overall readings they obtained in industrial work areas were 130 and 65 dB, respectively, while in most cases the observed noise levels were in the range of 85 to 105 dB. Sabine and Wilson, interestingly, seem to share the main concern of many managers at that time that intense industrial noise caused many workers to leave their jobs, or to be absent or take brief "vacations." There seems to be little concern with the effects of noise on hearing. In fact, early in the paper it is stated that ". . . there is no known method of selecting only those people who can successfully condition themselves to withstand intense noise." One hopes that the authors are merely repeating a commonly held misconception of the time, since, as we now know, most people's ears cannot be made to withstand intense noise, and in such conditions hearing loss will surely occur over a period of time. In the remainder of the paper Sabine and Wilson discuss the use of absorbing materials in factory spaces and the phenomena of reverberation and the "spreading effect" of sound. The authors suggest that with the use of absorbing materials "very substantial relief from annoyance is possible even under conditions where noise meter readings showed no significant reduction in noise level." It is interesting to note that forty years later acousticians are still debating the fact that the use of absorbing materials in industry seems to be more effective than measurements would indicate, particularly in the case of impulsive noise.

Paper 20 by Karplus and Bonvallet again gives the results of a study of the noise environment in U.S. industry. This study was conducted ten years later than Paper 19, and the authors visited forty plants. The results are given in considerable detail, and this paper probably represents one of the earliest and most comprehensive surveys of industrial noise made. The results are still of use today. It is of interest to note here that the authors mention the effect of noise on hearing as a prime concern and that noise levels are presented in octave bands and also overall (or "flat"). Loudness in sones are also presented, a practice that was common in the 1950s. The use of sones has been declining in the last twenty years.

REFERENCES

1. Giedion, S., 1948, *Mechanization Takes Command,* Oxford University Press, New York.
2. Lilley, S., 1965, *Men, Machines and History,* International Publishers, New York.
3. Klenn, F., 1959, *A History of Western Technology,* Scribner's, New York.
4. Oliver, J. W., 1956, *History of American Technology,* Ronald Press, New York.
5. Rosenberg, N., 1972, *Technology and American Economic Growth,* Harper & Row, New York.
6. Fosbroke, J., 1831, Practical Observations on the Pathology and Treatment of Deafness, No. II, *Lancet,* **19:**645–648.
7. Select Committee on Hand-Loom Weaver's Petitions 1835, in E. P. Thompson, *The Making of the English Working Class,* Vintage Books, New York, 1966, p. 307.
8. Deane, P., 1965, *The First Industrial Revolution,* Cambridge University Press, Cambridge, England.
9. Holt, E. E., 1882, Boiler-Maker's Deafness and Hearing in Noise, *Am. Otological Soc. Trans.,* **3**(1):34–44.
10. Barr, T., 1890, Injurious Effects of Loud Sounds on the Hearing, *British Med. J.* **2:**675.
11. Blum, J. M., et al., 1963, *The National Experience: A History of the United States,* Harcourt Brace Jovanovich, New York, p. 420.
12. Anon., 1913, Noise and Factory Efficiency, *Sci. Am.* **76:**189.

18

Reprinted by permission from *R. Philos. Soc. Glasgow Proc.* **17**:223–239 (1886)

Enquiry into the effects of Loud Sounds upon the Hearing of Boilermakers and others who work amid noisy surroundings. By THOMAS BARR, M.D., Surgeon to Glasgow Ear Hospital; Lecturer on Aural Surgery, Anderson's College; and Dispensary Surgeon for Diseases of the Ear, Western Infirmary, Glasgow.

[Read before the Society, 3rd March, 1886.]

IT is familiarly known that boilermakers and others who work amid very noisy surroundings are extremely liable to dulness of hearing. In Glasgow we would have little difficulty in finding hundreds whose sense of hearing has thus been irremediably damaged by the noisy character of their work. We have, therefore, in our city ample materials at hand for the investigation of this subject.

In the process of boilermaking, as most of you are aware, four different classes of men are engaged—riveters, caulkers, platers, and "holders-on." The riveter drives in with a large hammer the red-hot iron rivets for binding the plates together; the caulker hammers with a chisel the edges of the plates so as to ensure complete tightness; the plater forms the iron plates and arranges them accurately in position; while the "holder-on," stands inside the boiler holding a large hammer, the head of which he presses against the inner end of a rivet. These are not all equally exposed to loud sounds, and they differ, therefore, in the extent to which their ears are affected. The men who work inside the boiler, such as the "holders-on," are, of course, exposed to the loudest and most damaging sounds. Their ears are near to the rivet which is being hammered in by the riveter outside. The iron on which they stand is vibrating intensely under the blows of perhaps twenty hammers wielded by twenty powerful men. Confined by the walls of the boiler, the waves of sound are vastly intensified, and strike the tympanum with appalling force, while the vibrations from the iron pass directly through the bodies of the men to the delicate nerve structures in the inner ear. If, in such circumstances, we venture into the interior of a boiler, our first impulse is to hurry out, or to stop our ears with our fingers. We are conscious not merely of the sound waves, like blows, producing their terrible effects upon our ears, exciting therein sharp, painful, intolerable sensations, but our bodies seem to be enveloped in invisible, yet tangible waves which we actually feel striking against our heads and our hands. When I underwent this experience, I fortunately furnished myself with a couple of India-rubber plugs, and by carefully withdrawing and inserting them in the canals of my ears, I was able at pleasure to admit or shut out the fearful sound. Let no one who values his hearing perform such an experiment without similar precautions. After such an experience one is surprised that the delicate mechanism in the interior of the ears can retain its integrity for a single day under the action of these blows of compressed air. In order to experience the full effects of the noises in boilermaking, one must ensconce himself in one of the smaller interior chambers such as a " superheater," or flue, where the air-space is still more confined, while the plates which are being hammered are thin, and therefore give forth notes not only intensely loud but extremely shrill. Even men whose hearing has been blunted by years of exposure to the sounds of boilermaking are, I am told, forced in such circumstances to protect their ears with cotton-waste or such-like stopping. Amid the overpowering din, communications have generally to be made by pantomimic gestures, and when the foreman wishes to attract the attention of the men, he employs a shrill whistle like a policeman's. When my conductor at one moment, in the loudest and shrillest voice, spoke closely into the passage of my ear, the effect was not that of spoken intelligible words but that of acute pain as the sharp tones pierced my ear. No doubt this necessity for occasionally speaking loudly close into the ear must tend also to injure the hearing of these men.

I had the curiosity to take a phonograph into the interior of a flue while the riveters and caulkers were hard at work without and within. The manipulation of the phonograph in that situation was a somewhat difficult proceeding. . The results were not very satisfactory, for, while quite distinct indentations were produced upon the tinfoil, the reproduction of the sound was not effective. The indentations were small and closely arranged, indicating the great height of pitch of the notes, and contrasting with the large widely separated indentations caused by the human voice.

My enquiry included the examination of 120 men who were.

employed in boilermakers' sheds, but 20 of these being labourers who were not constantly working at the trade, I have excluded them, and have based my conclusions only on the examination of the 100 men who were directly engaged in the process of boiler-making. For the facilities kindly granted to me in the course of my investigations, I have to express my gratitude to my friend Mr. Kinghorn of the London and Glasgow Shipbuilding Company, to Mr. Kirk of Messrs. R. Napier & Sons, and to Mr. Jeffrey of Messrs. J. & J. Thomson.

The 100 men examined represented all ages, from 17 years, the youngest, to 67, the oldest. The average age was 34·93. The most serious results were found, as might have been expected, in the older men. The following shows the relative numbers at the various periods of life :—

Under 20 years, -	-	-	-	10
From 20 to 30, -	-	-	-	27
„ 30 to 40, -	-	-	-	21
„ 40 to 50, -	-	-	-	28
„ 50 to 60, -	-	-	-	10
„ 60 to 70, -	-	-	-	4
				100

The average number of years during which they had been exposed to the sounds of boilermaking was $17\frac{1}{2}$; the oldest had been at the trade for 54 years, and the youngest for three years.

The range of the enquiry embraced the four following points:—

1st.—Extent of the loss of hearing.
2nd.—Region of the Ear affected.
3rd.—Course of the deafness.
4th.—Prevention of the deafness.

I.—THE EXTENT OF THE LOSS OF HEARING.

This was determined by testing the air-conduction of sound, as in ordinary hearing, that is, when the waves of sound enter the outer canal of the ear, and are transmitted by way of the tympanum to the fluid of the labyrinth in the inner ear.

The following points were investigated:—

1. The power of hearing a simple tone such as the tick of a watch.

2. The power of hearing speech.
  (a.) Whisper at a yard distance.
  (b.) Moderately loud voice at a yard distance.
  (c.) Voice of a public speaker.

1. *The power of hearing a simple tone, such as the tick of a watch.*

I first tested the distance at which the tick of a watch was heard, the watch in question being heard when the hearing is normal 36 inches from the ear. The distance was accurately measured in each case with a measuring rule. The following table presents a view of the various results of this enquiry:—

  15 did not hear the watch in either ear on pressure.
  13 heard on pressure in one ear and not at all in the other ear.
  21 heard when the watch was in contact with, or pressed upon the ears.
  15 heard in one ear only on pressure or contact, and in the other at a distance varying in different men from half an inch to 20 inches.
  25 heard the watch on both sides at distances varying from half an inch to 10 inches.
  11 heard at varying distances of from 5 to 34 inches from the ear.
  ___
  100

In no single instance was the hearing normal so far as the tick of a watch was concerned. In about half the number of men, the watch was either not heard at all, or only on pressure or contact. The nearest to the normal was 34 inches and that was limited to one man and to one ear. Perhaps a more striking way of representing the extent of the loss of hearing in these 100 men, is to sum up the total number of inches at which the watch was heard by all the men, and then compare the result with the total number at which it should be heard by 100 men having perfectly normal hearing. The total number in normal hearing is found by multiplying the double of 36 inches (for the 2 ears) that is 72 by 100; this gives 7,200, which we shall regard as the standard number of inches at which the watch should be heard by 100 men with normal hearing. The total number heard by the 100 boilermakers was 704 inches, or only 9·36 per cent. of the normal hearing.

In order to draw a comparison between the hearing of these men and that of some other classes of the community, I have examined also 100 ironmoulders and 100 letter-carriers with reference to their powers of hearing.

Through the kindness of Mr. Campbell of the Hyde Park Foundry, and of Mr. Dawson of the General Post Office, I have been enabled to carry out these examinations. The ironmoulders are not especially exposed to very loud sounds, but their ears are menaced from another direction--namely, through the nasal passages. The middle ear which is the seat of most of the diseases of the ear, is really an offshoot from the nasal passage, the lining of which, in ironmoulders must almost constantly be coated with irritating dust and dirt. Besides, their exposure to great extremes of temperature must, by exciting nasal catarrhs, still further contribute to an unhealthy state of their nasal mucous membrane. Such an unhealthy state is very ready to invade the middle ear. We would therefore expect a small sum total of hearing for ironmoulders. Letter-carriers, on the other hand, from the comparatively healthy nature of their employment, enjoying, as they do, so much of the fresh open air, may be looked upon as living under circumstances pretty favourable to a healthy state of the organ of hearing. The following numbers give the results of the application of the hearing test with the watch to these 300 letter-carriers, ironmoulders, and boilermakers, including 600 ears :—

Total number of inches at which the tick of a watch should be heard by
100 Men having normal hearing,   7,200.
100 Letter-carriers (average age 30 yrs.) heard 5,694 or 79°/₀ of normal hearing.
100 Ironmoulders (,,   37½ ,,) heard 3,291 ,, 45¾°/₀   ,,   ,,
100 Boilermakers (,,   35 ,,) heard 704 ,, 9⅓°/₀   ,,   ,,

2. *The power of hearing speech:—*

In addition to the watch I also tested the hearing of these 100 boilermakers, with the voice. It is important, in order to form an accurate notion of the state of the hearing, not to limit oneself to such a test as the tick of a watch, or other such simple mechanical sound. There is not unfrequently in deafness a strange disparity between the power of hearing a watch and the power of hearing the voice. We meet with persons who hear the watch very badly, perhaps no farther off than an inch from either ear, and yet they hear quiet conversation so well, that their friends do not observe any defect of the hearing. We meet with others, on the contrary, who hear the watch as far as 20 inches from the ear, and yet they require a loud voice pretty near to the ear in order to hear so as to understand. One of the iron-moulders whom I examined presented this peculiarity very markedly. When I proceeded to apply the test of the watch, he said, "You need not try the left ear, for the hearing is quite gone on that side, I only hear with the right ear." On testing with the watch, however, I found, to the man's surprise that he heard its tick 3 inches from the left ear where he said there was no hearing, while on the right side, to which he believed his hearing was entirely limited, the watch was not heard at all even on pressure. When I tried him by means of speech, however, his first statement was found to be quite correct; he could hear speech only with the ear which did not hear the tick of a watch. In testing by speech I employed—

(a) A whisper at a yard distance, using such words as "twenty," "brother," "America," "forty," "house," "garden."—The use of isolated words is a more reliable method than that of complete sentences, as in the latter, the element of guessing disturbs the result.

In normal hearing, a whisper should be heard about 22 yards from the ear. At a yard distant from the ear it was found that *fifty-nine* men could not hear a whisper with either ear, or heard it very indistinctly; that *thirty-three* men heard it on both sides; while *eight* men heard it with one ear, but not with the other.

(*b*) A loud-spoken voice at a yard distance.—In normal hearing such a voice should probably be heard about 100 yards from the ear. *Thirteen* men did not hear so as to understand with either ear. *Ten* men heard on one side, but not on the other. *Seventy-seven* men heard such a voice at a yard's distance with both ears.

(*c*) A public speaker.—This particular form of enquiry, namely, as to their power of hearing a public speaker, seemed to me a pretty reliable way of ascertaining the extent to which the deafness of these men interfered practically with their social comfort or usefulness. *Twenty-one* men stated that they could not hear so as to understand any public speaker, however near they placed themselves to the platform or pulpit, hence they rarely or never entered a church or public meeting. *Fifty-four* men stated that they heard with difficulty in a church or at a public meeting, requiring to be near to the speaker, who must also be clear and distinct. Even in these favourable circumstances, they missed much of what was said. Only *twenty-five* asserted that they had no difficulty in hearing in whatever part of the hall or church they were placed.

Of the hundred ironmoulders whom I examined, *twelve* stated that they had difficulty in hearing in church or public meeting; while of the hundred letter-carriers, *eight* assured me that they heard with difficulty in church or public meeting.

I think these facts should be considered by those interested in the religious and social welfare of the people, and especially by public speakers and clergymen. I have been informed that two thousand five hundred men are probably actually engaged in this trade in Glasgow and its neighbourhood, besides those who may have left the trade, after having had their ears injured by the noise, and those labourers who are more or less associated with the same noisy work. Twenty-one per cent. of these are, as we have seen, practically debarred from church or public meeting; and although, no doubt, some excellent men, notwithstanding that they themselves hear nothing, attend church regularly for the sake of example to others, yet such are, I fear, few; while fifty-four per cent. are very much hampered in their enjoyment of church or public meeting owing to defective hearing. It might be worthy of consideration whether there should not be some arrangement made in the interests of boilermakers, so that this serious defect might be met by providing them with small places of worship, having good acoustic qualities, and supplied with speakers possessed of a strong voice and clear articulation. In the large, lofty ecclesiastical edifices which adorn our city, in the design or construction of which the question of hearing is, I fear, sometimes neglected in the calculations of the architect, the poor deaf boilermaker has no chance. Apart, however, from boiler-makers, the results of my examination of ironmoulders and letter-carriers—and these results bear out my observations of other classes—show that in the community there is a surprisingly large number who are more or less dull of hearing. Probably we are not far off the mark when we say that in every congregation or public meeting numbering 500 persons, 50 of these have some impairment of hearing. And of these 50 it may be said that if the speaker has not a clear and distinct mode of articulation, if he is not possessed of a fairly strong voice, they hear with difficulty, and do not follow the speaker with ease and advantage. When we consider that, owing to the strain of listening and the feeling that they are missing portions of what is said by the speaker, they soon become tired and discouraged in their efforts, and are apt to fall into a state of listlessness and inattention, it is not surprising that such persons are much tempted to remain at home. If my words reach any one destined for the clerical profession, or aspiring to success as a public speaker, I would urge him, in addition to his mental training, not to omit attention to the medium by which his mind has to communicate with the minds of his hearers —namely, the mechanical act of speech—I would urge him to cultivate such qualities as strength of voice, slowness, clearness, and distinctness of articulation, to avoid slurring over the consonant sounds, and the bad habit of lowering the voice at certain parts to an almost inaudible tone, last of all, to eschew, if possible, moustache and beard. These appendages are, I assure him, serious hindrances to the efforts which all persons who are dull of hearing instinctively make to read the facial movements of a speaker. Deaf ladies who wish to conceal their infirmity shun the society of moustached and bearded men. The reading of the lip and facial movements is universally practised by deaf people, and is a great help to them in understanding speech. I know persons who, when they visit the theatre, understand nothing of what is said on the stage until they have the face of the actor well in the field of an opera-glass. Hence, good vision is a great help to a deaf person—hence, also, the face of a public speaker should be well illuminated. In many pulpits there should be more light. Deaf persons hardly know their dependence upon their vision, and complain that their deafness is worse at twilight, not observing that their deficiency is aggravated by their imperfect vision.

A clergyman, therefore, cannot ignore the members of his congregation who are dull of hearing with safety to his reputation or popularity. I find that persons dull of hearing generally attribute their difficulty in hearing to the defects of the speaker's voice and articulation. In the privacy of my consulting room they often dilate very bitterly on the subject of " poor speakers." On the other hand, when they do make out fairly well what a speaker says, they are profuse and cordial in their expression of praise of the speaker. I may say, in this connection, that it is frequently the chief regret of a deaf patient that he or she cannot now hear the minister. I trust the minister will co-operate with the surgeon in assisting the deaf to hear.

II.—Region of the Ear Affected.

We now come to the consideration of what region of the organ of hearing is affected in persons exposed to loud noises. This necessitated the study of the following points :—

1. Bone-conduction of sound (when a sounding body is applied to the head).

   (*a*) Watch on temple.

   (*b*) Tuning-fork behind ear.

2. The power of perceiving notes of high pitch.
3. Hearing better in a noise.
4. Noises in the ear.
5. Giddiness.

1. *Bone-conduction of Sound.*—Supposing the ears were sealed up so that sound could not find admission by these channels, if a tuning-fork were made to vibrate, no sound would be heard, provided a space of air, however small, existed between the head and the fork. The waves of sound falling upon the head from the air would not be sufficient to throw the cranial bones into vibration, they would in great part be reflected from the surface of the head. But place the vibrating tuning-fork in contact with the head and its note would be heard resounding even more loudly than if the ears were open. In this case the vibrations of the tuning-fork are communicated to the bones of the head, and are transmitted to the osseous casing of the cavities of the ears, from which they pass to the endings of the auditory nerve in the labyrinth—this is what is termed *bone-conduction of sound.*

With the ears open, this mode of conduction is somewhat feebler than the ordinary method of hearing through the air. For we find that after a vibrating tuning-fork placed in contact with any part of the head has ceased to be heard, it will again become audible if transferred to a point opposite to, but not touching, the orifice of the ear. In a normal state of the hearing, therefore, sound waves conducted to the nerve of hearing by the bones produce a less effect than when conducted by the air. But, curiously, with many deaf people this is often reversed, and they hear much better by the bone-conduction. Without entering upon the more exact details of this matter, which would divert us too far from our proper subject, I will just say that as a general rule, liable, however, to exceptions, when in a deaf person the bone-conduction excites the nerve more readily than the air-conduction, we conclude that the obstacle to hearing exists in the parts of the ear external to the labyrinth—that is, in the outer ear or middle ear, the so-called conducting apparatus of the ear. When, on the other hand, in a deaf person, the air-conduction produces a greater effect on the nerve than the bone-conduction (the condition in health) we infer, although this is subject to more numerous exceptions than the other, that the labyrinth or nervous apparatus is the seat of the mischief.

In each of these hundred men, therefore, I tested the relationship between the bone-conduction and the air-conduction of sound.

For this purpose I employed a loudly ticking watch and a tuning-fork. In testing with the former, the watch was pressed upon both temples. The watch is only suitable where the deafness is very decided, otherwise it is heard by air-conduction as well as by bone-conduction, although placed on the bone, and this disturbs the result. It was thus applicable in 133 ears. In only 13 of these ears was it heard by bone-conduction, and thus in 120 ears the presumption was that some obstacle to hearing existed in the nerve structures. But the use of the vibrating tuning-fork yields the best results. In the case of each man I applied the vibrating tuning-fork (C) to the mastoid process, that is, the smooth bulging behind the ear; and when the sound had completely died away on that part, I instantly transferred it to a point opposite to, but not touching, the orifice of the ear. In 90 of the men the sound was audible for some time in the latter position after it had died away in the former. In these 90 deaf men, therefore, the bone-conduction was feeble as compared with the air-conduction, thus pointing to the nerve structures as the seat of the lesion. I employed likewise a second experiment as a confirmatory test. In each of these 90 men, after the sound had ceased on the mastoid process, I quickly transferred the tuning-fork to my own mastoid process (and I may say that I have reason to believe that my auditory apparatus is normal), and found that in every case the sound was heard by me for some time after it had ceased to be heard by the man under examination. This again proved defective bone-conduction, pointing still to diminished nerve perception. In the remaining 10 persons, representing 12 ears, the opposite results were found, the tuning-fork being heard for a longer time by bone—than by air—conduction. In these exceptional cases, both with reference to the tuning-fork and the watch, there was probably, in addition to the nerve lesion, some morbid state of the conducting apparatus, which, by intensifying the bone-conduction, compensated for the diminished perception of the nerve.

2. *Power of perceiving notes of high pitch.*—In connection with tests applied to the nervous structures of the ear, I examined twenty of the men with special reference to their power of perceiving tones of high pitch. The sounds in boilermaking which are probably most injurious to the ears are the notes which are extremely shrill, and which act with most damaging effect upon the short fibres of the basilar membrane in the cochlea—these short fibres being concerned especially in the perception of high pitch. For producing varieties of pitch, I employed a Galton's whistle, which consists of a silver tube about 2 cm. long, with a bore of 1 mm. Notes of various pitches are produced by altering the length of the tube by means of a piston which can slide in the tube. The more deeply the piston is pressed, the shorter the tube and the higher the notes. Notes from about 9,000 (9 mm.) vibrations to 28,000 (3 mm.) vibrations may be sounded in this way. Of the twenty men examined with the whistle I found the power of perceiving high notes strikingly diminished in every case. Above 9,000 vibrations per second the hearing gradually diminished, and few could hear 20,000, and then only very faintly. In health, it may be mentioned, the power of perceiving tones of high pitch is greater, as a rule, within limits, than that of perceiving tones of low pitch. In persons with normal hearing, notes as high as from 40,000 to 50,000 vibrations per second may be perceived.

3. *Hearing better in a noise.*—Another interesting point was made the subject of enquiry in these 100 men, namely, whether they heard speech better in the midst of a loud noise or during perfect quietness. It is a remarkable fact that many persons with very defective hearing hear better in a great noise, such as in a railway carriage, or in the presence of noisy machinery, than in a quiet place. The instance of this peculiarity first described was that of a husband who could be heard by his deaf wife only while the servant was beating a drum. Some try to explain such an apparent paradox, by pointing out that in a railway carriage the confined space, the nearness to the speaker, the elevation of the voice, and the close attention of the listener, may account for the apparently better hearing. The elevation of the speaker's voice must be taken in conjunction with the important fact that the general noise is not impeding the deaf

person's hearing as it is that of the normal subject. He is, so to speak, profiting by the loudness of the voice while he is not disturbed or distracted by the din around him. While I was travelling lately along with a friend afflicted with deafness, in a steamboat on the Clyde, my friend remarked that he rejoiced when the steamboat whistle brayed forth, because then the speaker raised his voice and his deafness was hardly detected. Nevertheless, it is unquestionable that, apart from these modifying influences, the hearing, in the case of some deaf persons, is actually improved in the presence of loud noises. The explanation, probably, is that in such cases the three small bones of the ear have become stiff and rigid, and have lost their wonderfully delicate powers of vibration in response to waves of sound, of moderate intensity, striking upon the tympanic membrane. When, however, they are shaken by strong coarse vibrations entering the ears, they are thereby, it may be conceived, rendered fit for the time being for transmitting ordinary vibrations from the human voice.

In regard to the men under examination, with a few exceptions, they reported that they heard better in *quietness*. On questioning carefully the few who asserted that they heard better in a noise I was satisfied that the peculiarity could be explained by the circumstances already mentioned, especially by the loudness of the voice, and the proximity of the speaker. If the explanation just offered of the cause of hearing better in a noise is a correct one, the absence of this phenomenon noticed in these men still further corroborates the view that the nerve structures are the seat of the lesion.

4. *Noises in the Ears.*—This is a very common symptom in the various diseases of the ear, in many of which indeed it is the main feature, and in a small number the only symptom for which relief is sought. I have been surprised at its comparative infrequency in these men. Only 8 men seemed to suffer permanently from these noises, and even in these the sensations were not of a distressing character. Thirty-four men, however, mentioned that they had at times noises in the ears; they were experienced chiefly after getting home at night, but were gone in the morning, or they were only present when they suffered from a cold. For a time after joining the trade, most apprentices are troubled with a sounding in the ears every evening for a few hours after leaving work. This comparative absence of persistent noises in the ears

of these men is in marked contrast with any 100 cases of ear disease which are treated at our Hospitals. For example, of 100 consecutive cases of ear disease affecting the conducting apparatus of hearing, treated at the Glasgow Ear Hospital, 78 were attended by a constant noise in the ear. Noises in the ear are probably in most cases due to pressure exerted directly or indirectly upon the terminal nerve structures in the labyrinth from causes external to the nerve, while the mischief in the boilermakers is probably of the nature of a degeneration of the nerve structures themselves.

5. *Giddiness* is now well recognised as a pretty frequent symptom of ear disease, being probably due, as in the case of noises in the ear, to pressure upon the nerve structures. Only one man stated that he had such severe giddiness as to amount to attacks of staggering, the attacks, not continuing for any length of time, and separated from one another by considerable intervals. In other fourteen men there seemed to be occasionally when stooping, or in the morning, when rising out of bed, some sensation of giddiness. But I attributed little importance to these sensations in their bearing upon the state of their ears. It may be said, then, that noises in the ear and giddiness have not been met with in anything like the frequency found in cases of ear disease attended by a like amount of impairment of hearing, when due to disease of the conducting apparatus of the ear.

In regard, therefore, to the question of the region of the ear affected when the organ is exposed to loud sounds, such as in boilermaking, the great preponderance of evidence (derived from the state of the bone-conduction, the diminished power of perceiving high notes, and the study of certain other symptoms) points to the *terminal nerve structures, probably in connection with the basilar membrane of the cochlea,* as essentially the seat of the mischief. There is no doubt, however, that in a certain proportion of cases, disease likewise exists in the middle and external ears, modifying the symptoms, just as disease of these parts exists in a certain number of all classes of the community.

III.—The Course of the Deafness.

In regard to the length of time intervening between their entrance upon the noisy employment and the commencement of the disturbance of hearing it was not easy to obtain accurate information. The statements of the men as to the duration of their deafness generally refer only to the time during which the loss of hearing had markedly attracted their attention. When closely questioned, most of them admitted that a certain degree of disturbance of their hearing began almost immediately after entering upon this kind of work, advancing in many cases slowly, but in some much more speedily to very decided deafness. When apprentice lads, with the keen sensitiveness of hearing which often characterises the young, are first put into these boilers, they are quite "stunned." The following descriptive terms were used in answer to the question how the noise first affected their ears:—"Bad," "very bad," "awfully bad," "fearful," "could hardly stand it;" and I am not surprised, from my own experience of the sound, to be informed that a number who enter upon this trade cannot stand it, and have to abandon it in favour of some other kind of employment. These apprentices said that even from the first they feel dull of hearing at night, and have a sensation of buzzing or confused din in their ears; these feelings, however, usually pass away in the course of a few hours, and on the following morning they feel quite well again. A few hours of absence from the noise apparently suffice at first to enable the irritated nerve structures to recover themselves. After a few weeks, or it may be months, if they are able to continue at the work, they appear to get accustomed, as they express it, to the noise; more correctly they have now diminished perception, probably due to partial paralysis of the nerve structures in the cochlea, and they become gradually less and less painfully and disagreeably affected by the noise. Some of the men went the length of asserting that they had reached such a state of torpidity to the sound as to be able to sleep in the interior of a boiler while all the riveters and caulkers were at work. Some, on the other hand, never reached the happy stage of being indifferent to the noises, and always felt it very unpleasant, even painful, when they had to work in the interior of a boiler, and they tried to avoid it. A number of the men mentioned how doubly painful the sensation was when, in consequence of the plates being thin, the sounds were of an extremely shrill character. I have mentioned that at first the dulness and noises in the ears pass off after a few hours of quietness. I found also, that even in after years when their hearing was markedly impaired, many of them thought that it improved when they were idle for a week or two, as at a holiday time. There seemed to be truth in these statements, for in a few cases where I was surprised at the comparatively small loss of hearing in men who had been long at the trade, I discovered, on enquiry, that they had at some time been absent from the trade for a year or two, perhaps away on a long voyage working on board ship. A number, however, stated that their hearing was not improved in the least by a period of absence from the noise.

The changes, therefore, which are produced upon the ear by the action of these loud noises seem always to tend at the earlier periods to pass away simply under the influence of rest to the organ; and even in the advanced stages there is frequently a tendency to distinct improvement on the removal of the irritating cause.

IV.—PREVENTION OF THE DEAFNESS.

The only other subject embraced in this investigation was as to the use by the men of means for shutting out the sound and protecting the ear. In answer to my enquiries, 28 men stated that they occasionally stopped up their ears with cotton or some such substance. Most of these only used such a preventive when they were working in the interior of a boiler while riveters were hammering outside, especially when they were put into a smaller cavity, such as a flue or "superheater." It was used in these circumstances, not with the object of preventing deafness, but to diminish the painful sensation experienced at the time in their ears. When severe dulness of hearing comes on, and the sensation is neither so painful nor so disagreeable, they usually discard the cotton plug. Most of the men seemed to be prejudiced against cotton plugs, chiefly owing to the notion that the use of cotton might make them liable to catch cold when they removed it at night; some, however, objected to this precaution because it interfered with their hearing, and others because it excited a disagreeable itchiness in the ear. But the main reasons for the neglect of such precautions were, I fear, want of appreciation of the value of hearing, want of forethought, and the conviction that dulness of hearing is an inevitable consequence of this kind of work. Speaking for myself, the interior of a boiler was quite intolerable unless my ears were plugged with India-rubber plugs, such as I show you here. When they were inserted in the orifices of my ears I was able with impunity to remain inside a flue while using the phonograph, although the noise was at its very worst. I observed, while the plugs were in my ears, that I was able to hear the voice of a speaker better than when the ears were open. Apparently, the dulling effects of closing the ears were more than compensated by the shutting out of the terrible and deafening clang. These plugs, which are hollow, and made of various shapes and sizes to suit various ears, were introduced by Mr. Cousins, of Southsea, and were originally intended for bathers in order to protect the drums of their ears when diving, and for this purpose they are very useful. As prepared by Mr. Cousins they are round, but I have, with his sanction, had such plugs made elliptical and oval in shape, which accords better with the form of the lumen of the orifice of the ear. I tried these plugs with some of the men both in the works of Messrs. Napier and of the London and Glasgow Shipbuilding Co. and they have so far reported favourably. They agree with me that they rather favour the hearing of speech than otherwise. In order to avoid irritating the canal of the ear, they should not be too large, and should be smeared with vaseline, which renders them effective in keeping out the sound without the necessity for very tight plugging. A plug of cotton well smeared with vaseline was, at my suggestion, also tried by some of the men. When the cotton is in the ear I found the sharp painful character of the sound was materially modified, giving place to a dull thud not particularly disagreeable. The India-rubber plugs are, however, more handy and effective. It would be well if apprentices when entering upon this kind of work were advised by the foreman to use such simple precautions for protecting the ears, for by these means it is probable that the injurious effects produced upon the ears by the intensely shrill notes would be materially obviated, while the protection which they would afford against the actually painful influences of the terrible din would prove a great comfort to the men.

Let me, in concluding, recapitulate the chief results of these enquiries :—

1. No one engaged in boilermaking for any length of time escapes injury to the hearing.

2. In about half the number of boilermakers the hearing power is so defective that the tick of a watch, which should be heard 36 inches from the ear, is either not heard at all or only on contact with the ear.

3. As compared with the normal standard, they possess only about $9\frac{1}{3}$ per cent. of hearing power.

4. As compared with men engaged in other occupations, their hearing power is extremely defective—for example, letter-

carriers have 79 per cent., and even ironmoulders 45 per cent. of the normal hearing power.

5. The extent of their defective hearing is also strikingly shown by the fact that three-fourths of their number could either not hear at all at a public meeting, or heard with difficulty.

6. In regard to the ironmoulders, 12 per cent. admitted difficulty in hearing in a public meeting, while of the letter-carriers 8 per cent. admitted the same difficulty.

7. In view of the fact that probably of those who attend church or public meeting 10 per cent. hear with some difficulty, it is of great importance that clergymen and public speakers should cultivate clearness and distinctness of articulation.

8. There is almost decisive evidence in favour of the view that the nervous structures of the inner ear are essentially the seat of the mischief in boilermakers' deafness. This evidence is derived mainly from the weak conduction of sound by the bones and the defective perception of tones of high pitch; but other peculiarities in regard to such symptoms as hearing better in a noise, noises in the ear, and giddiness, also support this view.

9. Lads when they join the business experience very painful sensations in the ear, and there is little doubt that the hearing power begins to suffer very soon after.

10. There is a tendency, at all events in the earlier stages, for the disturbance of hearing to diminish or pass away simply under the influence of absence from the noise, and even in the advanced stages there is sometimes a similar tendency.

11. The men have hitherto employed preventives in the form of cotton plugs, mainly in special and temporary circumstances.

12. Men who have recently joined the trade, and also many older men in whom the noises produce painful symptoms, would gladly employ some simple and efficient preventive or sound-deadener.

13. The india-rubber hollow cushion or plug promises to be useful, both for alleviating the painful effects of the noises, especially inside the boilers, and preventing the ultimate deafness.

19

Reprinted from *Acoust. Soc. Am. J.* **15**:27–31 (1943)

THE APPLICATION OF SOUND ABSORPTION
TO FACTORY NOISE PROBLEMS

Hale J. Sabine and R. Allen Wilson
The Celotex Corporation, Chicago, Illinois

WHILE the reduction of noise has been successfully achieved during the past years in buildings used for offices, hospitals, schools, and other common gathering places and in a variety of rooms used for many kinds of special purposes, it has remained for the current pressure of wartime demands for manufacturing products to focus any wide-spread attention to the problems found in factory work areas.

Perhaps the greatest single factor in bringing the factory noise problem into the open has been the mass employment of women, which began as a defense and war measure. Here were workers who, not knowing what they were expected to tolerate, were outspoken in their reactions to excessive noise conditions. Here were workers who felt that objectionable noise constituted sufficient reason to change jobs. In some industries, more or less regular absenteeism on the part of women employees led some executives to suspect that noise also had a bearing on this situation.

An interesting case in point is that of a textile mill where, during the past year, women have been brought into the weave sheds. New employees in this plant were given a rather extensive training course. Shortly after taking up the regular work as full time employees, a large percentage of these women left their jobs. By shrewd questioning the superintendent satisfied himself that much of this turnover was due to the reactions to plant noise. Since there is no known method of selecting only those people who can successfully condition themselves to withstand intense noise, he devised the following procedure as a test program. Before entering the training course, new women employees were given small tasks that kept them in or adjacent to the weave sheds. Those who were still on the job at the end of a full week were then entered for the training course. The results of this test were so conclusive and satisfactory that it is now standard practice. It is reported that between 40 and 50 percent of the new women employees undergoing this "trial" period now drop out after the third day and before the week end.

Another example where the problem appeared to be one of nervous fatigue as the result of exposure to noise, rather than difficulty in individual conditioning, was found in a munitions plant manufacturing .30-cal. and .50-cal. machine gun cartridges. Here it was observed that regular women employees were impelled to take brief "vacations" from their jobs at short intervals. Investigation by the plant health directors and engineers convinced them that noise was largely responsible for these absences. They further concluded that these rest periods were essential unless something could be done to modify noise conditions within the departments where these women worked.

There are other instances where noise has been a major factor in excessive turnover in men employees. Early in the growth of the defense program, a west coast factory found that it could not keep men in a department adjacent to a steel chipping operation until measures were taken to protect these men from the noise created by the pneumatic chipping hammers. A mid-west plant making a product vital to all automotive and mechanical equipment experienced the same situation in an automatic machine room.

In response to this increasing interest on the part of industrial hygienists, personnel officers, plant engineers, and others responsible for the efficiency and general welfare of factory workers, the present investigation was undertaken in order to determine what can be done with acoustical treatment to alleviate factory noise conditions.

The initial step in the survey was to determine the range of noise levels encountered in typical factories. Noise level measurements were made in 33 separate plants covering a wide diversity of industries and machine operations. Of all the readings taken in actual work areas, the highest was 130 db and the lowest 65 db. In the large majority of cases, the observed noise levels ranged quite uniformly between approximately

* Paper presented at the twenty-eighth meeting of the Acoustical Society of America, May 14–15, 1943, New York, New York.

85 and 105 db. In order to obtain comparative data on the noise output of various types of machines, readings were taken wherever possible at a fixed distance of 3 feet from the principal noise producing point of the machine. A few representative figures are given in Table I. The data shown should be considered only as illustrative, since in many cases the noise level may vary considerably depending on the type of work or material on which the machine is operating. It should also be noted that since the operator frequently is closer to the noise source than 3 feet, the level at his ear may be somewhat higher than the values shown.

One of the first questions which arose was whether the reductions in noise level of the order of 5 to 10 decibels theoretically attainable by means of acoustical treatment were enough to afford really worth-while relief when applied to the range of noise levels of approximately 85 to 105 db to which the average factory worker is exposed. Experience with a number of acoustical installations showed that very substantial relief from annoyance was possible even under conditions where noise meter readings showed no significant reduction in noise level. Corollary experiences revealed that it was often possible for an acoustically treated area to be more comfortable from the noise standpoint than an untreated area having actually a lower average noise level. From these observations it was soon concluded that factors other than the noise level contributed important elements to the over-all noise picture.

Analysis of the comments of workers indicated that these factors may be grouped under two headings, namely, reverberation and what may be termed the "spreading effect" of sound. Excessive reverberation seems to evoke a very definite feeling of annoyance apparently arising from the unnatural prolongation of an already disagreeable noise stimulus. Furthermore, the sustaining effect of reverberation on an intermittent or impact type of noise source greatly increases its interference with the understanding of speech, one of the most serious and frequent noise problems in factory areas. It must be noted, however, that when the noise sources are of a steady, continuous type, such as lathes or other rotary machines, the problem of reverberation is

TABLE I. Noise levels of various machines at distance of 3 feet.

Punch presses, various types	96–103
Headers	101–105
Drop hammers	99–101
Bumping hammer	100
Hydraulic press	130
Automatic riveters	95– 99
Lathes (average)	80
Automatic screw machines	93–100
Airplane riveting guns	94–105
Airplane propeller grinding	100–105
Cotton spinning	84– 87
Looms	94–101
Sewing machines	93– 96
Wood planers	98–110
Wood saw	100
Wire rope stranding machines	100–104
Ball mill	99

not involved and relief from this element of the total noise situation cannot be counted on.

Since reduction of the reverberation time is a primary effect of increasing the absorption in a room, it would be expected, and is borne out by experience, that the most effective use of acoustical treatment in alleviating an over-all noise condition is achieved in those areas where excessive reverberation is most obviously and unmistakably noticeable before treatment. Such areas are generally open, "one-piece" rooms having ceiling heights not less than about one-fifth the smallest floor dimension, and containing noise sources of a diversified and intermittent type at fairly wide spacings. A typical example is an assembly shop manufacturing corrugated metal culverts, housed in a room of 3000 square feet floor area with a 17-foot ceiling. The operations involve corrugating steel sheets, punching rivet holes, forming, riveting, and finally dropping the finished culvert on a concrete floor. It is easy to imagine the effect of an excessively high reverberation time on such noise sources. Analysis of the remarks of the employees and the plant manager after acoustical treatment pointed clearly to the reduction of reverberation as the chief factor in the relief experienced.

By spreading effect is meant the tendency of sound in a non-absorbent room to remain at a fixed level or decrease at a slow rate with increasing distance from the sound source, and the accompanying tendency for large amounts of reflected sound energy to arrive at the listening point from directions other than that of the sound source. The resulting illusion of distant noise sources being abnormally loud and close at hand, together with the inability to estimate easily the distance and direction of individual noise sources are responsible for the feelings of distraction and general uneasiness experienced by workers under these conditions. The spreading of sound is of course associated with excessive reverberation, although its effects are often audible under conditions where reverberation is not readily noticeable, as in the case of continuous, steady noise sources.

Reduction of this spreading effect is the second important action of acoustical treatment and is usually the most obvious and easily recognized change observed by workers. Typical remarks are "The noise of that machine over there stays where it belongs," or "Before, it seemed like every machine in the plant was making noise, but now I can only hear the ones right around me." Comments of this type incidentally revealed an interesting point, namely, that as far as symptoms of strain and fatigue due to noise are concerned, a machine operator is affected much less by his own machine than by the others in the room in spite of the fact that his own machine, being closest to him, sets up a higher noise level at his ear. This is especially true when the machine is operated by manual control, and also when the noise of neighboring machines is of an intermittent or irregular type, as from a battery of punch presses. The explanation probably lies simply in the fact that by directing attention to and controlling his own machine, the operator has a chance to "brace himself" against its noise. At any rate, it is undoubtedly this relative immunity of the average operator to annoyance caused by his own machine that makes possible the increase in working comfort actually realized through reduction of the spreading effect.

The amount of relief obtainable by cutting down the spread of noise is influenced greatly by the spacing and layout of the machines in a given area and by the type of noise they produce. The manner in which these factors affect the final result can be understood by considering the spreading effect in somewhat greater detail. It is well known that the introduction of absorbing material in a room reduces the intensity of distant noise sources more than the intensity of sources close to the point of observation. In order for absorption to be effective at any given position, as far as the ear is concerned, conditions must be such that before treatment the noise of distant sources is audible in the presence of nearby sources. In other words, no good will result from quieting distant sources if their noise is already completely masked by the noise from those in the immediate vicinity. This means that wide spacing of machines, differing qualities of noise from various units, and intermittent operation of the machines, are all conditions favorable to the attainment of good results. For example, treatment applied over punch press areas in which the machines are spaced at least 5 or 6 feet apart has proved a distinct benefit to the operators.

Situations where a number of continuously operating machines having identical noise characteristics are spaced very closely together afford negligible possibility of relief by the use of absorption for workers in the midst of the machines. If, however, only a part of the room is occupied by these machines and absorbing material is applied over the entire room, those workers regularly occupying the portion away from the machines will receive the full benefit of the treatment in cutting down the spread of noise into their area, and will experience the feeling of the noise being "pushed back" to where it belongs. The operators of the machines observe an immediate and pronounced drop-off in the noise in moving only a few steps from the edge of the machine area, and insofar as their duties permit they are afforded periods of relief from time to time which experience has shown to be very much worth while.

The increased facility in locating individual noise sources resulting from the reduction of the spreading effect of sound is a distinct advantage in the frequent cases where the operation of machines is gauged by ear. For example, the superintendent of a factory manufacturing wire cable reported that defective stranding by a machine is usually preceded by a warning "squeal." Before installation of acoustical treatment, time was consumed in determining which of the many machines in the room was at fault, but now, he says, he can "walk right to it."

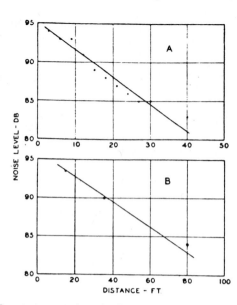

FIG. 1. Attenuation of noise level with distance from the source in large rooms with highly absorbent ceilings. (A) Room dimensions 100'×100'×9'. Slope of line, 0.35 db per foot. (B) Room dimensions 360'×560'×27'. Slope of line, 0.16 db per foot.

One important phase of the investigation was to establish, if possible, a theoretical basis on which to interpret and predict results from the use of absorption in industrial areas. The conventional method of calculating the noise reduction in decibels, arrived at by assuming that the average sound intensity in a room is inversely proportional to the total number of absorption units, has the advantage of simplicity, but usually has little or no physical significance when applied to actual conditions in factories, particularly in the case of very large areas. In the first place, the decibel reductions predicted by the theory refer only to the level of the generally reflected sound, and can, therefore, be realized only at points far enough away from the nearest noise source that the relative level of the direct sound is negligible. Since this distance is usually much greater than the normal spacing of machines, the calculated reductions cannot be expected under operating conditions inside of machine areas. In the second place, actual measurements made under conditions where the noise level due to a single source could be observed over a wide range of distances show that in rooms with highly absorbent ceiling treatment the intensity instead of levelling off to a constant

value at the distance from the source predicted by the theory continues to decrease at more or less a constant rate clear to the limits of the room. This means that the actual decibel reduction due to the addition of treatment instead of being a fixed quantity at all points beyond a certain distance, is continuously variable with distance, and that in predicting or specifying a decibel reduction, a corresponding distance must be specified.

These considerations, together with the fact indicated by experience that it is the reduction of the spreading effect of noise rather than reduction of noise level *per se* that is responsible for the results achieved by acoustical treatment, suggest that the effectiveness of a given acoustical installation may better be expressed in terms of the *attenuation of noise level with respect to distance from the source*. The higher this attenuation figure, the greater the reduction of noise from distant sources in relation to that from nearby sources, or, in other words, the greater the reduction in spreading.

While it has not been possible to carry out a study of attenuation characteristics detailed enough to afford a method of predicting results by calculation, nevertheless a number of useful observations of a general nature have been made:

(1) In rooms whose smallest floor dimension is at least several times the ceiling height, and having a highly sound absorbent ceiling, the intensity level due to a single source or concentration of sources decreases at a constant rate in decibels per foot over the entire area of the room, regardless of its size.

(2) This constant rate of attenuation varies inversely with the ceiling height. Attenuations of the order of 0.4 db per foot have been measured

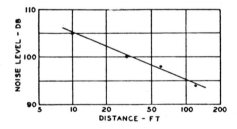

FIG. 2. Attenuation of noise level with distance from the source in a large room with untreated ceiling. Room dimensions 170'×200'×26'. Slope of line, 3 db per distance double.

for ceiling heights averaging 10 feet, and 0.2 db per foot for ceiling heights in the 20-foot range, the ceiling materials having absorption coefficients of at least 0.70. (See Fig. 1.)

(3) These attenuation rates are established at a distance from the source somewhat less than the ceiling height. Within this distance the intensity rises much more rapidly at a rate determined partly by the inverse square law and partly by the distribution of the noise source elements.

(4) The attenuation in the case of untreated ceilings is not only much smaller, but has a different characteristic, being of the order of 3 db per distance double. (See Fig. 2.)

All of these observations, and the latter in particular, suggest that in a large, low-ceilinged room, sound energy instead of being in a diffuse condition is contained largely in cylindrical waves which in travelling outward from the source are attenuated both by distance and by the absorption of the ceiling. The process, in fact, may be considered as analogous to attenuation in an absorbent-lined duct except that transmission takes place radially over a plane instead of in a single line.[1]

In cases where greater attenuation is needed than can be obtained by lowering the ceiling and increasing the absorption coefficient of the ceiling material, the use of baffles of absorbing material suspended from the ceiling is indicated. Although sufficient data have not been collected to permit exact design specifications for optimum performance, it would be expected that the closer the baffles are spaced and the lower the level to which they extend, the more effective will be the results.

[1] Cf. G. E. Morison, J. Acous. Soc. Am. 9, 244 (1938).

20

Copyright © 1953 by the American Industrial Hygiene Association Journal

Reprinted from *Am. Ind. Hyg. Assoc. Q.* **14**(4):235-263 (1953)

A Noise Survey of Manufacturing Industries

HENRY B. KARPLUS and GEORGE L. BONVALLET, Acoustics Research Group
Armour Research Foundation of Illinois Institute of Technology, Chicago

I. Introduction

RECENT interest in the effects of noise environment on hearing, speech communication, and work performance[1,2,3] has created a need for definitive data on noise conditions in American industry. Until recently, large variations in instrumentation and measurement techniques have existed which made comparison of the limited published data difficult or impossible. In the survey described in this paper, measurements were made in typical factories of most industrial categories. This article presents the data obtained, as well as an analysis of the survey from which some significant generalizations can be drawn.

II. Scope of the Survey

AN ATTEMPT has been made in this survey to study the noise conditions existing in a wide range of industries. Visits were made to 40 plants of widely different manufacturing industries. Measurements were made at about 600 different locations in the

Sound pressure levels of steady state noise conditions inside plants of various manufacturing industries were measured and analyzed in octave bands. Measurements were made where operating personnel were exposed to the noise. They were taken at about 600 locations in 40 different plants selected from a wide range of industries.

Both the total loudness and the highest octave band loudness for each set of measurements were computed, since these single number representations have a better correlation with undesirable features of noise than overall sound pressure levels.

Measurements are divided into those taken close to principal noise sources, called machine levels, and those further away, called area levels. Analysis shows that about 50% of the machine levels were between 90 and 100 db. with total loudnesses between 100 and 300 sones and those in the loudest octave band between 30 and 100 sones. Approximately 50% of the area levels were between 85 and 95 db. with total loudnesses between 50 and 200 sones and those in the loudest octave band between 20 and 50 sones.

A chart is given showing the approximate relation between overall level and total loudness. The loudness corresponding to a given overall level varies over a range of approximately two to one.

Noise levels are further analyzed according to the industry in which they were found. The noisiest machines were found in all industries concerned with metals, followed by the lumber industry. The noisiest areas were found in some of the metals industries, followed by the textile and the chemical industries. The lowest levels were found in the apparel industry. Noise levels of a few widely used tools are also shown. The loudest operations were chipping and riveting, especially when the work consisted of large metal plates.

A tabulation of the original data, showing overall levels, octave band levels, and loudnesses, is appended.

buildings of these plants, wherever operators were exposed to noise.

A. SELECTION OF FACTORIES AND MEASURING LOCATIONS:

As a guide for facilitating the selection of industries, use was made of the section "Manufacturing Industries" of the "Standard Industrial Classification Manual" issued by the U.S. Department of Commerce. This classification was modified somewhat for the analysis of the measurements in this survey. The classification used in this article, the types of plants visited, and the number of measurements made in each industry are shown in Table I. In some instances, different shops in the same plant were considered part of different industries. For example, the measurements taken in a department making small molded plastic components were classed with plastic products in the chemical industry and not with the machinery industry, where other departments of this plant were placed.

Most of the measurements here presented were taken specifically for this survey. In a few instances data from other sources were added. Interest was focused on those industries suspected of having noise problems, and therefore proportionately more plants in the metal industries were visited. This results in relatively more intense noises being measured.

Excluded from this survey were those locations where noise is incidental and not due to the work in progress. Thus the noise levels of offices, drafting rooms or warehouses were not included. Measurements were also confined largely to levels inside buildings. Noise levels of aircraft, traffic, and other general outdoor noises were not included. The noise of aircraft engines was also considered outside the scope of this project because, in general, these are tested in enclosures and operators are not exposed to the noise.

The statistical analysis of the data presented in this article gives equal importance to each measurement. No attempt has been made to weight readings with respect to the importance of an industry, the number of men exposed to a given level or the length of exposure. Very little is known about the effects of noise of short duration. Fatigue, interference with speech, discomfort, or hearing loss may depend differently on the

TABLE I.
CLASSIFICATION OF INDUSTRIES

Industry	Number of Measurements Machines	Areas
a. Food and Kindred Products	10	12
Meat Packing		
Candy Making		
b. Textile Mill Products	18	15
Spinning Mill		
Weaving Mill		
c. Apparel	5	3
Men's Work Clothes		
d. Lumber and Products	27	15
Millwork		
Lumber Mill of Paper Manufacturer		
Furniture (two plants)		
e. Paper, Printing, and Publishing	39	31
Paper Manufacture (two plants)		
Cardboard and Carton Making		
Publishing and Printing		
Newspaper		
f. Chemicals	21	21
Heavy Chemicals		
Paint		
Plastic Products		
Tire Making		
Petroleum		
g. Leather	13	23
Finished Leather		
h. Stone, Clay, and Glass	17	15
Soft Stone Cutting		
Tile		
Glass Containers (two plants)		
Concrete Aggregate		
i. Primary Metal Industry	37	30
Steel Foundry (three plants)		
Steel Mill		
Brass Foundry		
j. Fabricated Metal Products	81	58
Wire and Products		
Small Tools (two plants)		
Screw Machine Products		
Steel Pipe		
Machine Shops (in other groups)		
k. Machinery	29	25
Steel Plate and Boiler Making		
Heavy Power Machinery		
Internal Combustion Engines (two plants)		
Earth Moving Machines		
Electrical Appliances		
l. Aircraft	12	26
Aircraft Manufacturers (three plants)		

duration of a noise. When more information is available on such relationships, suitable weighting factors may be applied to these data. For this purpose, all the sound pressure levels and loudnesses have been tabulated in the appendix. It is expected, however, that the general conclusions would not be changed significantly.

B. MEASUREMENTS:

1. *Sound Pressure Levels:* Measurements of industrial noises were made with standard sound level meters and octave band

analyzers. Overall and octave band sound pressure levels were measured in decibels above the standard 0.0002 dyne per sq. cm. The overall level has been widely used in the past but is not adequate to describe a noise. Some measure of the contribution of different frequencies to the total noise is necessary. Since the noises encountered rarely contain high level pure tones, octave band analysis is suitable. All instruments used in the survey were calibrated in a random noise field by comparison with a laboratory standard. These techniques yield octave band sound pressure levels which may be relied upon to ± 3 db. Further details on methods of measurement may be found in an earlier article in this journal.[4]

2. *Machine and Area Levels:* Data are further subdivided into types of locations. Those taken close to the principal noise sources within two to five feet from a machine and in the vicinity of the operator are called machine levels. Other measurements were taken in places where men worked at some distance from the chief sources of noise and several machines could generally be heard. These have been termed area levels.

No correction has been applied to machine levels for the additional background because the factor of interest is the total noise to which the operator is exposed. In most cases, there was no significant contribution by the background except in two or three lower frequency octave bands. Where a machine did not stand out significantly above the background, it was considered to be too low to be measured and measurements of such a machine were considered area levels.

Where area levels were as high as machine levels because many similar large machines were operating quite close together, both machine and area levels are reported even though they are practically identical. The best examples of this were weaving sheds.

3. *Steady-State, Varying, and Transient Noises:* Commercially available sound measuring equipment is convenient for the study of steady noises. The precision of the meter is not the limiting factor and is quite satisfactory, because few noise levels remain the same from hour to hour or day to day. Many machines produce noises which

change somewhat with the phase of the operation. With the slow response setting of the meter, some noises were found to vary a negligible amount from the values reported, while in some cases deviations as great as ± 10 db. were observed. Variations of over 10 db. can be expected with different operations or when performing the same operation on different pieces of work. For example, chipping or grinding small castings is much less noisy than similar operation on large thin plates.

Rapidly varying noises like sharp intense impacts are not readily measured with standard sound level meters.[4,5] Because of its inertia, the needle of the indicating meter is unable to follow a very sharp peak and its maximum excursion depends on the attenuator setting. Of interest in impact noises are peak intensity and duration. Laboratory techniques are required to measure these. There is also no information available on the relatively undesirable effect of a transient noise of given peak value and duration compared with a steady noise. Again this may well differ according to whether fatigue, speech interference, or hearing damage is of interest. More fundamental research is required before transient and steady noises can be compared. In this analysis only steady-state noises are given. These include impact noises which last longer than about one-half second or repeat more frequently than approximately twice per second since sound level meters are satisfactory for these.

III. Analysis

THE OCTAVE band sound pressure levels read on the meter were corrected for known calibration errors and an overall level calculated.[4] This corresponds to the overall level which would be measured on a meter with the weighting switch in the "flat" or "C" position. From the corrected octave band levels, corresponding loudnesses in sones were obtained.[4,7] This conversion to loudness[6] takes into account the response of the average human ear to sounds of different frequencies. Accordingly, more emphasis is given to the high frequency bands.

By summing the loudnesses in sones in each octave band, a total loudness is obtained. This calculated total loudness correlates well with subjective estimates of relative loudness and is a better single number description of a noise than the overall level. The unit of loudness used here, the sone, unlike the decibel, is an arithmetic unit. Thus a loudness of 20 sones is twice as loud as one of 10 sones. This arithmetic scale has been found to be particularly easily understood by laymen.

Some of the undesirable effects of noise, such as deafening, appear to correlate with the level in narrow critical bands.[8,9,10,11] Consequently, the greatest loudness in any one octave band has also been given.

Reproduced in the appendix are all the measured octave band sound pressure levels, overall sound pressure levels, total loudnesses, and the loudnesses in the loudest band. In the following paragraphs, these values have been examined according to the relative number of times certain levels and loudnesses were encountered in this survey and how they were distributed among the various industries and among widely used tools.

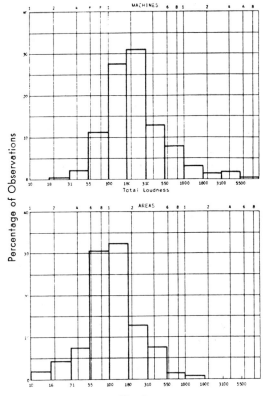

Fig. 2.
Distribution of total loudness.

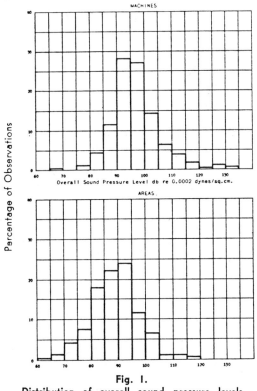

Fig. 1.
Distribution of overall sound pressure levels measured in industrial plants.

A. Distribution of Noises in Industrial Plants:

The distribution of all the overall levels, total loudnesses, and loudnesses in the loudest band encountered in this survey are shown in Figs. 1, 2, and 3. Machine and area levels are treated separately.

Inspection of these curves shows immediately that more than 50% of the machines measured produced overall noise levels between 90 and 100 db., total loudnesses between 100 and 300 sones, and loudnesses in the loudest band between 30 and 100 sones. Furthermore, very few machines (less than 6%) could be measured with overall levels less than 85 db., total loudnesses less than 50 sones, and loudnesses in the loudest band less than 20 sones. It should not be concluded that there are no machines quieter than these, but that machines producing noise levels lower than these figures were rarely noticed by plant personnel and are considered by them to be relatively

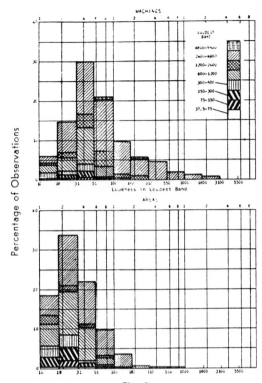

Fig. 3.
Distribution of loudness in loudest band.

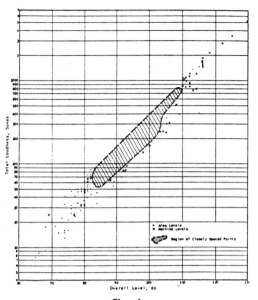

Fig. 4.
Relation between overall level and total loudness.

quiet in the usual factory background.

In general working areas, about 50% of the measurements showed levels between 85 and 95 db., total loudnesses between 50 and 200 sones, and loudnesses in the loudest band between 20 and 50 sones.

In addition, Fig. 3 shows by means of shaded areas the distribution of octave bands in which maximum loudness occurs. For example, Fig. 3-A shows that for machine noises having a loudness in the loudest band from 31 to 55 sones, the loudest band is in most cases the 2400-4800 cps. band. It is immediately apparent that the maximum loudness often occurs in the low

frequency bands when the loudness is low. For the louder noises, the fraction of measurements in which the loudest band is the 2400-4800 cps. band increases. Above 310 sones for machine levels and 180 sones for area levels, the loudest band was invariably the 2400-4800 cps. band.

The observations are summarized in a slightly different manner in Table II, where the median values, limits, and quartile deviations are shown. The lower limit gives lowest values observed. The lower quartile deviation is the value below which 25% of the observations were found, the median is the value below which 50% of the observations were found, the upper quartile 75%, and the upper limit gives the highest value observed. The extreme limits should not be interpreted too rigidly, since, undoubtedly, values outside this region can be found.

TABLE II.
LEVELS AND LOUDNESSES

Boundaries	Percent	Machines			Areas		
		Overall Level db.	Total Loudness Sones	Loudest Band Sones	Overall Levels db.	Total Loudness Sones	Loudest Band Sones
Lower Limit	0	66	4	1	60	2	1
Lower Quartile	25	92	130	31	84	65	15
Median	50	97	200	50	89	110	25
Upper Quartile	75	102	350	100	94	170	45
Upper Limit	100	133	6000	1700	120	1700	500

The relationship between overall level and total loudness for industrial noises has been plotted in Fig. 4. About 85% of the measurements are located in the shaded region. The other points show individual readings. From this chart the range of loudness likely to be encountered if only an overall level is known can be immediately read off on the ordinate. The range of loudness associated with a given overall sound pressure level is greater than 2 to 1. For example, if the overall is known to be 100 db., a loudness between 180 and 400 sones is to be expected.

B. OVERALL AND OCTAVE BAND LEVELS IN INDIVIDUAL INDUSTRIES:

The limits, quartile deviations, and medians of overall and octave band levels of individual industries are shown graphically in Figs. 5-16. More detailed distributions for three selected octave bands (150-300 cps., 600-1200 cps., and 2400-4800 cps.) are shown in Figs. 17, 18, and 19. These bands were chosen as typical for the low, middle, and high frequencies.

The spectra shown in Figs. 5-16 are all composite. The limits, medians, and deviations apply in each case to all the measurements taken in the particular octave band. Thus no machine had a spectrum exactly like any of the curves shown. None was as high as the upper limit in all octave bands.

The distributions in Figs. 17-19 show the percentage of the locations found having a level lower than that shown on the ordinate. It can be seen on these curves whether most measurements were close in value (nearly horizontal line) or whether wide variations were encountered (steeply inclined line). It also becomes apparent where limits were due to one or two isolated readings (steep slope near the 0 or 100% ordinate) and where many readings close to the limit were found.

The following observations may be made from a study of Figs. 5-19:

(1) Area levels are lower than machine levels. This is particular true for the aircraft industry. Here a few noisy operations are scattered at great distances in large plants so that area levels are quite low. In this instance, most of the area levels were taken in a different plant from the machine levels and only twelve machine levels were taken altogether. This is not a sufficient

sample from which to conclude that the machine levels in this industry are in general higher than in any other.

Fairly large differences between area levels and machine levels in the lumber and products industry may be explained by the fact that in most woodwork shops it is usual to find a large number of machines which are intermittently used, so that at any instant not many machines are operating. Consequently, the general area levels in this industry are lower in spite of the fact that machine levels are second only to the metal industries.

The chemical and the textile industries, however, show machine and area levels very close together. Conditions in these industries are in most cases very different from the aircraft industries. In these industries, it is usual to find many large machines very close together so that machine and area levels are frequently the same.

(2) High frequencies are less intense than low frequencies in area levels. For machine levels, a similar observation is found to hold for all but the textile industry and the industries concerned with work on metal.

(3) The difference between area and machine levels is greater at high frequencies than at low frequencies. This can be seen most clearly by comparing levels on Fig. 17 and Fig. 19. The high frequencies fall off more with distance.

Several factors contribute to this effect. High frequencies are usually radiated by small components of a machine, whereas low frequencies are effectively radiated by large surfaces. Very close to a noise radiator the intensity depends mainly on the amplitude of vibration of the source, whereas the general area level is a function of the total energy radiated and hence depends both on vibration amplitude and the radiating surface area. Consequently, the intense high frequency noises from small sources are concentrated near them whereas the low frequencies radiated by the larger sources are more uniformly distributed throughout the enclosed working area.

The general area levels are also reduced by absorption of sound. This takes place in the air and to some extent at room surfaces, even though they are not acoustically treated. High frequencies are diffused and ab-

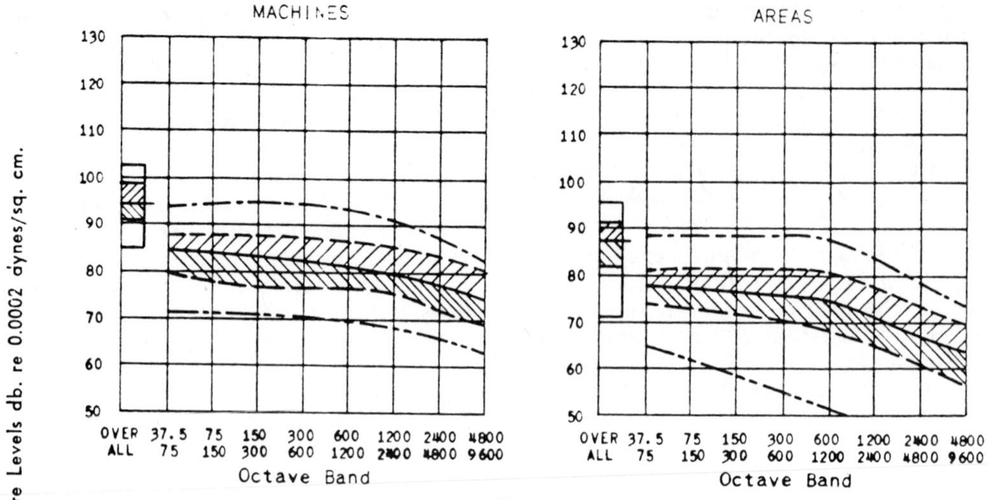

Fig. 5.
Sound pressure levels in the food industry.

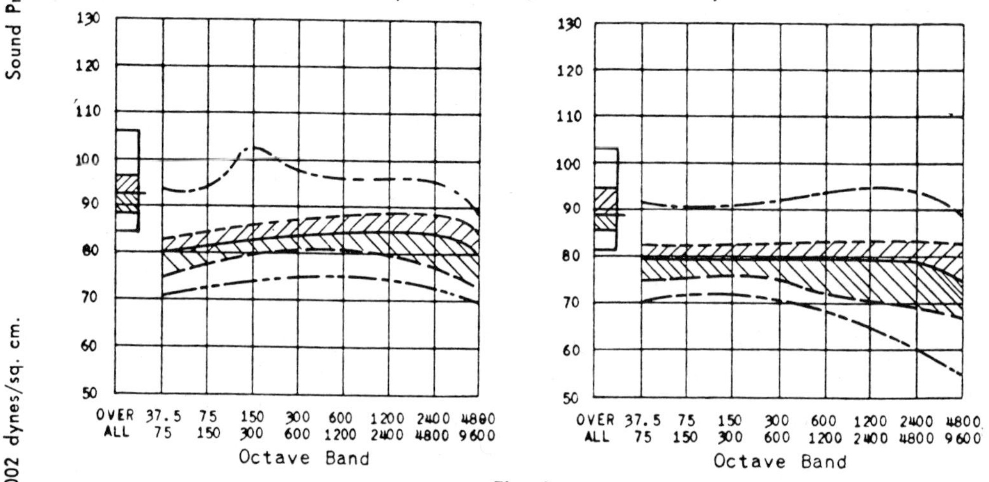

Fig. 6.
Sound pressure levels in the textile industry.

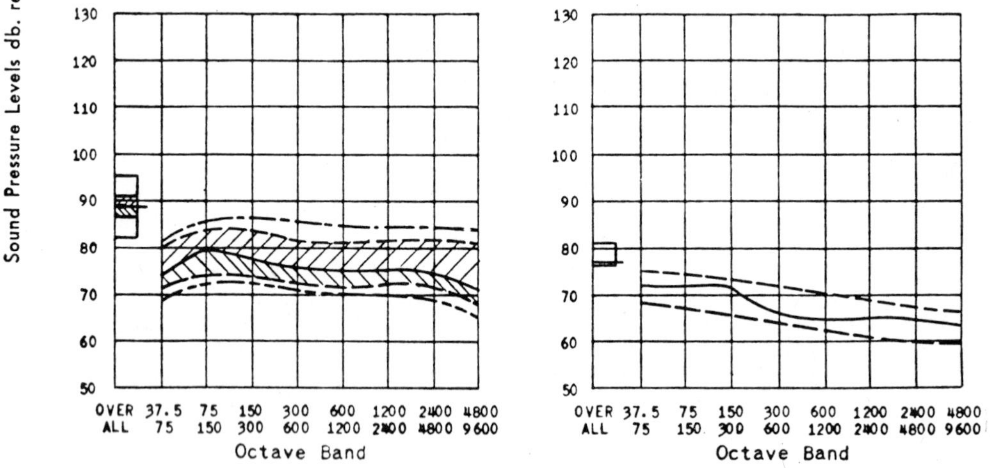

Fig. 7.
Sound pressure levels in the apparel industry.

Upper Limit ——·——·——
Upper Quartile Deviation — — — — —   Median ——————   Lower Quartile Deviation — — — — —
Lower Limit ——··——··——

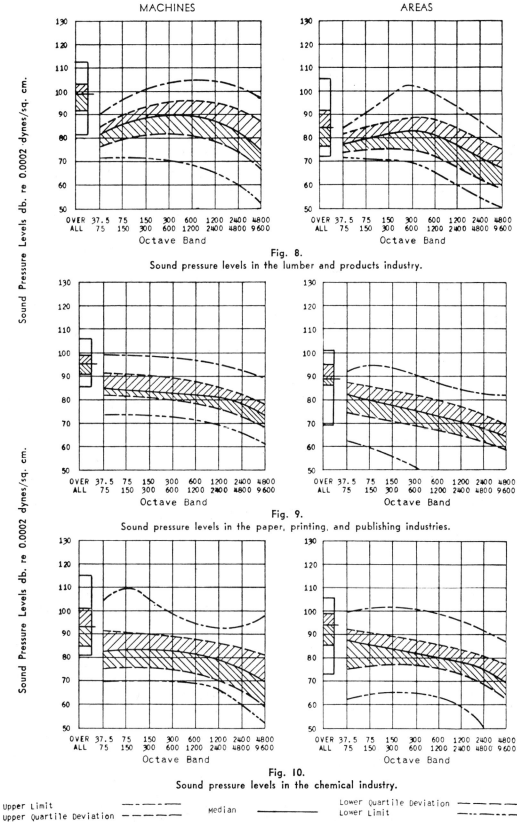

Fig. 8.
Sound pressure levels in the lumber and products industry.

Fig. 9.
Sound pressure levels in the paper, printing, and publishing industries.

Fig. 10.
Sound pressure levels in the chemical industry.

| Upper Limit | — — — · — | | Lower Quartile Deviation — — — — — |
| Upper Quartile Deviation — — — — — | | Median ———— | Lower Limit — · — — · — |

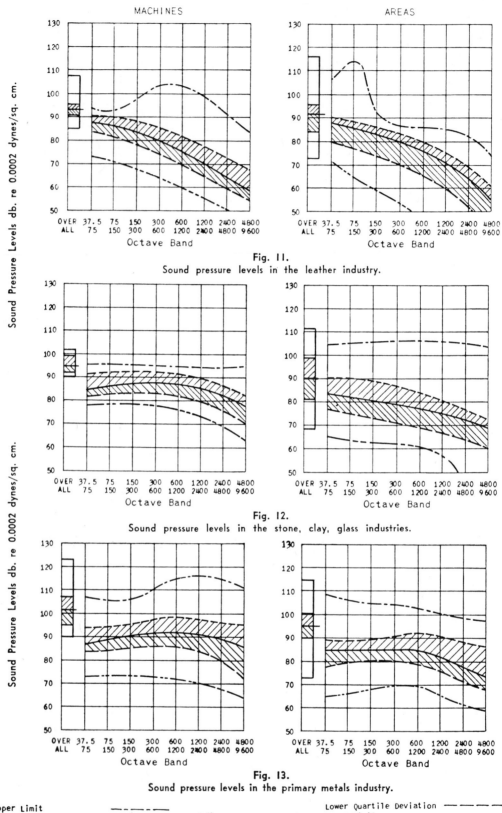

Fig. 11.
Sound pressure levels in the leather industry.

Fig. 12.
Sound pressure levels in the stone, clay, glass industries.

Fig. 13.
Sound pressure levels in the primary metals industry.

Upper Limit — · — · —
Upper Quartile Deviation — — — —   Median ——————
Lower Quartile Deviation — — — —
Lower Limit — · · — · · —

298

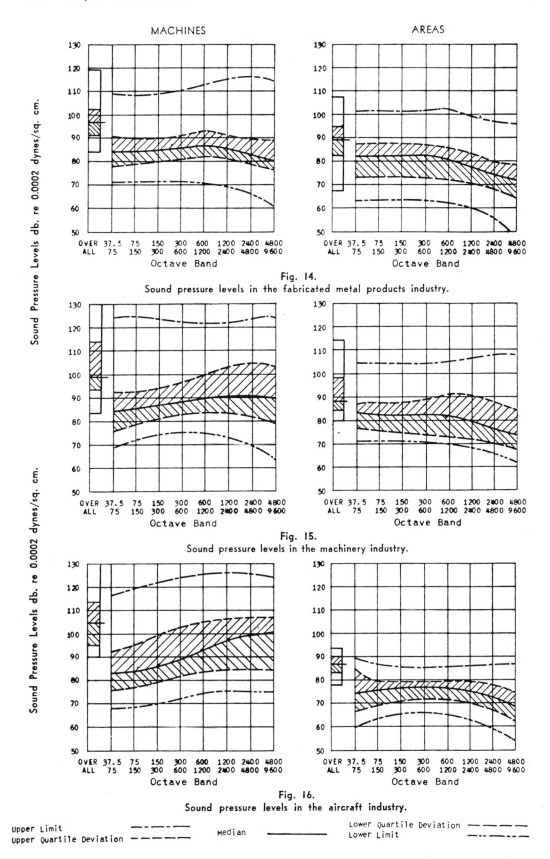

MACHINES

AREAS

Fig. 14.
Sound pressure levels in the fabricated metal products industry.

Fig. 15.
Sound pressure levels in the machinery industry.

Fig. 16.
Sound pressure levels in the aircraft industry.

Upper Limit
Upper Quartile Deviation   Median   Lower Quartile Deviation
Lower Limit

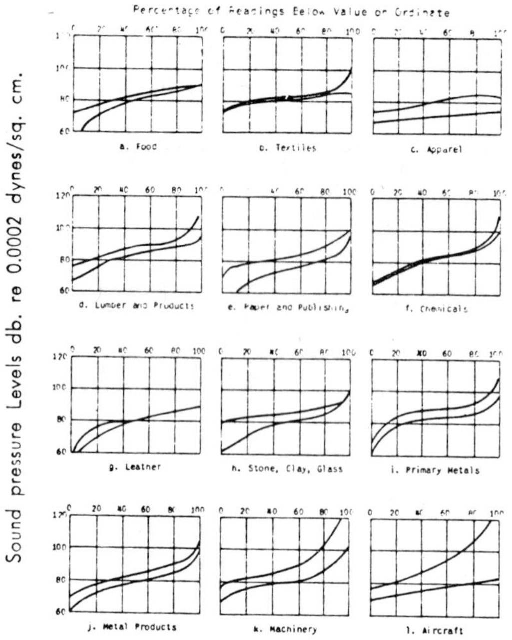

Percentage of Readings Below Value on Ordinate

Sound pressure Levels db. re 0.0002 dynes/sq. cm.

a. Food   b. Textiles   c. Apparel

d. Lumber and Products   e. Paper and Publishing   f. Chemicals

g. Leather   h. Stone, Clay, Glass   i. Primary Metals

j. Metal Products   k. Machinery   l. Aircraft

Fig. 17.

Distribution of sound pressure levels in the 150-
300 cps. octave band.

sorbed better than low frequencies, contributing further to the same effect. In
some plants, large quantities of materials
which absorb sound were found. Examples
of these were large piles of corrugated
cardboard for making boxes, or exposed
cloth in the apparel industry.

(4) The curves of Figs. 17-19 indicate the
range of observed values. The steep slopes
near the 100% lines indicate that the upper
limit of stone, clay, and glass industries
shown in Fig. 12 were due to very few observations. These were near a very large
machine making glass jars. Relatively larger numbers of observations close to the
upper limit are shown by less steeply sloping lines near the limit. This condition is
seen somewhat more often in the low frequency band shown in Fig. 17.

It may be well to point out again that
not much significance is to be attached to
the lower limit of machine levels; machines
with less noise exist but could not be measured.

C. LOUDNESSES IN INDIVIDUAL INDUSTRIES:

The loudnesses found in individual industries are plotted against the percentage of

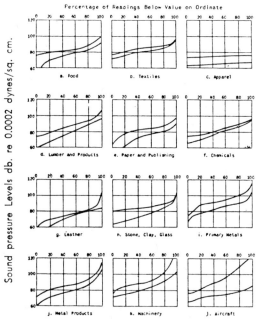

Percentage of Readings Below Value on Ordinate

Sound pressure Levels db. re 0.0002 dynes/sq. cm.

a. Food   b. Textiles   c. Apparel

d. Lumber and Products   e. Paper and Publishing   f. Chemicals

g. Leather   h. Stone, Clay, Glass   i. Primary Metals

j. Metal Products   k. Machinery   j. Aircraft

Fig. 18.

Distribution of sound pressure levels in the 600-
1200 cps. octave band.

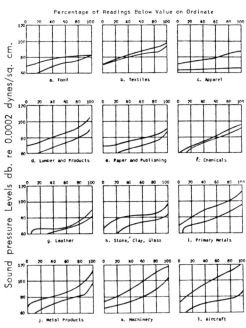

Percentage of Readings Below Value on Ordinate

Sound pressure Levels db. re 0.0002 dynes/sq. cm.

a. Food   b. Textiles   c. Apparel

d. Lumber and Products   e. Paper and Publishing   f. Chemicals

g. Leather   h. Stone, Clay, Glass   i. Primary Metals

j. Metal Products   k. Machinery   l. Aircraft

Fig. 19.

Distribution of sound pressure levels in the 2400-
4800 cps. octave band.

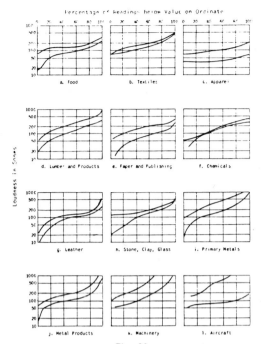

Fig. 20.
Distribution of total loudnesses.

measurements falling below this loudness in Figs. 20 and 21. Fig. 20 shows total loudnesses, and Fig. 21, loudnesses in the loudest band.

The curves generally follow a pattern similar to the distribution curves of octave band sound pressure levels, and especially to those for the 2400-4800 cps. octave band, as this is the most important one for determining loudness, particularly at the higher levels. As before, the metal industries are found to have the loudest machines with the lumber and products industry coming next. Relatively loud areas are found in the machinery, the primary metals, the textile, and the chemical industry.

D. SOUND PRESSURE LEVELS OF INDIVIDUAL MACHINES:

The octave band sound pressure levels of a few widely used machines are given in Figs. 22 to 33. They are presented approximately in order of decreasing loudness. The lowest, median, and highest levels observed are shown. All of the machines shown have sufficient uniformity that these three values are an adequate description.

The loudest operations encountered were chipping and riveting on large steel tanks

and aircraft assembly, shown in Fig. 22. These have been separated from similar operations on smaller or more massive pieces, like small castings or concrete, which have been grouped with other pneumatic power tools such as drills and wrenches in Fig. 23. It is seen here, and it is important to note, that the noise is much more a function of the work than of the tool. A chipping hammer used on large steel plates causes these to vibrate with large amplitude and to radiate noise effectively. Massive castings or concrete structures are less easily excited and radiate less noise.

The spectra of saws, Fig. 24, include a metal friction saw, circular wood saws, and stone saws. The upper limit in the four high bands is due to the friction saw. A log slasher was slightly lower, and a circular wood saw measured close to 90 db. in every octave band; stone saws were about 85 db. The lower limit is a measurement of a small cut-off saw on ½ x ¼ inch molding.

Fig. 25 shows planers which include operations on wood and stone. A wide range of intensities was found. Planing a flat surface on fairly soft stone was the least noisy, whereas planing a concave stone surface

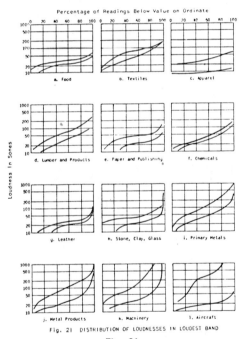

Fig. 21 DISTRIBUTION OF LOUDNESSES IN LOUDEST BAND

Fig. 21.
Distribution of loudnesses in loudest band.

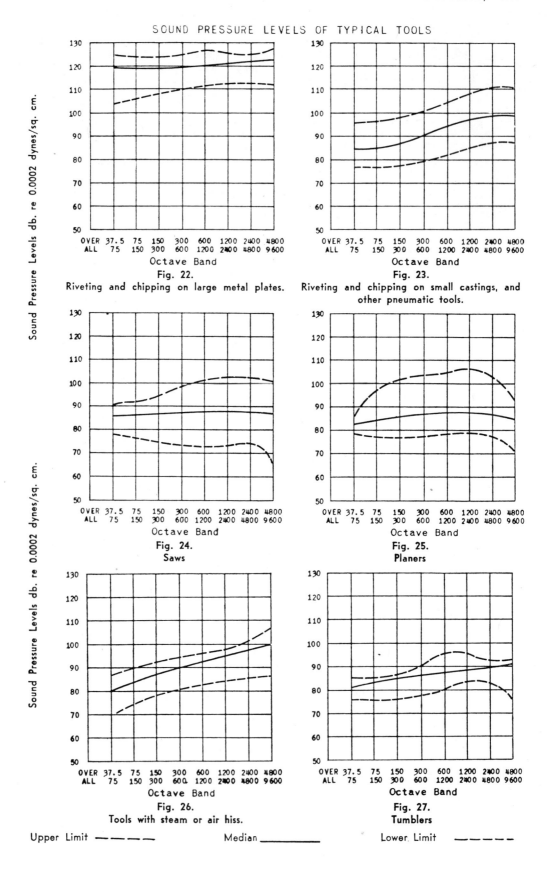

SOUND PRESSURE LEVELS OF TYPICAL TOOLS

Fig. 22.
Riveting and chipping on large metal plates.

Fig. 23.
Riveting and chipping on small castings, and other pneumatic tools.

Fig. 24.
Saws

Fig. 25.
Planers

Fig. 26.
Tools with steam or air hiss.

Fig. 27.
Tumblers

Upper Limit ‒‒ ‒‒ ‒‒   Median _____   Lower Limit ‒ ‒ ‒ ‒

SOUND PRESSURE LEVELS OF TYPICAL TOOLS

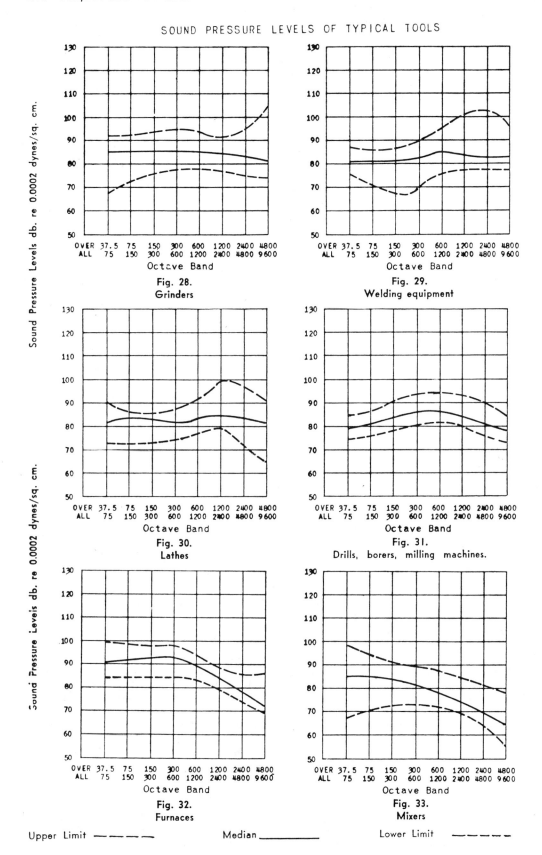

Fig. 28.
Grinders

Fig. 29.
Welding equipment

Fig. 30.
Lathes

Fig. 31.
Drills, borers, milling machines.

Fig. 32.
Furnaces

Fig. 33.
Mixers

Upper Limit — — — —   Median _____   Lower Limit   — — — —

was fairly close to the upper limit. Finish planing pine was close to the lower limit, whereas the normal planing operations of making a deep cut on hardwood were found close to the upper limit.

Steam and air hiss, as encountered in many operations, have upsloping spectra, as shown in Fig. 26. Upsloping spectra were also found in tumblers, Fig. 27, in which small castings are knocked together in a deburring operation. Grinding, welding, and machine shop operations like lathe work, drilling, milling, and boring are in general less noisy, as shown in Figs. 29-31. The high level shown in Fig. 30 was a screech produced periodically by a lathe taking a very thick cut.

Furnaces and mixers, Figs. 32 and 33, are examples of machines having downsloping spectra. Furnaces have fairly high overall levels, but are not usually judged to be very loud because of the low intensity in the high frequency region. Mixers were the least noisy class of machine listed here.

IV. Summary

STEADY-STATE noise levels, as they affect workers in manufacturing industries, were measured and analyzed in octave bands. Measurements made close to the principal noise sources were studied separately from those in general areas. The latter, though less intense, generally affect a larger number of workers.

Wide variations of noise levels were found. Analysis by industry showed that the noisiest machines were found to be those working on metal, followed by those working on wood. The noisiest areas were found in the machinery, the chemical, the stone, clay and glass, and the textile industries. Large differences between the machine and area levels were found in the aircraft, the machinery, and the lumber and products industries where machines were widely spaced or run intermittently, or where large amounts of material with sound absorbing properties were present. Little difference between the machine and area levels was found where many large machines were close together, as in the textile, the chemical or the leather industries.

The analysis showed that about 50% of the machine levels were between 90 and 100 db., with total loudnesses between 100 and 300 sones, and loudnesses in the loudest band between 30 and 100 sones. In general working areas, about 50% of the measurements were between 85 and 95 db. overall, total loudnesses between 50 and 200 sones, and loudnesses in the loudest band between 20 and 50 sones.

As a measure of an average or a most frequently occurring value, the medians have been computed. These varied for machines from 88 db. overall with a loudness of 100 sones in the apparel industry, to 103 db. with 500 sones in the aircraft industry. Medians for areas varied from 77 db. with a loudness of 30 sones in the apparel industry, to 93 db. with 170 sones in the chemical and the stone, clay, and glass industries.

In analyzing the data according to the type of machine or operation, most variations were relatively small. Many similar operations can be associated with typical noise spectra, and in several instances shown they did not differ by more than 8 db. from the median levels in octave bands. The most intense noises were found to be produced by pneumatic riveting and chipping, especially if the work consisted of large metal plates. The noise of these operations, as well as many others, depends not only on the machine but also on the work. Other factors affecting noise of machines are design, installation, and age.

An approximate relation between loudness and overall level has been given. These data indicate that if the overall noise level is known, the loudness can be estimated approximately to within limits of two to one. This corresponds to a change of level of about 8 db. For loud noises, the greatest contribution to the loudness was in the 2400-4800 cps. band in most cases.

V. Conclusion

A SURVEY of steady-state noise affecting workers in manufacturing industries has been made. Representative samples of a wide range of industries have been included. In a few industries more samples might have been considered desirable, but in the metal working industries, having the most severe noise problems, extensive data have been collected. The relative noise levels in various industries and for various operations have been discussed. These data should be of value in determining remedial measures. General area levels are less noisy than machine levels, so that the majority of workers are not exposed to very high levels.

It has been pointed out that more precise weighting would require consideration of the number of operators and duration of exposure. Further investigation of effects on exposure time and recovery period is needed.

VI. Acknowledgment

THE PROGRAM of which this is a part has been under the direction of DR. H. C. HARDY, Assistant Chairman, Physics Research Department, and D. B. CALLAWAY, Supervisor, Sound and Vibration Control Section. The cooperation of the Foundation Acoustics staff is acknowledged, particularly that of F. G. TYZZER, Senior Physicist, who assisted in the analysis and interpretation of the data.

References

1. NOISE: CAUSES, EFFECTS, MEASUREMENT, COSTS, CONTROL: Published proceedings of a Symposium. School of Public Health, The University of Michigan, Ann Arbor, February, 1952.

2. Problems of Noise in Industry. *A.M.A. Arch. of Indust. Hyg. and Occup. Med.*, February, 1952.

3. Proceedings of National Noise Abatement Symposia, Chicago. Vol. 1: October, 1950. Vol. 2: October, 1951. Vol. 3: October, 1952. National Noise Abatement Council, 9 Rockefeller Plaza, New York, N.Y.

4. G. L. BONVALLET: The Measurement of Industrial Noise. *Amer. Indust. Hyg. Assn. Quarterly*, 13:136, (September) 1952.

5. C. R. WILLIAMS, J. R. COX: Industrial Noise Measurement—Science or Art? *Proc. Third Natl. Noise Abatement Symposium*, October, 1952.

6. L. L. BERANEK. J. L. MARSHALL, A. L. CUDWORTH, A. P. G. PETERSON: The Calculation and Measurement of the Loudness of Sounds. *J. Acoust. Soc. Am.*, 23:261 (May) 1951.

7. F. MINTZ and F. G. TYZZER: A Loudness Chart for Octave Band Data on Complex Sounds. *J. Acoust. Soc. Am.*, 24:80 (January) 1952.

8. K. D. KRYTER: The Effects of Noise on Man. *Jour. Speech and Hearing Disorders*, Monograph Supplement No. 1, September, 1950.

9. L. L. BERANEK: Noise Control in Offices and Factory Spaces. (Industrial Hygiene Found. Am.) *Trans. Chem. Eng. Conf.*, 15th Annual Meeting, November, 1950.

10. H. O. PARRACK: Physiological and Psychological Effects of Noise. *Proc. Second Natl. Noise Abatement Symposium*, October, 1951.

11. H. C. HARDY: Tentative Estimate of a Hearing Damage Risk Criterion for Steady State Noise. *Jour. Acoust. Soc. Am.*, 24:756 (November) 1952.

Appendix

In the tables which follow, overall and octave band sound pressure levels as well as total loudnesses in sones and the loudness in the loudest band in sones are tabulated as determined from the field survey. Data are subdivided into types of locations. Those taken close to the principal noise sources within two to five feet from a machine and in the vicinity of the operator are called machine levels. Other measurements were taken in places where men worked at some distance from the chief sources of noise and several machines could generally be heard. These have been termed area levels.

It is emphasized that these data are indicative only and that variations in identical operations even in machines of similar or identical make may be large. Only in cases where measurements of the same type of operation were available, can statistical significance be attached to the data.

TYPICAL SOUND PRESSURE LEVELS AND LOUDNESSES
IN INDUSTRIAL PLANTS

NO.	INDUSTRY	OVER-ALL LEVEL db.	OCTAVE BAND LEVELS. db.								TOTAL LOUD-NESS SONES	LOUD-EST BAND SONES
			37.5 75	75 150	150 300	300 600	600 1200	1200 2400	2400 4800	4800 9600		
a	FOOD MACHINES											
1	Chop Cut Machine, Meat	100	98	91	90	89	90	84	80	84	230	44
2	Canning Machine, Meat	92	76	72	75	79	90	85	77	74	150	44
3	Vacuum Mixer, Meat	100	94	90	82	94	96	86	70	77	230	62
4	Sausage Machine	83	70	73	72	76	78	74	70	68	76	18
5	Bacon Presser	88	82	80	77	80	82	81	81	74	140	36
6	Ham Can Cleaner	90	85	86	82	77	78	77	78	79	130	28
7	Fudge Molding Machine	95	92	88	87	84	83	81	78	71	170	28
8	Candy Wrapper	100	100	84	75	76	79	80	82	76	160	40
9	Peanut Blanching Machine	95	92	87	89	83	83	81	81	75	180	37
10	Covering Candy with Nuts	92	86	86	85	84	81	79	77	73	150	26

TYPICAL SOUND PRESSURE LEVELS AND LOUDNESSES IN INDUSTRIAL PLANTS†

NO.	INDUSTRY	OVER-ALL LEVEL db.	OCTAVE BAND LEVELS. db.								TOTAL LOUD-NESS SONES	LOUD-EST BAND SONES
			37.5 75	75 150	150 300	300 600	600 1200	1200 2400	2400 4800	4800 9600		
	FOOD (Cont'd)											
	AREAS											
1	Sausage Kitchen	96	89	91	87	87	88	86	80	77	210	43
2	Meat Preparation Room	96	85	86	91	91	88	80	76	73	180	38
3	Bacon Slicing Room	83	78	74	74	75	76	72	70	70	76	16
4	Canning Area	74	66	66	66	66	66	66	66	66	41	10
5	Sausage Making Room	72	70	67	56	49	52	51	47	47	15	2.6
6	Chocolate Coating	91	82	84	84	83	82	79	73	64	130	24
7	Candy Mixing	87	83	79	83	73	70	63	62	55	67	17
8	Packing Department	82	79	74	71	69	73	72	74	68	72	20
9	Blanching and Roasting	92	89	84	86	82	80	80	73	70	140	26
10	Chocolate Melter and Mixing Pit	89	84	81	80	82	83	66	59	52	92	26
11	X-Ray Room	84	82	77	74	72	71	66	65	59	62	11
12	Covering Candy with Nuts	90	86	83	81	81	79	75	75	71	120	22
b	**TEXTILES**											
	MACHINES*											
1	Reducing Machines, 72 Spindle	93	76	78	80	81	88	85	85	78	180	55
2	Reducing Machine, 60 Spindle	95	78	82	80	84	87	92	90	85	180	89
3	Reducer Transfer Machine	89	76	78	76	78	82	85	83	78	150	44
4	Speeder Machine, Yarn Processing	97	76	79	84	85	89	93	93	85	290	110
5	Combing Machines	95	84	84	85	84	84	86	88	84	230	80
6	Pin Drafters	92	77	77	78	79	85	85	87	85	190	66
7	Preparers	97	81	84	87	90	90	92	91	85	300	99
8	Finishers	98	82	85	85	88	90	92	94	89	320	110
9	Spinning Frames	86	80	81	80	75	77	75	73	68	97	18
10	Twisters, Universal	87	81	83	82	79	79	78	78	74	120	28
11	Twisters, Cap Type	91	84	86	84	80	84	84	84	75	180	49
12	Ring Spinning	89	76	78	86	84	82	78	75	74	130	24
13	Carding Area	86	80	80	80	77	77	75	75	70	99	23
14	Combing Machines	86	77	74	75	76	80	81	81	78	290	99
15	Preparers	92	75	76	81	83	86	86	85	80	190	55
16	Woven Material Washing Area	103	80	84	103	93	85	75	70	66	190	70
17	Drawing Area	93	81	85	84	83	87	87	86	85	220	60
18	Weaving Shed	102	94	91	86	87	95	97	97	90	490	190
	AREAS*											
1	Combing Machines	95	84	84	85	84	84	86	88	84	230	80
2	Pin Drafters	92	77	77	78	79	85	85	87	85	190	66
3	Finishers	98	82	85	85	88	90	92	94	89	320	110
4	Spinning Frames	87	80	81	80	75	77	75	73	68	97	18
5	Ring Spinning	90	76	77	80	81	83	82	82	79	150	40
6	Reelers	82	75	74	74	72	74	73	73	64	87	19
7	Card Machines	86	80	78	78	75	75	76	76	73	100	25
8	Combing Machines	94	77	82	83	82	84	89	88	86	230	72
9	Dry Finish Area	87	79	84	80	78	75	73	66	56	85	15
10	Dye House Area, Noise of Preparer	85	79	75	76	75	76	77	76	72	94	24
11	Web Finish Area	87	79	81	78	78	82	74	66	57	92	24
12	Drawing Room	89	79	81	80	78	80	82	82	78	140	40
13	Spinning Machines	84	75	77	77	74	72	74	74	75	83	20
14	Storage Room, Machines on Floor above	95	92	90	85	81	85	77	69	62	150	30
15	Weaving Shed, Looms	101	92	90	85	86	94	96	96	89	450	170

*Levels in yarn processing areas are generally between machines, since this is the usual operator position. Many of the levels in this industry are typically area levels as well as machine levels and therefore are listed in the list of area levels
†See introductory paragraph at the beginning of appendix.

TYPICAL SOUND PRESSURE LEVELS AND LOUDNESSES IN INDUSTRIAL PLANTS†

NO.	INDUSTRY	OVER-ALL LEVEL db.	OCTAVE BAND LEVELS, db.									TOTAL LOUD-NESS SONES	LOUD-EST BAND SONES
			37.5 75	75 150	150 300	300 600	600 1200	1200 2400	2400 4800	4800 9600		
c	**APPAREL**											
	MACHINES											
1	Sewing Machine, Men's Garments	87	78	85	78	76	73	75	75	70	97	22
2	Button Holer, Garments	81	72	75	75	73	76	76	76	68	87	24
3	Fabric Cutter, Garments	91	75	85	81	75	81	85	85	86	180	55
4	Fabric Cutter, Garments	89	71	83	73	73	75	75	81	82	110	37
5	Sewing Machine, Garments	85	77	77	78	76	79	73	71	71	88	19
	AREAS											
1	Sewing Machines, Wearing Apparel	77	72	73	68	63	62	65	65	60	40	9.5
2	Button Holers, Wearing Apparel	77	70	72	73	66	64	63	63	56	40	7.9
3	Sewing Machines, Wearing Apparel	81	73	75	75	72	67	67	66	66	56	10
d	**LUMBER AND PRODUCTS**											
	MACHINES											
1	Wood Planer, ⅜″ Cut on 12″ x 16″ Pine Pieces	108	80	77	102	94	100	105	98	93	720	250
2	Circular Cut-off Saw, End Cut on 6″ x 2″ x 1″ Beech	106	83	80	82	88	89	87	104	102	720	400
3	Drum Barker, Tumbling of 2′ Long x 6″ Dia. Log Pieces	98	92	91	93	88	89	78	74	77	201	41
4	Slasher, Four 36″ Circ. Saws, 12″ Logs	99	90	90	88	91	96	88	89	92	340	80
5	Drum Sander, 36″, on Pine Millwork	92	84	87	83	86	81	77	70	66	130	27
6	Finishing Planer, 30″ on Pine Windows	90	84	83	82	85	84	80	79	77	150	31
7	Jointer, 16″ on Millwork	96	84	91	93	86	84	73	72	62	160	36
8	Drum Sander, 54″, Three 12″ Drums, Millwork	85	84	79	75	73	73	70	69	58	74	14
9	Shaper, Moulding Millwork	92	88	87	81	88	87	84	82	74	190	40
10	Double Surfacer, 30″ Rough Surfacing Pine	98	88	85	87	88	87	92	93	84	320	120
11	Cut-off Saw, 15″, 2″ x 4″ Pine	90	88	83	77	75	77	76	79	74	120	31
12	Planer, 30″, 4′ x 8″ x 2″, Oak Pieces	100	82	86	96	95	95	84	84	74	270	62
13	Sander, Three 14″ Drums, 10′ x 6″ Oak	90	83	82	83	82	83	77	76	72	130	26
14	Shaper, Pine Millwork	96	82	87	89	92	86	86	86	72	230	60
15	Sticker Moulder, Shaping Furniture Parts	101	85	89	88	91	96	97	92	84	400	120
16	Planer, Furniture Parts	112	84	88	104	102	107	107	98	100	940	310
17	Double-Ended Tenoner, Furn. Rails	99	81	89	96	88	84	89	89	85	280	80
18	Cut-off Saw, Comp. Air, Furn. Parts	107	85	84	88	101	94	100	99	100	660	240
19	Surfacer, 12″ x 6″ x 1″ Birch	104	85	86	96	96	101	95	95	85	490	150
20	Automatic Lathe, Birch Bed Posts	100	91	91	95	90	95	94	88	82	350	90
21	Jointer, Idling	99	80	86	98	92	80	76	74	68	170	51
22	Drum Sander, on Veneer Sheets, 8′ x 4′	98	85	86	90	92	95	85	84	78	250	62
23	Double Ended Tenoner, Furniture Components	96	78	86	92	92	84	78	90	74	240	89
24	Shaper, Furniture Components	87	78	79	82	76	76	79	79	73	110	30
25	Planer, Maple Furniture	103	78	81	96	91	99	98	95	85	480	150
26	Molder, Maple Furniture	101	76	83	88	97	96	94	92	85	380	110
27	Planer, Furniture Components, Hard Wood	101	82	87	88	92	95	95	93	90	390	120
	AREAS											
1	Mill Shop, Joiners, Planers, etc.	91	85	86	86	82	84	75	74	73	130	28
2	Carpenter Shop, Distant Noise Only	73	70	66	65	57	58	56	55	47	22	4
3	Log Sawing, Barking, Cutting Knots, etc.	99	90	93	91	89	95	81	76	72	230	62.3
4	Wood Finishing, Sanding, Planing, Jointing	85	84	75	74	74	71	69	67	57	67	12
5	Wood Finishing, Sanding, Planing, Jointing	83	82	75	73	70	65	60	56	52	45	8.5
6	Wood Finishing, Sanding, Planing, Jointing	87	82	77	83	78	77	70	66	52	84	17
7	Furniture Making, Planers, Jointers, Saws, etc.	101	82	86	95	96	95	92	86	78	320	72

†*See introductory paragraph at the beginning of appendix.*

307

TYPICAL SOUND PRESSURE LEVELS AND LOUDNESSES IN INDUSTRIAL PLANTS†

NO.	INDUSTRY	OVER-ALL LEVEL db.	37.5 75	75 150	150 300	300 600	600 1200	1200 2400	2400 4800	4800 9600	TOTAL LOUD-NESS SONES	LOUD-EST BAND SONES
	LUMBER & PRODUCTS—AREAS (Cont'd)											
8	Furniture Making, Planers, Jointers, Saws, etc.	94	79	85	91	86	84	83	81	73	180	37
9	Furniture Making, Lathes, Saws, etc.	103	80	82	90	103	89	83	81	73	290	80
10	Veneer Dept., Veneer Mach., Presses, Saws	81	76	77	72	71	66	65	64	54	49	9.2
11	Furniture Making, Planers, Shapers, Molders	97	78	86	92	92	84	78	90	74	240	89
12	Furniture Making, Planers, Shapers, Molders	84	75	78	80	75	71	69	68	62	71	14
13	Furniture Making, Planers, Shapers, Molders	94	78	80	88	88	89	78	74	70	150	41
14	Furniture Making, Planers, Shapers, Molders	91	78	80	84	82	85	83	80	75	150	34
15	Spraying Varnish, Air Exhaust, etc.	89	76	82	84	83	82	72	62	58	99	24
e	**PAPER, PRINTING, AND PUBLISHING MACHINES**											
1	Drum Barker, Tumbling Logs	98	92	91	93	88	89	78	74	77	200	41
2	Slasher, Four 36″ Dia. Saws, 50 Hp.	98	90	90	88	91	96	88	89	92	340	80
3	Wet Machine, Pulp Preparation	96	93	91	84	80	77	76	70	67	140	24
4	Stock Mixer, 300 Hp., Pulp Process	95	85	80	82	83	93	83	83	80	200	53
5	Paper Machine, Wet End, 400′ x 16′ x 12′	94	89	89	85	85	86	79	75	75	160	32
6	Calender, Clutch, and Gear Noise	97	84	94	88	85	91	87	82	83	230	47
7	Suction Pump, for Paper Machine	101	91	97	98	95	87	81	80	82	250	50
8	Printer-Slotter, Prints, Slots Cartons	98	90	91	89	92	92	92	88	80	300	75
9	Box Stitcher, Stitches Cartons	87	77	82	78	78	78	80	80	76	120	34
10	Folder, Automatic Taper (Cartons)	100	93	94	91	89	90	84	82	78	200	40
11	Folder, Automatic Taper (Cartons)	95	87	89	87	86	86	83	82	78	240	44
12	Corrugator, Double, Corrugated Cardboard	100	82	81	94	95	94	88	86	78	280	60
13	Corrugator, as 12 Cutting End	96	88	87	90	91	86	79	75	60	180	37
14	Corrugator, Single	89	80	81	83	82	77	71	71	70	100	21
15	Printer-Slotter, Prints, Slots Cartons	89	84	84	79	80	75	72	72	64	99	18
16	Trimmer Section, Pamphlet Stitcher	91	83	80	78	78	82	81	80	85	150	35
17	Folder, Large Sheet, Noise Due to Worn Parts	94	86	81	81	81	84	85	86	90	200	60
18	Folder, Large Sheet, Noise Due to Worn Parts	91	80	79	79	81	84	83	84	89	180	49
19	Shear, Idling, Noise Due to Worn Parts	103	88	86	82	85	92	97	86	89	320	120
20	Two-Color Press, 47″ x 71″	93	84	88	86	81	83	81	80	....	160	35
21	Two-Color Rotary Press, 45″ x 65″	92	86	88	84	82	80	76	76	80	140	25
22	Two-Color Press	92	83	83	84	83	83	81	80	85	170	35
23	Letter Press, Size 6/0	93	78	84	84	83	87	84	79	76	170	37
24	Four-Color Offset Photography Press	92	82	83	84	86	88	82	76	74	150	35
25	Ink Mill, Noise of Worn Parts	93	76	82	86	85	89	82	72	70	150	41
26	Cleaning Tank, Steam Noise	87	82	81	78	75	76	71	68	72	83	14
27	Printer-Slotter, Cardboard Boxes	99	90	91	89	92	92	92	88	80	300	75
28	Metal Stitcher	91	86	84	82	82	82	78	78	74	130	28
29	Combining Machine, Pasting Multi-Ply Sheets	92	85	86	86	83	86	83	73	76	160	32
30	Paper Machine, 300′ x 14′ x 12′	106	103	98	93	89	92	88	83	75	300	52
31	Paper Machine, 300′ x 14′ x 12′	97	95	91	82	81	81	80	79	77	160	30
32	Paper Machine, 300′ x 14′ x 12′	98	96	92	83	83	83	87	73	64	190	47
33	Pulp Grinder	105	102	95	91	95	96	94	90	78	420	90
34	Newspaper Press, Standard Type	102	94	99	94	94	90	84	82	75	280	45
35	Newspaper Press, Standard Type	103	95	101	96	94	89	87	84	77	310	51
36	Newspaper Press, Standard Type	102	94	98	94	92	87	80	76	68	230	42
37	Newsprint Rewinder	88	86	78	77	76	79	74	71	67	95	19
38	Linotype Machine	81	77	73	73	74	73	70	68	64	66	13
39	Air Ejector, Cleaning	83	78	71	71	70	67	68	76	74	69	24

†*See introductory paragraph at the beginning of appendix.*

TYPICAL SOUND PRESSURE LEVELS AND LOUDNESSES IN INDUSTRIAL PLANTS†

NO.	INDUSTRY	OVER-ALL LEVEL db.	OCTAVE BAND LEVELS. db.								TOTAL LOUD-NESS SONES	LOUD-EST BAND SONES
			37.5 75	75 150	150 300	300 600	600 1200	1200 2400	2400 4800	4800 9600		
	PAPER, PRNTG. & PUBLISHING (Cont'd)											
	AREAS											
1	Wet Machines, Paper Making	95	93	90	85	80	75	75	69	67	130	23
2	Beater Room, Pulping Machines	89	85	84	82	79	80	75	72	70	110	21
3	Paper Machine, 400' x 16' x 12'	94	89	89	85	85	86	79	75	75	160	32
4	Calenders	88	83	84	78	80	80	69	64	63	91	21
5	Paper Machine Auxiliary Equip., Pumps, etc.	96	88	93	90	85	88	77	75	72	170	29
6	Cardboard Box Folding and Taping	87	81	77	77	77	80	78	78	74	110	28
7	Corrugator	88	80	81	83	78	78	71	68	65	95	18
8	Folding Area, Paper Folders, Large Sheets	91	80	78	78	80	83	83	83	86	160	44
9	Cutting Area, Numerous Paper Shears	87	82	80	76	75	78	77	77	79	110	26
10	Pressroom, Two Color Printing	89	83	85	80	77	78	76	75	73	112	22
11	Offset Pressroom, Several Presses	71	69	62	58	55	57	55	52	51	18	3.7
12	Roto Press Room	68	65	60	60	56	54	55	55	52	18	4.3
13	Ink Room	87	76	80	79	79	79	75	72	70	97	19
14	Printing-Slotting Area	93	85	85	87	84	86	82	82	71	180	40
15	Metal Stitching	87	84	78	77	77	77	75	73	67	95	19
16	Shipping Room	76	75	62	60	59	61	60	59	45	26	5.8
17	Combining Area, Pasters, Mixers	93	85	86	86	83	86	83	73	76	160	33
18	Paper Machine, 300' x 14' x 12'	105	103	98	93	89	92	88	83	75	300	52
19	Paper Machine, 300' x 14' x 12'	97	95	91	82	81	81	80	79	77	160	31
20	Paper Machine, 300' x 14' x 12'	98	96	92	83	83	83	87	73	64	190	47
21	Beater Room, Pulping Machines	88	87	80	76	73	71	68	65	58	72	12
22	Proof Room, Noise of Message Conveyor Belt	75	74	66	62	61	59	52	48	40	21	4.3
23	Pressroom, Newspaper Presses	97	88	93	91	86	82	75	70	64	160	31
24	Paper Storage, Trucks, Conveyors	89	86	80	77	77	75	72	74	66	94	20
25	Mail Room, Paper Bundling Machines	84	79	79	71	76	69	64	66	64	61	10
26	Galley Room	83	78	77	77	68	67	67	66	60	57	11
27	Associated Press Room, Teletype Machines	77	74	71	69	60	59	52	49	46	24	4.3
28	Photography Department, Distant Noise	75	74	63	65	54	51	46	44	40	16	3.7
29	Stereotype Room, Working Printing Plates	92	85	89	81	80	75	76	69	58	110	21
30	Composing Room	81	77	73	71	72	67	65	62	56	47	9.9
31	Linotype Machines	81	78	74	72	71	69	67	67	62	56	11
	CHEMICALS											
	MACHINES											
1	Ink Mill, Noise of Worn Parts	93	76	82	86	85	89	82	72	70	150	41
2	Pigment Mills, Steel Ball Grinding	94	82	83	85	86	86	87	85	79	210	55
3	Pebble Mill	84	79	75	77	75	74	73	67	57	72	14
4	Mixer, Enamel Components	83	76	74	73	74	80	67	57	.....	62	21
5	Drum Cleaning	85	74	77	80	76	74	75	74	65	93	22
6	Pigment Mill, Enamel Grinding	82	69	72	76	75	75	73	69	56	70	14
7	Five-Roll Mill	113	105	110	109	101	89	82	79	80	430	102
8	Hammermill	104	98	93	92	97	95	94	92	96	460	110
9	Boiler Room, Blower	100	89	91	98	92	87	84	85	78	260	55
10	Compounding Roll, Plastic Components	94	87	86	84	84	90	84	79	77	190	44
11	Ball Mill, Plastic Components	99	82	82	82	84	93	95	93	92	320	130
12	Fabric Coating, High Temperature Type	81	72	71	68	74	76	69	69	67	61	14
13	Fabric Coating, Low Temperature Type	85	75	75	74	73	80	80	74	61	93	26
14	Fabric Coating, Low Temperature Type	82	71	72	73	73	74	80	64	54	72	26
15	Rewind, Fabric Beaming	93	88	89	84	80	76	75	69	58	110	20
16	Rotary Kiln	95	92	88	83	81	80	72	68	63	120	21
17	Rubber Mill, Tire Rubber	102	99	97	90	89	84	78	77	74	210	36
18	Tread Tubers, Making Tire Casings	97	94	92	86	82	82	81	86	70	210	60
19	Curing Press, Rubber Tire Processing	96	89	87	86	85	87	85	88	88	240	72

†*See introductory paragraph at the beginning of appendix.*

TYPICAL SOUND PRESSURE LEVELS AND LOUDNESSES IN INDUSTRIAL PLANTS†

NO.	INDUSTRY	OVER-ALL LEVEL db.	OCTAVE BAND LEVELS, db.								TOTAL LOUD-NESS SONES	LOUD-EST BAND SONES
			37.5 75	75 150	150 300	300 600	600 1200	1200 2400	2400 4800	4800 9600		
	CHEMICALS—MACHINES (Cont'd)											
20	Heavy Service Curing, Rubber Tire Processing	93	90	85	86	83	78	75	75	66	130	22
	AREAS											
1	Ink Mill Room	86	76	80	79	79	79	75	72	70	97	19
2	Pigment Grinding	92	80	81	84	85	85	85	83	78	180	44
3	Pebble Mills	79	76	70	70	71	68	68	62	47	47	9.6
4	Enamel Mixing	76	71	68	66	65	70	59	50	43	31	10
5	Paint Drum Cleaning	85	74	77	80	76	74	75	74	65	93	22
6	Enamel Component Grinding	74	66	66	68	67	66	63	58	46	33	7.1
7	Rotary Kilns, Heavy Chemical	95	92	88	83	81	80	72	68	63	120	21
8	Vibrator, Heavy Chemicals Conveyor	94	91	89	83	78	81	78	81	78	160	37
9	Mixer, Heavy Chemicals	100	97	92	90	89	88	87	78	72	220	38
10	Gas Compressor Room	94	90	89	80	82	85	80	78	74	160	30
11	Catalytic Cracking Area, Petroleum	105	91	93	101	101	99	90	78	66	350	82
12	Pulverizing Room, Ball Mill, Compounders	99	85	85	85	85	92	94	93	92	330	120
13	Plastic Molding Area	101	100	92	89	86	87	77	78	78	210	39
14	Rubber Mill Department	97	95	92	86	84	77	74	72	66	150	26
15	Tread Tubers Area	97	94	92	86	82	82	81	86	70	210	60
16	Tire Making	105	100	101	98	92	85	85	83	76	300	51
17	Tire Making	98	92	88	91	92	85	83	83	84	230	44
18	Curing Area, Large Curing Presses	95	88	86	84	83	85	84	87	85	220	66
19	Curing Area, Curing Presses, Misc. Equip	93	90	85	86	83	78	75	75	66	130	22
20	Tire Making	94	92	85	87	77	74	70	67	60	110	20
21	Rubber Ply Building and Cutting	95	92	88	87	84	85	80	76	68	170	30
g	**LEATHER**											
	MACHINES											
1	Fleshing Machine, Clipping Hide Irregularities	95	82	85	90	87	86	85	77	62	180	35
2	Hair Dryer	98	97	88	81	90	75	68	64	59	130	35
3	Lime Mixing Tub, Hide Processing Tank	88	84	83	80	77	71	66	60	51	73	14
4	Setting Out Machine, Leather Processing	92	86	85	83	83	81	77	72	65	120	23
5	Splitting Machine	91	85	82	81	81	78	80	81	70	140	37
6	Coloring Drums	97	89	93	90	86	86	77	70	61	170	33
7	Setting Machine. Noise of Hair Cutting	108	90	90	91	105	103	98	90	82	500	130
8	Plating, Preparing Leather Surface	94	91	86	85	82	81	72	66	56	130	38
9	Plating, Preparing Leather Surface	88	87	77	74	72	69	67	64	59	65	12
10	Fine Hairing, Smoothing Irregularities	93	91	84	77	79	77	68	64	57	94	18
11	Seasoning Machine	92	90	85	81	80	72	67	64	59	94	18
12	Spraying Machine	93	88	88	87	76	72	68	68	65	100	23
13	Measuring Machine, Area of Leather Piece	77	74	72	63	60	60	52	50	50	24	4.6
	AREAS											
1	Hide Storage	79	75	75	71	64	59	54	41	33	28	6.2
2	Loading Area	79	77	73	69	66	61	58	50	38	31	6.1
3	Hide Trimming, Manual	82	81	72	70	72	63	58	53	49	39	9.9
4	Fleshing, Clipping Hide Irregularities	93	83	84	87	86	84	80	75	70	150	28
5	Unhairing Machine	91	88	81	83	80	76	71	64	58	95	18
6	Tanning Department	95	92	88	88	83	79	70	65	54	130	25
7	Wringing	93	86	87	87	83	83	78	76	74	150	26
8	Setting Out Dept., Hide Smoothing	92	84	85	83	83	79	77	72	65	120	22
9	Shaving, Splitting	89	86	83	81	79	76	74	74	68	110	20
10	Drum Coloring	95	89	89	87	85	84	79	70	62	150	28
11	Hide Processing, Drying	92	88	85	85	82	79	74	70	64	120	21
12	Setting, Cutting off Hairs	96	90	90	89	87	83	82	82	74	200	40

†See introductory paragraph at the beginning of appendix.

TYPICAL SOUND PRESSURE LEVELS AND LOUDNESSES IN INDUSTRIAL PLANTS†

NO.	INDUSTRY	OVER-ALL LEVEL db.	OCTAVE BAND LEVELS, db.								TOTAL LOUD-NESS SONES	LOUD-EST BAND SONES
			37.5–75	75–150	150–300	300–600	600–1200	1200–2400	2400–4800	4800–9600		
	LEATHER—AREAS (Cont'd)											
13	Drying Loft	96	92	93	85	76	79	71	72	52	130	28
14	Power House	116	109	115	91	79	73	68	68	64	270	120
15	Plating, Preparing Surface	94	91	86	85	82	81	72	66	56	120	23
16	Staking, Stretching, Smoothing Hides	90	88	84	80	74	68	67	59	52	75	14
17	Buffing and Brushing, Unrelated Mach. Noise	96	94	91	83	81	81	78	73	62	150	34
18	Seasoning, Hide Preparation	93	91	90	78	72	68	63	60	54	84	23
19	Trimming, Manual	84	83	73	69	64	61	57	50	44	34	7.4
20	Leather Sorting, Distant Noises	68	66	59	55	53	51	36	34	24	9	2.3
21	Bundling, Packaging	76	74	67	67	58	58	54	48	47	22	4.3
22	Shipping Department	78	77	67	61	58	58	47	43	34	19	4.0
23	Lime Processing	89	88	77	75	70	69	65	63	57	55	9.2
h	**STONE, CLAY, GLASS** **MACHINES**											
1	Clay Crusher	101	97	91	94	94	91	82	76	65	230	47
2	Clay Tile Extrusion Machine	93	89	86	85	83	81	72	68	56	120	23
3	Packing Machine, Glass Jars	95	90	87	87	84	83	83	83	83	200	44
4	Bottle Forming, Glass	100	95	96	90	92	92	89	92	93	380	110
5	Mixer, Glass Components	100	95	93	91	92	91	78	72	75	220	47
6	Bottle Forming, Glass	98	87	89	89	92	90	89	88	83	290	72
7	Carving Stone, Pneumatic Chipper	93	83	85	86	87	84	80	79	75	160	31
8	Cutting Stone, 14″ Cut Off Saw	90	80	78	81	83	83	84	85	81	180	55
9	Cutting Stone Blocks, 48″ Saw	92	82	81	85	83	85	84	83	81	180	44
10	Stone Planer, Flat	90	82	79	82	82	82	83	80	72	150	34
11	Stone Planer, Circle 12′, Concave Surface	101	85	93	89	94	95	94	89	76	350	90
12	Crane (Above 300 cps Area Levels) 7½ Ton	99	92	94	87	84	82	81	80	73	190	34
13	Crane (Above 300 cps Area Levels) 15 Ton	90	82	81	81	80	82	82	83	74	150	44
14	Cutting Large Stone, 72″ Saw	97	82	81	86	84	88	95	90	82	450	100
15	Light Weight Concrete Aggregate	101	94	97	96	90	88	82	76	71	240	44
16	Clay Tile Kiln (Unloading)	99	82	78	88	90	91	92	81	91	290	75
	AREAS											
1	Clay Crushing	101	97	91	94	94	91	82	76	65	230	47
2	Clay Tile Extrusion Department	80	76	71	70	71	70	63	66	54	50	10
3	Boiler Room, Clay Products Plant	86	80	79	79	73	75	76	69	60	84	18
4	Tile Molding, Noise is Distant Machinery	93	92	85	79	70	67	63	61	55	74	20
5	Drying Room, Clay Tile	80	79	73	68	68	63	57	50	36	32	6.6
6	Clay Tile Kiln, Unloading	69	58	56	63	64	65	58	53	46	25	7.1
7	Glass Jar Mold Shop, Machine Repairs	96	96	85	81	75	77	76	76	76	130	24
8	Glass Jar Mold Cleaning Shop	100	99	93	87	84	84	82	78	74	200	34
9	Glass Jar Packing Department	90	87	84	81	80	77	72	74	72	110	19
10	Bottle Forming	112	98	101	102	103	104	104	104	103	1100	420
11	Batch House, Glass Jar Plant	98	96	90	88	88	85	72	71	64	160	30
12	Mixing Area, Glass Jar Plant	100	95	93	91	92	91	78	72	75	220	47
13	Stone Carving Area (Many Machines Contributing)	89	83	83	81	81	79	77	76	73	120	24
14	Stone Planing Area (Many Machines Contributing)	87	82	78	78	77	79	79	78	72	110	28
15	Near Crane (Many Machines Contributing)	90	75	74	80	80	82	86	83	74	170	40
i	**PRIMARY METALS** **MACHINES**											
1	Pneu. Chip., 5′ x ½″ Steel Tank, Weld. Groove	122	109	112	108	114	117	113	112	113	2800	1000
2	Ped. Grinder, 14″, on ⅜″ x 12″ Angle Iron	92	75	74	78	77	82	83	85	88	160	49

†*See introductory paragraph at the beginning of appendix.*

TYPICAL SOUND PRESSURE LEVELS AND LOUDNESSES IN INDUSTRIAL PLANTS†

NO.	INDUSTRY	OVER-ALL LEVEL db.	OCTAVE BAND LEVELS, db.								TOTAL LOUD-NESS SONES	LOUD-EST BAND SONES
			37.5 75	75 150	150 300	300 600	600 1200	1200 2400	2400 4800	4800 9600		
	PRIMARY METALS—MACHINES—(Cont'd)											
3	Furnace, 30' x 30' x 8', Oil, Ingot Heating	96	86	86	86	85	88	88	88	88	260	72
4	Furnace, 7' x 4' x 4', Oil, Heating	103	98	98	94	92	92	85	77	74	280	48
5	Wood Planer, 30" x 40", 16" x 12" x ⅞" Pine	108	80	77	102	94	100	105	98	73	720	250
6	Friction Saw, 46", on 4" x 4" x ¼" Steel Angleiron	107	88	91	89	95	100	101	101	100	760	300
7	Shakeout, 50 Sq. Ft., ½—1 Cu. Ft. Castings	114	110	109	102	105	106	99	97	90	790	160
8	Pneumatic Ram, 100 psi, Sand Mold Ramming	90	83	81	83	85	84	83	74	....	140	33
9	Shakeout, 40 Sq. Ft , ½—1 Cu. Ft. Castings	105	89	86	94	93	101	99	95	88	540	160
10	Tumbler, 36" x 72", Small Castings	100	78	78	86	91	95	95	92	92	390	105
11	Pneumatic Chipper, 100 psi., 1—3 Cu. Ft. Castings	110	88	90	95	96	97	106	104	106	1100	470
12	Push Up Machine, 14" x 5", Sand Molding	103	94	96	91	92	97	94	87	85	380	80
13	Core Blower, Making Sand Cores	116	86	84	86	87	108	101	110	114	1600	800
14	Core Draw Vibrator, Bench Cores	107	85	81	81	90	98	103	102	98	770	330
15	Shakeout, 10' x 10', 2—4 Cu. Ft. Castings	98	90	91	89	87	94	91	90	88	340	90
16	Pneu. Chipper, 100 psi, 1 Cu. Ft. Castings	102	90	90	90	89	94	95	93	92	410	140
17	Pneumatic Chipper, 100 psi, 4'—6' Castings	112	82	84	83	90	109	109	100	96	840	340
18	Air Hoist, 2000 Lb., Pneumatic	112	92	84	88	87	93	93	96	110	620	230
19	Stand Grinder, 36", Small Castings	96	88	87	88	86	92	83	81	81	180	50
20	Electric Furnace, 30 Ton, Early In Run	105	95	100	98	97	94	91	90	86	410	90
21	Sand Slinger, Air Rush and Sand Noise	102	90	96	94	94	94	91	87	88	350	69
22	Jolt Squeeze Machine, 13", Sand Molding	100	87	90	87	91	96	91	92	84	350	110
23	Shakeout, 8' x 5' x 3' Deep. 1 Cu. Ft. Castings	108	91	92	97	99	102	103	100	96	800	280
24	Pre-Mix Burner, 2" Outlet, Gas, Crucible Heating	94	91	89	87	77	78	78	76	77	140	24
25	Roughing Mill, 30' x 5' Rolls, on 4' x 2' x ½" Steel Plates	108	95	102	98	97	98	97	96	102	660	175
26	Leveler, 15' x 8' Area, for above Steel Plates	98	91	93	92	91	88	85	78	70	160	40
27	Decoiler, Uncoiling 2' Wide Strip Steel Coils	116	95	94	93	97	108	107	110	112	1700	800
28	Anneal. Furn., 120' x 60' x 40', Gas, Sheet Steel	101	98	84	91	95	84	77	71	70	190	48
29	Lectromelt Furnace, 60 Ton, Early In Run	103	94	98	97	93	93	81	76	70	270	54
30	Open Hearth, Five, Each 10' Wide	90	88	84	78	77	76	71	69	68	96	16
31	Scarfing, Acetylene Weld. Equip.	91	78	74	67	73	84	83	84	86	140	45
32	Bloomer Mill, Strip Steel	95	85	90	88	87	88	81	81	74	190	36
33	Strip Mill, Steel	97	88	84	82	79	88	85	90	94	270	90
34	Conveyor, Strip Steel	102	82	82	89	96	97	92	86	81	320	77
35	Furnace, 12' x 6' x 6', Heating Forgings	96	84	86	92	91	88	82	78	74	200	37
36	Forging Hammer, 3500 Lb., Steam Exh. Noise	109	86	88	93	91	92	95	102	107	750	330
37	Forging Manipulator, Air Exhaust Noise	101	85	86	90	92	95	92	90	95	370	90
	AREAS											
1	Chipping Area	102	84	85	86	89	91	96	97	101	510	190
2	Grinding Area, Noise Is From Chipping, etc.	92	75	74	80	79	87	87	79	84	170	47
3	Friction Saw Area, Noise Due To Distant Chipping	95	88	90	87	87	84	82	78	77	180	30
4	Shakeout Area, Noise Due To Shakeout	114	110	109	102	105	106	99	97	90	790	160
5	Pneumatic Ramming	91	80	81	83	85	84	83	74	72	140	33
6	Shakeout Area, Noise Due To Shakeout	99	85	86	91	93	95	93	90	86	370	90
7	Tumblers, Approx. 36" x 72", Small Castings	101	78	78	86	91	95	95	92	92	390	100
8	Shot Blast Room	98	82	82	85	87	91	91	91	92	320	100

†*See introductory paragraph at the beginning of appendix.*

TYPICAL SOUND PRESSURE LEVELS AND LOUDNESSES IN INDUSTRIAL PLANTS†

NO.	INDUSTRY	OVER-ALL LEVEL db.	OCTAVE BAND LEVELS. db.								TOTAL LOUD-NESS SONES	LOUD-EST BAND SONES
			37.5 75	75 150	150 300	300 600	600 1200	1200 2400	2400 4800	4800 9600		
	PRIMARY METALS—AREAS (Cont'd)											
9	Molding Area. Push-Up Machines, Shakeouts	103	94	96	91	92	97	94	87	85	380	80
10	Core Room, Core Molding By Machine	92	81	84	84	82	81	87	84	80	140	52
11	Core Room, Misc. Founding Equipment	90	85	82	81	81	81	78	77	74	130	26
12	Core Room, Core Draw Vibrators	90	82	81	81	84	83	76	75	71	130	28
13	Shakeout Area. Other Nearby Equip.	93	88	86	84	82	84	79	76	72	160	30
14	Casting Cleaning Area	97	88	88	88	88	91	85	88	82	250	66
15	Grinding Area, Noise of Handling Castings	93	86	86	84	82	88	77	80	81	160	38
16	Electric Furnace Area	105	95	100	98	97	94	91	90	86	410	90
17	Sand Slinger Area, Other Foundry Operations	93	85	86	87	85	82	74	70	68	130	26
18	Steel Pouring Area, Tumbling. Shakeout Noises	91	86	85	84	75	79	74	73	72	110	24
19	Roughing Mill Area, Handling ½″ Steel Plate	97	92	90	83	87	93	79	72	77	190	53
20	Steel Handling Area, Strip Coiling	87	78	78	79	78	85	74	65	62	68	30
21	Annealing Furnace Area	84	76	78	73	80	76	65	63	60	62	18
22	Lectromelt Furnace Area, Start of Melt	97	88	94	90	88	85	78	74	70	180	31
23	Lectromelt Furnace Area	92	80	90	84	81	78	75	72	70	120	23
24	Open Hearth Area	90	88	84	78	77	76	71	69	68	96	16
25	Scarfing Area, Misc. Distant Billet Handling	78	72	65	67	70	73	69	60	60	47	13
26	Bloomer Mill Area	95	85	90	88	87	88	81	81	74	190	36
27	Strip Mill Area	98	88	84	82	79	88	85	90	94	270	90
28	Conveyor	101	82	82	89	96	97	92	86	81	320	77
29	Drop Forge Area, Distant Operations	91	84	86	85	80	78	74	68	70	110	20
30	Blast Furnace Area	74	69	62	60	56	53	50	51	46	16	3.2
j	**FABRICATED METAL PRODUCTS MACHINES**											
1	Friction Saw. 46″, on 4″ x 4″ x ¼″ Steel	107	88	91	89	95	100	101	101	100	760	300
2	Milling Machine, Working Engine Block	93	76	77	87	88	87	82	78	74	160	33
3	Boring Machine	97	76	80	80	85	93	92	87	82	250	77
4	Pneumatic Press, Assembling Bearing on Shaft	100	74	76	88	85	94	91	94	96	370	140
5	Hob-grinder	91	78	78	78	75	74	79	84	88	150	48
6	Multiple Drill, On Piston Parts	98	78	81	91	93	95	85	77	75	220	64
7	Turret Lathe, Ram Type, On Piston Parts	104	84	84	83	79	87	102	101	93	620	300
8	Sprocket Grinder, Pneu., 1′ Dia., 36″ Castings	96	92	88	88	85	84	83	83	82	210	48
9	Hand Grinder, Pneu., 5″ Dia.. 24″ Castings	98	86	85	94	91	88	89	84	82	260	55
10	Stand Grinder, 36″, Small Castings	96	88	87	88	86	92	83	81	81	180	50
11	Nail Machine, Roofing Nails	110	96	96	94	97	106	102	101	98	850	300
12	Hoop Machine, No. 11 Ga. Steel Wire	101	97	90	87	85	98	87	84	81	270	66
13	Fence Wire Machine, Forming Steel Wire Parts	100	84	82	81	81	96	94	96	95	410	170
14	Wire Drawing Machine	95	84	88	88	85	90	81	76	80	180	44
15	Fence Machine, Forming Steel Wire Into Fence	92	79	81	81	81	91	83	81	82	170	46
16	Wire Drawing Machine	90	78	84	85	83	84	73	68	67	120	28
17	Turret Lathe, 3″ Hole In C. R. S., Noise of Worn Parts	91	81	85	78	80	84	84	82	81	160	37
18	Tracer Lathe, 4′ x 4″, Gear Noise, Cutting Squeal	95	92	85	83	87	86	81	79	74	170	32
19	Turret Lathe, On 3″ C. R. S.	92	82	74	79	82	83	79	87	84	170	65
20	Gear Shaper	88	72	75	78	81	83	81	79	76	130	30

†*See introductory paragraph at the beginning of appendix.*

313

TYPICAL SOUND PRESSURE LEVELS AND LOUDNESSES IN INDUSTRIAL PLANTS†

NO.	INDUSTRY	OVER-ALL LEVEL db.	OCTAVE BAND LEVELS. db.								TOTAL LOUD-NESS SONES	LOUD-EST BAND SONES
			37.5 75	75 150	150 300	300 600	600 1200	1200 2400	2400 4800	4800 9600		
	FABRICATED METAL PRODUCTS— MACHINES (Cont'd)											
21	Vertical Radial Drill, 2″ Dia. on Light Casting	93	77	79	78	79	89	85	85	82	180	55
22	Turret Lathe, Sprocket Cut	97	87	79	79	77	86	95	89	83	240	100
23	Arc Welder, on ½″ Steel Plate	86	84	76	80	76	81	78	76	78	120	24
24	Automatic Drop Hammer, Small Hardened Tools	111	97	94	95	100	106	103	105	104	1100	450
25	Automatic Punch Press, Air Blast Only, Small Tools	107	98	96	96	91	100	97	100	104	670	270
26	Pedestal Grinder, 36″, on Small Tools	106	76	78	76	81	97	90	98	105	530	220
27	Cutting Machine, on Hardened Tools	113	106	108	106	101	104	97	95	98	760	150
28	Sand Blast Machine, Four Position, on Hand Tools	118	87	86	88	94	100	108	113	115	2100	1200
29	Lead Pot Furnace, on Hand Tools	84	77	78	75	75	76	70	74	68	81	18
30	Lead Stirring Motor, Air Exhaust	110	82	81	80	81	87	103	107	103	1000	630
31	Pneumatic Vibrator, Removing Sand Mold Pattern	100	76	74	76	89	84	93	97	94	400	210
32	Core Blower, Sand Cores	102	82	80	78	88	98	89	93	99	370	120
33	Sand Muller, 1500 Lbs., For Mixing Sand	96	78	88	94	84	86	82	78	72	180	38
34	Tumbler, 6′ x 3′, Small Castings	97	85	83	82	82	91	91	90	90	280	90
35	Shot Blast, With Conveyor, Small Castings	100	86	88	87	87	98	88	84	83	270	76
36	Pre-Mix Burner, 2″ Outlet, Gas, Crucible Heating	94	91	89	87	77	78	78	76	77	150	24
37	Blower, For Cupola, 15″, Ducted Air	97	92	91	92	87	79	77	74	74	170	33
38	Milling Machine, Channel Cut, 10″ x 2″ x 2″ Steel	90	86	83	78	79	84	85	82	83	170	40
39	Pneumatic Hammer, 100 psi, Peening Operation	104	85	84	84	85	94	93	96	102	500	190
40	Butt Welder, Electric, on Tool Parts	91	88	76	70	79	88	84	83	83	160	40
41	Back Shear, Four-Cutter, for ⅛″ Steel, Operating	95	80	86	88	87	90	88	80	77	210	57
42	Roll Grinder, Truing 4′ x 30″ Roughing Mill Rolls	85	80	79	78	79	83	75	76	78	80	27
43	Shaper, Small Steel Parts	95	88	86	88	88	92	86	84	75	230	50
44	Corrugating Machine, on Sheet Steel	111	95	92	92	95	108	99	106	101	1000	500
45	Downspout Roll, on Sheet Steel	94	88	88	89	89	90	80	80	75	200	43
46	Corrugating Machine, Sheet Steel Riveting	114	102	98	98	95	112	107	104	101	1200	410
47	Liming Machine, Sheet Steel Processing	87	61	64	69	78	86	70	65	62	78	32
48	Automatic Screw Machine, Nut Blanking, 1⅛″ Stock	97	88	89	87	91	88	82	82	88	220	40
49	Chip Separator, Shake Table Type, Nuts and Bolts	95	87	88	87	88	89	82	80	87	200	37
50	Automatic Screw Machine, Blanking ½″ Brass Nuts	91	80	85	83	82	81	80	78	82	140	26
51	Hot Bolt Head Upsetter, on ½″ Bronze Bolts	111	90	90	88	90	107	107	96	102	840	310
52	Toggle Machine, Cold Header, No. 2—10 Screw Heads	101	89	88	96	93	94	90	86	87	240	63
53	Punch Press, 25 Ton, ⅜″ Brass Washers	101	90	90	90	92	98	90	87	93	350	72
54	Nut Former, ½″ Stock	97	91	89	91	85	87	81	80	85	210	35
55	Tumbler. Nuts, Bolts	95	82	84	84	83	86	86	86	91	220	60
56	Nut Tapper, ½″ Stock	94	78	80	83	83	86	86	88	87	220	73
57	Automatic Screw Slotter, ¼″ Bolts	96	78	82	84	87	92	88	87	88	250	67
58	Roll Threader, ¼″ Bolts	92	80	83	83	84	86	85	84	83	190	49
59	Spin Riveter, ¼ Hp, Electric, on Thin Steel Chassis	103	77	73	77	86	88	96	97	100	440	170

†*See introductory paragraph at the beginning of appendix.*

TYPICAL SOUND PRESSURE LEVELS AND LOUDNESSES IN INDUSTRIAL PLANTS†

NO.	INDUSTRY	OVER-ALL LEVEL db.	OCTAVE BAND LEVELS, db.								TOTAL LOUD-NESS SONES	LOUD-EST BAND SONES
			37.5 75	75 150	150 300	300 600	600 1200	1200 2400	2400 4800	4800 9600		
	FABRICATED METAL PRODUCTS— MACHINES (Cont'd)											
60	Radial Drill Press 4′ Arm, on Engine Block	90	84	81	81	81	86	79	77	74	140	33
61	Autom. Lathe. on 8″ Dia. Gray Iron Flange	88	76	75	74	76	78	86	76	76	120	44
62	Multiple Drill Press, 42 Spindle, on Engine Block	91	82	80	82	85	84	81	76	78	140	28
63	Vert. Tur. Lathe, 42″, on 5″ Thick Casting	88	76	75	76	81	82	82	76	71	120	30
64	Tumbler, 9 Cu. Ft., 3″ Steel Discs	87	76	77	75	76	79	82	79	72	110	30
65	Autom. Radiator Washers, Gas Fired Conveyor	97	84	84	89	94	89	80	74	66	180	45
66	Furnaces, Non-ferrous	105	85	95	104	92	90	75	69	64	230	74
67	Router, on Aluminum Stock	95	89	89	89	87	81	77	75	75	170	27
68	Hand Operated Pneu. Drill, on Aluminum Stock	92	86	85	80	75	76	86	84	77	170	49
69	Autom. Screw Machine, Cutting ½″ Bronze Bushings	94	77	77	76	81	84	87	89	90	210	73
70	Welded Tube Mach., Making Tube From Strip	91	84	87	82	75	77	72	72	58	100	18
71	Grinding Machine, on 2″ Metal Tubing	94	82	84	81	80	85	87	88	84	210	72
72	Electric Hand Grinder, 8″, on Steel Pipe	91	84	81	81	81	83	84	82	72	160	40
73	Grinding Machine, on 20′ x 4″ Pipe	99	86	85	82	94	95	89	86	82	280	62
74	Grinding Machine, on Steel Pipe	98	89	87	87	92	91	90	86	80	270	62
75	Grinding Machine, Metal Parts	91	82	82	81	83	83	81	84	78	160	49
76	Polisher, on 20′ Metal Tubes	95	84	85	88	87	85	84	89	88	240	80
77	Hand Grind, 4″ Electric, 1′ Steel Pieces	92	85	87	87	84	79	76	76	72	140	24
78	Pointing Machine, Steel Parts	103	98	99	96	90	88	78	74	64	240	45
79	Steel Conveyor, 30′ Steel Rods	107	98	93	93	96	102	104	100	93	790	270
80	Rotary Hammering Machine, 20′ x 2″ Steel Tubes	100	93	93	95	94	91	88	86	77	300	60
81	Pipe Mill, Processing Platform	95	89	87	87	86	89	88	86	78	240	60
	AREAS											
1	Machine Shop, Lathes, Presses, etc. (Steel)	85	71	75	76	77	80	77	75	72	97	22
2	Machine Shop, Lathes, Hob-Grinders	85	....	77	78	74	73	76	78	78	100	28
3	Metal Working, Grinding, Handling Castings	89	86	81	78	77	78	78	75	70	110	22
4	Grinding Area, Noise of Handling Castings	93	86	86	84	82	88	77	80	81	160	38
5	Nail Manufacturing Area, Nail Mach's.	103	86	88	88	91	98	96	95	92	460	300
6	Steel Wire Fence Manufacturing	102	84	82	81	81	96	94	96	95	410	170
7	Wire Drawing Area, Drawing Machines	95	84	88	88	85	90	81	76	80	180	44
8	Steel Wire Fence Manufacturing	94	79	81	81	81	91	83	81	82	170	46
9	Machine Shop, Lathes, Borers	83	78	71	75	74	77	71	70	67	74	17
10	Arc Welding Area, Steel Parts	84	78	76	76	73	76	76	75	63	88	22
11	Automatic Drop Hammer Area, 30 Mach's.	108	90	89	93	95	104	97	97	94	590	190
12	Automatic Punch Press Area, 12 Mach's.	105	98	96	96	91	98	95	94	98	520	130
13	Grinding Area, 36″ Machines, Small Tools	89	76	76	75	75	83	78	81	85	130	37
14	Grinding Area, 12 Machines, Var. Sizes, Steel	86	76	75	73	77	78	77	77	78	100	26
15	Milling Mach. Area. 20 Mach., Steel Pts.	86	76	75	75	75	82	76	73	68	94	24
16	Heat Treating Area, Lead Pots, Furnaces	83	76	75	74	75	76	69	65	66	67	16
17	Sand Molding Area, Small Tool Castings	83	79	75	76	69	71	69	67	62	61	11
18	Core Area, Noise Due to Air Blast	79	72	76	69	67	68	67	64	62	47	8.8
19	Shot Blast Area, Noise Due to Shot Machinery	88	82	80	82	79	78	72	70	70	97	18
20	Steel Pour Area, Tumbling, Shakeout Noises	91	86	85	84	75	79	74	73	72	110	19
21	Cupola Area, Cupola Down, Blower Noise	97	92	91	92	87	79	77	74	74	170	33
22	Tool Mfr. Area, Hum of Electric Butt Welder	86	83	77	73	71	76	76	70	65	81	18

†*See introductory paragraph at the beginning of appendix.*

TYPICAL SOUND PRESSURE LEVELS AND LOUDNESSES IN INDUSTRIAL PLANTS†

NO.	INDUSTRY	OVER-ALL LEVEL db.	37.5–75	75–150	150–300	300–600	600–1200	1200–2400	2400–4800	4800–9600	TOTAL LOUD-NESS SONES	LOUD-EST BAND SONES
	FABRICATED METAL PRODUCTS— AREAS (Cont'd)											
23	Rolling Mill Mach. Shop, Loc. and Dist. Noise	88	76	79	78	82	84	75	74	72	110	27
24	Rolling Mill Mach. Shop, Lathes. Drills	89	88	82	68	67	71	65	61	62	61	12
25	Corrugating Area, Sheet Steel, Sev. Mach.	100	96	90	85	86	97	87	76	77	200	72
26	Strip Steel Processing, Lining Area	82	62	63	60	60	82	70	58	60	50	24
27	Screw Machine Area, Bolts	93	76	74	84	84	85	82	82	88	160	40
28	Automatic Screw Machine Area, Nuts	87	78	79	79	79	79	75	75	76	100	22
29	Automatic Screw Machine Area, Small Parts	91	75	75	81	81	84	84	82	84	160	40
30	Automatic Screw Machine Area, Slotted Components	78	74	70	68	67	70	75	59	60	51	17
31	Light Riveting Area, Small Steel Chassis	80	76	70	70	69	66	68	66	69	51	10
32	Machine Shop, Drill Presses, Borers, etc.	88	83	78	77	79	80	75	74	74	100	21
33	Machine Shop, Drill Presses, Borers, etc.	84	77	75	72	74	75	76	72	73	82	18
34	Machine Shop, Drill Presses, Borers, etc.	87	80	78	79	81	82	79	75	71	120	24
35	Machine Shop, Drill Presses, Borers, etc.	85	75	74	74	76	80	76	74	71	92	21
36	Machine Shop, Drill Presses, Borers, etc.	83	76	75	74	75	75	74	72	68	77	18
37	Machine Shop, Drill Presses, Borers, etc.	85	80	78	75	75	75	75	75	70	92	22
38	Machine Shop, Drill Presses, Borers, etc.	85	74	77	75	77	77	75	75	74	91	22
39	Punch Press Department, Approx. 20 Mach.	93	80	83	83	83	88	85	83	80	190	44
40	Non-ferrous Foundry Furnace Area, 6—10 Furn.	105	85	95	104	92	90	75	69	64	230	74
41	Radiator Processing Area	97	84	84	89	94	89	80	74	66	180	45
42	Machine Shop, Screw Machines, Metal Working	87	75	75	76	81	80	79	80	76	120	34
43	Machine Shop, Lathes, Grinder, Saws	79	77	73	73	72	73	66	58	48	51	13
44	Storeroom, Tools, Materials	66	64	55	54	49	49	43	38	32	9	2.0
45	Pickling Area, Dist. Steel Handling Noise	88	84	82	80	79	76	68	62	62	82	17
46	Steel Fabrication Area, Handling, Cutting	91	88	85	84	80	81	71	68	63	110	23
47	Steel Fabrication Area, Handling, Cutting	93	86	87	87	86	81	78	73	68	140	27
48	Steel Fabrication Area, Handling, Cutting, Processing	101	92	95	94	96	87	78	72	62	220	38
49	Shipping Dept. (Steel Plant) Dist. Noise	76	70	71	70	67	62	54	52	58	29	6.6
50	Steel Finishing Dept., Lathes, Saws	84	80	78	77	74	71	63	59	52	58	12
51	Steel Polish and Grind Area	91	84	81	81	78	79	84	83	78	150	44
52	Steel Handling Area	94	88	89	87	84	83	82	78	70	170	30
53	Steel Parts Sawing Area, Handling Noise	92	88	85	83	82	79	73	68	68	110	21
54	Steel Processing Area, Handling Noise	96	89	89	88	87	88	82	79	72	190	38
55	Steel Working Area, Handling Noise	93	88	86	85	84	79	78	72	64	130	24
56	Power Plant Area, Alternators, Reciproc. Machy.	98	95	87	89	86	83	77	74	64	170	27
57	Shipping Department, Distant Steel Handling	85	80	80	77	70	67	64	63	52	56	11
k	**MACHINERY MACHINES**											
1	Engine Dynamometer Stand	99	96	88	88	84	82	90	89	90	290	79
2	Pneumatic Wrench, Engine Head Nuts	103	83	85	94	94	95	95	95	95	450	160
3	Milling Machine, Working Engine Block	93	76	77	87	88	87	82	78	74	160	33
4	Boring Machine, Working Engine Block	97	76	80	80	85	93	92	87	82	250	77
5	Pneumatic Press, Assembling Bearing on Shaft	101	74	76	88	85	94	91	94	96	370	140
6	Hob-Grinder	91	78	78	79	75	74	79	84	88	150	48

†*See introductory paragraph at the beginning of appendix.*

Typical Sound Pressure Levels and Loudnesses in Industrial Plants†

NO.	INDUSTRY	OVER-ALL LEVEL db.	37.5–75	75–150	150–300	300–600	600–1200	1200–2400	2400–4800	4800–9600	TOTAL LOUD-NESS SONES	LOUD-EST BAND SONES
	MACHINERY—MACHINES (Cont'd)											
7	Multiple Drill, on Piston Parts	99	78	81	91	93	95	85	77	75	220	64
8	Turret Lathe, Ram Type, on Piston Parts	105	84	84	83	79	87	102	101	93	620	300
9	Welder, 600 amp., on Steel Castings	110	86	86	82	89	98	105	106	102	1000	470
10	Pneu. Riveting Hammer, 20' x 8' Dia., x ¾" Steel Tank	131	121	124	123	122	121	121	118	122	5400	2000
11	Pneu. Chipper, 24' x 4' Dia., x ½" Tank	127	96	92	116	107	118	121	122	122	5900	3000
12	Pneu. Chipper, 10' x 8' Dia., x ½" Tank	126	106	107	108	113	118	117	116	124	4100	1500
13	Pneu. Chipper, 12' x 8' Dia., x ½" Tank	129	122	120	119	118	120	119	120	122	5400	2400
14	Hand Grinder, Electric, 12", 8" Thick Steel Plate	97	69	72	75	93	90	87	88	88	240	73
15	Arc Welder, on ½" Steel Plate	86	84	77	80	76	81	78	76	78	120	24
16	Pneu. Air Hoist, 4000 Lb.	102	82	88	84	77	88	85	90	101	320	90
17	Pneu. Wrench, on Engine Nuts	105	78	78	78	85	96	100	97	100	550	200
18	Int. Comb. Engine Test, 150 Hp. Each	102	86	90	98	91	94	92	90	90	370	90
19	Steel Plate Shear, for 6' x ¼" Steel, Idling	98	92	88	87	90	93	87	78	76	230	53
20	Back Shear, Four-Cutter, For ⅛" Steel, Operating	95	80	86	88	87	90	88	80	77	210	57
21	Engine Lathe, 60", on 40' x 1' Steel Shaft	94	79	82	86	90	87	81	71	69	160	35
22	Engine Lathe, 60", on 30' x 1' Steel Shaft	92	86	85	86	77	87	79	65	66	140	37
23	Gas Welder, No. 6 Tip, Cutting 15" Steel	95	83	84	83	84	86	87	86	91	240	66
24	Gas Engine Test, 600 Hp.	99	94	95	90	86	87	86	86	84	270	60
25	Gas Engine Test, 200 Hp.	116	104	107	100	99	100	103	107	113	1440	570
26	Stud Driver, Pneu. ⅜" Engine Studs	95	82	81	80	89	88	85	87	87	230	66
27	Automatic Radiator Washer, Gas-Fired, Conveyor	97	84	84	89	94	89	80	74	66	180	45
28	Riveting Jig, Airplane Wing Assembly	116	88	94	96	97	104	108	111	113	1800	900
29	Internal Combustion Engine Test, 1500 Hp.	116	101	103	107	111	110	107	106	104	1400	500
	AREAS											
1	Shot Blast Room, Small Castings	98	82	82	85	87	91	91	91	92	320	100
2	Engine Dynamometers	97	94	87	85	84	82	85	84	85	210	49
3	Machine Shop, Lathes, Presses, etc.	85	71	75	76	77	80	77	75	72	97	22
4	Machine Shop, Hob Grinding Area	85	77	77	78	74	73	76	78	78	100	28
5	Assembly Shop, Grinding, Pneu. Chipping	91	86	84	83	79	80	81	79	74	150	31
6	Pneu. Rivet. Area, Large Steel Plate Tanks	105	85	81	90	91	99	99	96	98	550	170
7	Pneu. Chip. Area, Cleaning Steel Tank Welds	113	92	95	94	103	102	107	108	108	1400	630
8	Hand Grinding Area, Steel Plate	85	80	78	79	77	76	69	66	61	77	16
9	Arc Welding Area, Steel Parts	84	78	76	76	73	76	76	75	63	88	22
10	Pneu. Air Hoist, 4000 Lb.	84	76	76	77	75	76	74	75	72	89	22
11	Int. Comb. Eng. Testing, 150 Hp. Each	88	85	85	74	72	76	73	68	67	88	16
12	Earth Moving Equipment Assembly	91	85	86	81	82	84	75	71	71	120	28
13	Steel Plate Shear, for ¼" Steel, Idling	98	92	88	87	90	93	87	78	76	230	53
14	Engine Lathe, 60", on 1' x 40' Steel Shaft	90	77	80	85	85	84	78	70	67	120	28
15	Engine Lathe, 60", on 1' x 30' Steel Shaft	88	81	84	81	76	77	72	64	62	87	17
16	Compressor Test, 100—300 Hp. Int. Comb. Eng.	105	100	99	95	92	95	92	93	91	450	120
17	Engine Test, 50 Hp. Int. Comb.	109	98	107	100	92	95	88	87	82	400	74
18	Machine Shop, Lathes, Drills, Borers	87	83	82	72	74	77	75	73	67	91	19
19	Machine Shop Area, Lathes, Drills, etc.	85	81	80	70	70	72	70	66	66	65	12
20	Punch Press Room, Electrical Parts, Steel	86	78	76	77	79	80	79	79	78	120	31
21	Punch Press Room, Electrical Parts, Steel	90	81	81	80	81	82	84	83	83	170	44

†*See introductory paragraph at the beginning of appendix.*

TYPICAL SOUND PRESSURE LEVELS AND LOUDNESSES IN INDUSTRIAL PLANTS†

NO.	INDUSTRY	OVER-ALL LEVEL db.	OCTAVE BAND LEVELS. db.								TOTAL LOUD-NESS SONES	LOUD-EST BAND SONES
			37.5 75	75 150	150 300	300 600	600 1200	1200 2400	2400 4800	4800 9600		
	MACHINERY											
	AREAS (cont'd)											
22	Punch Press Room, Electrical Parts, Steel	97	83	85	84	89	89	90	90	89	290	89
23	Punch Press Room, Electrical Parts, Steel	100	86	86	86	92	91	93	93	91	360	120
24	Int. Comb. Eng. Test Area, 1500 Hp.	104	86	92	97	100	96	96	91	85	440	110
25	Radiator Processing Area	95	84	84	89	94	89	80	74	66	180	45
1	**AIRCRAFT MANUFACTURE**											
	MACHINES											
1	Pneu. Riveting Gun, Sub-assembly	133	118	123	122	120	129	123	125	127	5800	1700
2	Pneu. Riveting Gun, Sub-assembly	125	118	115	116	112	113	114	115	120	3400	1400
3	Pneu. Riveting Gun, Wing Assembly	106	84	81	85	93	90	101	103	104	750	380
4	Riveting Hammer, Fuselage Assembly	108	102	101	101	98	99	98	94	103	690	150
5	Pneumatic Drill, ⅛" Holes in Aluminum	91	86	76	76	75	85	82	84	85	170	49
6	Bumping Hammer, on Thin Metal	112	86	94	103	108	109	107	104	101	1200	420
7	Circular Saws, Cutting Metal	105	69	70	82	86	88	86	98	103	460	220
8	Rivet Bucking, Wings	105	82	86	88	96	94	100	101	96	670	300
9	Rivet Bucking, Fuselage	90	75	78	80	93	85	82	80	88	180	42
10	Router, on Aluminum Stock	95	89	89	89	87	81	77	75	75	170	27
11	Hand Operated Pneu. Drill, Aluminum Stock	92	86	85	80	75	76	86	84	77	170	49
12	Riveting Jig, Wing Assembly	116	88	94	96	97	104	108	111	113	1800	900
	AREAS											
1	Routers, Riveting	92	88	86	83	80	81	76	76	70	130	24
2	Aluminum Machining	88	86	79	77	74	73	69	68	54	73	12
3	Sub Assembly, Light Riveting	88	86	78	77	76	74	71	70	62	81	14
4	Sub Assembly, Distant Riveting	91	89	83	79	78	73	70	64	60	89	15
5	Sub Assembly, Distant Riveting	93	89	86	82	80	81	84	74	67	140	30
6	Wing Assembly, Distant Riveting	91	83	79	82	79	80	80	85	82	160	55
7	Fus. Assem., Dist. Riveting	88	86	77	76	76	76	73	74	65	90	20
8	Fus. Assem., Dist. Riveting	86	84	77	74	73	72	67	61	54	64	11
9	Sub Assembly, Dist. Riveting	86	85	77	74	72	71	66	62	53	60	10
10	Assembly, Distant Riveting	86	84	77	77	74	73	70	67	56	75	14
11	Assembly, Distant Noises	89	87	79	80	76	76	73	69	57	90	16
12	Assembly, Distant Noises	89	86	78	81	82	78	74	70	60	98	19
13	Assembly, Distant Noises	88	84	83	80	78	71	68	60	52	77	15
14	Assembly, Distant Noises	85	84	75	75	71	67	62	58	45	50	9
15	Small Parts Assembly	88	84	82	79	78	72	70	68	62	83	15.9
16	Wing Rivet. Bucking Area	83	69	73	73	74	74	69	72	70	70	18
17	Stamping, Shears, Grinding	83	73	75	75	75	74	73	72	70	77	18
18	Riveters, Routers, Grinders	89	74	81	82	83	74	78	77	76	123	26
19	Wing Assembly, Distant Riveting	84	62	68	72	72	76	73	78	77	95	31
20	Wing Assembly, Distant Riveting	83	67	72	74	73	77	78	76	65	88	24
21	Main Assembly, Distant Riveting	81	68	72	71	71	74	75	75	65	75	22
22	Fuselage Assembly, Distant Riveting	84	68	71	72	78	77	77	76	70	91	24
23	Router Area, Distant Riveting	92	78	78	81	85	84	84	83	82	170	44
24	Small Parts Assembly, Distant Riveting	86	73	75	76	76	78	79	79	78	110	31
25	Small Parts Assembly, Distant Riveting	82	64	72	71	72	74	75	74	71	75	20
26	Wing Dept., Sm. Saws, Aluminum	80	67	68	67	67	68	70	76	75	64	24

†*See introductory paragraph at the beginning of appendix.*

Part VI

NOISE MEASUREMENTS AND
INSTRUMENTATION

Editor's Comments
on Papers 21, 22, and 23

There are several reasons for measuring noise. In the case of machines these include: (1) identification of noise sources on a machine, (2) comparison of the noise of different machines, (3) determination of whether the noise of a machine is within prescribed limits, and (4) determination of the noise level of a machine at a certain distance. In cities, where it is necessary to measure the noise in the community caused by aircraft, surface transportation, factories, and other sources, measurement equipment and techniques can become quite complicated and differ from those used with single machines.

Lord Kelvin once said: "If you can measure that of which you speak, you know something of your subject, but if you cannot measure it, your knowledge is unsatisfactory" [Paper 6]. In the last century most assessment of noise was completely subjective; one exception might be the work of Barr (Paper 18), where people's hearing was examined by measuring the distance at which they could just hear a ticking watch. Even in the present century, until the 1920s, noise measurements seem to have been mainly unreliable and often unrepeatable. A multitude of different methods were being used and developed. Before this time, empirical methods were mainly used to determine noise sources and devise noise control solutions and no quantitative measurements of noise levels existed.

There are several good reviews of early noise measurement methods [1, 2]. Snook [1] reviewed methods in use to identify sources of automobile noise in 1925. The methods used then can be divided into two groups: contact devices and noncontact devices. Contact devices included the following. A listening-stick consisting of a small dowel-rod was held between the teeth of the observer and placed in contact with vibrating surfaces. Resonant chambers of wood were sometimes mounted on the end of the dowel-rods. These chambers modified the vibrations but apparently made low-frequency vibrations audible. Acoustic stethoscopes, much like a physician's, were also frequently used. There were also simpler stethoscopes consisting solely of a rubber hose of about one-inch diameter. One end was pressed against the vibrating device and the other against the ear. Several other slightly more complicated types of mechanical or electrical contact devices were also in use [2], including the geophone, a mass on flexible supports in a small cavity connected to the ears by rubber tubing; a contact microphone made of a rubber bag filled with granular carbon with two carbon electrodes; an inertia type of microphone; and an electrical stethoscope, which used a telephone receiver to which was attached a rubber frustrum and a strip of steel. The signal from the electrical stethoscope was amplified electrically and fed to the ears.

These early contact devices suffered from several defects. Some had mechanical resonances and thus a poor frequency response. Most relied on the human ear as the receiver, and it was difficult or impossible with most to obtain a permanent record or chart of the noise. Most of these devices have now become obsolete, although the stethoscope is still sometimes used. In the last twenty or thirty years various other devices have been perfected to measure the vibration of machines, including: displacement, velocity, and acceleration transducers. For a recent review of these devices see Ref. 3. All these transducers are still in use, and many make use of the phenomena of capacitance, reluctance, or strain. The voltage generated is normally amplified; it can be frequency-analyzed and a permanent record can be obtained. The most common type of vibration transducer in present use is probably the piezoelectric accelerometer. It has the advantage of good, wide frequency response, low weight, and rugged construction.

Although contacting and noncontacting gauges have been and still are used to measure the vibration of parts of machines in order to locate noise sources, some sources are aerodynamic in origin and are not caused by vibration. In these cases and for measurements of community noise, it has been necessary to develop other methods of measuring airborne noise. Several of the methods that came into use in the 1920s have persisted to the present day; they have been reviewed by Snook and Free [1, 2].

At first the audiometer, which had been developed primarily to test hearing and deafness, was adapted to measure the loudness of noise in the mid 1920s. It became widely used for community noise surveys, such as the famous New York noise surveys [4, 5] until it was superseded by such other devices as the sound level meter in mid 1930s. Since it was used widely in early community noise surveys [4, 5, Paper 24, 6] and automobile noise measurements, it is probably of considerable interest to describe the principle of the audiometer. The instrument in its various forms worked on the principle of masking. The apparatus consisted of a buzzer to produce a standard noise, an attenuator, and a telephone receiver. The person making the noise measurement listened simultaneously to the noise from the source being measured and to the buzzer noise from the telephone receiver of the audiometer. The attenuator of the audiometer was then adjusted until the buzzer noise was just masked by the noise being measured. The attenuator of the instrument was calibrated independently so that its setting could be used to indicate the loudness or sound pressure level of the noise measured. Although the audiometer was very useful, it had several disadvantages: it was heavy and cumbersome; a permanent record of the noise was not obtained; and if the frequency of the noise being measured differed too much from the predominant frequency of the buzzer, the results became inaccurate [2]. Some audiometers had several buzzers that produced masking sounds of different frequencies in order to try to overcome this frequency problem.

In order to make the equipment lighter, a similar but much simpler approach was developed in the early 1930s. This approach using tuning forks was used by Davis in England to measure street noise [7], and by Bassett and Zand in the United States to measure aircraft noise [Paper 6]; it was also used in the New York survey. The method involved striking a tuning fork in a standard way, placing it close to the ear, and then using a stopwatch to find the time it took for the sound to become inaudible (that is, masked by the noise being measured). Different pitch tuning forks could be used with noises of different frequency, and the time to inaudibility correlated very well with audiometer measurements.

In the mid-1920s and early 1930s, another type of instrument was developed for noise measurement [1, 2, 8, 9]. This instrument was known at first as an acoustimeter [2] or noise meter [Paper 6], but soon became known, when it became portable, as a sound level meter, a name that it retains today. As the cost, size, and weight were reduced, sound level meters soon became valuable tools for use in many noise measurement problems and superseded most of the other devices already discussed. They consist essentially of a microphone, one or more electrical amplifiers, various filters, attenuators, a rectifier, and a meter, connected in series. Their development was made possible by advances in electrical circuit design and amplification in the 1910s and by the perfection of suitable high-quality microphones. The development of the condenser microphone in the period 1915–1920 was perhaps the real breakthrough that made the sound level meter and similar accurate laboratory measurers of noise possible, and most credit should perhaps go to Wente. Wente published papers in 1917 and 1922 describing a condenser microphone that he had developed. Both papers have been reprinted in the companion Benchmark volume, "Acoustical Measurements," edited by Harry B. Miller and so they will not be repeated here. Since the early 1900s, small improvements have since been incorporated, but the essential design is contained in Wente's first paper. This paper was published in the *Physical Review,* and it is interesting to note that two of the nine members of the editorial board in 1917, E. Buckingham and W. C. Sabine, are also well known for their contributions in acoustics. The condenser microphone did receive some competition later from piezoelectric (initially Rochell-salt) and moving-coil microphones. However, the former suffered from stability problems (from temperature and humidity changes) and the latter from poorer frequency response. The most stable microphone for laboratory measurements has remained the condenser microphone. For a recent review of noise measurements and instrumentation, see Ref. 10.

In addition to the efforts just discussed to improve methods of measurement of noise of vehicles and noise in the community, parallel efforts were being made to improve laboratory methods of sound and noise measurements. Sabine made some early measurements of reverberation time, absorption, sound transmission, and sound pressure in his sophisticated Riverbank Laboratories. Some of his papers are reprinted in the companion Benchmark volume [11]. Sabine obtained surprisingly accurate, repeatable results using organ pipes and stop watches [12].

Absorbing materials are used both in the control of reverberation in auditoria and for absorption of noise. Paris was active in several areas of noise control in the 1920s in England. The coefficient of absorption of a material was (and still is) measured at normal incidence (for convenience) and in reverberant conditions (closer to practical use). Paper 21 by Paris was one of the first to discuss the relationship between measurements of absorption coefficient obtained by these two methods and is still a useful paper.

The noise of machines today is still measured in laboratory conditions in two basic types of rooms: anechoic and reverberation. The anechoic, or "dead", room, attempts to simulate free-field conditions where there are reflections. In such rooms, which are ideal for identifying sound sources on machines, the directivity of the source is preserved and the inverse square law should apply. Churcher described such a room built at Metropolitan Vickers Company in England in the early 1930s. However, this room had flat absorbent walls. Bedell described a similar anechoic room built in the mid 1930s in the United States [14]. Most anechoic rooms built today have walls constructed of many absorbing wedges. Meyer's paper (Paper 22) is perhaps the first to describe measurements of the absorption coefficient of various nonhomogeneous wall linings, in particular of pyramid and other shapes, which are still in use today. In addition, Meyer's paper describes the construction of a very large anechoic room (882 m^3, or 28,000 ft^3) built with wedges in the late 1930s in Germany, probably the largest in use at that time. Similar anechoic rooms with wedges were soon built in other countries (see, for instance, Beranek and Sleeper [5], who built three such anechoic chambers early in the 1940s at the Cruft Laboratory at Harvard University, the largest being 2044m^3 [72,200 ft^3]).

The acoustical properties of anechoic chambers using absorbing wedges or pyramids on the walls is usually superior to those with flat absorbing walls, unless the absorbing walls are very thick. Olson described an anechoic chamber with good properties, but the flat lining made of blankets mounted on the walls was eight feet thick [16]. Such a design consumes much working space.

Reverberation rooms are extensively used in measurements of the noise of machines. With a measurement of the space-averaged sound pressure level and the reverberation time in the room, the sound power of a machine is easily determined. The sound power is a very useful quantity since it can be used to determine the sound pressure level in different acoustical environments. Meyer wrote one of the first papers (Paper 23) to discuss the measurement of the sound power of machines using the reverberation room method.

The reverberation room method of sound power determination is extensively used, and its use seems to be increasing. The use of reverberation rooms was recently reviewed in Ref. 17.

It is impossible to review all the recent developments in instrumentation and measurements in this brief commentary, or to decide which are most important, but it would be improper not to comment at all. In the last thirty or forty years, such instrumentation as electronic amplifiers, filters, and so on, has been improved greatly. The more recent use of the transistor has reduced the size, weight, and cost of many components. Analog filters have been used for many years for frequency analysis of sound [3]. However, real-time analyzers have been developed in recent years. These devices use the "time-compression" principle and can use either analog, digital, or hybrid (analog and digital) techniques of analyzing data. Fast-Fourier Transform (FFT) Digital Fourier Analyzers have also become widely used in the last five years. They make use of the Fast-Fourier Transform (FFT) algorithm, which was published by Cooley and Tukey in 1965 [18]. This algorithm is much faster than conventional digital Fourier transform algorithms and has made the digital FFT analyzer a useful efficient means of signal analysis. Perhaps it would not be fair to give Cooley and Tukey all the credit for this algorithm since other similar algorithms were published earlier. However, they presumably "rediscovered" this algorithm at a time when it could be utilized by the new electronic instrumentation that was becoming available, and their paper is undoubtedly a significant contribution.

Another development worthy of mention is the direct measurement of sound intensity. Intensity can, in principle, be measured either on the surface of a vibrating machine using an accelerometer and a microphone [19, 20, 21] or in the fluid medium itself. The idea is not new; Rayleigh's disc, which dates back to the last century, can be used to measure intensity, although the device is rather fragile and impractical. In 1932 Olson took out a patent on a sound intensity meter [22], and several other workers have continued work on measurement of sound intensity since, including Clapp and Firestone [23] and Schultz [24, 25] and Fahy [26]. Although such sound intensity meters are still in the research stage and are only just beginning to be produced by instrument manufacturers, it appears that they are now sufficiently developed to give reliable results, although some engineering judgment must still be applied to their interpretation. Some recent applications of intensity measurements include the direct measurement of the sound power of machines such as diesel engines by Chung [27] and others [28], the measurement of fan sound power in ducts [29], and the direct measurement of sound transmission through structural walls in buildings, machinery, and aircraft using intensity devices [30, 31]. A recent conference was devoted entirely to the measurement of sound intensity, and the entire book of proceedings is useful [32]. Other recent methods to determine noise sources are reviewed in Ref. 33. Measurements of and instrumentation for community noise and aircraft are becoming more complicated now that measures such as L_{10}, L_{90}, L_{eq}, L_{dn}, and PNdB are becoming widely used.

REFERENCES

1. Snook, C. H., 1925, Automobile Noise Measurement, *Soc. Auto. Eng. J.* **17**(1):115-124.
2. Free, E. E., 1930, Practical Methods of Noise Measurement, *Acoust. Soc. Am. J.* **2**:18-29.
3. Crocker, M. J., and A. J. Price, 1975, *Noise and Noise Control,* Vol. 1, CRC Press, Cleveland, Oh., pp. 99-172.
4. Free, E. E., 1926, Measurements of the Street Noise in New York City, *Phys. Rev.* **27**:507.
5. Free, E. E., 1926, How Noisy Is New York?, *Forum* **75**(2):xxi-xxiv.
6. Lemon, B. J., 1925, Glimpses of Balloon Tire Progress, *Soc. Auto. Eng. J.* **16**(2).
7. Davis, A. H., 1930, Measurements of Noise by Means of a Tuning Fork, *Nature* **125**:48-49.
8. Hunt, J. H., and G. F. Embshoff, 1925, Some New Electrical Instruments for Automotive Research, *Soc. Auto. Eng. J.* **16**(4):444.
9. Firestone, F. A., 1926, Technique of Sound Measurements, *Soc. Auto. Eng. J.* **19**:461-466.
10. Six articles printed in Special Issue, 1977, Noise Measurements and Instrumentation, *Noise Control Eng.* **9**(3):100-162.
11. Northwood, T. D., 1977, *Architectural Acoustics,* Benchmark Papers in Acoustics, Dowden, Hutchinson & Ross, Stroudsburg, Pa., 428p.
12. Siekman, 1978, More on Sabine (letter to the editor), *Sound Vib.* **12**(2):8-9.
13. Churcher, B. G., 1933, Acoustics Laboratory of Metropolitan Vickers Company, *Engineering* **135**:563.
14. Bedell, E. H., 1936, Some Data on a Room Designed for Free Field Measurements, *Acoust. Soc. Am. J.* **8**:118-125.
15. Beranek, L. L., and H. P. Sleeper, 1946, The Design and Construction of Anechoic Sound Chambers, *Acoust. Soc. Am. J.* **18**(1):140-150.
16. Olson, H. F., 1943, Acoustic Laboratory in the New RCA Laboratories, *Acoust. Soc. Am. J.* **15**(2):96-102.
17. Eleven articles printed in special issue, Noise Measurement Facilities and ANSI SI-21-1972 Qualification, 1976, *Noise Control Eng.* **7**(2):48-114.
18. Cooley, J. W., and J. W. Tukey, 1965, An Algorithm for the Machine Calculation of Complex Fourier Series, *Math. Comput.* **19**:90.
19. Hodgson, T. H., 1976, Investigation of the Surface Acoustical Intensity Method for Determining the Noise Sound Power of a Large Machine in situ, *Acoust. Soc. Am. J.* **61**(2):487-493.
20. McGary, M. C., and M. J. Crocker, 1981, Surface Intensity Measurements on a Diesel Engine, *Noise Control Eng.* **16**(1):26-36.
21. McGary, M. C., and M. J. Crocker, 1982, Phase Shift Errors in the Theory and Practice of Surface Intensity Measurements, *J. Sound Vib.* **82**(2):275-288.
22. Olson, H. F., 1932, System Response to the Energy Flow of Sound Waves, U.S. patent no. 1,892,644.
23. Clapp, C. W., and F. A. Firestone, 1941, Acoustic Wattmeter, an Instrument for Measuring Sound Energy Flow, *Acoust. Soc. Am. J.* **13**:124-136.
24. Schultz, T. J., 1956, Acoustic Wattmeter, *Acoust. Soc. Am. J.* **28**(4):693-699.
25. Schultz, T. J., P. W. Smith, Jr., and C. I. Malme, 1975, Measurement of Acoustic Intensity in Reactive Sound Field, *Acoust. Soc. Am. J.* **57**(6):1263-1268.
26. Fahy, F. J., 1977, A Technique for Measuring Sound Intensity with a Sound Level Meter, *Noise Control Eng.* **9**(3):155-163.
27. Chung, J. Y., J. Pope, and D. A. Feldmaier, 1979, Application of Acoustic Intensity Measurement to Engine Noise Evaluation Proc. SAE Diesel Engine Conference, *Soc. Automot. Eng.* Paper 790502.

28. Reinhart, T. E., and M. J. Crocker, 1982, Source Identification on a Diesel Engine Using Acoustic Intensity Measurements, *Noise Control Eng.* **18**(3):84–92.

29. Roland, J., M. J. Crocker, and M. Sankbakken, 1982, Use of Acoustic Intensity Measurements to Evaluate the Sampling Tube Method of Measuring In-Duct Fan Sound Power, *Am. Soc. Heat. Refrig. Air-conditioning Eng. Trans.* **88**(pt. 2):

30. Crocker, M. J., P. K. Raju, and B. Forssen, 1981, Measurement of Transmission Loss of Panels by the Direct Determination of Transmitted Acoustic Intensity, *Noise Control Eng.* **17**(1):6–11.

31. Wang, Y. S., and M. J. Crocker, 1982, Direct Measurement of Transmission Loss of Aircraft Structures Using the Acoustic Intensity Approach, *Noise Control Eng.* **19**(3):80–85.

32. *Proceedings of the International Congress on Recent Developments in Acoustic Intensity Measurement*, 1981, Centre Technique des Industries Mechaniques (CETIM), Senlis, France.

33. Crocker, M. J., 1977, Identification of Noise from Machinery, Review and Novel Methods, *Inter-Noise 77 Proc.*, E. J. Rathe, ed., International Institute of Noise Control Engineering, pp. A201–211.

21

Reprinted from *Philos. Mag.* **5**:489–497 (1928)

On the Coefficient of Sound-absorption Measured by the Reverberation Method. By E. T. PARIS, D.Sc.[*]

THERE are several methods of measuring the coefficients of sound-absorption for different materials [†]. In the reverberation method of W. C. Sabine a specimen of the material to be tested is mounted inside a reverberation-chamber, and the coefficient of absorption (defined as the ratio of the difference between the incident and reflected energy-fluxes to the incident energy-flux) is deduced from the observed effect of the presence of the specimen on the rate of decay of sound in the chamber. According to the theory of reverberation, the specimen receives sound at all angles of incidence from 0° to 90°.

There are, however, other methods of measuring sound-absorption in which plane-waves of sound are made to impinge on a specimen at some definite angle of incidence. For example, in Watson's experiments a beam of sound (produced by means of a source at the focus of a paraboloidal reflector) was directed on to the specimen to be tested. In this method the angle of incidence is determined at will by suitably arranging the relative positions of the specimen and the reflector. Again, in the stationary-wave method sound-waves fall normally on the specimen, the coefficient of absorption being deduced from observations on the interference pattern in front of the specimen.

In general the coefficient of absorption of a given substance is a function of the wave-length of the sound and of

---

[*] Communicated by the Author.

[†] A description of the methods at present in use is given by Davis and Kaye, 'Acoustics of Buildings,' chap. v. (Bell, 1927).

the angle of incidence, so that the value of the coefficient obtained by the reverberation method cannot be expected to agree with that obtained by one of the other methods mentioned above.

In the present paper an attempt is made to correlate the coefficient of absorption of a substance, measured at some particular wave-length and expressed as a function of the angle of incidence, with the coefficient of absorption determined by Sabine's reverberation method for the same wave-length.

The coefficient measured by Sabine's method will be called the "reverberation coefficient of absorption" and denoted by α_r. The coefficient of absorption for waves incident at an angle θ will be denoted by $\alpha(\theta)$.

The theory of reverberation as developed by Sabine, Jäger, and others is a departure from classical methods in theoretical acoustics, and involves certain assumptions regarding the distribution and behaviour of sound-energy in an enclosed space in which reverberation is occurring. Accounts of the theory and of the assumptions made have been given in recent years by E. A. Eckhardt[*] and E. Buckingham[†]. The question of the validity of the assumptions involved will not be discussed here, but attention will be confined to examining the result of introducing the coefficient $\alpha(\theta)$ into the theory of reverberation with the object of finding a relation between $\alpha(\theta)$ and α_r.

The rate at which sound-energy strikes the walls of a room in which reverberation is occurring is found as follows[‡]. Let dS be an element of area of the wall of a room in which the acoustical energy-density ρ is uniform, and let dV be an element of volume at a distance r from dS in a direction making an angle θ with the normal n to dS (fig. 1). Of the energy $\rho\,dV$ contained within dV at any instant, that portion will ultimately strike dS which is moving within directions included within the solid angle

$$d\omega = dS \cos \theta / r^2$$

subtended by dS at dV. This fraction is $d\omega/4\pi$, so that the amount of energy inside dV which will ultimately strike dS is

$$(\rho\,dV\,dS\cos\theta)\,/\,(4\pi r^2).$$

The total amount of sound-energy that falls on dS in one

[*] Journ. Franklin Inst. cixv. p. 799 (1923).
[†] Sci. Paper of the Bureau of Standards, No. 506 (1925).
[‡] *Cf.* Buckingham, *loc. cit.* p. 201.

second is obtained by integrating this expression throughout a hemisphere of radius a described about dS as centre, a being the distance traversed by sound in one· second. Thus, using polar coordinates r, θ, ϕ (fig. 1) so that

$$d\mathrm{V} = r^2 \sin\theta\, dr\, d\theta\, d\phi,$$

we have for the rate at which sound-energy falls on dS :

$$\frac{\rho\, d\mathrm{S}}{4\pi} \int_0^a dr \int_0^{2\pi} d\phi \int_0^{\pi/2} \sin\theta \cos\theta\, d\theta = \frac{\rho a\, d\mathrm{S}}{4}. \quad . \quad (1)$$

Fig. 1.

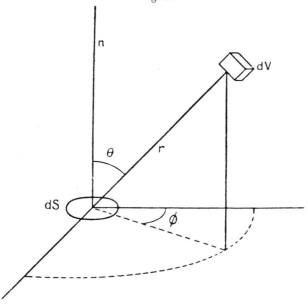

The rate at which sound is absorbed at dS is

$$\alpha_r\!\left(\frac{\rho a\, d\mathrm{S}}{4}\right), \quad . \quad . \quad . \quad . \quad . \quad (2)$$

where α_r is the reverberation absorption-coefficient appropriate to the surface of which dS forms a part.

Another expression for the rate at which sound is absorbed at dS can be formed by making use of the coefficient $\alpha(\theta)$. Thus the rate at which energy coming from $d\mathrm{V}$ is absorbed at dS is

$$\alpha(\theta) \,.\, (\rho\, d\mathrm{V}\, d\mathrm{S} \cos\theta) / (4\pi r^2),$$

and hence the total rate of sound-absorption at dS is

$$\frac{\rho\,dS}{4\pi}\int_0^a dr \int_0^{2\pi} d\phi \int_0^{\pi/2} \alpha(\theta)\sin\theta\cos\theta\,d\theta$$

$$=\frac{\rho a\,dS}{2}\int_0^{\pi/2}\alpha(\theta)\sin\theta\cos\theta\,d\theta. \quad . \quad (3)$$

By equating the expressions (2) and (3) for the rate at which sound is absorbed at dS, we find

$$\alpha_r = 2\int_0^{\pi/2}\alpha(\theta)\sin\theta\cos\theta\,d\theta. \quad . \quad . \quad . \quad (4)$$

This equation provides a means for finding α_r when $\alpha(\theta)$ is known. It is clear, however, that in general the determination of α_r in this way would be very laborious, for it would be necessary to find $\alpha(\theta)$ for all values of θ between 0 and $\pi/2$ (probably with apparatus of the type used by Watson) before the integration in (4) could be effected. Incidentally, equation (4) shows that a simple determination of the coefficient of absorption at normal incidence, that is $\alpha(0)$, by stationary-wave or other small-scale testing apparatus, supplies insufficient data for calculating the reverberation coefficient α_r.

There is, however, a class of substances for which the difficulty of finding $\alpha(\theta)$, and thence α_r by means of equation (4), is not great. The substances referred to are the porous acoustic tiles and plasters used in architectural acoustics. For such materials it is possible to determine a quantity called the "acoustical admittance per unit area of reflecting surface" from which the coefficient of absorption at any angle of incidence, $\alpha(\theta)$, can readily be calculated [*].

Substances of this type absorb sound by virtue of the fact that during reflexion there is a movement of air into and out of the pores of the material. If the pores are small this in-and-out movement of air is accompanied (owing to viscosity) by the degradation of sound-energy into heat. If Φ is the velocity-potential due to incident and reflected sound-waves at the surface of a porous medium of the kind indicated, and q is the volume of air per unit area of surface moving periodically in and out of the pores, then $q = \Omega\Phi$, where Ω is the "acoustical admittance per unit area of reflecting surface."

An ideal substance of the type contemplated would consist of a flat solid wall perforated by a great number of pores,

[*] Proc. Roy. Soc. A, cxv. p. 407 (1927).

bounded by surfaces everywhere perpendicular to the flat face of the wall. The characteristic feature of such a reflecting surface is that sound-waves are produced only in the pores and not in the solid material itself. Also the wave-motion in the pores is not transmitted internally from one part of the reflecting surface to another, so that Ω is independent of the angle of incidence. Acoustic tiles and plasters, while not conforming precisely to this specification, appear to do so to an extent which justifies the employment of an " acoustical admittance " (assumed independent of the angle of incidence) in connexion with the investigation of their acoustical properties.

It can be shown * that the coefficient of absorption of a porous substance of the kind under consideration can be expressed in terms of acoustical admittance per unit area as follows :—

$$\alpha(\theta) = 1 - \left| \frac{\cos\theta - a\Omega}{\cos\theta + a\Omega} \right|^2, \quad \cdots \quad (5)$$

where a is the velocity of sound, and θ, as before, is the angle of incidence. If Ω is separated into its real and imaginary parts, say $\Omega = \Omega_1 + \iota\Omega_2$, then

$$\alpha(\theta) = \frac{4a\Omega_1 \cos\theta}{\cos^2\theta + 2a\Omega_1\cos\theta + a^2(\Omega_1{}^2 + \Omega_2{}^2)}. \quad \cdot \quad (6)$$

By substituting this value for $\alpha(\theta)$ in (4), we obtain

$$\alpha_r = 8a\Omega_1 \int_0^{\pi/2} \frac{\cos^2\theta \sin\theta \, d\theta}{\cos^2\theta + 2a\Omega_1\cos\theta + a^2(\Omega_1{}^2 + \Omega_2{}^2)}. \quad (7)$$

Let

$$x = \cos\theta, \quad A = a^2(\Omega_1{}^2 + \Omega_2{}^2), \quad \text{and} \quad B = 2a\Omega_1.$$

Then, since $4A - B^2 > 0$,

$$\alpha_r = 4B \int_0^1 \frac{x^2 \, dx}{x^2 + Bx + A}$$

$$= 4B \left[x - \tfrac{1}{2}B \log_e (x^2 + Bx + A) \right. $$
$$\left. + \frac{B^2 - 2A}{\sqrt{(4A - B^2)}} \tan^{-1}\left(\frac{2x + B}{\sqrt{(4A - B^2)}} \right) \right]_0^1. \quad (8)\dagger$$

---

* Proc. Roy. Soc. A, cxv. p. 418 (1927).
† Peirce, ' Short Table of Integrals,' p. 9 (Ginn & Co.).

Or, putting in the values of A and B,

$$\alpha_r = 8a\Omega_1 \left[1 - a\Omega_1 \log_e \frac{(1+a\Omega_1)^2 + (a\Omega_2)^2}{a^2(\Omega_1{}^2 + \Omega_2{}^2)} \right.$$
$$+ \frac{a^2(\Omega_1{}^2 - \Omega_2{}^2)}{a\Omega_2} \left\{ \tan^{-1}\left(\frac{1+a\Omega_1}{a\Omega_2} \right) \right.$$
$$\left. \left. - \tan^{-1}\left(\frac{a\Omega_1}{a\Omega_2} \right) \right\} \right]. \quad (9)$$

By means of (2) the reverberation coefficient can be calculated for any porous reflecting surface for which the acoustical admittance per unit area is known.

For example, a determination was made, by means of stationary-wave apparatus, of the acoustical admittance of "Akoustolith" tile (supplied through the Building Research Board of the Department of Scientific and Industrial Research) at 512 vibrations per second Details of the apparatus employed and the method of finding "acoustical admittance" have been given elsewhere [*]. The value obtained for Ω was $0.0641 - \iota \times 0.0643$, and inserting this value in (5) and (9), we find that $\alpha(0) = 0.23$ and $\alpha_r = 0.35$. Thus, according to this calculation, the reverberation coefficient is considerably greater than the coefficient at normal incidence [†].

The calculated value of α_r agrees well with that quoted by Watson [‡] for Akoustolith tile at 512 vibrations per second, viz. 0.36. Davis and Kaye [§], however, quoting results obtained at the Building Research Station of the Department of Scientific and Industrial Research by a reverberation method, give the surprisingly low value of 0.19 for the coefficient at the same frequency, this being less than the coefficient at normal incidence. There appears, therefore, to be some uncertainty as to the true value of the reverberation absorption-coefficient of this substance, and no conclusion can safely be drawn as to whether the calculated value of α_r is or is not confirmed by experiment.

The absorption of sound by an ideal porous substance,

[*] The apparatus is described in Proc. Phys. Soc. xxxix. p. 274 (1927); the method of finding Ω is given in Proc. Roy. Soc. A, cxv. p. 413 (1927).

[†] The value of 0.23 for $\alpha(0)$ is less than that given in an earlier paper, viz. 0.26 (Proc. Phys. Soc. xxxix. p. 281, 1927); this may be due to the fact that the specimen used in the present test was an old one and had suffered some damage which had been repaired with plaster of Paris.

[‡] 'Acoustics of Buildings,' p. 25.

[§] 'Acoustics of Buildings,' p. 116.

presenting a flat surface to the incident sound, and perforated
by a large number of similar channels bounded by surfaces
perpendicular to the face, was the subject of a theoretical
investigation by Rayleigh *. It has been pointed out in a
previous paper † that, in the special case when the pores are
so long that the vibrations within them are sensibly extin-
guished before the stopped ends are reached, the acoustical
admittance per unit area (derived from equations given by
Rayleigh) has a specially simple form, being proportional to
the fraction of the area of the reflecting face which is
occupied by the pore-openings. If σ is the perforated and
σ' the corresponding unperforated area,

$$a\Omega = \sigma/(\sigma+\sigma'). \quad \ldots \ldots \quad (10)$$

The coefficient of absorption at an angle θ is, by (5),

$$\alpha(\theta) = \frac{4a\Omega\cos\theta}{(a\Omega+\cos\theta)^2}, \quad \ldots \ldots \quad (11)$$

which becomes unity when $\cos\theta = a\Omega$, so that there is
always some angle of incidence, given by $\cos\theta = \sigma/(\sigma+\sigma')$,
at which total absorption occurs. With $B = 2a\Omega$ we have,
instead of (8),

$$\alpha_r = 4B\int_0^1 \frac{x^2\,dx}{(x+\tfrac12 B)^2}$$
$$= 8a\Omega\left\{\frac{1+2a\Omega}{1+a\Omega} - 2a\Omega\log_e\frac{1+a\Omega}{a\Omega}\right\}. \quad (12)\ddagger$$

Some numerial results obtained by means of (11) and (12)
are given in the following table and serve (in the absence of

$\sigma/(\sigma+\sigma')$.	θ_m.	$\alpha(0)$.	α_r.
0·05	87 08	0·18	0·30
·1	84 15	·33	·49
·2	78 28	·56	·72
·3	72 32	·71	·84
·5	60 00	·89	·94
·7	45 34	·97	·95
·9	25 50	·997	·93
1·0	00 00	1·000	·91

* Phil. Mag. xxxix. p. 225 (1920); Sci. Papers, vi. p. 662.
† Proc. Roy. Soc. A, cxv. p. 413 (1927).
‡ Peirce, 'Short Table of Integrals.' p. 6 (Ginn & Co.).

data concerning substances actually employed for absorbing sound) to indicate how a reverberation coefficient may differ from the coefficient at normal incidence. The first column gives the value of $a\Omega$, or $\sigma/(\sigma+\sigma')$; the second column gives the value (θ_m) of the angle of incidence at which total absorption occurs; the third and fourth columns give the values of $\alpha(0)$ and α_r calculated from (11) and (12) respectively.

If the figures in the last two columns are plotted against

Fig. 2.

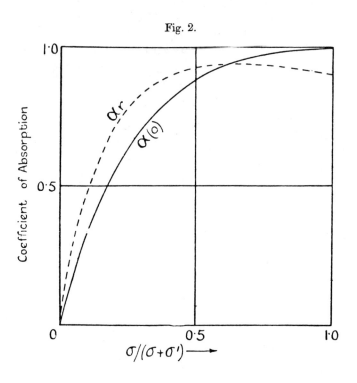

$\sigma/(\sigma+\sigma')$ as in fig. 2, it is seen that $\alpha_r = \alpha(0)$ when $\sigma/(\sigma+\sigma') = 0.65$ approximately, α_r being greater than $\alpha(0)$ when $\sigma/(\sigma+\sigma')$ is smaller than this value and less than $\alpha(0)$ when it is greater. The rise in α_r as $\sigma/(\sigma+\sigma')$ is increased from 0 to 0.3 is very rapid, but little can be gained by making the proportion of perforated area greater than 0.3.

The reverberation coefficient passes through a maximum value of about 0.95 when $\sigma/(\sigma+\sigma')$ is between 0.6 and 0.7, that is when θ_m is between 45°·5 and 53°·1. This result is in accordance with the fact that the amount of energy incident at an angle θ on an element of surface is proportional to $\sin 2\theta$, and is a maximum when $\theta = 45°$. It is therefore reasonable that α_r should be a maximum when θ_m, the angle at which total absorption occurs, is in the neighbourhood of 45°.

In conclusion, it should perhaps be emphasized that the results obtained in this paper are dependent on the assumption made in the theory of reverberation as to the behaviour of acoustical energy in an enclosed space. Interference is entirely ignored. The justification for this assumption is that it provides a theoretical basis for the reverberation formulæ which have been used with such striking success in auditorium acoustics. It remains to be seen whether its employment for establishing a relation between reverberation coefficients of absorption and those obtained by other methods will be justified by experimental results.

The investigation described in this paper was undertaken in connexion with experiments on sound-absorption carried out in the Acoustical Section of the Air Defence Experimental Establishment with the aid of funds provided by the Department of Scientific and Industrial Research on the recommendation of their Physics Research Board.

Summary.

A formula is deduced from reverberation theory by means of which the reverberation coefficient of absorption of a material can be calculated if the coefficient for plane waves is known for all angles of incidence between 0 and $\pi/2$. In the case of certain porous substances the reverberation coefficient can be found if the acoustical admittance per unit area is known. It is shown that in general the reverberation coefficient differs from the coefficient at normal incidence. Some numerical values for an ideal porous substance are given by way of illustrating the results obtained.

Biggin Hill, Kent,
November 1927.

22

MEASUREMENT OF THE SOUND POWER
OF NOISE SOURCES

Erwin Meyer and Paul Just
*Information Release from the Imperial Office of
Telegraph Technology*

*These excerpts were translated expressly for this Benchmark
volume by colleagues of Malcolm Crocker, Purdue University,
from "Messung der Gesamtenergie von Schallquellen" in*
Z. Tech. Phys. **8**:309-316 (1929).

1. INTRODUCTION

In order to measure the sound-field produced by a source, one
normally uses measuring devices that are sensitive to the amplitude of
either acoustical pressure or particle velocity. The recorded values
are valid only for the point in the room where the measurement was
made. From the pressure or the particle velocity, the flow of acoustic energy at the point of measurement can be calculated using known
formulas. But if one wishes to determine the total acoustic energy*
radiating from a source, only one method is available. Measurements
must be made at many points spherically distributed about the source,
the sound power at each point must be calculated from the measurements, and the sound power of each point summed to give the total
radiated energy.* Besides the fact that this process is quite tedious,
it is imprecise when used on strongly directional sources or on
sources composed of several individual noise sources.

The acoustical problem posed here is the same as in optics in
photometry where one tries to determine the flow of light energy
resulting from a lamp. Here one can also determine the total amount
of light radiated by making measurements with a photometer at very
many places in the room. But in photometry there is a much easier
means - an instrument, called the Ulbricht-sphere, [1] which provides
the integral value directly. The goal of the present experimental
investigation is to determine, whether in acoustics a measurement
technique is possible that is analogous to the measurements with the
Ulbricht-sphere.

2. THEORETICAL FOUNDATION

If one wishes to determine the sound power emitted by a source
with a single measurement, it is necessary to insure that all the
sound waves radiated from the source contribute to the sound level at

---

* [<u>Editor's</u> <u>Note</u>: Literally "energy" but "sound power" is meant.]

the point of measurement. This can only be accomplished...by the use of reflective walls.

In the following we assume a completely closed room. The complete sound absorption is A, where

$$A = \sum_{i=i}^{n} a_i S_i ,$$

S_i represents individual surfaces in the room and a_i represents the sound absorption coefficient of each surface, S_i. As suggested by W.C. Sabine, the value of a that corresponds to 1 m^2 of open window area in the room is set equal to unity. Using the assumption that the walls have no curvature so that the distribution of sound in the room is uniform, W. Jäger [2] has shown that a noise source with constant sound power L produces, in the steady state, a constant energy density E throughout the room. This energy density is given by the equation:

$$E = \frac{4L}{A \cdot c} ,$$
(1)

where c is the speed of sound.

The energy density increases with decreasing sound absorption A, and it is independent of the room volume. This equation is valid only for small values of absorption. One can see that when the absorption approaches 1 in a strongly damped room, the energy density cannot remain a constant since, as in free space, the amplitude of sound decreases with increasing distance from the source. Constant energy density is only made possible by the numerous reflections from the walls. The lower the energy loss is through absorption, the closer the strength of the reflected waves approach that of the incident and the more accurate equation (1) becomes. The uniform energy density arises from the summation of many reflected sound waves in the room.

[Editor's Note: Material has been omitted at this point.]

The aforementioned assumptions made in Jäger's theory mean that for our purpose the test room should have certain properties: It must not have curved surfaces and thus the best room to choose is one with rectangular corners. The sound absorption should be as low as possible. If equation (1) is to be used to determine the sound power L of a source in the room, one must measure E and A experimentally. Before we examine Jäger's theory more closely, we will examine methods of measuring E and A in the next two sections.

3. DETERMINATION OF SOUND ABSORPTION

Equation (1) is valid for the steady state, but the aforementioned theory also predicts behavior in the nonsteady state, such as when the sound source is switched on or off, that is the initial

unsteady sound field and the reverberation. We will consider only the
case of reverberation after the source has been switched off. If E is
the steady state energy density, E is the instantaneous value of
energy density, V is the volume of the room in m^3, and A is absorption
as above, Jäger's theory [3] gives

$$E' = E \cdot e^{-\frac{cA}{4V} \cdot t}$$

(2)

If we set $E'/E = 10^{-6}$ and call the corresponding reverberation time T,
we get

$$A \cdot T = 0.16 \cdot V.$$

(3)

Equation (3) can be used to find the absorption A of a room with a
known volume when the reverberation time T is measured. This measurement must be made at several frequencies because the absorption and
thus T are frequency-dependent. To accomplish this, a method
developed by the authors was used [4]. In place of the usual subjective measurements an objective graphical process is used. One does
not use a pure tone sound because the standing waves that result give
a very irregular reverberation measurement curve. A warble tone is
used where the noise contains all frequencies within two frequency
limits [5]. A bandwidth of 100 Hz has proven adequate for both high
and low frequency measurements. The sound is produced with a frequency generator and loudspeaker. A rotating condenser on the generator varies the frequency between the upper and lower limits about ten
times per second. A distortion-free microphone and amplifier coupled
to an adequately quick responding galvanometer are used for the sound
measurements. The position of the galvanometer, which corresponds to
the sound pressure seen by the microphone, is photographically
recorded. Illustration 1 shows a sample measurement. The upper edge
of the graph contains the time scale; at time t = 0 the noise source
is switched off, and the curve shows the decaying reverberation. When
the deflection of the galvanometer drops to very low values, the gain
of the amplifier is automatically multiplied by ten so that the
further decay of reverberation appears on a magnified scale. When the
data from the curve in Illustration 1 are plotted with a log scale
ordinate, one must get a straight line according to theory. Illustration 2 shows this plot. From it, the above defined reverberation time
can be found to be T = 1.7 sec. This method has already been successfully applied to find the reverberation times of rooms such as radio
sound studios. It can also be used to measure the absorption coefficients and the applicability of various materials that may be used as
sound damping materials on walls in noisy areas such as factories for
noise reduction purposes [6]. For our purpose, the absorption coefficient A can be derived from the measured reverberation time T using
equation (3). Experiments were conducted in two nearly empty rooms of
about 130 m^3 volume that happened to be available for measurements.
The rooms differed in the materials of the walls and floor as well as
in geometric form and window area. Illustration 3 shows values of A

as a function of frequency for both rooms. A gives the number of square metres of open window area equivalent to the absorption in the room. The measured values of A were confirmed by the subjective impression of listeners engaged in conversation in the room. Room a has very little low frequency absorption, so it sounds dull and hollow whereas room b has a much brighter tone color. Both rooms are very lightly damped; the average absorption coefficient a per m^2 varies from 0.02 to 0.15 for the very highest frequencies.

4. DETERMINATION OF ENERGY DENSITY

A device that directly measures the energy density at some point in the sound field does not exist, so the energy density must be calculated from measurements of pressure or particle velocity. In this investigation the pressure amplitude was measured at first using a Reiss microphone. For the purpose of calibration, a metal plate (electrode) with many holes for sound transparency was placed in front of the membrane, which was covered with a thin foil to make it a conductor. In order to obtain the frequency response, it is only necessary to apply a dc-voltage and a constant ac-voltage between the membrane and cover (electrode) for all frequencies. Then a constant (electrostatic) force is applied to the membrane for all frequencies. The Reiss microphone, a carbon microphone, has a disadvantage that makes it a poor choice for acoustic measurements. Its sensitivity and frequency response undergo constant small changes so that the microphone must be calibrated over and over again. For that reason, a condenser microphone was built with a low frequency circuit that offers just as good a frequency response, a very high sensitivity, and stability of calibration.

The frequency response for both microphones is shown in Fig. 4. The condenser microphone K.M. was calibrated in the same way as the Reiss microphone R.M. [7]. The curves show the combined response of the microphones and amplifier, since a three- or four-stage amplifier is needed. The amplifier gain is adjusted to the desired value by a potentiometer, the resulting current is measured with a thermocouple or with a rectifier and a dc galvanometer.

In order to measure the pressure amplitude it is necessary not only to have the relative frequency response, but the entire system must be calibrated against an absolute. This was done for a number of frequencies between 100 and 10,000 Hz with the help of a Rayleigh disc. If a dc-voltage V_0 and an ac-voltage $V_1 \sin \omega t$ is applied between the membrane and the electrode ($V_1 \ll V_0$, V_1 effective voltage), this results in a pressure acting on the membrane:

$$p = C \cdot V_0 \cdot V_1.$$

The constant C can be determined by finding a pressure amplitude and values of V_0 and V_1 that result in the same reading on the amplifier. If one measures the voltages V_1 and V_0 and measures the pressure p with a Rayleigh disc, C can be calculated. Since C is determined by geometric constants such as membrane area, distance from membrane to cover, and so forth, it is a constant and so a knowledge of V_0 and V_1 is sufficient for all further calibrations.

Given the effective pressure amplitude in dyne/cm^2, the energy density E is

$$E = p^2/c^2 \cdot \rho.$$

Using $\rho = 0.0012$ g/cm^3 and c = 340 m/sec, the sound power L in microwatts is given by solving equation (1) for L

$$L = 6.1Ap^2 \ [\mu W]. \tag{4}$$

If it is desired to measure the sound power developed by a loudspeaker as a function of the input frequency, an appropriate signal generator is required. The system used was developed earlier and offers the advantage of a constant sine wave voltage output at all frequencies. It consists of two high frequency generators, one of which operates at a constant frequency while the frequency of the other is varied using an adjustable condenser. The difference in the high frequency signals provides the low frequency output. A schematic view of the loudspeaker test apparatus, which was similar to the one used previously [8], is given in Fig. 5. The frequency response of the speaker was automatically recorded on film in such a way that the graphic records and the adjustable condenser were driven by the same motor. A small secondary condenser, driven by a separate motor was connected in parallel with the adjustable condenser. It varied the output frequency by ± 50 Hz about 10 times per second, producing a warble tone in order to prevent standing waves in the recording room.

5. CONTROL EXPERIMENTS

[Editor's Note: Material has been omitted at this point.]

Because of the geometric reflection of sound waves, the ideal recording room should be rectangular, as near cubic as possible, with no irregular surface or corner shapes. The dimensions must be at least large enough to be comparable with the wavelength of low frequency sound. The sound absorption should be kept to a minimum.

For such a room, equations (1) and (4) predict a uniform energy density in the room. The energy density will depend on the sound power of the source and the absorption of the room. Since equation (1) is derived using a number of simplifying assumptions, it is necessary to check by experiment whether it is valid for the room being tested.

The first experiment checked whether the energy density in the room was uniform. A constantly working loudspeaker is the best sound source for this purpose. The loudspeaker may not, as already mentioned, produce a pure tone sound because of the standing waves that would result in tremendous variations in the measured values of energy density [10]. As in the case of the determination of reverberation time, one must use a narrow band warble tone. A condenser microphone is used to record the sound. As a person carries the microphone around the test room the output is recorded on photographic film. Fig. 6 shows some of the results of the experiment for various frequencies as the microphone was moved all about the room. The recorded pressure amplitudes were read off at constant intervals and converted to logarithmic values for display in Fig. 6. One can see that the energy density in the room is adequately uniform to allow practical experiments to be conducted assuming uniformity. Only in close proximity to the noise source was an increase in pressure amplitude noticeable. In all following experiments the microphone was not facing the loudspeaker, which was always turned to face a wall, the ceiling, or a corner of the room.

The next experiment examined the effect of the location of the noise source within the room and the effect of the room itself. To answer these questions two positions of microphone and speaker were tried in both of the test rooms and the frequency response of a Rice-Kellog speaker was measured in each case. The experimental setup shown in Fig. 5 was used and the recorded pressure amplitudes were converted to sound power in microwatts using equation (4). The absorption characteristics of the two rooms a and b are given in Fig. 3. The results of the four frequency response measurements are shown in Fig. 7. It is clear that the results are largely independent of microphone position or test room and mainly show the loudspeaker characteristics. The standard deviation between the values in Fig. 7 is about 30%, which is quite satisfactory for acoustical energy measurements.

The final control experiment involved measuring the total sound power originating from the source using two methods--by the method just described and by a point-by-point measurement of the sound field. For this purpose a simple horn-type loudspeaker was used and the sound field was measured point-by-point using a Rayleigh disc. The frequency used was a warble tone centered on 800 Hz. The point-by-point measurement and summation gave a total sound power of 36 milliwatts while the integral method described above yielded 30 milliwatts under ideal conditions. For practical acoustical purposes this is a useful accuracy. The fact that the integral method gives a lower value of sound power is no accident--the variation is presumably caused by the fact that the pressure amplitude is not exactly constant in the whole room. In the area near the source the amplitude is higher, and this is not taken into account. One can also work in reverse and, as in photometry, use a noise source that has been calibrated by a point-by-point measurement to "calibrate" the test room. In this case the absorption measurements described above would not be necessary.

6. LOUDSPEAKER MEASUREMENTS

We will now go on to consider some applications of the method described above. As mentioned in the introduction, measurements of pressure or particle velocity amplitude were made at a single point in front of the loudspeaker. In order to measure an average value in the circumferential direction, the speaker can be rotated rapidly on a vertical axis and the amplitude recorded by an instrument of slow response [12]. Now, using the method described here, the sound power can be measured directly. This is particularly important in the case of speakers that consist of more than one sound source where each source is used for the reproduction of sounds within a given frequency range.

In order to record the frequency response of a loudspeaker it is necessary to set some variable constant. The most practical method is to hold constant the ac-voltage on the control cathode of the amplifier that drives the speakers. In this way the response of the amplifier tubes, the speaker, and the frequency generators are brought into the experiment. The measurements were carried out as shown in Fig. 5 with a noise bandwidth of ± 50 Hz. The use of a noise band is necessary for two reasons. First, to prevent standing waves from forming in the room, and second, to eliminate the effect such waves would have on the sound production of the loudspeaker. Using a bandwidth of sound has the disadvantage of rounding off sharp resonances of the speaker.

Figures 8a and 8b show the frequency response of the Rice-Kellog speaker in terms of sound power (8a) and p^2 (8b), were p is the pressure amplitude directly in front of the speaker. The differences between the two are quite small at low frequencies where the sound waves have a spherical form, but at high frequencies considerable differences exist. At high frequencies the speaker acts as a very directional source, so that high pressure amplitudes are measured directly in front of the speaker despite a low level of overall sound power. The curves in Figs. 8a and 8b can be considered as extreme cases when dealing with practical problems. In general, the listener would hear sound similar to that given by the sound power response curve, although somewhat modified by the absorption characteristics of the room [13].

Figure 9 gives the frequency response curves of six sample loudspeakers. The response is given in terms of microwatts of sound power. The ac-voltage supplied to the tube amplifier was between 3 and 8 volts. Curve (a) shows an older type electromagnetic horn speaker, (b) is a newer electrostatic speaker, (c) is an electrodynamic type, and (d) is an electromagnetically driven plate loudspeaker. A second speaker of the same kind as (d) is shown in both (e) and (f). In case (e), a tube of low internal impedance (RE 604) was used, while in (f), a tube with very high internal impedance was used (RE 164d). Plots (e) and (f) confirm the well known fact that high impedance tubes favor the higher frequencies, because the apparent impedance of the speaker rises with increasing frequency.

The results shown in Fig. 9, as well as response curves measured for many other available speakers, show that the ideal in sound reproduction, constant sound power at all frequencies, has by no means been reached.

Given a loudspeaker's sound power output, it is fairly simple to calculate the speaker efficiency. The only other information needed is the input power, which can be determined, for example, by using the three-voltmeter technique. The following efficiencies as a function of frequency were calculated for an electrodynamic speaker:

TABLE 1

Hertz	100	200	400	800	1600
Efficiency	0.09%	0.23%	0.35%	0.9%	0.7%

The highest efficiency is about 1%. It can be clearly seen from the values in Table 1 that the radiation resistance increases with frequency.

Efficiencies were measured for a few other loudspeakers at the frequency which gave the maximum sound pressures. For an electromagnetically driven plate speaker of high sound power, the efficiency reached 3.5%, for an electrostatic speaker (with throttle and a 1,000 V dc static charge) 2%, and for an electromagnetic horn-type speaker at resonance, 7%.

7. OTHER NOISE SOURCES

If one desires to measure the sound power of a source that develops a complicated spectrum or band of sound according to the method described here, the problem arises that the sound absorption in the test room is frequency dependent. In principle it is possible to construct a room with constant sound absorption, since materials are available that have different absorption characteristics. They can be combined in a room to get uniform absorption at all frequencies [14]. Since such a room was not available for experimental purposes, an average value of absorption for all frequencies was used for the experiments described below. Otherwise the experimental method was the same as before.

First the sound power of conversation was measured. For conversational speech with four people, values between 3 and 14 microwatts were recorded. These results compare well with values reported by Crandall, who stated 10 microwatts as the average value for the English language [15].

The sound power of a few musical instruments was also measured. In order to prevent standing waves, trills and runs were played, and the players moved about constantly in the room. Strong fortissimo on the violin gave 100 microwatts, while a cornet played as loudly as possible gave 50 microwatts. These examples show that the sound power of musical instruments has often been underestimated. It is said that several million trumpeters would be needed to produce one horsepower of acoustic energy, while in fact, "only" 15,000 men would be enough.

Measurements of the kind described in this paper are quite simple to carry out. It is planned to use this technique in the field of building acoustics. One can get a more precise definition of <u>sound transparency</u> than before, in that the sound power striking the wall and the (radiated) power are measured. [<u>Editor's</u> <u>Note</u>: It is not clear in the German text what the author means. Presumably the term <u>sound</u> <u>transparency</u> means <u>transmission</u> <u>loss</u> and the <u>striking</u> and <u>radi-ated</u> sound power are equivalent to the <u>incident</u> and <u>transmitted</u> power, respectively.]

SUMMARY

In the present paper a method of determining the sound power emitted by a source is described. The measurements are carried out in a completely closed room having a minimal sound absorption (reverbera-tion room). The absorption characteristics are determined using a newly developed objective experimental method of determining the reverberation time. The sound pressure amplitude in the room is meas-ured using an electrostatically calibrated microphone. Given measured values of pressure and absorption, the sound power of a source can be calculated using Jäger's theory. The validity of this measurement technique has been confirmed by experiment. As examples for the use of the method, the frequency response and efficiency of several loudspeakers are given, and sound power levels of human speech and a few musical instruments are also given.

Submitted on July 6, 1929.

NOTES AND REFERENCES

[<u>Editor's</u> <u>Note</u>: Only the Notes and References cited in the text are included here.]

[1]   Compare B.R. Ulbricht, The Spherical Photometer (Das Kugelpho-tometer), Berlin, 1920.

[2]   W. Jäger, Zur Theorie des Nachhalls, Wienr Akad. Ber. Mathem. Naturw. Klasse 120. Abt. 2a, 1911.

[3]   The reverberation process in rooms of certain geometric form has recently been dealt with in a more rigorous mathematical manner than previously by E. Waetzmann and K. Schuster in "Uber den Nachhal in geschlossenen Räumen", Ann. d. Phys., 5. Folge, Bd. 1, S. 671, 1929. There are disagreements with Jäger's theory,

but for cubic and cylindrical rooms with little damping the differences are relatively small, so they have been neglected here. For the calculation of sound power these differences play no role, since they apply to equations (1) (2), and (3) in the same way.

[4]   Erwin Meyer and Paul Just, "Zur Messung von Nachhalldauer und Schallabsorption," Elektr. Nachr. Techn. (ENT) 8, (1928), 293.

[5]   Recently several warble tones useful for architectural and room acoustics have been recorded and are available as phonograph records from Parlophonplatten. See Zeitschr. f. techn. Physik, 10 (1929) 148, vol. 4.

[6]   See Erwin Meyer and Paul Just, "Über Messungen an schalldämpfenden Materialen," Telephgraphen u. Fernsprechtechnik (TFT), 18, (1929), 40.

[7]   The electrostatic calibration described here considers only the frequency characteristics of the membrane and not the disturbance in the sound field caused by the microphone body. These disturbances are smaller for smaller microphones. Theoretical work on the refraction from a microphone body has been done on spherical microphones with sound incidence perpendicular to the membrane. [See St. Ballantine, Phys. Rev. 32 (1928), 998].

[8]   Martin Grützmacher and Erwin Meyer, Eine Schallregistriervorrichtung zur Aufnahme der Frequenzkurven von Lautsprechern und Telephonen, ENT. 4 (1927), 208.

[10] Erwin Meyer and Paul Just, ENT, l.c (1928), 296, Table 1.

[12] E. Gerlach, "Eine registrierender Schallmesser und seine Anwendungen, Zeitschr. f. techn. Phys. 8 (1927), 515.

[13] See J. Wolff, Proc. of Inst. of Radio Eng. (1928) 16, 1729 and L.G. Bostwick, Bell Syst. Tech. Journ. (1929) Vol. 8, 135.

[14] Nearly constant absorption for all frequencies was accomplished in a radio recording studio made with thin wood panels and cloth curtains. As shown in a recent experiment, wood panels absorb waves, while curtains hung close to the wall absorb the higher frequencies.

[15] C.F. Sacia, "Speech Power and Energy, Bell Syst. Journ. 4 (1925), 627.

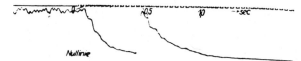

Figure 1. Reverberation curves

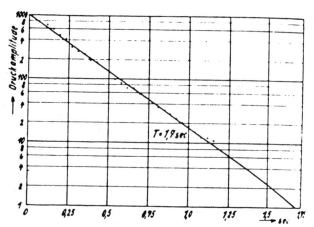

Figure 2. Reverberation line (curve of Fig. 1 plotted on a log-log scale)

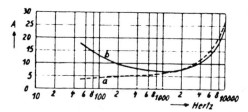

Figure 3. Sound absorption of two experimen rooms as a function of frequency

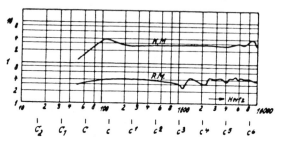

Figure 4. Frequency responses of the Reiss microphone and condenser microphone

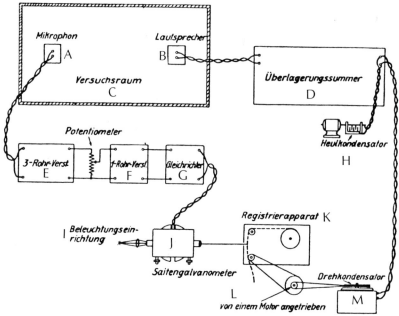

Figure 5. Apparatus for recording the frequency response of loud-speakers. A-microphone; B-loudspeaker; C-test chamber; D-summer; E-3-tube amplifier; F-1-tube amplifier; G-rectifier; H-warble condenser; I-light source; J-galvanometer; K-recording apparatus; L-both driven by the same motor; M-constant frequency source. Note: The galvanometer bends a beam of light in proportion to the applied voltage.

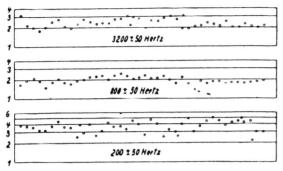

Figure 6. Energy density at various locations in the test chamber

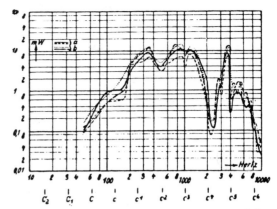

Figure 7. Frequency response curves of the Rice-Kellog loudspeaker, measured in two different rooms "a" and "b" with various microphone and speaker locations

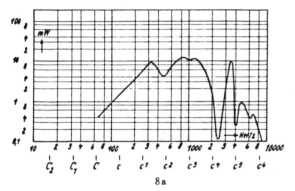

8 a

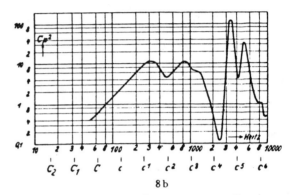

8 b

Figure 8. Frequency response of the Rice-Kellog speaker: a) sound power, b) pressure amplitude at the microphone location

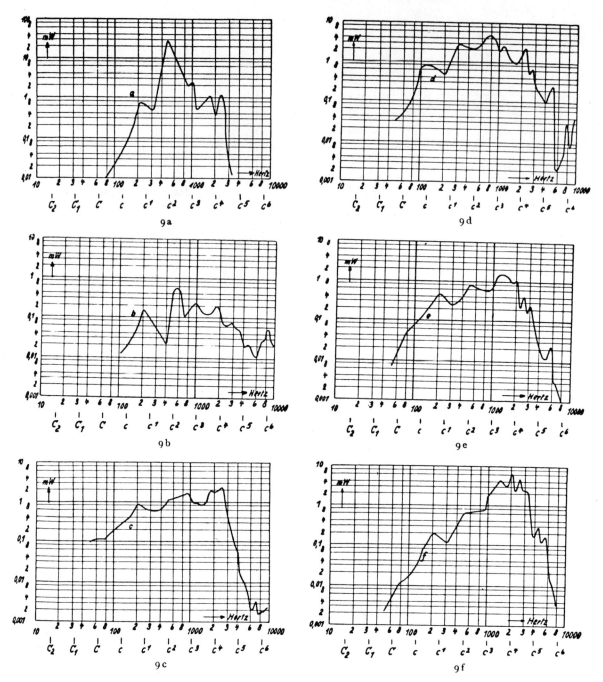

Figure 9. Frequency response curves of various loudspeakers

23

A NEW SOUND ABSORPTION ARRANGEMENT OF HIGH EFFECTIVENESS AND THE CONSTRUCTION OF AN ANECHOIC ROOM

Erwin Meyer, Gerhard Buchmann, and Arnold Schoch

*Communication from the Institute for Schwingungsforschung
of the Technischen Hochschule, Berlin*

*This excerpt was translated expressly for this Benchmark
volume by colleagues of Malcolm Crocker, Purdue University,
from "Eine neue Schallschluckanordnung hoher Wirksamkeit
und der Bau eines schallgedämpften Raumes" in*
Akust. Z. **5**:352–364 (1940).

INTRODUCTION

In acoustical measurement technology, propagating waves have an eminent significance. Calibrations of both sound sources and sound receivers require them. Measurement devices like sound pressure and noise level meters are tested with propagating waves. Different countries have standards to establish the nature of the sound fields used in the calibration procedures for electroacoustic devices. In the domains of room as well as building acoustics, propagating waves play a great role, for example in the determination of the reflection from or the transmission through building materials.

1. CONSTRUCTION OF PREVIOUS ANECHOIC ROOMS

Propagating sound waves are produced if sound reflection is prevented in the neighborhood of the sound source. Equivalent free space for acoustic transmission, even when disturbing external effects like weather and wind are not in question, is also disturbed in many cases by the ever present surface reflections. Closed rooms are hence usable only if the walls are sufficiently sound absorbent or sufficiently far from the location of the measurement. If we denote by E_D the energy density in the direct sound field at distance r from a point source of sound and by E_R the mean energy density of the reflected portion of the sound, in accordance with acoustical theory of rooms we have the relation:

$$E_R/E_D = - \frac{16 \, \pi r^2}{S \, \ln(1 - a)} .$$

Here a is the average sound absorption coefficient of the walls and S is the total area of the wall surface. Propagating waves may be said to prevail if the quotient E_R/E_D is small, that is, if either the sound absorption coefficient is almost 100% or if the wall area S is great, that is, the volume of the room is itself great. It is also obvious that sufficiently close to the sound source (r small) the wall reflections do not produce serious disturbance.

The problem of creating an anechoic room therefore involves finding absorbing treatments of high efficiency over a large frequency range. Only porous materials are suitable here, their absorption coefficients as many theoretical and experimental investigations have shown depend on the three parameters: porosity, acoustic flow impedance, and thickness of the treatment. The acoustic impedance of a thick layer of a porous substance is given by the expression

$$W = \frac{\rho c}{P} \sqrt{1 - \frac{iz}{\omega \rho}}$$

in which ρc denotes the wave impedance of the air [the specific acoustic impedance for a plane wave], P is the porosity, z the flow resistance of the material and $i = \sqrt{-1}$. For the case of large absorption coefficient, the impedance of the material and the specific acoustic impedance of the wave must be approximately equal. We see that this is the case, if the porosity is approximately unity, and the flow resistance z is sufficiently small compared with the mass impedance $\omega \rho$ [ω is the angular frequency of the wave]. Cotton wool satisfies this condition and hence has been used often for anechoic rooms [1]. Its porosity is almost 100% and its flow resistance amounts to only a few cgs units (acoustic ohms). It is a consequence of the small flow resistance that particularly at low frequencies the layer of treatment must be extremely thick in order that the reflection at the hard boundary of the layer should play no role. In addition to this normally-used construction, another familiar one employs a combination of hanging layers consisting of 16 flannel and muslin curtains [2]. The corresponding absorption coefficients lie (in the range from 100 to 1000 Hz) between 95% and 99%. Here the flow resistances of each layer are respectively 15 ohms and 6 ohms.

2. DEVELOPMENT OF A NEW SOUND ABSORPTION SCHEME

The absorbing materials mentioned above, in particular cotton wool, have various practical disadvantages. They are not fireproof and they are also comparatively expensive. In room acoustics other available porous materials can be used. Among these, rock wool takes an outstanding position. It is made by atomizing, in a steam jet, the molten minerals occurring in blast furnace processes. It is completely fireproof and is favorably priced. Rock wool has a smaller porosity than cotton wool (i.e. about 90%) and a flow resistance between 80 to 200 ohm according to the method of preparation. The acoustical impedance especially for the lower frequencies lies essentially above the ρc for air, so that a homogeneous layer of uniformly packed rock wool will reflect more sound than a corresponding layer of cotton wool. On the other hand, the large flow resistance provides the advantage that the portion of the sound penetrating the material is strongly absorbed. The absorption per unit length is given by the expression:

$$\frac{\omega}{c}\sqrt{-\frac{1}{2}+\frac{1}{2}\sqrt{1+\left[\frac{z}{\omega\rho}\right]^2}} \ .$$

This increases with the frequency and the magnitude of the flow resistance.

In materials of this kind the principal task is to get the sound into the material, or in other words, to provide a matching of the wave impedances. We achieve the matching through a gradual transition from air to the material by constructing the latter in cone, pyramid, or cylindrical form; the sound penetrates into the homogeneous layer through a transition layer with gradually increasing cross-section. By such an arrangement we accordingly alter the porosity as well as the flow resistance gradually from the values of the medium that is 1 or 0 respectively to the values of the absorbing material [3]. As soon as we set up this principle we encounter a whole series of problems on the length and shape of the transition in connection with the properties of the chosen absorption material. The theoretical handling of this is difficult, demanding a solution of the wave equation with very complicated boundary conditions [4]. Hence we seek at first a solution of the problem along experimental lines.

It is necessary to describe briefly the measurement scheme with which small reflection coefficients are measurable. As is well known, it consists of a Kundt tube that is closed at one end by the device to be investigated. At the other end is located either the sound source or a strong absorbing material. In the latter case the sound must be admitted to the tube from the side. From the sound pressure P_{max} at the pressure antinode and the sound pressure P_{min} at the pressure node, the reflection coefficient (with respect to amplitude) is given by

$$P_{max}/P_{min} = (1+R)/(1-R) \ .$$

The sound pressure is measured by a movable microphone (a small Rochelle salt microphone) and recorded with an amplifier by a tube voltmeter or level recorder. It is of great importance that the Kundt tube shall have completely rigid walls. For the measurements, two tubes of square cross-section are used. The one is used for large scale samples and for frequencies up to 400 Hz. It possesses a cross-section of 40 x 40 cm^2 and a length of 15 meters. The second is only 14 x 14 cm^2 in cross-section and 5 meters long. It can be used up to 1000 Hz. Above the frequencies given, the first cross-modes exist in the tube and this hinders the development of a plane wave. Both tubes were built from brick walls of 25 cm thickness and with a concrete "floor" and "ceiling". The inner coating was constructed with particular care and made as impervious as possible. With this construction a difference of up to 40 dB between the sound pressure levels at the nodes and the antinodes was achieved in the large tube, using an acoustically hard sample at one end.

In room acoustics it is customary in the calculation and measurement of reverberation time to define the absorption coefficient a for absorbing materials, that is, the ratio of "non-reflected energy to the incident energy." In this system of measurement the values of the materials commonly used range from 90% to 100%. But it is more suitable for (absorbing) materials in anechoic rooms to define instead the amplitude reflection coefficient R. Thus R = 30% means a = 91%, R = 10% means a = 99%, and R = 3% means a = 99.91%.

We pointed out above that, for porous materials, the reflection coefficient becomes lower when the porosity is increased. The suggestion therefore is that the porosity may be increased by not packing the absorbing material too uniformly tight, but by placing it to form channels aligned in the direction of the sound propagation. We then attain a sound absorbing scheme consisting of many parallel sound-dissipating channels: many experimentally well-confirmed theories of sound propagation inside channels are available [5]. Sound attenuation in the channels is greater the smaller the channel diameter and the smaller the wall impedance. For example, a rock wool coating of 5 cm thickness with a channel 10 cm in width has in the frequency range from 100 to 1000 Hz a sound attenuation from 5 to 50 dB/m [6]. A channel length of 1 meter at a mean frequency of about 300 Hz will be sufficient to provide an attenuation of the reflected wave of 10% of the incident wave. Smaller channel widths have still more favorable relations.

Figures 1 and 2 confirm the predictions mentioned above. For comparison, in Fig. 1, homogeneous materials without special surface treatment have been measured, namely a rock wool plate (Ravolit plate) of the Berlin firm of L. Mandt Gypsum Works with porosity of 90%, flow resistance of 80 ohms perpendicular to the plane of the plate, acoustic impedance normal to the plate for high frequencies about $3\rho c$ and secondly a dense layer of loose rock wool. Both layers are 1 m in thickness. The measurement tube was that used for low frequencies. The plates of 5 cm width were set close together so that the complete cross-section of the tube was filled. Eight plates were arranged next to each other. We note that in all three curves the reflection coefficients are still comparatively high but decrease with increasing frequency. The impedance plays a role up to about 300 Hz.

Figure 2 shows the influence of the separation of the layers for the measurements corresponding to curves 1 and 2. Instead of 8, only 4 or 6 rock wool plates respectively are distributed over the cross-section of 40 x 40 cm^2; so that in the first case, 3 channels of 7 cm width each are present, and in the other, 5 channels of 4 cm width each. Correspondingly the reflection coefficients for the low frequencies are higher in the first case than in the second, whereas the values for the higher frequencies are reversed.

A combination (curve 3) of a dense packing of 50 cm thickness with a distributed arrangement (as in curve 1) of likewise 50 cm thickness gives lower reflection values. The corresponding sound absorption coefficients are around 96%.

The next step is to allow the channels to begin with a broad opening at the entrance and to have their cross-section decrease gradually or otherwise expressed, to have the channel width increase towards the opening. Figure 3 provides examples of this. This refers to pyramidal or wedge-shaped rock wool elements 1 m in length. Curve 1 gives the values for the rock wool pyramids packed manually that are covered with a gauze covering; they are packed so that the flow resistance is about 100 ohms, and the porosity is 90%. The base area of the pyramids is 13 x 13 cm^2, so that in the large measuring tube 9 elements can be fitted. The second curve refers to a wedge-shaped arrangement that was cut out and composed from 5 cm thick compressed rock wool plates (manufacturer Monolan). The wedges, which had a base area of 5 x 13.5 cm^2, were stacked in many layers on top of each other. The wall behind this arrangement was in each case sound-reflecting (5 cm thick lacquered wooden plate); in the arrangement of curve 1 there was in addition a brick wall of 6 cm thickness inserted between the wooden plate and the rock wool. At 100 Hz the reflection coefficients already have values of about 10% and at yet higher frequencies, drop to values of a few percent. That, incidentally, represents the limit of accuracy of the duct method. Thus, an absorption coefficient of 99% was obtained at 100 Hz and yet higher values of more than 99.9% were measured in the higher frequency range.

This effect is not only confined to rock wool material, as curve 3 shows. This curve was recorded for pyramids of fiberglas. The shape of the curve is similar, although the material was not packed with the right density. Therefore the flow resistance and the porosity were not adequately chosen and the low absolute values for the reflection coefficient could not be obtained in the low frequency range.

After the principle of gradual transition has been proved to be essential for sound absorption, questions arise concerning the length and the shape of the transition. In order to decide the first question, transition lengths were used in which the cross-sectional area increases linearly with the depth. The absorption elements are thus wedge shaped.

One must expect very clear behavior wherever the layer thickness of the material existing behind the transition is so great that the amplitude of the wave reflected from the back end is sufficiently small. For the experiments, plates of rock wool with bitumen are used, which can easily be cut into any desired shape. A 40 cm thick compact layer with the end of the duct open, served as an absorbing layer. The measurements were made in the small measurement duct and could hence be extended up to frequencies of about 1000 Hz. The wedge shaped transition piece consisted of wedges of 15, 30 and 60 cm length. The measurement results are presented in Fig. 4.

The matching (of impedances) is approximately reached at frequencies 600, 300 and 150 to 200 Hz, respectively. For the lower tones the insufficient thickness of the absorbing material and perhaps also the fact that the wave impedance of the material has a considerable reactive component, causes a disturbance. One can say that under experimental conditions the transition length must amount to about 1/4 to 1/3 of the wavelength. In addition to this measurement, two more sets of experiments were carried out. The first set was performed with a constant length of the rectangular absorption layer and with variations in the ratio of the length of the wedge shaped transition piece to the length of the completely packed rectangular piece. In the second set the length of the wedge shaped transition piece was varied while the total volume of the absorption material was kept constant. The end of the duct was in each case acoustically hard (5 cm thick iron plate). Figure 5 gives the results of the first series of measurements; Fig. 6 shows those of the second series. Here also one recognizes the decisive influence of the transition length, which must be about one third of the wavelength. For low frequencies resonance phenomena again play a part owing to the insufficient thickness of the absorption element. If the element is designed correctly, small values of the reflection coefficient can be obtained over a particularly wide frequency range. The reflection coefficient for the arrangement 4 in Fig. 6 (layer dimensions 25 and 90 cm) for example, has a value of less than 10% at frequencies higher than 90 Hz and a value of only a few percent at frequencies higher than 250 Hz. As far as the shape of the transition piece is concerned, the measurements presented show that a linear as well as a quadratic increase (pyramids) of the cross-sectional area of the absorption material gives the same desired performance.

Shapes of the wedge shaped transition piece that have an area increase that is greater than that of the second power, such as an exponential shape, achieve equal reflection coefficients in the higher frequency range, using the same total length of absorption material, but have higher values in the low frequency range. This is because the total quantity of the absorption material is smaller.

Figure 7 confirms what was just said. The layer of absorbing material behind the transition interval is 40 cm thick and is adjacent to the open end of the duct. The cross-sectional area of the front edges of the exponential wedge depends on the workability of the material and has a width of 1/2 cm to 1 cm. The reflection coefficients of both the exponential wedge and the linear wedge are practically equal above 500 Hz within the limits of measurement accuracy.

All the shapes discussed so far exhibit a point or an edge at one end. Some investigations have been made to examine the opposite case, where the shape is blunt. These shapes produced smaller reflections in the low frequency range than the type with a uniform distribution of material, but in the high frequency range the values obtained were considerably higher than those of the sharp type.

All the measurements in the Kundt tube were performed with normal incidence sound waves. With this type of arrangement it is expected, however, that the absorption for oblique incidence sound waves would also be extraordinarily good because the sound waves must take a longer path to reach the wall. An investigation of this subject was not carried out since it is of interest only for low frequencies and in this range great problems in the measurements will be encountered.

3. THE STRUCTURE OF THE ANECHOIC ROOM

The preceding investigation shows that one can manufacture effective sound-absorbing equipment that, at the same time, has practical advantages over previously known materials. This gives a new foundation for the building of a very good anechoic room. But the room must possess as well as the property of a high absorption capability of its walls, the further property of being sufficiently large, especially whenever one wishes to measure low tones with high accuracy, or whenever one wants to investigate a large mechanism. The large scale is required with respect to the wavelength of the low frequency waves and secondly because of the well-known feature of a sound absorbing layer to damp effectively sound waves at grazing incidence [7]. One must be able to position the sound source and measurement equipment far enough away from the layer so that the sound propagates as if in a free space. Finally, yet a third fundamental question about the planning of an anechoic room is important, namely the question whether the room should be asymmetrical. It is clear that along the lines or surfaces of symmetry, noise reflection from the walls that are still present are in phase and hence can produce great irregularities in the sound field. One can get around these difficulties by choosing the positions for the measurements so that they do not coincide with these symmetry positions of the room. This is a much easier procedure than to build the room somewhat assymetrical, which makes construction difficult. Thus, the room was planned with parallel, rectangular walls and has therefore the advantage that the reflection sites are easily found for a noise source placed in any particular position.

Figure 8 shows the floor plan and the elevation of the anechoic room. The dimensions are: length 16 m, width 11 m, height 9 m. This gives a volume of 158 m^3. In front of the anechoic room is a small laboratory of floor area 32 m^2. Figure 9 shows an outside view of the building.

The main problem was to manufacture all the absorbing materials. It is obvious to consider a procedure where the wedge or pyramid shaped elements are manufactured automatically. Manufacture by such methods was suggested to the relevant industries and was partially successful. However, there is still some work to be done on development in this area. Nevertheless, in order to cover the room nothing else is possible but to have the absorbing material manufactured by hand and at the same time to take the manufacture under one's own direction. It was decided to choose pyramid-shaped elements. This shape was made in such a way that the pyramidical part rested on a rectangular base with square cross-section. The following dimensions were chosen: base area 15 x 15 cm^2; total height, 1 m; height of the square part, 15 cm. This is a similar shape to that described in Fig. 4, Curve 1.

Rock-wool elements hold together well when densely packed because of interlocking of fibers. In order to give a durable form the elements were covered with a very thin gauze, and a finish applied for stiffening. Thin wooden slats were inserted into the rock-wool elements to achieve a higher rigidity. The absorbing elements placed on the side walls of the room each had a wooden beam that was pointed in front and that penetrated the entire element. At the base area these beams were attached to a wooden triangle that was located diagonally. Fastening eyes were also fixed to this triangle. The ceiling pieces had only a short wooden slat with two small transverse slats; the fastening eyes were here fixed to the wooden slat. The pyramids to be placed on the floor were given a wooden strip with a diagonal slat for stiffening. In each case a perforated cardboard disc was fixed at the end of the square-shaped part of the absorbing elements. Therefore, also the ducts of the perforated bricks that were used for the construction of parts of the walls of the room became effective as sound absorbers. The pyramids were suspended by small hooks screwed into the ceiling, but on the side walls the hooks were nailed into place. In order to simplify this work, the masonry of the inner wall was not made from solid bricks, but from perforated bricks with holes normal to the wall surface [8]. Wooden plugs were driven into the holes of the bricks and the hooks were driven into these plugs.

The production of the absorbing elements proceeded in the following way: (Figures 10a and 10b). The gauze correctly cut to shape is first of all placed into the mold shell that was used to form the elements. Then a quantity of mineral wool is weighed out and poured in and compressed together with the wooden suspension frame work that carried the fastening eyes. The cardboard disc was already in the right position.

The free ends of the gauze are not sewn or glued, but held together with pins that proved to be the most time-saving method. In the last step, the base area of the absorption elements were wrapped up with a precision-cut piece of gauze. The absorbing element weighs, including the wooden frame, from 2 up to 2.2 kg, so that it gives a weight per square meter of 90-100 kg [9]. The fastening of the absorbing elements was carried out using a scaffolding equipped with an electric elevator (Fig. 11). [A partial list of materials used in the room construction follows.] An area of 840 square meters was covered. For this, about 32,000 absorbing wedges were necessary.

They contained 64 tonnes of mineral wool, [10] 18,000 square meters of gauze, wooden slats with a total length of approximately 30 km, 24,000 hooks and about one million pins. The absorption material elements were manufactured in eight shell molds in different rooms. Each room had a separate air-exhaust system because of the dust annoyance. With 11 workers it took 4 1/2 months to manufacture the absorption elements and cover the room.

Figures 12a and 12b show two views of the anechoic room in which particularly the grating for walking in the room should be noticed and also the construction of the door. The grating consists of a central pathway 1 meter wide, of very narrow mesh, and of a fairly broad-meshed section (mesh-size 25 x 25 cm^2), which covers the remaining floor area. The grating is on the same level as the floor of the laboratory since the floor of the anechoic room is 1 meter below that of the laboratory. The door has been made especially soundproof and moves on rollers. Figure 12a shows the measurement apparatus with sound generator and movable microphone. Several pieces of tackle for attaching acoustical equipment are provided that can be operated partly from the entrance door, and partly from a room above the ceiling. Fixed installations are carefully avoided with the exception of the grating, which is absolutely necessary and that has elements that can easily be removed. The room contains a ventilation system with two ducts whose walls have similarly received a sound-absorbing covering. The one-year experience showed that no dust-nuisance exists caused by the absorbing mineral wool.

The floor of the room is likewise sound-damped. In the planning stage it was intended that a completely reverberant floor should be installed and the noise sources should be built into it as well as the sound receiver. In this case one has a known sound source radiating into a half-space. Because many noise sources, for example large loudspeakers or musical instruments prevent installation of a completely reverberant floor, one covering was chosen for the entire room. However, in order to allow measurements with this other method, it is planned to cover the grating with strong varnished plywood which, except for the low frequencies, works as a good acoustic reflector. The dimensions of the room with walls completely covered amount to 14 x 9 x 7 = 882 m^3. Somewhat less than half of the original room volume has thus been lost through installation of the sound-absorbing treatment.

4. MEASUREMENTS OF SOUND PROPAGATION IN THE ANECHOIC ROOM

The acoustical impression upon entering the room is extraordinarily great, as so, incidentally, is the optical impression. The room is completely without reverberation. On entering, one has, as in all anechoic rooms, a peculiar feeling of pressure in the ears, which disappears when one stays in the room for a while. The decline in the sound-intensity with increasing distance from the source is subjectively very striking. In measurements, this decrease of intensity makes it necessary to use considerably greater amplification, compared with other rooms. Other effects are also noteworthy. If speaker and listener stand at the opposite ends of the central strip in the room with a number of other persons between them, communication between them is not possible because of the obstruction to the direct sound wave(s) and the increased noise level caused by the presence of these people. A wooden or metal plate of 1/2 square meter area gives a

distinct echo if there are short noise impulses like hand-clapping or
something similar.

The acoustical properties of the room can only be ascertained by
measurement. The usual method is to produce a known sound-field and
to determine the interference due to reflections from the wall. A
spherical sound-field was used in the present experiments. In order
really to produce spherical sound waves, in practice the noise source
must be of small dimensions and therefore it radiates a small amount
of sound power. To cover the frequency range, three different sound
sources are utilized: up to 200 Hz, a dynamic cone loudspeaker, the
backside of which was closed off by a small, heavy-walled wooden box;
from 200 up to 2000 Hz a pipe of about 5 cm diameter, the open end of
which represents the sound source, and at whose other end there was a
dynamic cone loudspeaker. Above 2000 Hz a high frequency loudspeaker
system is used whose small cone is replaced by a tube of 1 cm diameter. The measuring microphone, a small condenser microphone that is
as omnidirectional as possible is suspended from a wooden pole of
about 1 meter in length. The top end is drawn on rollers along a
wooden rail 9 meters long, like an overhead railway. During the movement of the microphone away from the sound source, the sound pressure
is amplified and read out on a Neumann graphic level recorder. It
must decrease proportionally to $1/r$, where the distance is signified
by r. In order better to recognize deviations from this relationship,
the position of a potentiometer on the amplifier is altered synchro-
nously with the movement of the microphone in such a way that the
amplification is varied proportionally to r. Thus, in an ideal spherical field the voltage signal and thus the reading on the graphic
level recorder remain constant. Figure 13 shows examples of these
measurements. The whole chart indicates a microphone traverse length
of almost 9 meters. Apart from a short initial part, during which an
over-compensation of the amplifier causes a disturbance, the trace
which has a dynamic range of 25 dB, is quite flat and smooth for the
middle frequencies. Within measurement error no deviation, from the
$1/r$ rule can be shown, in the whole range of measurement of 9 meters.
The results do not hold too well for low and for high frequencies.
For the low frequency tones, residual sound reflections from the wall
cause interferences. For the high frequency tones the source of the
irregularities lies in the measurement apparatus itself. By way of
example, covering the beam carrying the microphone by a felt strip
produced a real improvement. If the measuring apparatus is nearer to
the grill, then the disturbances for the higher tones become likewise
greater. The effect of a noise-reflecting plate, one square meter in
area, set up normal to the microphone track is shown in Fig. 13. A
striking interference effect occurs at 1000 Hz.

If the noise source and the microphone track are near a wall
area, there appears an additional damping, as was already shown above
in the conditions for room size. This effect was likewise proved
through some recordings (Fig. 14). The loudspeaker and microhone were
placed directly over the grating of the floor area. For the low and
middle frequencies, one obtains an average decay of from 1.2 to 0.9
dB/m. The propagation of high frequencies is, in addition, disturbed

by the lattice grating.

In this way, measurements were carried out in several places in the room. Figure 15 shows some results. The moving track was about 4.40 meters over the grating and was about 3.50 meters distant from one of the side walls. The greatest fluctuations in dB are indicated as coordinates in Fig. 15. These are presented separately for distance ranges from the noise source of 0-3 m, 3-6 m, and 6-9 m. For most acoustic measurements a measurement accuracy of about 1 dB is sufficient. This means roughly a 10% change in amplitude. Such an accuracy is attained in the region 0-3 m for nearly all the frequencies that occur. Above 300 Hz the fluctuations even for the largest distances are considerably smaller, amounting to only a few percent. This slight wavelike quality of the sound field does not merely, as Fig. 15 proves, reach only to about 1000 Hz, but right up to the highest frequencies, as long as the measuring apparatus is not interfering with the sound field.

Similar results as in Fig. 15 were obtained also in other parts of the room. Especially favorable results were obtained whenever the noise source is [sic] placed in a corner of the room and the measuring track goes out of the room at an angle. Then the first reflections are associated with much longer paths compared with the direct sound. The improvement at low frequencies is considerable. In this way, the greatest deviations from the 1/r law along the whole 9 m long measurement track amounted to the following: for 60 Hz, 2-3 dB; for 75 Hz, 2 dB; for 120 Hz, 3 dB; for 200-300 Hz, less than 1 dB. The variations above 300 Hz did not change compared to the values of Fig. 15.

Yet a second procedure for testing the room acoustics of an anechoic room was applied; the short pulse tone method [12]. Short wave impulses are emitted and displayed on an oscilloscope together with the echoes which arrive later. Figure 16 gives a selection of the measurement results for different frequencies and for different distances from the noise source. An electrodynamic conical loudspeaker with a sound screen served as the noise source in the present case. ["Sound screen" is a literal translation, probably meaning "baffle".] The noise receiver was the abovementioned condenser microphone. Apart from the lowest frequencies, no echoes at all could be confirmed.

With the experiments presented, it is shown that the room completely fulfills the expectations and that it is eminently suitable for all acoustical measurements requiring progressive sound waves.

[Material has been omitted at this point.]

NOTES AND REFERENCES

[1] The anechoic room built in 1932 for the Institute for Vibration Research has this sound absorbing material. See W. Janovsky and F. Spandöck, Akust. Z, 2(1937), p. 322.

[2] E.H. Bedell, J. Acous. Soc. Am. 8, (1936), p. 118.

[3] This seems to contradict L. Cremer's results of his theoretical investigations, published in Ebektr. Nachr. Technik 12 (1935) p. 362. However, Cremer examined arrangements with slowly increasing flow resistances, whereas a gradual decrease of porosity is important for a good matching [of impedances] of absorbing material.

[4] Lord Rayleigh, Theory of Sound, Vol. II, p. 96 and Vol. II, p. 235.

[5] L. Cremer, Akust. Z. 5 (1940), p. 57 and P.M. Morse, J. Acous. Soc. Am. 11 (1939), p. 205.

[6] W. Lippert, Akust. Z. 6 (1941), Vol. 1.

[7] G.V. Békésy, Z. Techn. Phys. 14 (1933), p. 6.

[8] The large area of perforated bricks (490 sq m compared to the total room area of 840 sq m) together with the wooden ceiling, resulted in a remarkably short reverberation time of the empty room (2 sec). Interesting also was the observation that with small noise impulses a whistling sound occurred, which was attributed to the decay (of sound) in the numerous resonators that were formed by the perforated bricks.

[9] The walls are designed for a load of 100 kg/m^2 applied with a lever arm of 80 cm.

[10] The large amounts of rock wool required were graciously provided with very high quality by the two German firms Deutsch Eisenwerke AG., Gelsenkirchen; and Deutsche Patentwärmeschutz AG., Dortmund-Hörde.

[11] We have to thank the firm of W. Genest G. Ltd., Berlin for placing at our disposal an experienced worker for the installation of the absorption material.

[12] F. Spandöck, Ann. Phys. 5, 20 (1934), p. 328.

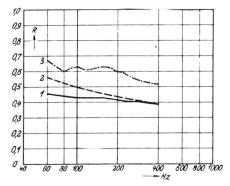

Figure 1. Reflection coefficients for uniformly dense rock wool layers of length 1m. 1. Eight rock wool layers held together with pitch (Ravolit plates) packed densely with end of tube open. 2. Eight rock wool plates packed densely with end of tube acoustically hard. 3. Uniformly loose packing of rock wool, with tube end acoustically hard.

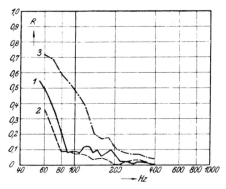

Figure 3. Reflection coefficients of three different absorption material layouts with smooth transitions, all 1m thick. 1. Handmade rock wool pyramids bound with gauze. 2. Wedge-shaped sheets of rock wool, cut from 5 cm thick sheets and installed in opposed positions. 3. Fiberglass pyramids.

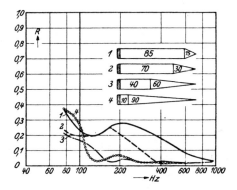

Figure 5. Dependence of the reflection coefficient on the ratio of wedge length to backing length at constant overall length (1m).

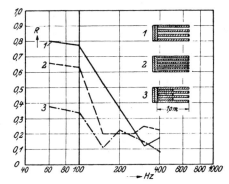

Figure 2. Reflection coefficients for channelized absorption scheme, 1m long. 1. Four plates distributed over the tube cross-section 40 × 40 cm². 2. Six plates distributed over the tube cross-section 40 × 40 cm². 3. Mixed system: four plates 50 cm long over the cross-section 40 × 40 cm². Dense packing over 50 cm length.

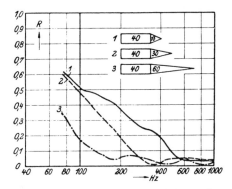

Figure 4. Dependence of the reflection coefficient on the length of the wedge-shaped transition portion of the sound absorbing material. The termination impedance was kept constant. Each wedge was backed by 40 cm of rock wool. The wedge lengths: 15cm (1), 30 cm (2), 60 cm (3).

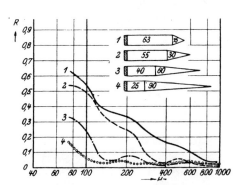

Figure 6. Dependence of the reflection coefficient on the ratio of wedge length to backing length with the amount of sound absorbing material kept constant.

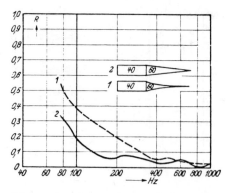

Figure 7. Comparison of the reflection coefficients of two different wedge shapes. Total length of wedge and backing 1m, backing length 40 cm, end of tube open. 1. Exponential wedge shape. 2. Linear wedge shape.

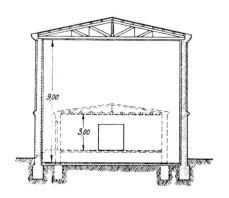

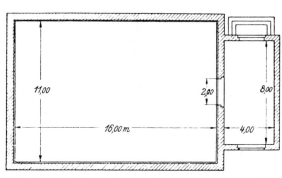

Figure 8. Floor plan and cross-section of the anechoic room

Figure 9. External view of the anechoic room

362

Figure 10.   Manufacture of the sound absorbing elements

Figure 11.   Mounting of the sound absorbing elements in the anechoic room

Figure 12.   Views of the completed room.

363

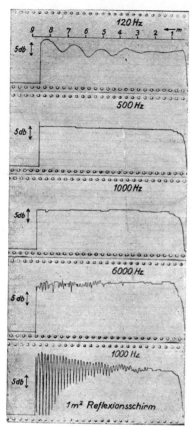

Figure 13. Examples of measurements made to check the 1/r relationship in a spherical sound field. a) 120 Hz, b) 500 Hz, c) 1000 Hz, d) 6000 Hz, e) reflection effects at 1000 Hz produced by a one square meter hard surface positioned perpendicular to the microphone path.

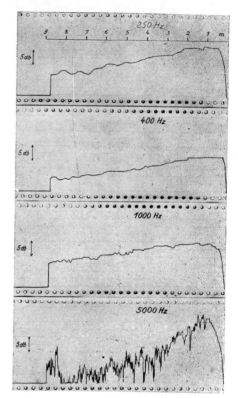

Figure 14. Absorption of sound waves traveling parallel to the absorbing floor of the room at 250, 400, 1000 and 5000 Hz

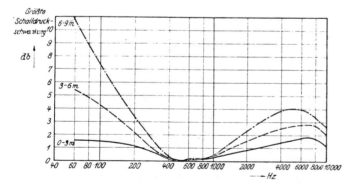

Figure 15. Greatest deviation from the 1/r law of a spherical sound field as a function of frequency and distance from the source. The measurement path was along the length of the room.

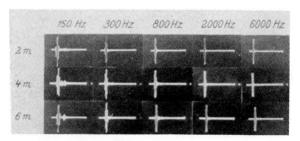

Figure 16. Reflection measurements using sound pulses of different frequencies at varying distances from the source.

Part VII

COMMUNITY NOISE

Editor's Comments
on Papers 24 Through 27

Noise has obviously been a social problem ever since people began living together in communities. The ancient Greek city of Sybaris, in which the noise of potters and tinsmiths was a nuisance, was mentioned in the Introduction. Bernadino Ramazzini described a similar situation in more recent times in *De Morbis Artificum* (Diseases of Workers), published in 1713 [1]. In the chapter on diseases of coppersmiths he states, "In every city, e.g., Venice, these workers [coppersmiths] are all congregated in one quarter and engaged all day in hammering copper. . . . from this quarter there arises such a terrible din that only workers have houses and shops there; all others flee from that highly disagreeable locality."

With the rapid growth of cities and industrialization in the last few centuries, manufacturing and industrial noise has become an increasingly troublesome community problem, particularly in the compact cities of Europe, where factories are often situated close to residential areas. However, undoubtedly a more serious source of community noise is surface transportation. As already discussed in Part I, the noise of wheeled vehicles was recognized as a problem in ancient Rome. Certainly the advent of motorized vehicles, steam engines, and rail vehicles in the last century has made transportation noise the major source of noise in most cities. In some parts of large cities near major airports, aircraft noise can be a serious problem as already discussed in the commentary in Part II.

The first well-documented survey on community noise is the report *City Noise*, which was published in 1930. This is the report of a Noise Abatement Commission appointed by Dr. Shirley W. Wynne, Commissioner of Health of New York City. The purposes of the survey and report were (1) to measure and tabulate general city noises to produce a basis for future comparison; and (2) to determine the magnitude of specific noise sources in order to furnish a basis for recommendations for improvement. The whole report is of interest, but because of its length, an extract is reprinted here as Paper 24. Of interest is the fact that this section on noise measurement, prepared by Fletcher (who was later to become President of the Acoustical Society of America), Beyer, and Duel, records noise levels from different sources obtained with an early instrument, a noise meter, which was remarkably like present-day sound level meters (see Fig. 1). The noise meter employed a weighting network (Fig. 2) that is similar to the A-weighting filter. The sound pressures were given using the (for then) relatively new decibel scale (p. 381). Such a scale had been in widespread use for only three or four years and still had not been universally accepted (see Paper 1). However, note that the root mean square reference pressure used in Paper 24 was not the one commonly used today of 0.00002 Pa. (For further discussion on reference pressures see p. 105.) An audiometer was also used to check the results from the noise meter (see p. 378). It is particularly interesting to look at the recommendations made in 1930. It was found that the volume (amount) and composition of street traffic mainly determined the outside noise at any point in the city, and it was recommended that a goal of making motor trucks as quiet as automobiles should be adopted. It is sad to note that in the United States and Britain and in many other industrialized countries, surface traffic is still the major source of noise fifty years later and that the governments of several of these countries still have as their goal quieting heavy trucks so that they are no noisier than automobiles.

Paper 25 by Bonvallet reports a survey of city noise in Chicago that was initiated in 1947. The findings were similar to those of the New York survey of 1930. Bonvallet found that traffic noise was the main source, affecting many more people than did industrial noise. The measurements were made in octave bands, and the recommendation was made that the noise level in the 400-800-Hz octave band should be used for monitoring or regulation purposes. The Chicago survey eventually led to an amendment to the Chicago Zoning Ordinance, as described in Paper 26 by Hardy. Hardy's paper describes the Chicago noise ordinance and a method to predict community noise. It is interesting to note that the ordinance initially controlled only industrial noise and that octave bands (not to be exceeded) rather than A-weighted sound levels were specified. Later, vehicle noise was also included in the ordinance. The Chicago noise ordinance was important since it seems to have been the first in the United States to specify maximum quantitative levels for manufacturing industries and vehicles. Earlier city ordinances normally incorporated qualitative provisions banning noise that was a nuisance. Such qualitative ordinances are difficult if not impossible to enforce.

Although ordinances such as the Chicago ordinance are, in principal, easy to enforce, they are rather simplistic because the levels set do not take into account other factors known to be important, such as existing background noise, season, time of day, character of noise (impulsive or continuous, pure tone or broad band), type of neighborhood, previous exposure, and so on. As a result of studying several case histories of noise complaints, Stevens, Rosenblith, and Bolt produced a scheme for the *composite noise rating,* to try to allow for these additional factors. Paper 27 describes this scheme, which is perhaps the first attempt to assess community reaction to noise. The composite noise rating, CNR, is no longer widely used, although similar schemes that allow for these factors—existing background noise, season, time of day, and so on—are used in Britain [2] for evaluating whether noise complaints from neighborhoods near industries are justified. A similar recommendation has been adopted by ISO [3], although the U.S. delegation voted against its adoption. The British and ISO procedures [2, 3] have used A-weighted sound levels rather than octave bands of noise.

Many other schemes for rating community noise have been proposed since the CNR, such as L_{eq}, noise pollution level, traffic noise index, day/night sound level (L_{dn}), and so on. Schultz has reviewed many of these schemes [4], many of which are useful not in the regulation of community noise, but rather in determining the likely impact of a new highway or factory on a nearby community. Such schemes may also be used to decide the effect on the whole population of a country of reducing the noise emission of surface transportation vehicles or aircraft. Such a discussion is beyond the scope of this book, and the reader is referred to a recent chapter by Kessler [5] on this topic. Eldred [6] and Von Gierke [7] have discussed the day/night sound level, L_{dn}, the unit that was favored by the Environmental Protection Agency.

REFERENCES

1. Ramazzini, B., *De Morbis Artificum* (Diseases of Workers), trans., University of Chicago, 1940.
2. British Standards Institution, 1967, Method of Rating Industrial Noise Affecting Mixed Residential and Industrial Areas, *British Standard B. S. 4142.*
3. International Organization for Standardization, 1971, Assessment of Noise with Respect to Community Response, *ISO Recommendation* 1996.
4. Schultz, T. J., 1982, *Community Noise Rating,* 2nd ed., Elsevier, New York.
5. Crocker, M. J., and F. M. Kessler, 1982, *Noise and Noise Control,* vol. II, CRC Press, Boca Raton, Fla., pp. 181–219.
6. Eldred, K. M., 1974, Assessment of Community Noise, *Noise Control Eng.* **3**(2):89–95.
7. Von Gierke, H., 1975, Noise—How Much Is Too Much, *Noise Control Eng.* **5**(1):24–34.

24

Reprinted from pages 111–123 of *City Noise*, E. F. Brown, F. B. Dennis, Jr., J. Henry, and G. E. Pendray, eds., Noise Abatement Commission, Department of Health, City of New York, 1930, 308p.

Noise Measurement

DR. HARVEY FLETCHER, *Chairman*
Committee on Noise Measurement Survey

PROFESSOR ALBIN H. BEYER
DR. ARTHUR B. DUEL

AT the first meeting of the committee, held on November 7, 1929, its purposes were outlined as follows: (1) The general city noises will be measured and tabulated to furnish a basis of comparison for any future condition of supposed improvement or deterioration. (2) The magnitude of noises from specific sources will be measured to furnish a basis for recommendations for improvement.

Although some survey work had been done previously, it was agreed that the results available were inadequate, so a complete noise survey was undertaken to measure out-of-door noise in the city. For this purpose, a motor truck with apparatus and a crew of observers was sent throughout the city during the period from November 1929 to May 1930. The results of this survey were reported by Dr. R. H. Galt in a preliminary report dated March 21, 1930, already submitted to the Commission, and in a final report dated June 18, 1930, which is attached. As will be seen from this report, the noise levels were measured at 97 different outdoor locations in New York City. The average noise levels at these locations ranged from about 45 decibels to about 80 decibels. At each of these stations the noise levels varied up and down through a range of about 20 decibels. The highest level encountered in the streets was 101 decibels, which was due to a riveter 35 feet away from the measuring instrument. The minimum level obtained was 38 decibels, in a residential section at 4:30 A.M.

It is particularly interesting to note that the average noise levels taken over a 5 to 10 second interval vary more in a 20 minute period than the levels averaged over a 20 minute interval vary in a 24 hour period. For example, the 20 minute averages taken at 48th Street near 8th Avenue varied from 53 to 63 decibels, or a variation of only 10 decibels, during the 24 hour period, while the small interval average varied usually more than 20 decibels.

It will be seen from the results that the intensity of the noise due to street traffic is directly proportional to the number of vehicles of a given kind passing in a given time. In most cases the outdoor noise is principally due to traffic, so if one knows the number of passenger automobiles and the number of trucks passing per minute it is shown how one can estimate the noise level. This should be useful when desiring to reduce the noise in any locality below a definite level.

The noises from some of the specific sources studied produced maximum levels as follows:

MAXIMUM NOISE LEVELS FROM SPECIFIC SOURCES

Hammering on steel plate (almost painful)	113	decibels
Automobile Horn	102	"
Riveter	101	"
Subway	97	"
Blast of Explosives	96	"
Steamship Whistle	94	"
Elevated Train	91	"
Motor Truck	87	"
Lion Roaring	87	"
Steam Shovel	86	"
Police Whistle	83	"
Street Car	83	"
Passenger Automobile (noisy)	83	"
Radio Loud Speaker	81	"
Thunder (1 to 3 miles)	70	"
Passenger Automobile (quiet)	65	"
Church Bells	61	"

Through the courtesy of the joint subcommittee on Development and Research of the National Electric Light Association and the Bell System, the results of their work on indoor noises were communicated to the Noise Abatement Commission by the chairman, Mr. O. B. Blackwell. Attached is a report, dated June 13, 1930, by Mr. R. S. Tucker which gives a report of this work as it pertains to New York City. The noise was measured at 70 indoor locations. As will be seen from Fig. 1 of the report, the range of noise levels averaged over about five minutes found in business offices, factories, department stores, etc. was from 32 to 72 decibels, while in residences it was from 22 to 45 decibels. In general the noise level in a third story room facing the street is about 10 to 15 decibels lower with the windows open, and 20 to 25 decibels lower with the windows closed than that on the street, when the street noise is the predominant source. However, in the majority of cases it was found that the predominant noise arose from sources inside rather than outside the room. One of the specific sources of noise, which was studied in some detail in the measurements conducted by this committee, is the automobile horn. A report dated June 20, 1930, by Dr. J. C. Steinberg gives the results of this investigation. Thirty-three different types of automobile horns which were sent in by various manufacturers were tested for noise level, and a frequency-intensity spectrum was obtained for the sound emitted by each. The noise levels produced at 23 feet ranged from 70 to 102 decibels, the average being 90 decibels. Judgment tests were made also, both with respect to the objectionableness of the horns for such reasons as "frightening effect," "harsh or raucous tones," etc., and with respect to their ability to override noises simulating street noise in New York.

It appears that 42 per cent of the horns are louder than is necessary to override the loudest average noise levels existing on New York streets, 45 per cent had about the right loudness, and 13 per cent were not loud enough. The louder horns were judged to be very objectionable. The sound waves of all the horns are made up of fundamental frequencies and overtones, of different intensities. The sound waves for the most objectionable horns—motor driven type—had both harmonic and inharmonic overtones; those for the least objectionable horns—vibrator type —had harmonic overtones only. In most cases, the presence of inharmonic overtones causes the sound to have a "harsh" or "raucous" character.

The results of noise measurements contained in the attached reports are not to be regarded as exhausting the field of valuable noise studies. From more extensive investigations of specific noise sources and conditions, much might be learned of practical value in noise abatement. The committee has attempted merely to collect enough data to attain in a broad way its original objectives, and thus to establish a sound basis for the estimation of future changes in noise conditions and for immediate recommendations which are made as follows:

Recommendations Concerning Noise Out-of-Doors

1. AUTOMOBILE HORNS

    I. The level of the emitted sound at the reference distance of 23 feet should be about 88 to 93 decibels. Levels in excess of this are unnecessary and objectionable. Levels less than this are insufficient to override the noise due to heavy street traffic.

    II. Fundamental frequencies between 200 and 300 cycles per second appear satisfactory. The overtones should be harmonics of the fundamentals. The energy should preferably be distributed approximately uniformly in frequency, with just enough energy in the higher frequency ranges to secure directive properties. A relatively large amount of energy in the higher frequency ranges causes the sound to have a sharp and disagreeable character. The presence of inharmonic frequencies in the sound wave causes the sounds to have a raucous character.

    III. Since an automobile horn must produce a rather loud sound in order to be useful as a signalling device, the use of horns should be severely restricted. In particular, the sounding of the horn should be prohibited when an automobile is one of a large number blocked in the street.

2. TRAFFIC CONTROL

    Since the volume and the composition of street traffic largely determine the out-of-door noise at any particular place, all efforts should be made to keep traffic volume down and to restrict the type of traffic to that of the least noisy character. This can be done by traffic laws, by restriction of building heights to keep down congestion, and by multiplying thoroughfares. If it is particularly important to have quiet in a certain neighborhood, for instance near a school, a hospital, or a court, traffic should be so routed in that section of the city as to reduce the volume to a minimum in the immediate neighborhood under consideration and to keep commercial traffic away.

3. AUTOMOTIVE VEHICLES

Other than horn sounds, the noises produced by a motor truck, bus, or automobile are due chiefly to engine, gears, brakes, chains, tires, and rattling loads; of these the engine exhaust is generally the most noisy. Adequate mufflers should be required, and the use of muffler cut-outs should be prohibited in the city. Power should be transmitted by gears, not by chains. Loose frames, windshields, tailboards, and loads should be prohibited. The noise of changing gears should be reduced. For the present, the goal should be to make motor trucks as quiet as the present passenger automobiles, and to keep all automobiles as quiet as the less noisy half now are. These improvements could be attained by the proper construction, care, inspection, and use of the vehicles and would reduce the average street noise level at most places in the city by 5 to 10 decibels.

4. ELEVATED AND SURFACE ELECTRIC LINES

From the standpoint of noise on the street, elevated lines should be replaced by subways, and surface street cars by buses. These changes would reduce the maximum noise at most places on the streets involved by 7 to 10 decibels, and by 2 to 3 decibels, respectively.

The noise produced by an elevated train or a surface car may be reduced appreciably by so constructing the rails and rolling stock that sudden jolts and impacts are eliminated. A street car passing over a track crossing causes a noise about 6 decibels greater than that for ordinary track; and a street car moving fast over ordinary track causes about 7 decibels more noise than does a similar car moving slowly over the same track. Street cars should not be allowed to move rapidly over track crossings, and no crossings should be located in the immediate vicinity of a hospital or a school.

5. STREET SURFACE

Since the noise due to a horse-drawn vehicle on an asphalt street was found to be about 12 decibels less than that for a street paved with cobblestones, a great improvement will result from the introduction of proper street surfacing in districts where horse-drawn vehicles are numerous.

By careful and constant repair, street surfaces should be kept free from irregularities which cause jolts to passing vehicles.

6. OTHER SOURCES OF NOISE

All vehicles, wagons, trucks, and carts, whether propelled by motors, drawn by horses or pushed by hand, should be provided with rubber tires, which should almost universally be pneumatic.

The ordinance already passed restricting the use of loud speakers on streets should be enforced; and attention should be given to loud speakers in homes which disturb persons living nearby.

The measurements indicate that the noise from a boat whistle in the harbor frequently attains such a level that conversation is interfered with in rooms as much as a quarter of a mile distant. A noise of this intensity occurring at night will often interrupt sleep. It should be possible to mitigate this nuisance by restricting the duration and the frequency of occurrence of whistle blasts; and it is not unlikely that measurements of the sound intensity required for proper signalling by boat whistles in the harbor would indicate that reductions of 10 decibels or more could in many instances be made. If the sound levels now caused by boat whistles are necessary under sea-going conditions, provision should be made for producing lower levels in the harbor.

Although the riveter as a source of street noise is less frequent and usually more remote than the motor vehicle, nevertheless the noise of riveting proved by measurement to be one of the most intense found in the survey, and it is generally recognized as one of the most annoying in type. Moreover, when a building is being constructed close to an occupied building, the occupants of the latter may for a considerable time be subjected to an extremely intense noise from a source much closer than the riveter usually is to the street. Consideration should be given to the possibility of reducing the radiated sound by mechanical or acoustical damping, and to the possibility of replacing riveting under some or all conditions by electric welding.

Considerable reduction of the high noise levels measured for steam operated pile drivers, shovels, and other machines used in construction and excavation should result from the substitution of electrical power for steam.

Adequate mufflers would largely remove the noise produced by stationary engines of the explosive type, such as are used in air compressors, pumps, etc.

7. NIGHT NOISES IN RESIDENTIAL SECTIONS

The audibility of a noise, and hence its annoyance, depends upon the background of other noises. The measurements indicate that this background level on a busy street by day may be 35 decibels greater than on a residential street at night; hence a noise which during the day might be inaudible on a busy street may be very disturbing at night in a residential district. Noise restrictions should therefore be far more severe in residential sections of the city at night than in the entire city by day. Since noise sources are less numerous at night, such regulations will restrict relatively few activities.

8. POLICE DUTY AT NOISY LOCATIONS

In view of the extremely great noise at certain locations such as at the intersection of Canal Street and the Bowery, consideration should be given to the length of hours of duty for policemen at such places.

Recommendations Concerning Noise in Buildings

1. CHARACTERISTICS AND PLANS OF BUILDINGS

By proper treatment of wall surfaces and spaces between walls, and by proper selection of room furnishings, the transmission of sound may be reduced and its absorption increased. To diminish the cost, such treatment should, if possible, be included in the original designs.

To isolate noise, subdivide places of work; for example, have several one-desk offices rather than one office with several desks. Glass partitions should extend to the ceiling. Special rooms should be provided for conferences and interviews. Manual workers should be kept away from thinkers; computing machines and type-writers should be in different rooms from desk workers. Noisy factory operations should be segregated from others.

Windows and ventilators should be so designed as to reduce sounds coming from without.

2. TREATMENT OF INDOOR NOISE SOURCES

In addition to isolation, already mentioned, indoor noise sources may in many instances be quieted by improvement in design, with especial consideration of noise. For example, the exposed surface of rotating parts in electric motors may be made less irregular, reducing the motor whine; electric fan noise may be reduced by proper motor selection and blade design; impact of metal parts in machines of many types may be eliminated. The communication of vibrations from machine to walls, shelves, and other effective sounding boards may be reduced by mounting machines upon massive, disconnected pillars, and in some cases by absorptive pads.

Noise Survey of the City of New York Report of Street Noise Measurements

I. *Introduction*

At the start of the survey, two principal objectives were assigned: to obtain a quantitative measure of the outdoor noise of New York; and to determine the major sources of intense and prevalent outdoor noises, and to measure and analyze the contribution of each source.

It is considered that these objectives have now been attained. The extent to which this has been achieved will be made clear by this report, which presents in greater detail the results summarized in the preliminary report of March 21, 1930, and includes certain later measurements. The results remain substantially as given in the preliminary report; but it has been possible to incorporate an improvement in the interpretation of the noise meter measurements, based upon recent laboratory tests, which indicate that the entire scale previously employed in reporting data should be shifted slightly. By this change, for example, a noise level given in the preliminary report as 72 decibels is now given as 70 decibels, the entire scale of levels as measured by the noise meter being reduced by 2 decibels. The measurements of deafening remain unaltered.

II. *Extent of the Survey*

Noise measurements have been made at 138 stations in the City of New York, distributed as shown in the following table:

TABLE 1			
Borough of	Locations Out-of-Doors	Locations Indoors & In Subways	Repetitions	Total Stations
Brooklyn	8	0	0	8
Bronx	9	1	0	10
Manhattan	75	8	32	115
Queens	5	0	0	5
Grand Total	97	9	32	138

The 97 outdoor locations represent a wide range of noise conditions, from the quiet of a remote residential street to the din of the main highways; a wide range of traffic conditions, from no traffic to 129 automobiles per minute; and a wide range of geographical conditions, from the open terrain of the campus of New York University to the shut in canyon of lower Broadway.

Measurements were made near excavation and construction work; in a subway; beside elevated lines; on a bridge over East River; on the pier and simultaneously in an office when the whistle of a departing steamship was blown; on the streets near two schools and two hospitals; in rooms of a hospital and of a place of business, and on the adjacent streets; on "Radio Row;" and under other special conditions.

Most of the measurements were made between the hours of 9:30 A.M. and 5:30 P.M. At each of three locations, however, tests were made at intervals throughout a period of twenty-four consecutive hours.

Of a total of about 10,000 observations out-of-doors, 7,500 indicate the aggregate effect of all sources of noise at the particular place and time, without isolation of specific sources of the noise; in about 2,500 cases, not only was the measurement recorded, but the source of the noise was identified. Thus in about 200 instances the noise due to a motor truck was measured; the sound of a police whistle was measured about 60 times; of an automobile horn, 80 times; of squeaking brakes, 10 times; and so forth. Altogether, the noise produced by each of about thirty individual sources was thus measured, at stations widely distributed over the city. In addition, some hundreds of observations were made upon specific sources at especially chosen stations of a less general character.

At several stations the aggregate noise from many sources was subjected to a frequency analysis by means of band-pass filters. The same process was applied to the noise from each of several separate sources.

The measurements at most of the stations indicate not only the intensity of the noise, but also its deafening effect upon the human ear, in three different frequency regions. Among other results of these auditory observations, it is possible to specify the intensity which a policeman's whistle or an automobile horn must attain in order to be audible in the loudest traffic noise, or in average traffic.

A special study of automobile horns, from the standpoint of

Photograph by *Bell Telephone Laboratories*

Two methods of measuring noise were used as a check on each other—the audiometer on the left measuring the "deafening" effect and the noise meter on the right recording the noise level directly in decibels.

their effectiveness and desirability as acoustical signalling devices in traffic, has been described in another report.[1]

III.   *Apparatus Used In Noise Measurements*

The measuring apparatus used in the survey consists of two instruments, a noise meter and an audiometer, the former yielding a purely physical measurement, the latter a measurement which involves the organs of hearing of the observer.   The action of the noise meter, of which a simplified diagram is given in Figure 1, starts with a microphone which picks up the sound wave in the air and produces an electrical counterpart of it; the electrical wave is then amplified and rectified, finally operating a

---

1. "Report of Measurements on Automobile Horns," by J. C. Steinberg, appearing in this volume.

meter and causing the pointer or needle to move along the meter scale to a point determined by the intensity of the sound wave.

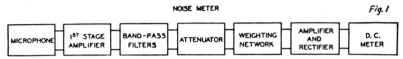

NOISE METER   *Fig. 1*

As Figure 1 shows, the noise meter is supplied also with an attenuator and a weighting network. The attenuator enables the observer to control the amplitude of the electrical wave by large steps, so that when a very loud sound is encountered, which would drive the needle off the scale of the meter, a known amount of attenuation may be inserted, bringing the needle back on the scale. The weighting network is designed to render the noise meter more sensitive to those single frequencies to which the human ear is more sensitive, and vice versa, so that the pitch of a tone will automatically affect the meter in the same way in which it affects the ear; while this reproduction of the action of the ear is by no means complete, owing to the very complicated characteristics of the ear, it represents an improvement upon any arrangement which neglects the ear entirely. The relative attenuation introduced by the network at various frequencies is approximately as shown in Figure 2, which is based on the curve of equal loudness of the normal ear at a sensation level of 30 decibels.[2]

In addition to the parts already mentioned, the noise meter contains a set of band-pass filters shown in Figure 1, any one of which may be omitted from the circuit, or inserted, by throwing a switch. The action of such a filter is to suppress any wave of which the frequency, or pitch, does not lie in the particular frequency region passed by the filter. One filter passes only waves of low pitch, having frequencies less than 500 cycles per second; another passes only the region, or band, from 500 to 1500 cycles; a third filter passes the band from 1500 to 3000 cycles; the fourth passes all frequencies above 3,000 cycles. In most of the observations of the survey, these filters were not employed, since the aggregate noise level was usually the quan-

2. H. Fletcher, "Speech and Hearing," p. 230.

tity to be measured, rather than the amount of noise associated with some particular frequency band. Under some circumstances, however, it was desired to know how a complex noise was made up of the various simpler components; whether the pitch was predominantly high or low; and so forth. For such measurements the filters were employed, with results to be given later.

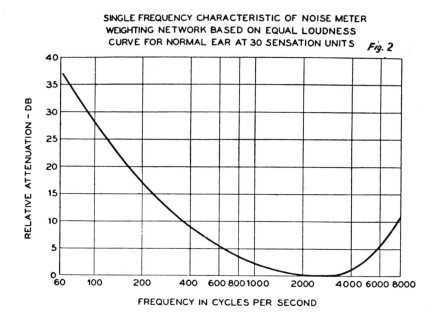

SINGLE FREQUENCY CHARACTERISTIC OF NOISE METER
WEIGHTING NETWORK BASED ON EQUAL LOUDNESS
CURVE FOR NORMAL EAR AT 30 SENSATION UNITS   *Fig. 2*

Attention should be called to one additional feature of the instrument, the time constant of the noise meter. When a noise begins suddenly, rising at once to a certain intensity and then remaining steady, the indicating needle of the noise meter does not instantly move to a steady position, but rises gradually, requiring two seconds to reach approximately a steady position, for one instrument used, and five seconds for another instrument. When the noise ceases, the needle takes a similar time to settle back to zero. Consequently, if a noise fluctuates rapidly and repeatedly, varying from a low level to a high level and then down and up again, as outdoor noises often do, the noise meter needle will not follow the noise level exactly, but will round off the peaks and valleys. The range of noise levels actually occurring will thus often be greater than the range of noise meter readings. The discrepancy is greatest when an isolated noise maximum or minimum occurs of very short duration, as for the noise due to an explosion, and becomes negligible when the maximum or minimum is sustained over a period of two to five seconds or longer, as for the sound of a passing elevated train.

The audiometer employed in the survey to measure the deafening effect of noise has been described elsewhere.[3] The instrument contains a phonograph which produces a test tone of the type usually termed a warble, the pitch varying rapidly up and down through a certain interval. Three such tones were used, one of low pitch, one of middle pitch, and one of high pitch. The phonograph produces each warbling wave, in turn, electrically; the electric current passes through a control or attenuator and is applied to a receiver held at the observer's ear. By means of slots in the receiver cap, the noise is admitted to the ear, as well as the test tone from the receiver diaphragm. The observer then adjusts the attenuator until the test tone is just audible in the presence of the noise. From this setting of the attenuator, and from a similar setting made in the laboratory in a noiseless room, the observer can tell how much his ear was deafened by the noise, for each of the three test tones. If a plot is made showing the amount of deafening plotted against the frequency characteristic of the test tone, the resulting curve is termed the audiogram of the noise.

IV. *The Decibel Scale of Noise Levels*

In ordinary conversation, a noise is often described as loud, or intense, with no attempt to state exactly how loud, or how intense, it is. When a more exact statement is required, the ordinary person will describe a noise as more intense than some other well-known noise. This method of describing by means of a comparison is the basis of the decibel scale which is used in reporting the noise measurements of the survey.

[*Editor's Note:* Material has been omitted at this point.]

---

3. Journal of the Acoustical Society of America, Vol. 1, Oct. 1929, p. 147.

Levels and Spectra of Traffic, Industrial, and Residential Area Noise*

G. L. BONVALLET

Armour Research Foundation of Illinois Institute of Technology, Chicago, Illinois

A survey of city noise in the Chicago area was initiated in 1947. This report describes traffic, industrial, and residential area noise which was investigated as a part of the work. Levels of traffic noise were 35 to 45 db, 45 to 65 db, and 65 to 75 db in the 400–800-cps band for light, average, and heavy traffic conditions, respectively. Industrial noise ranged from 50 to 60 db in the same band for fifty percent of the cases measured. Ninety percent were below 65 db. A limiting spectrum for noise which is considered not objectionable is presented. Residential area noise ranged from 38 to 47 db in the mentioned band for fifty percent of the cases measured. At night and for winter conditions, traffic noise was 10 db lower in the mentioned band than in the daytime. At night, industrial area noise dropped to levels of existing traffic conditions, and in winter it was lower by about 5 db in octave bands mainly because factory windows were closed. At night, residential area noise was 5 to 10 db less in octave bands than during the day, and for winter conditions there was a drop of 6 to 8 db in octave bands due to the modified character of distant traffic.

INTRODUCTION

A SURVEY of noise in the Chicago area was initiated in 1947 under the sponsorship of the Armour Research Foundation of Illinois Institute of Technology and the Greater Chicago Noise Reduction Council. The work is being done in the public interest with a view not only toward investigating noise conditions, but also toward formulating a basis for tolerable levels. The program includes a study of noise of mass transportation and other vehicles, and noise in traffic lanes, in industrial areas, and in residential areas. Phases on vehicle noise and automobile horn noise have been published.[1,2] The work described here includes the investigation of traffic, industrial, and residential area noise.

It is the opinion of the group engaged in this work that complete octave band data are of more value in describing the noise than the single over-all level. However, it would be desirable to have a single number to describe the loudness of noises such as these. Present methods of evaluating loudness, referred to in the first article published,[1] are unsuitable for rapidity and simplicity. As a result of numerous listening judgments on these and other mechanical sources of noise, it was found that loudness and objectionable quality of noises such as these could be correlated better with levels in one of the middle octaves, such as the 400–800-cps band, than with over-all levels. It is believed that the 400–800-cps band level represents a good compromise; the relatively high energy content at the lower frequencies is not so objectionable, and high frequency energy is absorbed to a greater extent by intervening structures and air.

The test equipment consisted of the sound-level meter and its associated octave band filter which constitute the Western Electric Type RA-361 Noise Analyzer.

All readings were made on the flat network. Several thousand readings were taken at several hundred stations in various parts of the city. In general, the measurements were made during the usual business hours of 8:00 A.M. and 5:00 P.M., and most of the data were taken under summer conditions when factory windows were open. Some information also was gathered at night and under winter conditions.

The present report is divided into three parts: (a) traffic noise, (b) industrial noise, and (c) residential area noise. This order suggests itself because it is well known that the noise of traffic affects many more people than noise of industry. This is confirmed easily by comparing population density near traffic routes and near industrial areas.

TRAFFIC NOISE

Traffic in the City of Chicago includes many types of vehicles, and the resulting noise conditions vary widely. The main types of traffic encountered may be divided into the following three categories:

1. Heavy traffic conditions such as (a) in busy "downtown" commercial areas, (b) at intersections of mass transportation and other vehicles including "L" trains, older street cars, and motor trucks, and routes of the same vehicles with heavy traffic, and (c) street corners where motor coaches accelerate.

2. Less noisy or average areas with (a) conditions similar to the above but with less traffic, (b) routes of relatively heavy or fast automobile traffic, (c) PCC type street car and electric trolley bus routes where motor truck traffic is not heavy, (d) motor bus routes exclusive of starting zones.

3. Light traffic conditions where no mass transportation vehicles, relatively few automobiles and trucks pass, or where these vehicles generally are at low speeds. Residential areas and commercial streets and intersections with little traffic comprise this category. For the more quiet conditions in this group the background either is distant traffic or is unidentifiable. Residential areas with industrial noise, of course, are not in this

---

* Presented in part as Paper No. 34 at the 37th Meeting of the Acoustical Society of America, New York, New York, May 7, 1949.

[1] G. L. Bonvallet, J. Acoust. Soc. Am. **22**, 201 (1950).
[2] D. B. Callaway, J. Acoust. Soc. Am. **23**, 55 (1951).

category, since the traffic noise is lower than the industrial noise and cannot be measured.

Such a division of traffic conditions is necessarily arbitrary, but for the purpose is quite useful. Because of the varied nature of the conditions no attempt beyond the above was made to estimate vehicle traffic density. Other published material[3,4] has referred to similar traffic and related vehicle noise.

In measuring traffic noise a compromise was necessary regarding the conditions to be measured. With relatively heavy traffic the level varies with time over a smaller range than when vehicles pass intermittently. The level at the moment one or more vehicles are passing the measuring station is not an average condition in most cases. Such maximum levels of vehicle noise were presented in the earlier report. Similarly, when no vehicles are operating in the area of the measurement station, the level is a minimum. A rather wide range of levels exists between these two conditions.

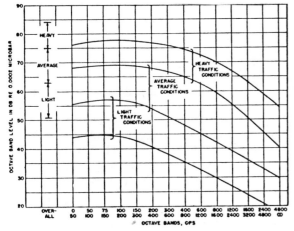

Fig. 1. Average noise levels of heavy, average, and light traffic conditions. Readings were made with vehicles at 25 to 50 feet for heavy traffic, and at 100 feet or more for average or light traffic.

The average level used in this report is a reading which, with a variation of plus or minus a few decibels, describes a particular traffic condition approximately midway between the above maximum and minimum. It is important to point out the difficulty of making meaningful measurements because of the subjective factor involved. Readings were made only when a representative number of vehicles were about 100 ft away, except in the case of heavy traffic in which case they may have been from 25 to 50 ft away. The measurements were made at the curb which was about 10 ft from traffic lanes.

Figure 1 shows the range of over-all and octave band levels in the three types of traffic conditions. The curve for light traffic was obtained from Fig. 4 for residential area noise with little or no traffic. Although the back-

[3] *City Noise* published by the Noise Abatement Commission of the City of New York (The Academy Press, New York, 1930).
[4] D. F. Seacord, J. Acoust. Soc. Am. 12, 183–187 (1940).

ground is termed unidentifiable in such quiet areas which are far from industrial zones, generally it actually is traffic noise at a distance.

Investigation of conditions at night and in winter indicated that the character and density of traffic were changed to the extent that complete measurements would require much more time and care than the results would justify. For reduced traffic at night, reductions in the 400–800-cps band were about 5, 10, and 15 db for light, average, and heavy traffic conditions, respectively. For winter traffic without snow the levels were about the same as for night conditions. With a blanket of snow on the ground a further drop of about 2 to 3 db in over-all and middle octave levels was found, with a somewhat greater drop at higher and a smaller drop at lower frequencies.

INDUSTRIAL NOISE

Measurements were made at over a hundred places in various industrial areas. In general, two conditions were encountered. These were: (a) places where a particular industrial noise being measured was above the background of industrial, traffic, or unidentifiable noise or was measured in traffic lulls, and (b) places where the background noise, generally from traffic, was always above any industrial noise. In a few cases the industrial noise being investigated was below a background of industrial noise in the same vicinity. In each measurement, the judgments of at least two persons were employed in the determination of the proper category. Measurements were made in the street, at the curb, or on the sidewalk at an average distance of about 25 ft from the industrial property line. The readings were taken without regard to proximity of residences.

It is important to point out that far more industrial areas are noisy as a result of noise of motor trucks, trains, street cars, and other vehicles than because of the industrial noise itself. It is significant, however, that in many cases the traffic noise near an industrial plant may be identified with operation of that plant.

The levels in the 400–800-cps band above and below the background are indicated by the two small distribution curves in Fig. 2. This figure shows the percentage of cases in each of the 5-db ranges. For instance, about 17 percent of the cases were in the 60–65-db group. In order to use the data which were below the background, a combined curve was approximated as shown. By checking the properties of the related ogive curve of the combined data on both rectangular coordinate and probability paper, the new curve was found to represent a reasonable, normal distribution.[5] Although it does not necessarily follow that data of this type follow a normal distribution law, it is interesting to learn that such appears to be the case. It is seen that about 90 percent of the levels in the 400–800-cps octave band were 65 db

[5] A. G. Worthing and J. Geffner, *Treatment of Experimental Data* (John Wiley and Sons, Inc., New York, 1943).

or less. Only about 10 percent were above 65 db. It may be seen also that about 50 percent of the cases were in the range from 50 to 60 db.

The levels of noises which were above the background are plotted as dots in Fig. 3, and the levels where the industrial noise was below the background are plotted as circles. Although the plotted data show more cases with industrial noise above than below the background, actually in most cases the industrial noise was below. There were hundreds of places where data were not taken since industrial noise was below the background. On the other hand, a reasonable amount of time was spent finding industrial sites where the noise was above the background in order to make measurements.

The upper curve in Fig. 3 is drawn through 65 db in the 400–800-cps band and represents a typical spectrum of industrial noise. It is the considered opinion of the group engaged in this survey that this curve divides industrial noises into two groups: those with octave band levels below this curve which were considered not objectionable, and those with octave band levels above and which were considered objectionable, as judged from the existing background conditions. With further study and measurement this type of curve might be considered a limiting spectrum for unobjectionable noise in certain areas. As mentioned, 90 percent of the cases were below this curve. Based on cases which were investigated but not measured, it is believed that a figure of 95 or 96 is more reasonable as the percentage of unobjectionable cases.

The lower curve in Fig. 3 was obtained from Fig. 2, and indicates that 10 percent of the measurements were below about 45 db in the 400–800-cps band. Actually, the percentage of noises in the city below this curve is much higher.

Investigations of industrial areas made after the usual business hours indicate that noise conditions are not so intense because of reduced traffic in addition to cessation of industrial operations. Since noise at night in many industrial areas is due only to night traffic, the levels which are to be expected are those either of traffic or of the even lower background of residential areas. Such levels would be the same as in night conditions mentioned in the previous section on "Traffic Noise."

The main cause of reduced levels in winter is the closing of factory windows. The data indicate that with no snow, levels are lower in winter than in summer by about 5 db in over-all and octave band levels. Sufficient information was obtained with a blanket of snow covering the ground to give a preliminary indication of a further drop of about 2 db in over-all and middle octave band levels, with a somewhat greater drop at the higher and a lower drop at low frequencies.

RESIDENTIAL AREA NOISE

Noise in residential areas consists of local or distant traffic noise, industrial noise not originating in the area,

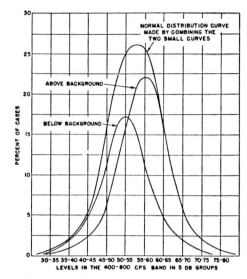

FIG. 2. Percentage distribution curves of industrial area noise levels in the 400–800-cps band. Measurements of noise conditions which were above the background are combined with those below to obtain a probable distribution of noise in both conditions. About 90 percent of the readings are below about 65 db.

noises of children at play, local activities such as carpentry and other trades, vendors, dogs, and other miscellaneous sources. In some areas residences may be found adjacent to industrial plants. There are many residential areas on street car and motor truck routes and on routes of heavy, noisy automobile traffic. Even homes on otherwise inherently quiet boulevards often are disturbed by motor coaches, especially in the offensive accelerating condition which is worse because of the modern torque converter. Steam and diesel engines and "L" trains operate in or near many sections where people live. A discussion of these noise sources is further complicated by railroad switching and whistling, and airplanes in flight. These noises are not further described here, although data on these were taken and

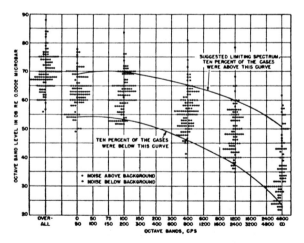

FIG. 3. Average noise levels in industrial areas. The approximate distance to the property line in each case was 25 ft. About 90 percent or more of the industries were below the upper curve which is a suggested limiting spectrum.

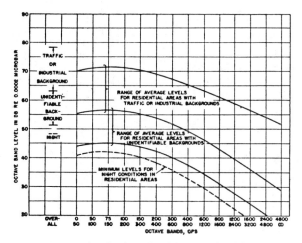

FIG. 4. Average levels of residential area noise. The upper two curves define residential area noise conditions which are intense due to nearby industrial or traffic noise. The area below is for more quiet conditions where the background is more or less unidentifiable. The lowest curve indicates minimum levels at night.

used to obtain the residential area curves. Fortunately, modern mass transportation and other vehicles are being designed to be quieter, and operating agencies are replacing the noisy and outmoded equipment as rapidly as possible.

The greater part of the residential neighborhoods are mile-square areas with relatively heavy traffic routes along the borders and with somewhat lighter traffic routes further subdividing these areas. Noise produced by passenger automobiles usually is not objectionable if the speed is limited. The new PCC street cars are a great improvement over their predecessors and generally are not objectionable. However, the same routes are traveled by noisy motor trucks of many types. There are certain areas in each square mile, therefore, which have more objectionable noise than others which are further from traffic routes.

Under such conditions as these it is extremely difficult to make meaningful measurements. A so-called "average" level is almost meaningless. However, a sufficient number of them becomes valuable. The average level is a measurement of average background conditions and is made when there is no noise due to nearby sources such as children, vendors, etc. as mentioned in the first paragraph in this section.

In Fig. 4 curves of over-all and octave band levels in the various residential areas are shown. The upper two curves define noise conditions which are intense due to nearby industrial or traffic noise. The area below is for more quiet conditions where the background is more or less unidentifiable. The background under these quiet conditions is due primarily to distant traffic rather than to industrial activity. There are, of course, areas where the levels are lower than the data shown.

The data in this figure do not represent a percentage distribution of quiet locales and areas which have reasonably more traffic or industrial noise. Rather, it is

weighted with noisy cases. The total area of residential zones having relatively quiet conditions is much greater than that which has objectionable traffic and industrial noise, hence most people enjoy relative quiet.

There are over 1500 route miles of tracks and lanes used by mass transportation vehicles. This figure excludes railroads, some truck routes, and most automobile routes. On the other hand, there are over two hundred square miles within the corporate city limits, and about one-fourth of this is industrial area. Much of the area zoned for industrial use, however, is vacant or contains very low level noise sources. It can be seen, then, that random noise in residential areas is due to traffic to a much greater extent than to industrial activity. This is even more importantly the case at night when people desire quiet.

A sufficient number of observations in residential areas were made at night to obtain significant data. The lowest levels measured are indicated by the lowest curve in Fig. 4. The range extends from this curve upward into the region of levels in relatively quiet residential areas, a reduction from daytime levels of 5 to 10 db in most octave bands.

Some little time was spent in investigation of winter conditions. Although more data would be desirable, the measurements given here indicate the approximate conditions. These suggest a reduction of about 6 db in over-all and 6 to 8 db in middle octave band levels for winter conditions without snow. With a blanket of snow on the ground a further reduction of about 2 db in over-all and middle octave band levels was found, with a greater reduction at higher and a smaller one at lower frequencies.

GENERAL NOISE CONDITIONS

An attempt to present a general picture of city noise is given in Fig. 5. Over-all and octave band levels for traffic, industrial, and residential area noise are shown. These data give a general summation which is descriptive of the conditions encountered and therefore are suggestive, within limits, of what might be expected under similar conditions. The curves which indicate heavy and average traffic are identical to those of Fig. 1.

In the case of industrial and residential area noise it was believed to be of most value to show curves which indicated the range for 50 percent of the cases, that is, one-fourth of the cases measured had higher levels and one-fourth had lower levels than the indicated range. For industrial noise the octave band levels are taken from Fig. 2, rather than from Fig. 3. The suggested limiting curve for industrial noise and the minimum night-time curve for residential area noise are again presented for completeness.

It has been found valuable for some purposes to reduce the traffic and industrial noise levels to the approximate conditions which might exist just outside of residences. The levels therefore may be lowered to correspond to an increased distance of about 25 ft. Reductions of 5±3 db for the over-all level and 8±3 db for the 400–800-cps band are proper for such purposes. It should be pointed out that all measurements are difficult, and all distances observed in the measurements are approximate with wide variations. As a result the reductions mentioned here are approximate at best.

CONCLUSIONS

In spite of the difficulty of obtaining levels of noises such as these, traffic, industrial area, and residential area noises have been measured. Having made thousands of such measurements, curves plotted from them are meaningful.

The most prevalent city noise unquestionably is that of traffic. The most prevalent source of noise in industrial areas also is that of traffic. In many cases the traffic noise in an industrial area is that due to related traffic, such as the motor trucking identified with a particular plant. In residential areas, the so-called unidentifiable background usually can be identified as noise of distant traffic.

There probably is little question of the advantage of octave band data over over all levels. It is the opinion of workers in this survey that loudness can be correlated better with levels in one of the middle bands such as the 400–800-cps band than with over-all levels. Levels in this band represent a good compromise between levels in lower and higher octave bands.

Measurements of traffic noise ranged from 35 to 47 db in the 400–800-cps band for light traffic, from 47 to 65 db for average traffic, and from 65 to 75 db for heavy traffic conditions.

Industrial area noise levels ranged from 40 to 80 db in the mentioned octave band, with 50 percent of the cases in the range of 50 to 60 db. It is believed that 90 percent of the cases measured and probably 95 to 96 percent of industrial noises in the city may be considered not objectionable. A limiting spectrum for industrial noises is suggested such that levels in octave bands which are lower than this curve may be considered as not objectionable.

Residential area noise generally is due to distant traffic. It ranged from 37 to 49 db for areas with relatively unidentifiable background noise, and from 48 to 65 db in the 400–800-cps band for residential areas with backgrounds of industrial noise or average to heavy traffic. The range for 50 percent of the cases in quiet areas with unidentifiable background was 38 to 47 db.

At night, traffic noise dropped by about 5, 10, and 15 db in the 400–800-cps band for light, average, and heavy traffic conditions, respectively. In winter the character and density of traffic causes a drop in levels which was approximately the same as these figures. With a blanket of snow covering most of the ground a

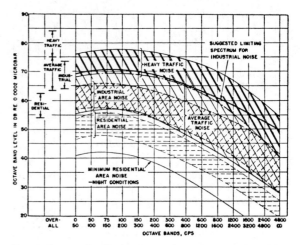

FIG. 5. Average levels of traffic, industrial, and residential area noise. The data for industrial area and residential area noise are shown as a range including 50 percent of the measurements. The curves for traffic conditions are those of average and heavy traffic. The suggested limiting spectrum for industrial noise and the curve of minimum levels in quiet residential areas for night conditions also are indicated.

further reduction of 2 db in the 400–800 cps octave band level resulted, with a greater drop at higher and a smaller drop at lower frequencies.

Industrial noises are reduced at night, but many of the industrial area levels depend on traffic, hence night conditions are those of night traffic in those areas. In winter, industrial noises are reduced mainly because of the closing of factory windows. Reductions of about 5 db in the 400–800-cps octave band were found, with a greater drop at higher and a smaller one at lower frequencies. With a blanket of snow covering the ground a further reduction of 2 db in the same octave band was found, with a greater drop at higher and a smaller one at lower frequencies.

Residential area noise at night dropped below daytime levels from 5 to 10 db in most octave bands, generally as a result of reduced distant traffic. Winter conditions caused a drop of about 6 to 8 db in most octave bands because of the modified character of distant traffic. Measurements of the effect of snow were meager but indicated a reduction of about 2 db in middle octave bands, with a somewhat greater reduction at higher and a smaller one at lower frequencies.

ACKNOWLEDGMENT

The writer acknowledges the encouragement of Dr. H. A. Leedy, Director of Armour Research Foundation, and the guidance of Dr. H. C. Hardy, Supervisor of the Acoustics and Vibrations Section. F. G. Tyzzer provided valuable criticism and suggestions. The work was carried out with the assistance of F. Mintz, D. B. Callaway, H. H. Hall, and C. T. Lorie of the Foundation staff.

ENGINEERING AND ZONING REGULATION
OF OUTDOOR INDUSTRIAL NOISE

Howard C. Hardy
Consultant in Acoustics
Chicago, Illinois

WE think of the words *noise control* usually in the sense of the application of acoustical engineering practice. However, in a broader sense these words can mean some form of legal regulation. This paper treats both aspects of the subject as it applies to the control of industrial noise in the community. Occasionally, communities do pass noise regulation laws, and there are some very peculiar ones in the statute books. Experience teaches us that merely passing a law is not enough unless (1) the law is a reasonable and realistic one, (2) there is a capable supervision and enforcement agency,

* An address before the Third West Coast Noise Symposium in Los Angeles, November 13, 1956.

and (3) there is a straightforward, practical, and economical approach to a solution to the noise problem. That is why the word *engineering* has been placed ahead of *zoning regulation* in the title.

The author has had the privilege of participating in the preparation of the noise regulations in the Proposed Comprehensive Amendment to the Chicago Zoning Ordinance. This has been previously described in an earlier issue of NOISE CONTROL.[1] In this amendment the philosophy was introduced for the first time of the use of legislative performance standards for industrial nuisances. Previously, the approach had been to set up zoning restrictions based on *black lists* of industrial processes, with different degrees of re-

striction often based on obsolete practices or the prejudices of the lister.

To put it another way, a manufacturer is no longer banned from a light manufacturing district because he has a punch press, provided his plant does not exceed certain limits on noise. There are similar restrictions on vibration, air pollution, use of explosive materials, heat and glare, and other activities or uses which can be considered nuisances in the neighborhood. This is a much more realistic approach to zoning because in the great majority of cases, and in all those in which good acoustical engineering practice is used, the presence of a punch press cannot be determined by observations at the property line of the plant.

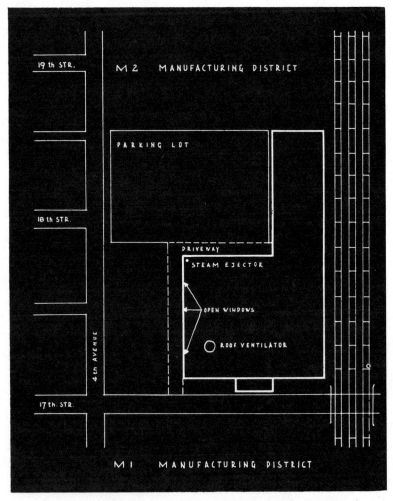

FIG. 1. Hypothetical factory in an M-2 manufacturing district near a residential district. Sound sources near the homes are open windows, a roof ventilator, and a steam ejector.

The new Chicago ordinance provides that "at no point on the boundary of a Residence or Business District shall the sound-pressure level of any individual operation or plant (other than the operation of motor vehicles and other transportation facilities) exceed the decibel levels in the designated octave bands shown below."

Octave Bands (cps)	Along Residence District Boundaries	Along Business District Boundaries
Below 75	72	79
75 to 150	67	74
150 to 300	59	66
300 to 600	52	59
600 to 1200	46	53
1200 to 2400	40	47
2400 to 4800	34	41
Above 4800	32	39

The same standard is applicable to both light and heavy manufacturing. In practice, however, a heavy manufacturing district is nearly always separated by buffer districts of light industry, railroads, or natural boundaries, so that the distances traversed by the sound from heavy industry are generally much greater.

It should be remarked that these numbers have been carefully worked out based on previous good practice and what has been accepted by civic and industrial leaders *in the city of Chicago.* Quoting from my previous articles, "It should be emphasized that although this limiting spectrum is believed to fit Chicago conditions, *it would be a mistake to apply these numbers arbitrarily to industrial areas in other sections of the country.*"

Why Have a Noise Performance Law?

From the point of view of an industrialist, a law such as the one quoted above may look at first like another of those hampering restrictions which will put a great financial burden on business. It should be remarked, however, that all but

a small minority of industry are now complying with these standards. The Armour Research Foundation completed a large metropolitan noise survey in Chicago in 1948.[2] From the data of this survey it is estimated that not over one percent of the industrial plants exceed the limits set in the ordinance, although these numbers are recognized by acoustical engineers to be quite low, much lower, for instance, than in the case of traffic noise.

Since industry has had in the past the false reputation of being a noise maker, most of the industrial district administrators in Chicago and the Chicago Association of Commerce and Industry have supported the measure to bring the few bad neighbors into line. The octave-band levels stated in the ordinance are for new construction; old establishments are given ten years to comply.

There is another reason for having the standard. Many companies have been confronted with time-consuming complaints and sometimes legal action in which noise has been mentioned. Not all of these protests have been valid. A few have been in the category of economic harassment. There was no assurance before now that if noise control measures were instigated, the protests would be discontinued. With a noise *speed limit,* both parties now know just how they stand in such matters.

Enforcement of Zoning Performance Code

The Chicago Plan Commission, in drawing up this new ordinance, realized from the first that it is very important to set up realistic means for enforcement. The new law provides for an Office of Zoning Administration and a Board of Appeals for this office. The Zoning Administration is expected to hire an adequate technical staff to administer the noise performance code and the many other performance codes in the new zoning ordinance amendment. It will be the task of such groups to draw up manuals of operation. Each applicant for a building permit will be

asked to fill out a questionnaire, from which the engineers in the Office of Zoning Administration can estimate whether there will be a potential nuisance from noise or any physical cause. On noise there will be questions such as "What is the location of ventilation blowers?"; "If there is a cooling tower, where is it located?"; "What will be the location of all heavy equipment over 20 tons?"; and instructions such as "Designate all open windows."

It is expected that the Office of the Zoning Administrator will furnish advice (some of which will be in pamphlet form) to the architectural engineers who request it to aid them in furnishing certified building plans.

Other Civic Regulations Which Influence Industrial Noise Control

It should not be thought that the above are the only ordinances that affect community industrial relations in the city. Most cities have laws regarding noise during construction and provide that pile driving, riveting, and operation of pneumatic equipment be used only during certain working hours if the operation is in certain zoned districts.

Of course, most city noise is comprised of traffic noise, and this applies also to the operation around industrial areas. Regulations of traffic noise can be found in vehicular laws and the building or zoning regulations. The Chicago Zoning Ordinance provides for adequate parking around the plant area. All loading berths must be at least 50 ft from any property in a residential district unless completely enclosed by building walls. (There are also restrictions on the size and nearness of building walls.) The larger the plant, the larger the parking and loading areas; and these can often be used to advantage in providing isolation space between the plant and residential areas.

In the over-all industrial plan of the city the layout of railroads and highways, trucking depots, and freight transfer points can be

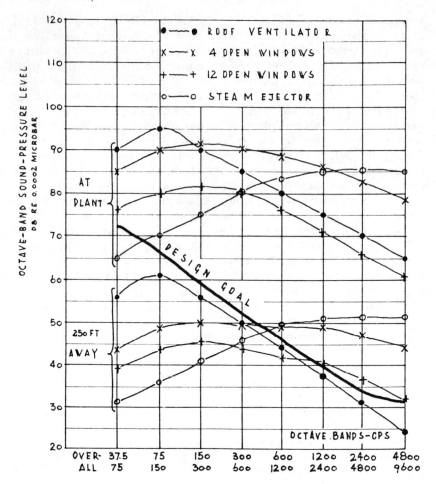

FIG. 2. Octave-band noise levels of various noise sources as measured at the plant and as computed 250 ft away. The design goal is shown as a solid line.

contributors to the noise on this side of the plant, namely (1) a roof ventilator, (2) open windows, and (3) a steam ejector which goes off occasionally and is definitely disturbing. Closer examination shows that the noise from four of the windows is much greater than that from the others because there is a heavy machine near the opening.

A sound survey is made; the data obtained are shown on lines A to D in Table I and in the curves in Fig. 2. This is insufficient data to enable determination of which contributes the most noise near the residences. To do so, one must make an analysis for each frequency band similar to that illustrated in Table II for the 600–1200-cps band only. The over-all data for the four sources are shown in the first column. It is found by observation that the sound is not spread uniformly in space but beamed slightly—in the case of the windows, towards the street in question; in the case of the ventilator, upwards away from the street. The data corrected for beaming are shown in the second column.

The next step is to determine how much sound power is radiated in the prescribed direction. Translated from decibels into sound intensity, the sound-pressure level read on a sound-level meter is a measure of the sound power flowing from a unit area. Knowing the effective radiating area, one can compute the total sound power radiated from this and the sound-pressure level at the prescribed distance.

Figure 3 helps us to do this. Entering the graph at the prescribed sound-pressure level, we read the intensity in watts per square foot. In the case of the roof ventilator, this is 0.6×10^{-5} watts. If the sound were radiated uniformly in all directions, it would be spread over a sphere of 5-ft radius, which has an area of $4\pi(5)^2$, or 315 sq ft. The effective power radiated is, therefore, $315 \times 0.6 \times 10^{-5}$, or

Method of Analysis to Be Followed by Plant Engineers

The remainder of this article will discuss briefly the method of analysis used by engineers to evaluate neighborhood noise problems. (See also reference 3.)

The steps which the plant engineers should take are somewhat as follows:

1. Determine a design goal: how much noise in different octave bands will be acceptable. (When there are city ordinances like those described above, this is easy. Otherwise, previous experience of the company or other companies *in similar neighborhoods* should be followed.)

extremely important in minimizing transportation activity and its noise-producing effect.

2. Determine what noises exist or will exist (nearly always there is more than one noise source).

3. Measure each sound source as to level and direction.

4. Compute the total power radiated in the pertinent directions.

5. Compute the intensity levels which will exist near residential areas.

To illustrate this, consider the case of the hypothetical factory shown in Fig. 1, which is being surveyed to determine whether it exceeds the noise limits given above. Observation 250 ft away at the property line along a residential street indicates that the levels are higher than allowed. Listening indicates that the plant is the offender. Observation further indicates that there are several possible

Table I. Noise Data for a Typical Industrial Plant

Octave Bands (cps)	Octave-Band Levels (db)							
	37.5 75	75 150	150 300	300 600	600 1200	1200 2400	2400 4800	4800 9600
A. Roof ventilator (at 5 ft)	90	95	90	85	80	75	70	65
B. Four open windows (total of 50 sq ft)	85	90	92	91	88	86	82	78
C. Twelve open windows (total of 150 sq ft)	75	80	82	80	76	72	66	60
D. Steam ejector (at 5 ft)	65	70	75	80	83	85	85	85
E. Same as A (at 250 ft, computed)	56	61	56	50	44	38	31	24
F. Same as B (at 250 ft, computed)	43	48	50	49	49	49	47	44
G. Same as C (at 250 ft, computed)	38	43	45	44	42	40	36	31
H. Same as D (at 250 ft, computed)	31	36	41	46	49	51	51	51
Design Goal	72	67	59	52	46	40	34	32

Table II. Typical Computations to Obtain Data of Table I

Noise source	Octave-Band Level 600–1200 cps	Beaming Correction	Corrected for Beaming	Intensity (watts per sq ft)	Radiating Area (sq ft)	Effective Sound Power (watts)	SPL at 150 ft	Required Treatment (db)
Roof ventilator	80	(−2)	78	0.6×10^{-5}	315	0.0019	44	—
Four windows	88	(+3)	91	11.7×10^{-5}	50	0.0059	49	>3
Twelve windows	76	(+3)	79	0.8×10^{-5}	150	0.0012	42	—
Steam ejector	83	—	83	2.0×10^{-5}	315	0.0063	49	>3
					Total	0.015	53	>7

Table III. Noise Data for Jet-Engine Silencer

Octave Bands (cps)	Octave-Band Levels (db)							
	37.5 75	75 150	150 300	300 600	600 1200	1200 2400	2400 4800	4800 9600
1. Design goal (2000 ft)	72	67	59	52	46	40	34	32
2. Inside test cell, afterburner on (no acoustical treatment)	146	150	152	150	148	146	145	143
3. Beaming at top of stack	−1	−3	−6	−8	−10	−12	−14	−16
4. Air attenuation (2000 ft)					−1	−3	−5	−8
5. Inverse square attenuation	−52	−52	−52	−52	−52	−52	−52	−52
6. Level at 2000 ft without quieting	93	95	94	90	85	79	74	67
7. Required noise reduction	21	28	35	38	39	39	40	35
8. Allowed level at top of stack	125	122	117	112	109	107	105	108
9. Allowed level at 100 ft	98	93	85	78	73	69	65	65

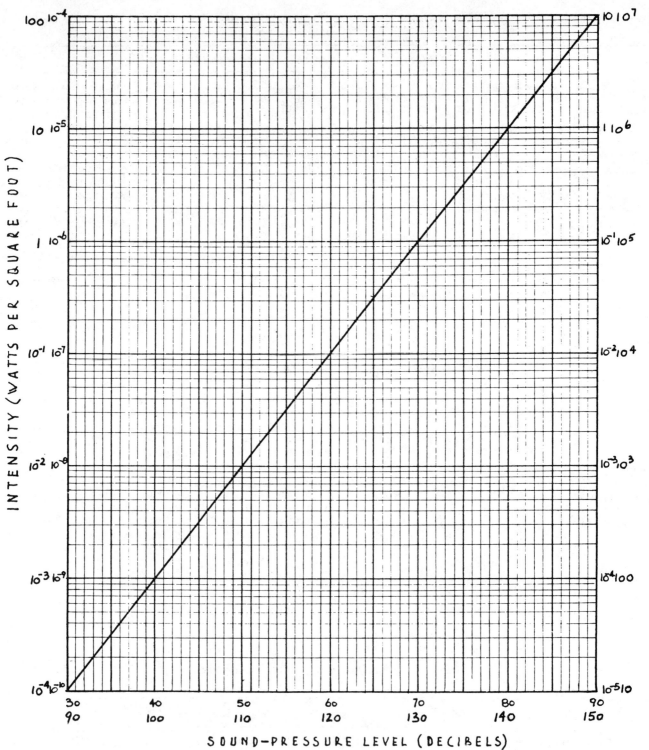

FIG. 3. Relation between intensity and sound-pressure level. The same curve can be used for computing the power required to obtain a given level.

0.0019 watt. Using the right-hand vertical scale, we find that this would be 53 decibels at 90 ft, or 44 db at 250 ft. (Without a logarithm table this can be obtained by an Inverse Square Law calculation: $(250/90)^2 = 7.7$; from Fig. 3, a ratio of 7.7 to 1 gives 9 db.)

The total effective power for all four sources is given in the fifth column of Table II. The steam ejector is found to give the most intense sound in this band. The sound from the four noisy windows, the roof ventilator, and the twelve less noisy windows is of lower intensity. Since 46 db is the desired limit, the amount of quieting needed is also indicated.

The data for all the bands are summarized in the lower part of Table I and in Fig. 2. From this we see that although the roof ventilator had the highest SPL as originally measured, it is the only one of the sound sources which is under the design goal. Also, it is seen that only the high frequencies need to be reduced at the other sources. This is the region of the spectrum which is most economical to quiet. The 15-db reduction required for the steam ejector can be obtained by the installation of a simple muffler. Closing the four windows or putting a simple sound trap on the windows will obtain the reduction needed for them. Closing two of the twelve windows will satisfy the condition for the other source.

It can be seen, then, that with comparatively simple and economical measures a noise problem can be brought under control. Incorrect or offhand engineering procedures would probably have resulted in quieting the wrong sources too much and the right sources too little. In such cases, good acoustical engineering analysis saves a lot of money.

Analysis of Jet-Engine Noise Problem

Good engineering and good economics have been proven in many cases in the silencing of jet engines. The following example shows how one obtains the specifications for the treatment of an engine test cell.[4] The noise spectrum of a

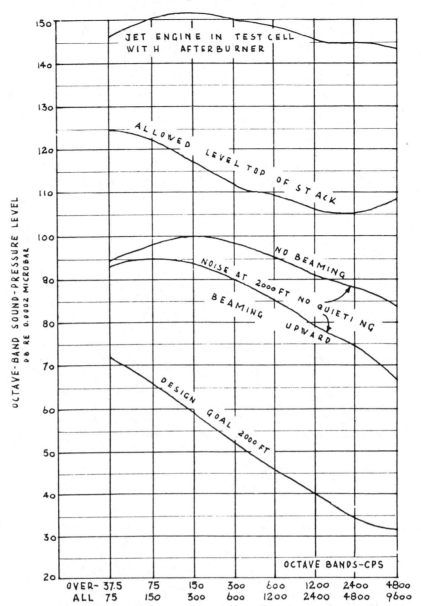

Fig. 4. Noise spectrum applicable to a jet-engine test facility.

large jet engine with afterburner is given in the upper curve of Fig. 4 and the second line of Table III. The design goal of line 1 is the same as used before. If the exhaust is directed upwards, a considerable part of the energy will be beamed in that direction, especially at high frequencies. The effect is shown in line 3 of the table and in Fig. 4. Also, at high frequencies there is some absorption of sound in the air.

The inverse square attenuation is calculated on the basis of an exhaust opening of 18 ft square. This is equivalent in area to a sphere of radius given by $4\pi r^2 = 18^2$, or an

effective radius of 5 ft. The Inverse Square Law attenuation is $10 \log (2000/5)^2 = 52$ db and can be read from Fig. 3, if desired. The level that would exist at 2000 ft without quieting is given in line 6, and the required noise reduction for one cell can be obtained by subtracting line 1 from line 6.

Two other spectra are given which may be convenient for specification writing or for subsequent evaluation and monitoring. If the design goal has been reached, the allowed level at the top of the stack will be that given in line 8. More convenient for measurement is that at 100 ft given in line 9.

It will be noticed that the levels at 100 ft are considerably below those causing hearing damage and are not high enough to prevent speech communication at close quarters.

Many constructions have been devised for quieting these engines the required amounts. One of the most economical is the use of turns acoustically lined with a material of high acoustical absorption, but able to withstand the severe turbulence and temperatures involved. The large tests cells for the Ford Aircraft Division, which are shown in Fig. 5, are examples of such structures.

In these structures the exhaust gases with cooling air may have a volume flow as high as a million cubic feet per minute. The gases are passed through the exhaust labyrinth shown, the walls of which are lined with an efficient absorptive material specially designed to withstand excessive heat, heavy vibration, and high turbu-

lence. By this means, the noise is reduced on the order of 50 db, approximately 10 db more than required in the analysis above. A similar labyrinth quiets the intake path through which the combustion and cooling air pass.

Conclusion

The example of the silencing of the jet engine, the most formidable sustained noise source developed by man, is included to show that there are no industrial plant noise problems which cannot be silenced adequately if management so desires. Experience indicates that costs may run anywhere from $100 to $1 000 000. The cost per decibel may be anywhere from $10 to $2000. One can only make a wild guess at how much has been spent in the United States on outdoor noise control, but it must be on the order of $100 000 000.

This is a large enough figure to warrant some deliberation among civic authorities and plant engi-

neers. It will be helpful if each community can work out for itself some practical noise control performance standards. These will, no doubt, encourage more industries to become quiet neighbors. The engineering techniques are available and many companies are putting them into practice. There is no reason why we cannot keep up with the increased audio output of our modern industrial technology; maybe we can even lower it.

References

[1] H. C. Hardy and G. L. Bonvallet, "Proposed new city zoning performance standard for noise in Chicago," NOISE CONTROL 1, No. 6, pp. 14–17 (November 1955).

[2] G. L. Bonvallet, "Levels and spectra of traffic, industrial, and residential area noise," J. Acoust. Soc. Am. 23, 435 (1951).

[3] H. C. Hardy, "How to control industrial noise—Part 1, Background data in acoustics for the plant engineer," Plant Engineering, pp. 98–101 (November 1956).

[4] H. C. Hardy, "Design characteristics for noise control of jet-engine test cells," J. Acoust. Soc. Am. 24, 185–190 (1952).

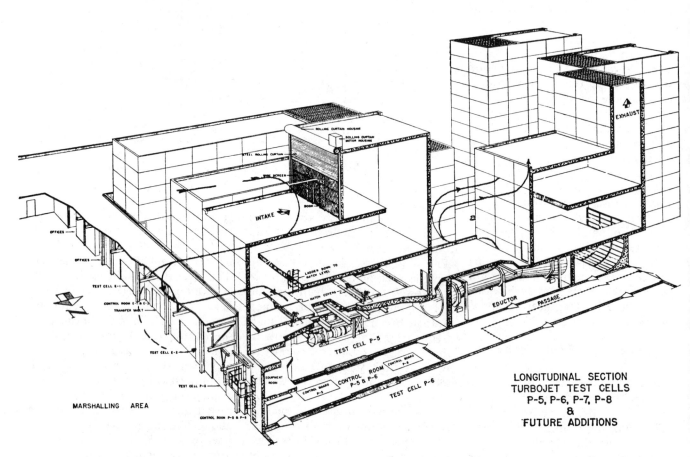

FIG. 5. The jet-engine test cell design of the Ford Aircraft Division. Exhaust gases are cooled by air and water, then quieted by passing through an exhaust labyrinth lined with acoustical material. Similar structures quiet the intake and cooling air paths.

Copyright © 1955 by the Acoustical Society of America

Reprinted from *Noise Control* 1:63-71 (1955)

A Community's Reaction to Noise: Can It Be Forecast?

K. N. STEVENS, W. A. ROSENBLITH, AND R. H. BOLT

NOBODY—least of all the authors of this paper—will try to tell you how quiet your neighborhood should be. In a country like ours, these matters are not decided by fiat. We do it another way —we find out how much quiet, or lack of noise, people like. This job calls for sensible yardsticks, yardsticks that will help us to assess a noise and to rate a community's reaction to noise.

How did we arrive at the scheme that we are about to outline? We studied a number of actual communities. In each one we measured all the things about the noise that seemed to matter. We looked at the way the people had reacted to the noise: mild squawks, vigorous complaints, or what. Then we tried to find a relation between the noise in all its aspects and the observed reaction of the community. We summarized these findings in a set of charts and tables and applied the scheme to forecast community reactions in some new situations. By and large the method worked. From this we are hopeful that it may help to set design goals for the noise control engineer who is given the job of bringing "peace and quiet" to a neighborhood.

The basis of such a scheme is the assumption that people will continue to behave in the future as they have in the past. There will be small drifts, but on the whole the social group will be stable in its responses to noise. Drastic events may upset this stability, but our assumption is probably a reasonable one, so long as we keep alert to signs of possible change. A noise that was once taboo may in time become acceptable. The opposite may also happen. A particular noise may at some time be linked with a disastrous event; thereafter it becomes a warning signal, and the community becomes much less tolerant of it.

What is noise? The only generally accepted definition of noise specifies it as sound unwanted by someone in some context. Hence, we should not be surprised that the same amount of sound energy may be considered noise in a block of hospital buildings, while it may pass unnoticed in the neighborhood of supermarkets and a drive-in theater.

If we were to list the factors that make for pleasant residential living, we would encounter insistence on cleanliness, absence of pollution and excessive traffic, and quiet. Though people differ in their tastes and in the emphasis they put on these requirements, we would not go too far wrong if we assumed that the majority would like to be rid of sources of noise that are not in their control. These opinions or value judgments may be expressed more forcefully by suburbanites than by tenement dwellers, but the desire for "peace and quiet" is accepted as an ideal by most Americans.

Stimuli and Responses

That people react differently to different sounds is a truism. We can relieve some of its obviousness by adding that people may react differently to identical sounds and may behave similarly in response to sounds that are physically quite different.

How can we best understand man's behavior in response to sounds? That depends to some extent on what aspects of man's behavior we are interested in. We have no reason to assume that a man will react similarly to a word, a cry, a whistle, the whoosh of a jet plane, the bark of a dog, or the dripping of a water faucet. Outside of the laboratory we may well ask ourselves if it is indeed

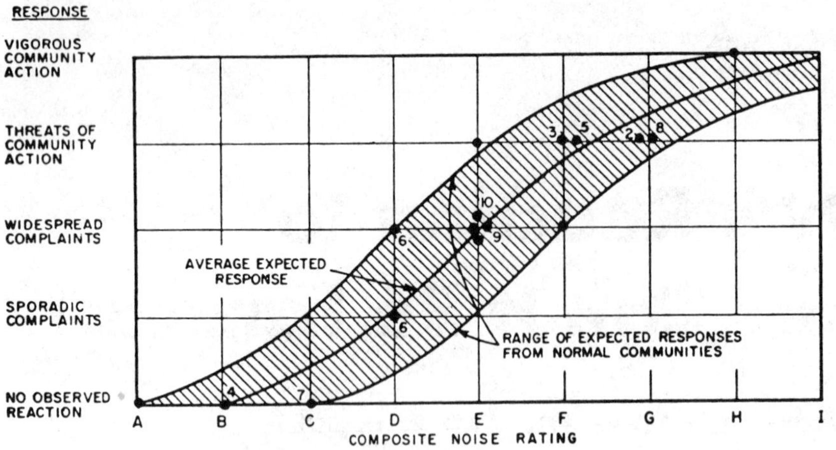

FIG. 1. The wide curve shows the range of responses that can be expected from communities exposed to noises of increasing severity. The center curve is the average response. Community response is assessed along the ordinate. Each point represents a case history of neighborhood reaction to noise; the numbered points refer to the case histories in Table III.

the stimulus that determines a person's or a group's response.

What do we mean by "the stimulus" in everyday situations? Is it enough to state the sound pressure levels in various octaves, or must we also add that this sound is produced by a railroad train that has been passing a dozen times a day for the past twenty years? The dictionary tells us that whatever affects us, excites us, or goads us into action is a stimulus, but such a definition is of limited value to us. The more technical definition used by experimental psychologists tells us that stimuli are events in man's environment. More precisely, the experimenter generally designates as stimuli a particular category of events in the environment that he has under his control. Under these circumstances an acoustic stimulus becomes a certain amount of vibratory energy presented to an observer at selected time intervals.

Next we need to consider what we are going to call a "response." A man, or for that matter a guinea pig, behaves in many ways: he breathes, his heart beats, he perspires, he looks about, he moves, he talks, and he may even complain. Obviously, we must select certain aspects or characteristics of the organism's total activity and identify them as responses. The organism has a whole repertory of different responses to a given stimulus. When we make an arbitrary

choice, we must not forget that by focusing on one type of response we cannot assume gratuitously that other responses will follow along and accommodate us by being similarly related to the stimuli with which we are concerned.

It is often assumed that we can predict the nuisance value of a particular noise stimulus from its loudness. Even if we disregard the fact that many noises are by no means steady (which makes it difficult to assess their loudness even by the most ingenious schemes), we are still faced with a number of paradoxes. Is the loud noise produced by the surf more objectionable than the proverbial dripping faucet or the whine of the bloodthirsty mosquito? Most people would agree that the surf sound should probably not be classified as a noise. As a matter of fact, it may be regarded as a pleasant sound, in a class, perhaps, with the sounds emitted by your own hi-fi equipment (but not with those generated by your neighbor playing Stravinsky's "Rites of Spring" when you think it's nap time).

These examples should be enough to indicate just how circular these everyday stimulus-response relations actually are. Under the circumstances it would be unreasonable, except in extreme cases, to entrust to an ordinary sound-level meter the job of telling you whether your neighborhood is really quiet enough.

This may sound rather discouraging. How then do you go about setting up a realistic criterion for neighborhood noise based on community reactions? Laboratory data on the annoyance that is experienced (or at least reported) by paid subjects listening with earphones to pure tones, or even bands of noise, are clearly of limited value.

A realistic approach might involve the following steps: (1) observe the way in which communities react to known noise conditions; (2) supplement these data where possible by answers to carefully worded questionnaires; (3) use these indexes of behavior to set up a tentative response scale, and finally, try to develop a computational method that is not too involved in order to calculate what we might call the *effective stimulus*.

Why are we suddenly referring to the *effective stimulus* instead of just "the stimulus"? People in a community live in an environment so complex that it would be foolish to attempt to specify it completely. So, when we are interested in a particular type of response behavior, we have to make a choice and select certain variables on the stimulus side that seem to have greatest descriptive and predictive value. In laboratory experiments we often describe the stimulus as the energy change in the environment at certain frequencies and for a specified length of time.

In the community noise problem we have to broaden our definition of the effective stimulus still further. The effective stimulus is not just what is happening "here and now," but includes such factors as the noise levels to which the community has been exposed in the past and the number of times the particular acoustic events have occurred. The nature of the source that produces the particular noise is sometimes an important and occasionally the most important factor. We have, therefore, developed the concept of the *composite noise rating* as a description of the effective stimulus.

This evaluation of the noise stimulus is, of course, inadequate to account for all the nuances of the stimulus that derive from the context or connotation. Many factors may modify the reaction of a community. People may adjust to a new level of sound because it means more business, or their livelihood, or because they have lost certain fears that accompanied the acoustic event when it was new.

A Tentative Yardstick for Community Reactions

The reaction to a given noise may vary greatly from person to person in a community. Some people complain at the slightest provocation; others do not express annoyance even under quite severe noise exposure. Since it is difficult to make a reliable prediction of how one person in a community will respond to a noise stimulus, we shall focus our attention on a large group of residents, or a *living community,* and talk about the response of the group rather than the response of individual persons in the group. For a given noise stimulus we expect the responses of different communities to exhibit less variation than the responses of individual persons within the communities.

How do we measure and specify the response of a community? Can we set up a scale on which to measure the disturbance exhibited by a community when exposed to noise? Such a scale is proposed here, but we should note that the methods currently available for measuring community reactions are much less precise than the techniques used to measure certain aspects of the stimulus.

The response scale we propose is shown on the ordinate of Fig. 1.

At the low end of the scale is the region where no reaction is observed. The people in the community are not sufficiently disturbed to complain to those responsible for the noise or to the municipal authorities. Many of the residents probably do not notice the noise, but others may be somewhat disturbed. Careful questioning or observation of an insider would bring the attitudes of these people into the open.

The next point on the scale, "sporadic complaints," describes

the situation in which some residents in the community are sufficiently disturbed to voice their opinions to those responsible for the noise, by means of telephone calls, letters, or the like. However, the complaints are not, for the most part, persistent. If a substantial number of residents in the community were to complain, and if some of the complaints were persistent, the point on the scale marked "widespread complaints" would be reached.

The term "threats of community action" describes a more severe condition in which large numbers of persistent complaints and threats are voiced. Groups might organize in an effort to bring about legislation or other restrictive action against those responsible for the noise. "Vigorous community action" describes the condition in which community action is strong enough to force the offenders to limit drastically or cease their operations.

The points on our response scale are not so well defined as we might wish. It is a relatively simple matter to measure the intensity of the noise that prevails in a community with a meter. It is simple also to obtain pertinent information on the time schedule of the noise, the background noise, and so forth. Our information on the community response, however, is gleaned from comments on the number of telephoned complaints and the number of letters of complaint, and from impressions of the severity of the disturbance voiced by the complainers. A carefully planned and executed opinion survey of communities exposed to noise would give much more precise data on the response. Such surveys are rarely made, however. For the present, when we evaluate case histories of communities disturbed by noise, we must rely primarily on counting the number of unsolicited complaints in order to assess the response of a community. We recognize that such data are often ill-defined and vague and that the frequency of the complaints and their severity cannot always be clearly separated.

The Composite Noise Rating (CNR)

What do we mean by the noise stimulus in a neighborhood? We have already noted that the stimulus cannot be described simply as the noise intensity measured at a given time. In order to predict the response of a community from measurements or computations of the noise stimulus to which the community is exposed, we must incorporate several characteristics in addition to intensity in the computation of the stimulus. Our objective here is to find a combination of the various physical aspects of the noise that will yield an adequate composite description of the *effective stimulus* in terms of a single rating, which we call the composite noise rating.

Noise Level Rank

Of primary importance in the determination of the community response are the over-all level and spectrum of the noise. We shall assume that the spectra are given as sound pressure levels in octave bands of frequency and are measured out of doors in the vicinity of residences. We assume further that the values are obtained by averaging over a reasonable time interval and over a reasonable number of locations in the community.

Figure 2 shows a family of curves that define the *noise level rank.* The ranks are designated by the letters from **a** to **m**, in ascending order, that is to say, **f** is a higher rank than **b**, but not necessarily three times **b**. Each rank denotes the area between two neighboring curves. At the low end is rank **a**. The lower boundary of the rank would be the average threshold of hearing for octave bands of noise. The highest rank is **m**, and the upper boundary of this rank represents a noise spectrum in which people can communicate only by shouting in each other's ears at a distance of a few inches.[1, 2] This choice of scale implies that a noise that is inaudible does not contribute to annoyance, and that a noise in which it is virtually impossible to communicate, even by shouting, may be treated as socially unaccept-

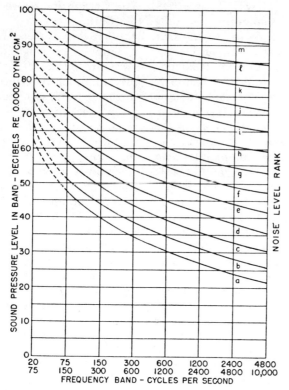

FIG. 2. Family of curves used to determine the noise level rank for residential noise. The spectrum of the noise is plotted as sound pressure levels in octave bands of frequency. The highest zone into which the spectrum protrudes is designated as the noise level rank.

able, whether or not questions of annoyance are involved.

To determine the level rank of a noise, we superimpose the measured or calculated octave band spectrum of the noise onto Fig. 2. The noise level rank is given by the highest area into which the spectrum protrudes in any band. This procedure implies that the noise level in a single octave band can sometimes determine the noise level rank uniquely. In effect, we are stating that different frequency bands contribute independently to the shaping of the response, and that the effects of different bands are not additive. The data of the case histories presented in the Appendix are not inconsistent with this assumption.

Where do the curves in Fig. 2 come from? As a starting point in the construction of these curves, we drew on the results of certain laboratory experiments on loudness. If two stimuli are identical in every respect except intensity (that is to say, if they have the same temporal character, the same background noise, etc.), it is reasonable

to assume that the sound that people judge to be the louder will elicit the more severe community response. Consequently, the boundary lines between ranks in Fig. 2 have been selected so that stimuli of equal noise level rank would be judged, in general, to be equally loud (except at very low and very high frequencies, as is noted below).

Loudness data have not been reported for continuous spectrum noise below 100 cps, and consequently the slope of the curves at low frequencies is an extrapolation. At high frequencies the curves are drawn slightly below the curves of equal loudness, because laboratory data on annoyance indicate that high-frequency noise is usually judged to be more annoying than noise of lower frequency at the same loudness.[3, 4]

The distance between adjacent curves in the middle and high-frequency range is approximately 5 db. The steps are somewhat smaller at low intensities and somewhat larger at high intensities. At low frequencies the spacing be-

tween curves is decreased, as the bunching of the equal-loudness contours for bands of noise at low frequencies would suggest. Several considerations dictate the selection of 5 db as a suitable discrete step for the specification of noise level rank. From previous experience, we believe that the range of variation usually encountered in the reactions of a community to a given noise is so wide that a change of noise level of less than 5 db would not produce a significant change in the general pattern of reaction to the noise. Also, the fluctuations of the noise levels both in time and in space are often as high as 5 db, and it would be unrealistic, therefore, to specify the levels with greater precision.

Determination of the noise level rank is the point of departure in the evaluation of the CNR of a noise stimulus, and we shall turn now to other stimulus properties that affect the community response. Important characteristics of the effective stimulus other than the average level and the spectrum fall into four general categories: background noise, temporal and seasonal factors, detailed description of the "packaging" of the noise, and previous exposure. Quantitatively, we account for these properties by means of correction numbers which we apply to the noise level rank. For example, a correction number of + 1, i.e., one rank upward, applied to a noise level rank of d yields e. The evaluation of the correction numbers is discussed in the following paragraphs, and a summary of all correction numbers is presented in tabular form following the discussion.

Background Noise

When we talk about a noise stimulus in a community, we are generally focusing our attention on noise originating from a particular *source*. Some of the sound energy reaching a community may originate from other sources. The sound originating from these sources is called "background noise." Generally the residents accept this background noise as a part of their daily environment, and the noise does not disturb them particularly, i.e., they don't react to it, or they have adapted to it. It is clear, however, that the background noise must be considered as a factor modifying the "effective stimulus."

It may happen, for example, that the noise from a particular source is masked by the background noise in one community, but is much more intense than the background noise in another community. The two communities will respond quite differently to these two stimulus situations. In a sense, the background noise level plays the role of a reference level with which the noise under consideration is compared.

In order to take into account differences in background noise, we measure the spectrum of the average background noise in octave bands of frequency, and we plot this spectrum in Fig. 3. The figure is divided into zones labeled with correction numbers from − 3 to + 2. The zone in which the major portion of the noise spectrum lies designates the correction number to be applied for background noise. The spectrum shape of the curves in the figure is an average spectrum derived from many measurements of background noise in different localities. Table I indicates the type of locality in which each level of background noise is often found.

Temporal and Seasonal Factors

When the offending source of noise operates more or less uniformly and continuously over an appreciable period of time, the noise stimulus may be described adequately simply by the sound level in octave bands of frequency. Usually, however, the noise source does not radiate sound continuously but operates on some sort of time schedule. For example, it may operate only between the hours of 9:00 A.M. and 5:00 P.M., or it may be heard in a residential area for only three or four 20-second periods each hour (as is the case when three or four aircraft per hour pass over a community). How do we correct for such irregular time schedules in our computation of the CNR? Our approach is outlined in the following discussion.

TABLE I. Correction numbers to account for daytime ambient noise levels in typical neighborhoods. On the average, the correction numbers should be increased by one for nighttime conditions.

Neighborhood	Correction number
Very quiet suburban	+1
Suburban	0
Residential urban	−1
Urban near some industry	−2
Area of heavy industry	−3

Day or Night.—During the daytime, many people are away from their residences and do not hear the noise. Residents who stay near their homes are often engaged in activities that are not greatly disturbed by moderate noise levels. In the evening and at night, however, the noise tends to interfere with relaxation and sleep. We expect, therefore, that a noise of a given level rank will produce a more severe response if it occurs at night than if it occurs only in the daytime. Empirical evidence suggests that we apply to the level rank a correction number of − 1 if the noise source operates only in the daytime. No correction number

is applied if the source operates at night (say between 8:00 P.M. and 8:00 A.M.).

Repetitive Character.—In the discussion that follows, we shall be concerned primarily with noise sources that operate on a more-or-less regular schedule every day of the week. We shall give only brief consideration to schedules that are less regular.

If a noise source operates only during a certain fraction of the time each day, the community response will, in general, be less severe than the response to a continuous stimulus. To account for this restricted time schedule in our description of the stimulus, we apply a negative correction number to the level rank. At present, quantitative evaluation of the influence of repetitiveness is not well established because field data are not available for a wide range of conditions. Preliminary data seem to indicate, however, that the correction number is a function of the percentage of time the noise source operates within, say, an 8-hour period. Our experience indicates that the correction number is reason-

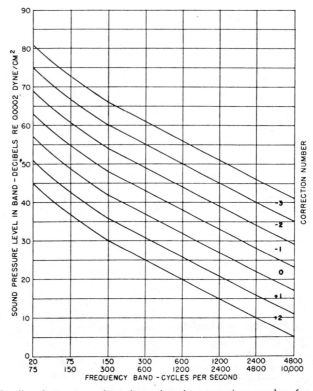

FIG. 3. Family of curves used to determine the correction number for background noise. The spectrum of the ambient background noise is plotted as sound pressure levels in octave bands of frequency. The zone in which the major portion of the noise spectrum lies designates the correction number to be applied for background noise.

ably independent of the particular schedule of operations but is concerned only with the total number of minutes of operation within the period. For example, if the source operated for a total of 30 minutes each day, the correction number would be the same whether these 30 minutes occur within one hour or whether the source operates for only 4 minutes each hour.

We have derived an empirical relation between the percentage of time the source operates and the correction number, and this relation is given in Fig. 4. This relation and the assumptions that govern its use should be regarded as tentative, subject to modification as more data become available.

Additional correction numbers should be applied if the noise source does not operate on the same schedule each day. If operations are restricted to one or two days a week, an additional correction number of −1 seems to be required.

Winter or Summer.—In northern climates, there is a marked difference in people's living habits in winter and summer. In the winter, almost all activities associated with residential living are carried on inside well-insulated houses, usually with the windows closed. In the

TABLE II. List of correction numbers to be applied to noise level rank to give composite noise rating.

Influencing factor	Correction number
1. Background noise (see Fig. 3)	+2 to −3
2. Temporal and seasonal factors	
a. Daytime only	−1
Nighttime	0
b. Repetitiveness (see Fig. 4 and text)	0 to −6
c. Winter	−1
Summer	0
3. Detailed description of the noise	
a. Continuous spectrum	0
Pure-tone components	+1
b. Smooth time character	0
Impulsive	+1
4. Previous exposure	
None	0
Some	−1

summer, however, residents are frequently out of doors, and windows are left open day and night. Consequently, for a given level of noise measured outside, the stimulus to which residents are exposed is greater in summer than in winter. To account for this difference, we apply a correction number of −1 if the source operates only during the winter and no correction number for summer or year-round operation. In warm climates no correction is applied for either winter or summer.

Description of the Noise

Spectrum Character.—A noise spectrum that contains audible pure-tone or single-frequency components is apparently judged to be more annoying than a spectrum that is reasonably continuous. This deduction is based on engineering experience with noise spectra of both types. If, for example, the sound-pressure level in an octave band reaches the noise level rank d by virtue of the contribution of a single-frequency component, we propose that a correction number of +1 be applied, i.e., that the level rank of the noise be raised from d to e. The implication is that, all other things being equal, the level of a pure tone must be about 5 db below the level of a continuous spectrum noise in the same octave band to produce the same neighborhood reaction.

Peak Factor.—A noise that is rea-

sonably continuous in time, at least for a few seconds or more, is apparently judged to be less annoying than an impulsive noise, for example, the sound of a drop forge or gun shots. Experience indicates that a correction number of +1 should be applied to the level rank if the noise is impulsive. At present no firm definition of impulsive noise in quantitative terms is proposed, and some judgment is required to distinguish between impulsive and continuous noise.

Previous Exposure

Experience has shown that residents of a community differ from one another in their ability to adapt to an intruding noise after repeated exposures. For example, people near a railroad can become accustomed to the noise even though they may have shown some reaction during the first few days of exposure. The noise of an occasional aircraft overhead is now accepted by most people, and they may therefore be considered to be adapted to this sound. This adaptation is unstable, however. An accident or near accident may upset the community sufficiently to warrant on-the-spot reconsideration. No correction should be applied to the noise level rank if an intruding noise is a new one to which the residents have not been exposed previously. If there has been some previous exposure to the noise (or to noise of a similar type),

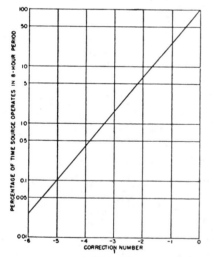

Fig. 4. Proposed correction numbers for repetitiveness of the noise when the source operates on a reasonably regular daily schedule. To a first approximation, the correction number is dependent only on the percentage of time the source operates in an 8-hour period, and not on the particular time schedule within the period.

a correction number of − 1 is proposed.

Summary of Computation of the Effective Stimulus

Table II summarizes the various physical characteristics of the stimulus and indicates the quantitative influence of each in the form of correction numbers that must be applied to the noise level rank to obtain the composite noise rating. In order to avoid confusion, we usually designate noise level ranks by lower case letters and composite noise ratings by capital letters. The noise level rank and the CNR are identical when (1) the acoustic environment is similar to that of a suburban community, (2) the noise source operates at night, (3) the noise source is continuous, (4) the noise source operates in the summertime, (5) the noise spectrum is continuous, (6) the noise has a uniform short-time character, and (7) there has been little previous exposure of the community to the noise.

Empirical Stimulus-Response Relations

Up to this point we have proposed a scale on which to measure the response of communities to a noise stimulus, and we have described a scheme for specifying that stimulus. The next question is: is it possible to find a unique relation (or one that can be defined in statistical form) between the stimulus and the response? This question can be answered only after a study of the reaction of residents in a number of communities that have been exposed to various amounts of noise under sufficiently different circumstances.

We have examined a number of case histories of noise in residential communities. In each of these case histories we have reasonably reliable measurements of the noise stimulus and of the reaction of the community to the noise. The available data have been summarized in Fig. 1. The composite noise rating is plotted as the abscissa and the response scale is plotted as the ordinate. For each case history we compute the CNR of the noise stimulus according to the procedure outlined above. From the reaction of the community we find the appropriate level on the response scale. And from these estimates of stimulus and response we obtain a point on Fig. 1. We proceed in the same manner for each case history.

The reaction of some communities to the noise was rather strong. In these instances the points in Fig. 1 lie near the upper end of the response scale. In others, there was only a mild reaction, or perhaps no reaction at all. Thus Fig. 1 summarizes data and experiences for a wide range of values of both responses and noise stimuli. Table III summarizes the calculations for some typical case histories.

The points in Fig. 1 cluster around a mean line labeled "Average Expected Response from a Normal Community." There is a certain amount of spread of the points about the mean, but most

TABLE III. Summary of typical case histories of response to noise in residential areas.

No.	Description of facility and noise	Level rank	Background noise	Day or night	Repetitive character	Winter or summer	Spectrum character	Peak factor	Previous exposure	Composite noise rating	Observed community response
	1	2	3	4	5	6	7	8	9	10	11
1	Large wind tunnel in Midwest, jet engine operating	h	+1	0	0	−1	0	0	0	H	Municipal authorities forced facility to shut down
2	Large wind tunnel in Midwest, no burning	f	+1	0	0	0	0	0	0	G	Vigorous telephone complaints and injunction threats
3	Exhaust for air pumps, factory in industrial area	j	−3	−1	0	0	+1	0	−1	F	Lodging house owner entered complaints with operator of factory and with local Department of Health
4	Engine run-ups at aircraft manufacturing plant	e	0	−1	−1	0	0	0	−1	B	No complaints reported
5	Aircraft in flight one mile from airport	l	−1	0	−4	0	0	0	−1	F	Vigorous complaints by letter and telephone; one community attempted to prevent passage of aircraft
6	Aircraft in flight four miles from airport	j	−1	0	−4	0	0	0	−1	D	Sporadic complaints in some communities, widespread complaints in others
7	Aircraft engine manufacturing plant: test cells	f	−1	−1	0	0	0	0	−1	C	No complaints reported for daytime operation
8	Transformer noise from power company	f	+1	0	0	0	+1	0	−1	G	Injunction threats
9	Large fan at power company	e	0	0	0	0	+1	0	−1	E	Residents complained consistently, company took steps to reduce noise
10	Weapons range, intermittent firing	l	−1	−1	−5	−1	0	+1	0	E	Vigorous complaints from nearby residents

of them lie within a shaded area labeled "Range of Expected Response from Normal Communities."

We have observed previously that the response of a community is not determined solely by the physical characteristics of the noise stimuli. There are other factors that are not necessarily directly related to the stimulus but that tend to influence the general attitude of the community. In most of the communities that we studied, it was not possible to examine the general attitudes or biases independently or to account for them in quantitative fashion. Consequently, we expect a certain spread in the data, like the spread on the graph in Fig. 1. For example, if a community is near a noisy factory and if there are good public relations between the residents and the factory management, the response may be less severe than the average line in Fig. 1. If, on the other hand, the factory makes no attempt to reduce the noise, and no regard is evidenced for the feelings of the

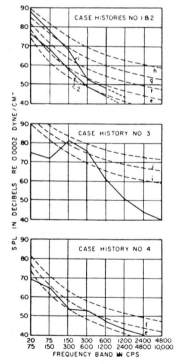

FIG. 5. Measured spectra of intruding noise for four of the case histories listed in Table III. The dashed lines indicate the boundaries of the noise level ranks from Fig. 2. In each case, the noise level rank is the highest rank into which the noise spectrum protrudes.

residents, the reaction may be more severe than the expected average.

Another factor difficult to account for in our evaluation of the effective stimulus is the connotation of the noise. What message does the noise source convey to the people in the community? People may respond quite differently to the same noise, depending on the meaning it has for them. Perhaps the noise brings to the community a message of danger; perhaps it brings a happy message. If an aircraft has recently crashed in the vicinity, the response to the sound of aircraft overhead will probably increase sharply, even though our instruments detect no change in the stimulus. On the other hand, the noise of aircraft returning from a safe mission may bring feelings of joy to the people in a community.

These and other factors may result in different reactions in communities that are exposed to the same noise stimulus or to noise stimuli with the same CNR.

From study of the case histories, however, we believe that a majority of cases would lie within the shaded region shown in Fig. 1. Only in a small percentage of cases, say five to ten percent, would we expect the response to lie outside the shaded region.

The data summarized in Fig. 1 indicate that we can, within certain limits, establish a relation between stimulus and response. We must recognize, however, that there will be a range of uncertainty in the expected response of a community to a given stimulus so long as we are unable to measure and account for factors that have little if any relation to the physical stimulus.

Conclusion

We have presented a scheme for computing the effective noise stimulus to which a community may be exposed and for measuring the response of the community to noise.

On the basis of the twenty-odd case histories we have studied, we feel that the scheme presented here will, within the statistical range of variation indicated by the shaded area in Fig. 1, predict the response

of a community to a given noise situation in most cases. However, we should like to regard the method as a framework within which new data and experience on neighborhood reaction to noise can be gathered. We may find that the new data provide further validation of the scheme, or we may find that the computational procedure requires modification, perhaps with the introduction of additional correction numbers.

We started by asking whether a community's reaction to noise could be forecast. Perhaps you're disappointed because we have not been able to give you a clear-cut answer—we have not said how many decibels it takes to make people squawk how loudly. It just doesn't seem to be that simple. The answer involves decibel readings, spectra, background noise, and time schedules, and includes some factors that we cannot even measure with the customary instruments of the physicist or the engineer. The scheme that we have presented attempts to take account of the factors that seem to be important. Armed with this arsenal of meter readings, correction numbers, and a yardstick for complaints, we have been able to quantify certain aspects of engineering experience. We need to know whether this scheme will work when *you* try it. For a long time noise forecasting is going to be like its older and more established brother, weather forecasting. We may wish the weather man were more accurate in his predictions, but we would not want to be without him.

APPENDIX
Summary of Case Histories

The empirical relation shown in Fig. 1 was derived from a number of case histories of neighborhood reaction to noise. Pertinent data from some of the case histories are summarized in Table III and Fig. 5. The columns of the table give information about both the stimulus and the response.

The first column indicates the type of facility that generated the noise. Column 2 gives the noise level rank, which is obtained by plotting the octave band spectrum of the noise in Fig. 2 and noting the highest rank into which the noise spectrum protrudes. To illustrate this procedure, Fig. 5 shows spectra for four of the cases. In columns 3 to 9 are listed the various correction numbers for each case history, following the scheme outlined in the text. The composite noise rating, obtained by applying the correction numbers to the noise level rank, is given in column 10. In column 11 there is a brief description of the community re-

sponse, from which we select an appropriate point on the response scale of Fig. 1.

Each of the ten case histories listed in Table III is represented by a numbered point in Fig. 1. The lines representing the average and the range of expected responses in Fig. 1 are based on these ten points, plus a number of points derived from other case histories.

References

[1] L. L. Beranek, *Acoustics* (McGraw-Hill Book Company, Inc., New York, 1954), see section on Speech Interference Levels.

[2] W. A. Rosenblith and K. N. Stevens, *Handbook of Acoustic Noise Control* (Wright Air Development Center, June, 1953), *Vol. II,* "*Noise and man,*" WADC TR 52-204. Available from Department of Commerce, Office of Technical Services, Washington 25, D. C., attention Technical Reports Services. This handbook also contains a preliminary discussion of the scheme presented in this article.

[3] K. D. Kryter, "The effects of noise on man," J. Speech and Hearing Disorders, Monograph Supplement 1 (1950).

[4] I. Pollack, "Loudness as a discriminable aspect of noise," Am. J. Psychol. 62, 285–289 (1949).

AUTHOR CITATION INDEX

SUBJECT INDEX

409

412

413

About the Editor

MALCOLM J. CROCKER was born in Portsmouth, England in 1938. He obtained a B.Sc. in aeronautical engineering in 1961 and an M.Sc. in noise and vibration studies in 1963, both from Southampton University. From 1957 to 1961 he completed a five-year apprenticeship with the British Aircraft Corporation. From 1963 to 1966 he worked on noise and vibration problems in space vehicles at Wyle Laboratories in Huntsville, Alabama.

After obtaining a Ph.D. in acoustics from Liverpool University in 1969, Dr. Crocker returned to the United States to teach acoustics at Purdue University, where he was a professor of mechanical engineering until 1983. He was named Assistant Director/Acoustics of Purdue's Ray W. Herrick Laboratories in 1977, where he also conducted research on vehicle and machinery noise. Dr. Crocker was general chairman of Inter-Noise 72 and Noise-Con 79. He is a fellow of the Acoustical Society of America and a member of INCE/USA. He has published widely in acoustics and noise, including over 200 technical articles and several books. He was a founding director of INCE/USA in 1971 and of International/INCE in 1974. Dr. Crocker was the 1981 president of INCE/USA and since 1973 he has been the editor-in-chief of *Noise Control Engineering Journal*.

In 1983 Dr. Crocker came to Auburn University, Alabama where he is professor and head of the department of mechanical engineering. Dr. Crocker is continuing an active research program at Auburn in engineering acoustics and noise and vibration control.

415

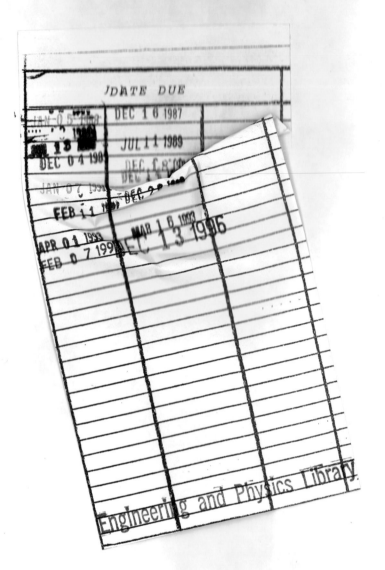